Second Edition

NUCLEAR FORENSIC ANALYSIS

Second Edition

NUCLEAR FORENSIC ANALYSIS

Kenton J. Moody

Nuclear Chemistry Division, LLNL
Livermore, California, USA

Patrick M. Grant

Forensic Science Center, LLNL
Livermore, California, USA

Ian D. Hutcheon

Glenn T. Seaborg Institute, LLNL
Livermore, California, USA

CRC Press is an imprint of the
Taylor & Francis Group, an **informa** business

Front cover image:

The atmospheric nuclear explosion was the Truckee shot of the Dominic test series, conducted over Christmas Island in 1962 by the Livermore Lab. It was a B-52 verification airdrop of a parachute-decelerated XW-58 warhead for the Polaris A-2 missile. The 210-kt air burst detonated at ~ 7000 feet.

Back cover image:

Neutron radiograph of an Afghan scam billet. A number of strange, disk-like objects were supported in the center of the device, and the neutron radiography revealed a low-Z object immediately above them that had not been visible by x-ray.

CRC Press
Taylor & Francis Group
6000 Broken Sound Parkway NW, Suite 300
Boca Raton, FL 33487-2742

Printed on acid-free paper
Version Date: 20141027

International Standard Book Number-13: 978-1-4398-8061-6 (Hardback)

Library of Congress Cataloging-in-Publication Data

Moody, Kenton James.
Nuclear forensic analysis / Kenton J. Moody, Ian D. Hutcheon, Patrick M. Grant. -- Second edition.
pages cm
Includes bibliographical references and index.
ISBN 978-1-4398-8061-6 (hardcover : alk. paper) 1. Criminal investigation. 2. Smuggling--Investigation. 3. Nuclear activation analysis. 4. Nuclear nonproliferation. I. Hutcheon, Ian D. II. Grant, Patrick M. III. Title.

HV8073.M762 2015
363.25'9336--dc23 2014034195

Visit the Taylor & Francis Web site at
http://www.taylorandfrancis.com

and the CRC Press Web site at
http://www.crcpress.com

This work is dedicated to the memory of Professor W. Frank Kinard, College of Charleston, for his many years of friendship and collaboration. A foremost scholar and educator in general nuclear chemistry for over 41 years, Frank influenced and inspired many contemporary nuclear forensic analysts. His recent loss to the nuclear science and education communities is significant, and he will be greatly missed.

Contents

Preface

In the shock and outrage following the attack on the World Trade Center on September 11, 2001, a previous attack on the Center nearly a decade earlier is often overlooked. On Friday, February 26, 1993, a massive explosion occurred in the public parking garage beneath the north tower, killing six persons and injuring more than a thousand. The site quickly became a crime scene, and forensic investigators recovered sufficient evidence to ultimately arrest and convict six Islamic extremists, each of whom was given a prison sentence of 240 years.

The blast tore a 30-m hole through several concrete sublevels, causing more than $500M in damage. The explosive device was eventually deduced to consist of 600 kg of urea nitrate, some nitroglycerin, azides of magnesium and aluminum, compressed hydrogen gas, and interestingly, sodium cyanide, all contained in a rented Ryder van. Some initial evaluations of the incident estimated that the explosive charge could have delivered greater than 1 ton of high-explosive yield, implying a device whose mobility would certainly be at issue. At the Lawrence Livermore National Laboratory, those of us working on the evolution of the discipline that would become modern nuclear forensic science asked ourselves whether the device could have been a "failed nuke"; that is, an improvised nuclear explosive device that failed to work efficiently and derived but a relatively trivial amount of yield from the nuclear materials (tons rather than kilotons). In the New York of 1993, this possibility was not seriously considered, and substantial time elapsed before anyone interrogated any of the explosive debris with radiation-survey instrumentation.

But in the world two decades removed from the initial World Trade Center attack, a world in which the smuggling of nuclear materials in Europe and the discovery of uranium enrichment equipment in Iraq during the first Gulf War have made front-page news, it is unlikely that investigation of an explosion of this magnitude today would not be accompanied by radioactivity-screening instrumentation. However, it remains unclear whether crime scene investigators would possess sufficient knowledge and training to collect and preserve materials that could be productively interrogated by nuclear forensic analysis (NFA).

Forensic science has several practical working definitions. A dictionary (Merriam-Webster.com) will tell you "belonging to, used in, or suitable to courts of judicature," whereas an introductory text by Saferstein (*Criminalistics*, 9th edn., 2007) is more restrictive: "in its broadest definition, the application of science to law." A street-level law-enforcement investigator might (less formally) think of forensic efforts as "a beacon pointing the way out of the sewer of criminal sludge." Not all applications of forensic science necessarily result in formal judicial procedure, however. Accident investigations, nonproliferation assessments, treaty verification, and intelligence operations often also rely on forensic science without the resulting data ever seeing the light of day. What is true about all forensic investigations, though, is the fact that the questioned situation is not the result of a well-controlled experiment in which all initial conditions are known. On the basis of our experiences, we offer here a definition that appears to encompass all of the situations described above: "forensic science is the application of any appropriate technical or sociologic discipline to narrow the limits of informed conjecture." The common thread among all of these definitions is the emergence of forensic science as an important element in efforts to solve problems, while at the same time ensuring that uniform standards of impartiality and justice are met.

Both nuclear chemistry and forensic science can trace their roots to the late nineteenth century, and both became mature disciplines during the twentieth century. Many of the pioneers in both fields were contemporaries: F. Galton published the methodology for classifying fingerprints in 1892, and H. Gross published the first manuscript describing the application of scientific techniques to criminal investigation in 1893. W.C. Roentgen discovered x-rays in 1895, and H. Becquerel's chance discovery of the radioactivity of uranium minerals followed in 1896. J.J. Thomson first identified the electron in 1897, and the Curies discovered the elements polonium and radium in 1898. A.L. Hall published the treatise that was the beginning of scientific ballistics in 1900, and the structure of the atom and the typing of blood were elucidated during the first decade of the twentieth century. Indeed, the ultramicrochemical expertise of forensic pioneer P. Kirk made important contributions to the U.S. Manhattan Project to develop the first atomic bombs. However, although both disciplines have coexisted for more than 100 years, the common ground uniting them into NFA has emerged only within the last 20–25 years.

This book originated from our involvement in the development of this hybrid discipline from its earliest days. During the early 1990s, when the smuggling of nuclear materials was first becoming a reality, we developed and adapted analytic tools based on first-principles nuclear science and an understanding of the standards of forensic science. We then applied these tools both to known materials, obtained from the U.S. weapons complex through legitimate means, and to interdicted materials obtained by the activities of law-enforcement agencies. In formulating the content of this book, we often asked ourselves: "Had it been available a priori, what information would have saved time and energy in generating the results of the forensic analyses?" We wrote the manuscript to capture the foundations of the discipline as it evolved into mature science, while it could still be encompassed in a single text, as well as to provide pertinent, multidisciplinary information within a single volume. The target audiences for the principles of NFA are upper-division, undergraduate college students; graduate students with a firm foundation in the physical sciences; experienced forensic scientists having a desire to learn the nuances of radioactive materials; and first-responders likely to encounter radionuclear samples in the field and tasked to preserve their forensic signatures. It was not our intention to write a textbook, but rather a collection of information that could be used as enrichment material in a college course or professional seminar. We focus particularly on the chemical, physical, and nuclear aspects associated with the production or interrogation of a radioactive sample. We try to present this material with minimum mathematical formality, using terminology that is perhaps not jargon-free, but that is at least consistent. As far as we know, this book is a seminal effort in the field.

To validate the core material treated in the text, we have included a number of nuclear forensic case histories. The intent is to demonstrate that the study of nuclear forensic science is not merely academic exercise but has contemporary application to real-world investigations. We have attempted to select cases with the greatest educational value, but the least political sensitivities; the production and control of nuclear materials is, after all, a politically charged and sensitive subject. In the interest of economy, we have omitted the forensic investigation of most of the submitted "scam" nuclear items, as well as our first case involving actinide material: the seizure of a bottle of uranium yellowcake by local authorities from a high school student, after its procurement at a county swap meet.

Coming as we do from reasonably diverse backgrounds and technical perspectives, each of us would like to individually acknowledge colleagues that contributed to our professional development and some of the successes that we have had. We would not become scientists without the influence of our teachers, who introduce us to the scientific method and instill in us an enthusiasm for research.

One of us (KJM) would like to thank K.S. Robinson, T.G. Hughes, R.M. Martin, and G.T. Seaborg for opportunities to explore the various aspects of the chemical sciences; A. Ghiorso for showing that, although nothing replaces hard work, a well-developed scientific intuition can help guide an analysis; and E.K. Hulet for demonstrating the necessity of selecting the most important problems on which to work. The efforts of L. Cronk and R. Wikkerink in isolating, handling, and packaging very radioactive materials in the early days of the NFA program are also gratefully acknowledged.

Attempting to acknowledge all of the contributors to the gradual metamorphosis leading one of us (IDH) from his beginnings as a physicist examining radiation damage in lunar rocks and soils, to a geochemist dating the oldest objects in the solar system, to a forensic scientist, is a tenuous proposition at best, doomed certainly to incompleteness.

A few individuals must, however, be recognized for their salient contributions: P.B. Price and J.H. Reynolds instilled a deep curiosity in the wonders of the solar system just beyond immediate reach; Toshiko Mayeda and R.N. Clayton introduced the wonders of isotope anomalies and asked in exchange only for "hard work" (Tosh's favorite words) and elegant science; and G.J. Wasserburg drove home the importance of asking difficult questions and demanding credible answers.

This metamorphosis has been a gradual process, spanning more than 30 years. My current engagement with nuclear forensic science owes a particular debt to Sid Niemeyer for his early recognition of the importance of this emerging discipline, and for his ceaseless efforts to establish nuclear forensics as a core mission in the Departments of Energy and Homeland Security; more recently, thanks to S. Prussin for his commitment to scientific excellence and the refusal to seek easy solutions to complex problems.

PG gratefully acknowledges his graduate-school mentors at the University of California, Irvine for superb guidance: F.S. Rowland, G. Miller, and V. Guinn. During my Los Alamos National Laboratory days, I was most fortunate to learn from, and work with, an expert radiochemistry staff, particularly H. O'Brien,

P. Wanek, B. Erdal, J. Gilmore, K. Wolfsberg, M. Kahn (University of New Mexico), and D. Miller (Cal-State Northridge). A similar windfall occurred next at Livermore, and particular thanks go to T. Baisden, R. Torres, F. Kinard (College of Charleston), J. Meadows, D. Sisson, D. Phinney, D. Dye, A. Mode, and (of course) the co-conspirators in NFA, Ken and Ian.

Staff and affiliates of the Livermore Forensic Science Center (FSC) have routinely conducted outstanding forensic analyses over many years and have succeeded in distinguishing the Center as one of accomplished excellence. Their work is notably reflected in the case studies of this book, and these scientists include R. Whipple, A. Alcaraz, C. Koester, G. Klunder, P. Spackman, A. Williams, B. Hart, D. McAvoy, J. Haas, R. Randich, T. Fickies (California Department of Justice), R. Russo (Berkeley National Lab), N. Gharibyan, A. Love, J. Bazan, T. Tillotson, and D. Eckels. I was exceptionally privileged to work closely with the first two directors of the FSC—both extremely gifted, personable, and genuinely cool guys: Brian Andresen and Glenn Fox. More recently, eternal gratitude to Brad Hart for returning to the FSC as its fourth director and rescuing it from a death spiral, and to Dawn Shaughnessy for generously engaging me in her fundamental research group for a career capstone. Special thanks to Jim Blankenship of the FBI for his support over many years, and for his steady focus, despite myopic incursions by other agencies, on what the law-enforcement aspects of nuclear forensics should entail.

And finally, to three generations of Grants: D.F. Patrick, Mary, and Cathy; Joni, Lori Lyn, and Crystal-Lyn; Lori Anne, Michael, Kevin, and Mari—the greatest thank-you—for the diverse facets of life often taken for granted, and basically... for everything.

Acknowledgments

Each of us, however, would also like to thank CRC Press and its editorial staff, particularly Becky (McEldowney) Masterman, who first approached us with the idea for this book. We are likewise grateful to staff of the LLNL Office of Classification and Export Control for their informal security-classification reviews of the manuscripts prior to publication: R. Grayson (first edition) and A. Volpe (second edition). Professor S. Prussin read the first edition completely and provided detailed suggestions and corrections, many of which were incorporated in the second edition.

We are indebted to the true pioneers of nuclear forensic analysis (NFA): that fraternity of cloistered nuclear scientists who followed the lead of E. Fermi and H. Anderson after the end of World War II (see Chapter 1). Working in postwar secrecy, imposed isolation, and relative anonymity, these National Laboratory and Intelligence Community scientists played important roles in averting a third global conflict and (in the West) winning the Cold War without mass fatalities or nuclear winter. In the process, they laid the foundations for the modern embodiment of NFA. Their accomplishments and resulting contributions during this singular era in history have been too little appreciated.

Beginning as we did during the immediate post-9/11/2001 era for the 1st edn, the book took considerably more time and effort than any of us expected, often at the expense of family interactions. So we conclude this preface by modestly and gratefully acknowledging the patience and understanding shown by our families during this endeavor.

Authors

Kenton J. Moody earned a bachelor of science in physical chemistry from the University of California at Santa Barbara in 1977. He attended graduate school at UC Berkeley, where he studied with Nobel Laureate Glenn Seaborg (the discoverer of plutonium) and received a PhD in 1983 with a specialty in actinide radiochemistry. Following a two-year stay at the Gesellschaft fuer Schwerionenforschung (a German accelerator laboratory), he joined the Nuclear Chemistry Division at Lawrence Livermore National Laboratory (LLNL) in 1985, where he has performed extensive diagnostic radiochemical measurements in support of various national security programs.

He is a technical leader for the application of nuclear and radiochemical techniques to problems in national security and the U.S. nuclear stockpile. He also performs basic research on the heaviest elements (the latter in collaboration with physicists at accelerator laboratories in the former Soviet Union). In addition to numerous classified reports detailing the performance of nuclear explosive devices, he has coauthored more than 100 refereed journal publications in the subject areas of the decay properties of the heaviest elements, nuclear reaction mechanisms, fission, and nuclear structure. He has codiscovered six chemical elements and more than four dozen heavy-element isotopes.

Ian Hutcheon is currently the deputy director of the Glenn Seaborg Institute, the Chemical and Isotopic Signatures group leader in the Chemical Sciences Division, and a distinguished member of the technical staff at the Lawrence Livermore National Laboratory. Prior to this position, he was a senior research associate in the Division of Geological and Planetary Science at the California Institute of Technology in Pasadena.

His technical training is in physics and geochemistry: he received an A.B. at Occidental College in 1969 and a PhD in physics from the University of California at Berkeley in 1975. He then spent two years as a post-doctoral fellow and five years as a senior research associate in the Enrico Fermi Institute at the University of Chicago.

He has authored over 170 publications in peer-reviewed journals in the areas of secondary-ion mass spectrometry, the early history of the solar system, and nuclear forensic analysis. He also serves on the review panels of the NASA Cosmochemistry

Program and the Sample Return Laboratory Instruments and Data Analysis Program. He is a member of the American Geophysical Union, the Meteoritic Society, and the Microbeam Analysis Society.

Patrick M. Grant earned BS (1967) and PhD (1973) in chemistry from the Santa Barbara and Irvine campuses, respectively, of the University of California. He worked in radiochemistry and nuclear medicine at Los Alamos National Laboratory for eight years, and was an associate group leader for medical radioisotope research and production. He then spent two years in the oil, gas, and minerals industry at Chevron Research Company. He has been a staff member at Livermore National Laboratory since 1983, serving as the deputy director and special operations and samples manager of the Forensic Science Center.

Grant has also held positions as a senior nuclear reactor operator and as an adjunct university professor of chemistry. He has served as a subgroup member of the U.S. National Security Council's Coordinating Committee on Terrorism and is a member of Livermore's Select Agent Human Reliability Program. He was also a member of three FBI Scientific Working Groups for forensic aspects of radioactive materials, chemical terrorism, and all-WMD analytes.

In addition to numerous classified and law-enforcement reports, he has authored or coauthored more than 120 refereed publications in the open literature in the subject areas of chemistry, physics, nuclear medicine, thermodynamics, spectroscopy, forensic science, incident analysis, and laser science. He has won the Health Physics Society's Silverman Award in radiobiology and a Department of Energy Award of Excellence. He is a fellow of the American Academy of Forensic Sciences and a member of the editorial board of the *Journal of Forensic Sciences*. One of his unclassified investigations, a scientific explanation for the Riverside Hospital Emergency Room "Mystery Fumes" incident [*Forensic Science International* **87**:219–237 (1997) and **94**:223–230 (1998)], was extensively highlighted in the popular media and is now appearing in fundamental forensic science textbooks. Additionally, a significant contribution to the interpretation of the bullet-lead evidence in the long-standing conspiracy debate over the assassination of U.S. President John F. Kennedy was published by the *Journal of Forensic Sciences* [**51**:717–728 (2006)].

1 Introduction

It's 8:15;
That's the time that it's always been.
We got your message on the radio—
Conditions normal and you're coming home.

Enola Gay,
Is mother proud of Little Boy today?
This kiss you give—
It's never ever gonna fade away.

Enola Gay
Orchestral Manoeuvres in the Dark (OMD)
© EMI Virgin Music Ltd

The Manhattan Project, effected by the United States during World War II, forever changed the technical, social, and political framework of the world. The endeavor produced a nuclear industry with several diverse facets, with the principal ones today being weapons, power, and medicine. The focus of this book is on the science, techniques, and analyses of nuclear materials and their near environments for information pertinent to nuclear incident investigations by law enforcement and intelligence agencies. The main applications of such forensic analyses are in instances of nuclear smuggling, nuclear terrorism, nuclear extortion, and nuclear arms proliferation. To those ends, the weapons and power aspects of the nuclear revolution are the most important areas for consideration.

1.1 NUCLEAR MATERIALS

The primary component of established nuclear weapons is uranium or plutonium metal, suitably enriched in a fissile isotope (^{233}U, ^{235}U, or ^{239}Pu, respectively). These nuclides are incorporated in a definition of Special Nuclear Material (SNM), which comprises any substance enriched in ^{233}U or ^{235}U or containing any of the isotopes $^{238-242}$Pu. Further categories of weapons-related material are Other Nuclear Materials (e.g., ^{237}Np, ^{241}Am, ^{252}Cf, ^{6}Li, T[^{3}H], D[^{2}H]) and Source Materials (e.g., natural U, Th, D-38 [U depleted in ^{235}U]). Uranium-235 is additionally grouped into the areas of high-enriched uranium (HEU: $>90\%$ ^{235}U), intermediate-enriched uranium (20%–90% ^{235}U), and low-enriched uranium (LEU: $<20\%$ ^{235}U). An HEU composition of importance in the US nuclear weapons program is termed Oralloy (Oy) and, unless another enrichment is specifically given, refers to U metal enriched to a nominal 93.5 weight-% (wt.%) in ^{235}U. Similarly, Pu is considered weapons grade if the nuclidic makeup is at least 93% ^{239}Pu, with $\leq 7\%$ ^{240}Pu, and is termed reactor fuel grade if

TABLE 1.1
Approximate Critical Masses of Special Nuclear Material Metals for Two Different Configurations

Material	Bare, Isolated Sphere (kg)	Fully Tamped (Reflected) Sphere (kg)
^{235}U	52	17
^{239}Pu (α phase)	10	4
^{239}Pu (δ phase)	16	6
Civil Pu	13	—
^{233}U	15	6

more than 8% of the substance is ^{240}Pu. The Pu used for civilian power production in mixed-oxide (MOX) fuels is nominally 50%–60% ^{239}Pu and 25% ^{240}Pu.

The explosive yields of nuclear weapons are given in metric kilotons (kt) or megatons (Mt) of TNT equivalent. One kiloton corresponds to the detonation of 10^9 g of high explosive (HE), with the exothermic release of 4 TJ of energy. The complete nuclear fission of 1 kg of ^{235}U or ^{239}Pu releases about 17 kt of TNT equivalent.

The minimum quantity of fissionable material necessary for a nuclear explosion is termed its critical mass, and the value depends on the specific isotope, material properties, local environment, and weapon design. The critical masses of several SNM metals and geometries are shown in Table 1.1. For a hypothetical nuclear weapon, U must be enriched to about 80% ^{235}U or higher, and about 25 kg of HEU would be necessary for an implosion bomb. Similarly, a weapons-grade Pu device would require about 4 kg of the fissile material in a perfected bomb model. However, it is also possible to weaponize civil Pu. Although the increased ^{240}Pu concentration in this material makes a weapon less efficient and reliable, the United States successfully tested such a device at less than 20 kt in 1962.

Two less popular pathways to a fission bomb are through the production of ^{233}U or ^{237}Np. The former is generated in a nuclear reactor by irradiation of fertile ^{232}Th source material to produce the fissile product. The Th fuel cycle uses naturally abundant targets that require no special enrichment or other extraordinary tactics to breed ^{233}U. After suitable neutron irradiation and an appropriate radioactive decay interval, chemical separations (reprocessing) isolate ^{233}U from the Th target and other species. The Th cycle generates but one-third of the waste volume as that of conventional U reactors, and it is nonproliferative in the sense that it makes no usable weapons-grade Pu. The Th–^{233}U production system is discussed in more detail in Chapter 3.

Neptunium-237 is also fissionable and can be configured to produce a nuclear explosion, although less conveniently, as it requires high-energy neutrons to fission. It forms in irradiated reactor fuel and can be isolated from reprocessing streams and high-level waste. Although it does not qualify as SNM or source material, the International Atomic Energy Agency (IAEA) is monitoring separated ^{237}Np through voluntary agreements with applicable countries.

By far, the development of nuclear weapons has relied predominantly on the production of HEU or weapons-grade Pu for fission primaries. The military uses of HEU are both weapons and naval propulsion reactors; the principal civilian uses are research and test reactor fuel, targets for medical radioisotope production, and propulsion reactors for Russian icebreakers. HEU is optimal bomb material because primitive, gun-assembled devices are feasible, whereas with Pu, neutrons from the spontaneous fission of ^{240}Pu would preinitiate any such similar designs. More technically sophisticated, implosion-assembled primaries are consequently necessary for Pu. Thus, for example, Iraq concentrated its limited technical resources on the production of HEU in a covert nuclear weapons program.

Weapons Pu results from the irradiation of ^{238}U in a nuclear reactor, under tailored conditions of neutron bombardment and postirradiation decay, followed by radiochemical reprocessing. (A major difference between weapons-grade Pu production and nuclear power generation is that, for a ^{240}Pu content of less than 10%, a reactor bombardment of 7 or fewer months is necessary; economic power production, in contrast, dictates fuel irradiation for about 4 years for optimum performance.) A Pu program is relatively straightforward from a materials production viewpoint, but it requires more difficult weapon designs and is also difficult to shield from detection by national technical means. The production of enriched U, in contrast, is costly and energy intensive. Although several methods for isotope enrichment are in commercial operation or under advanced research and development, the most widely used are gaseous diffusion of UF_6 (used by the United States and France) and high-performance gas centrifugation of UF_6 (used by, e.g., Russia, Europe, and South Africa).

Of the other technologies, such as chemical exchange, laser isotope separation, and electromagnetic separation, the last deserves passing mention for historic reasons. Developed under the Manhattan Project, electromagnetic separation of U isotopes in calutrons was ultimately abandoned by the West in the late 1940s as too energy demanding. However, it was a method implemented by Iraq, whose use of this technique was revealed to the rest of the world only through on-site inspections by the United Nations after the Gulf War of 1991.

1.2 NUCLEAR POWER AND Pu PRODUCTION

Large nuclear reactors use U of various enrichments to produce electric power, propel vessels, and create diverse radionuclides, including Pu. The 433 commercial nuclear power reactors worldwide at the end of 2012 were generating 370 gigawatts electric (GW_e), or approximately 13%–14% of the world's electric power, with an overall conversion efficiency of about 30%. More than 100 of these commercial LEU reactors are in the United States although, until recently, the last American reactor order had been placed in 1973. [However, in 2012, the US Nuclear Regulatory Commission approved a standardized and certified power reactor design (Westinghouse AP1000 Generation III+), following which Southern Nuclear Co. was granted a construction and operating license for new, 1100-MW_e units at the Vogtle and Summer power stations.] In addition, the US Navy operates more than 80 submarine and surface combat ships fueled by HEU.

The use of nuclear reactors to produce electric power has been out of favor in the United States and other countries for several decades. However, fears of destructive global warming and oceanic pH reduction by the generation of CO_2 and airborne black-carbon particles through consumption of fossil fuels had prompted a modern revival of nuclear energy. That revitalization was recently stalled by international public angst and diminished confidence in nuclear power fostered by the 2011 catastrophe at Fukushima Daiichi, and the deceleration was abetted by discoveries of vast deposits of cheap shale-gas reserves. In addition to general slowed growth and increased licensing delays, some nations (e.g., Switzerland, Italy, and Germany) made the extreme political decision to give up nuclear power entirely.

Yet, nuclear reactors are currently the only viable decarbonized energy supply. To avoid adverse and indisputable global consequences of extraordinary proportion, mankind must eliminate or seriously curtail carbon emissions from energy generation via fossil fuels [1]. Indeed, even burning natural gas (predominantly CH_4) reduces such releases, in comparison to combusting coal, by only approximately a factor of 2. [But when escaped into the environment, CH_4 has ≥20× the influence of CO_2 as a greenhouse gas.] Political decrees of "no more nukes" serve principally to increase dependence on fossil fuels, exacerbate import expenses, amplify CO_2 and carbon-particle emissions, decrease energy-mix diversity, raise worries about energy security, and increase the costs and difficulty of combating climate change.

The lessons learned from the complete renovation of the US licensing system for the Westinghouse AP1000 reactor appear to define the path to the future for nuclear power, that is, a new era of standardized design to resolve every nuclear safety issue for all new construction, with an added design concept of the small modular reactor (SMR). The SMR is forecast for production by the early 2020s and would generate power on the order of ≤300 MW_e. By constructing the plant in a controlled factory environment, thereby minimizing the appreciable time and expense of production at a remote job location, subsequent on-site assembly may be accomplished within 3 years. Further, the forecast costs are \$0.5–1B, rather than the ~\$5–7B for a 1000-MW_e nuclear power reactor. Increased passive safety systems and underground designs would be explored, and SMRs could operate individually or in clusters for the more extensive capacities required by utilities. Supplemental SMR modules may be added at future times as energy demands increase.

At the present time, though, a variety of extant nuclear reactors exists for numerous applications. These include high-temperature designs cooled with gaseous He (HTGR), liquid-metal-cooled (Na or Pb) fast breeder reactors (LMFBR), heavy-water units (e.g., CANDU), accelerator-driven reactors, and pebble-bed modular reactors. A brief review of the history of power reactor technology, with concepts for future designs, was published 10 years ago [2]. However, the predominant nuclear reactor type for the generation of electric power today is the light-water reactor (LWR), which is typically fueled with nominally 3%–4% LEU. The two common variants of the LWR are the boiling-water reactor (BWR) and the pressurized-water reactor (PWR). Existing operating power reactors worldwide are 19% BWRs and 63% PWRs, with the remaining 18% divided among heavy-water, graphite-moderated, gas-cooled, and liquid-metal-cooled reactors of various models. At this writing, future power reactors on order or under construction will comprise 83% PWRs, 7% heavy-water units, and 6% BWRs.

A commercial LWR typically contains about 100 t of ceramic fuel pellets and operates at 300°C. Design output power is approximately 3–4 gigawatts thermal (GW_t) and, at about 30% efficiency, results in electric output of approximately 1 GW_e. The lifetime of an LWR is about 40 years, but a given fuel rod resides in the reactor for only 4 years, and about 20% of a core is replaced annually. The approximately 20 t of highly radioactive spent fuel discharged each year is then stored on-site in cooling ponds for some years.

The Pu production of nuclear power reactors is considerable. The nominal 1-GW_e reactor burns about 1 t of fissile material per year and results in the production of about 200 kg of Pu (i.e., sufficient material for a bomb approximately every week). Each year, power reactors around the world generate about 70 t of Pu.

1.3 NUCLEAR WEAPONS AND THE COLD WAR

The Manhattan Project culminated in the production of three atomic bombs [3,4], two of which were deployed against Japan to expedite the end of World War II. At 8:15 on August 6, 1945, the American B-29 bomber, *Enola Gay* (named after the pilot's mother), dropped a gun-assembled U device on the city of Hiroshima. The weapon, designated *Little Boy*, weighed about 4100 kg and contained approximately 65 kg of 84%-enriched ^{235}U. Only 1% of the ^{235}U fissioned, but the explosive yield was 13 kt and resulted in the deaths of 140,000 people by the end of 1945. Three days later, an implosion-assembled Pu bomb, *Fat Man*, was dropped on Nagasaki. That device consisted of a 6-kg solid sphere of greater than 95% ^{239}Pu, with a total weight of 4900 kg. It was 20% efficient, gave a yield of 21 kt, and caused 70,000 fatalities by the end of 1945. Nuclear weapons have not been used in armed conflict since.

What ensued, however, was a Cold War competition and nuclear arms race between the United States and Soviet Union (USSR) that dominated both superpowers until the early 1990s. Following the detonation of a fission bomb by the USSR in 1949, both countries embarked on accelerated programs of research and development for designs of novel, more efficient, and more powerful nuclear weapons. Advances in primary technology led to hollow-shell pits, boosting, and variable "dial-a-yield" mechanisms. Pure fission bombs could thus approach yields of 500 kt with optimum designs. Beyond those efforts, moreover, fusion weapons were also pursued. Both superpowers developed and tested immensely more powerful multistage weapons containing thermonuclear fuels such as DT and ^6LiD. The first successful US hydrogen bomb, exploded in 1952, was an unwieldy, cryogenic, unweaponizable, proof-of-concept device. The yield of *Ivy Mike* on Eniwetok atoll in the Marshall Islands was 10 Mt. The USSR followed with their first thermonuclear device in 1954, and the largest nuclear bomb exploded to date was a 60-Mt shot in 1961 by the Soviet Union.

At the same time, significant efforts were also focused on improving the safety and security of nuclear weapons. They were driven in the United States by a classified 1958 RAND report that used Bayesian statistics to evaluate the probability of an accidental or unauthorized nuclear detonation by the escalating Cold War strategic bombing posture [5]. These improvements were technical [e.g., development of insensitive HE, fire-resistant pits, and Ga-stabilized Pu alloys], operational [permissive action links (PALs), two-man arming rule, and arming only over enemy territory], and sociologic [psychological testing of military personnel controlling nuclear warheads and open-literature publication of technical safety articles in scientific journals].

Along with the three other declared nuclear powers (Great Britain, France, and China), the United States and the USSR conducted numerous tests of various devices in the atmosphere and underground. Primarily at test sites in Nevada and the Marshall Islands, the United States executed more than 1000 nuclear tests between 1945 and 1992 (>800 underground and >200 atmospheric). The US South Pacific test series occurred during 1946–1962 and resulted in more than 65 atmospheric nuclear detonations. Estimates of the nuclear programs of other nations have the USSR conducting more than 700 total tests, France more than 200, England and China more than 40 each, and more recently, India, Pakistan, and North Korea on the order of a half-dozen tests apiece. In all, more than 2000 nuclear tests exploded worldwide between 1945 and 1998, with more than 500 detonations taking place in the atmosphere. Atmospheric testing by the nuclear powers distributed large quantities of fission products and actinides, including 5000 kg of ^{239}Pu, throughout the biosphere as worldwide fallout.

The ensuing buildups of nuclear warheads by the five historic powers were awesome, expensive, illogical, and scary. A single 475-kt device would decimate an area of 125 km^2 from the explosive blast alone. Yet, including its total nuclear force of strategic, tactical, and reserve weapons, the United States peaked at a total of approximately 31,000 weapons in 1967 and spent $6 trillion on its military nuclear program between 1940 and 1995. Similarly, the USSR produced 45,000 warheads by 1986, becoming the dominant component of the highest historic global stockpile inventory of 70,000 nuclear weapons. To appreciate the scope of destruction that these weapons could cause, consider the following simplified calculation: Assuming that all 45,000 Soviet weapons had an average yield of 500 kt, then 5.6 million km^2 of Earth's surface could be destroyed by the Soviet nuclear arsenal alone. This number corresponds to more than 60% of the total land and water mass of the United States. Given that blast is but one lethal component of a nuclear explosion, the devastation would be even more severe if other significant prompt factors, such as thermal effects and massive doses of neutron-, x-, and γ-radiations, were also considered. In addition, other grim effects, more delayed in time, include dangerous radiation exposure levels to any surviving population and the downwind fallout. These escalations in destructive power occurred within Cold War strategies of launch-on-warning and mutually assured destruction (appropriately "MAD"). The latter consisted of the complete annihilation of at least 40% of an adversary's population and 70% of its industry.

In total, the United States produced 100 t of weapons-grade Pu and 994 t of HEU for its weapons program. The Soviet Union produced comparable numbers. As of 2011, the US war-reserve military stockpile had been reduced to about 5000 warheads, 40% of which were operational strategic weapons. The stockpile is spread over the triad of bomber aircraft, intercontinental ballistic missiles (ICBMs), and submarine-launched ballistic missiles (SLBMs). Each component of the US strategic triad provides independent MAD. Current appraisals of Russia indicate 8,000–10,000 warheads, with 2,400 of them strategic. France has perhaps 300 stockpiled nuclear weapons, China 240, and Britain 220, while present estimates for Israel, India, and Pakistan are in the ballpark of 80–100 each.

1.4 NUCLEAR TREATIES AND NONPROLIFERATION PROGRAMS

At its peak in 1962, atmospheric testing by the superpowers generated about 72 Mt of fission yield as fallout in a single year. The next year, the United States, USSR, and Britain agreed to the Limited Test Ban Treaty, which prohibited nuclear explosions within Earth's atmosphere, the oceans, and in outer space. In 1970, the Treaty on the Nonproliferation of Nuclear Weapons (NPT) entered into force. Signed by a total of 190 nations (including Iran), it is the most extensively observed arms control treaty in history and deals primarily with horizontal proliferation. However, the nuclear weapon states of Israel, India, Pakistan, and North Korea (which withdrew in 2003) are not party to this agreement, and Iran is noncompliant. For the most part (with France and China sporadic exceptions), nuclear tests after 1963 have been conducted underground. Additional limitation by the Threshold Test Ban Treaty of 1974 restricted such underground shots to less than 150 kt.

In 1991, after the fragmentation of the Soviet Union, the Republic of Kazakhstan closed the Semipalatinsk nuclear test site. The next year, the United States initiated a self-imposed moratorium on nuclear testing, conducting the last such experiment (*Divider* in Operation Julin) at the Nevada Test Site in September 1992. Finally, the five historic nuclear weapon states—and almost all other countries—signed a Comprehensive Test Ban Treaty (CTBT) in 1996. The CTBT contains both horizontal and vertical nonproliferation aspects and has been signed by 183 nations and ratified by 158. However, the CTBT can enter into force only after it has been ratified by all 44 countries that possessed either nuclear weapons or nuclear reactors in 1996, and 8 (China, Egypt, India, Iran, Israel, North Korea, Pakistan, and the United States) have not done so.

Professional government negotiators continue to accomplish various other nuclear arms control agreements, treaties, and antiterrorism measures. These have become too numerous to treat in worthy fashion here but would include, for example, the START efforts, Global Threat Reduction Initiative, Global Proliferation Security Initiative, the bilateral (United States and Russia) Pu Management and Disposition Agreement, Global Initiative to Combat Nuclear Terrorism, Convention on the Physical Protection of Nuclear Material, a Fissile Material Cut-Off Treaty, nuclear weapons–free zone treaties, and so forth. Thus, for example, the New START Treaty of 2011 limits both the United States and Russia to a cap of 1550 deployed strategic nuclear weapons and 5000 total warheads by 2018. It also established counting formulas for deployed and stored reserve warheads, but did not at all address tactical weapons.

Prior to New START, in 2002, Presidents George W. Bush and Vladimir Putin signed the Strategic Offensive Reductions Treaty (*aka* the Treaty of Moscow), which would reduce strategic nuclear arsenals by two-thirds, to 1700–2200 warheads for each country, by 2012. This treaty was remarkable from a diplomacy standpoint. Negotiated for the United States by Secretary of State Colin Powell within 6 months, the Treaty of Moscow comprised three pages, five articles, and 475 words. In contrast, START I was 700 pages of excruciating detail that required more than a decade to negotiate.

In 2009, US President Barack Obama labeled nuclear terrorism the most pressing and extreme danger to global security. Indeed, Osama bin Laden, remaining al-Qaida terrorists, and other Muslim extremists believe that possession and use of unconventional weapons, including weapons of mass destruction (WMD), is a religious duty. The United States hosted the first Nuclear Security Summit in Washington, DC, during 2010. Attended by 47 nations, among the agreements reached were either the complete shutdown or the conversion to LEU operation of the approximately 120 worldwide HEU-fueled research reactors; the collection and protection of all HEU in civilian possession; the transfer of nuclear material inventories to existing weapons states; and the concurrence of the United States and Russia to each destroy 34 t of excess weapons-grade Pu. By 2011, 72 research reactors in 33 countries had been so converted or shut down, and in 2012, Austria became the 22nd nation to remove all HEU from within its borders.

The second Nuclear Security Summit took place in South Korea in 2012 and was attended by 53 nations. Progress was gauged on the 2010 goal of securing all vulnerable nuclear materials worldwide by 2014, and dialogues on security considerations of radiologic sources were conducted.

In early 2013, the Obama administration proposed further reductions to the US nuclear stockpile. The current level of 1700 deployed strategic warheads, to reduce to 1550 by 2018 via New START, would be further decreased to roughly 1000 or so as a cost-savings measure. Facing likely Congressional opposition, rather than negotiate a new arms-control treaty, President Obama could choose a course of announcing unilateral cuts or of reaching informal agreement with Russian President Putin for mutual reductions within the New START framework. This latter tactic would not require ratification.

Technology advances contribute to global nuclear nonproliferation as well. An International Monitoring System (IMS) configured under the CTBT consists of an array of 337 seismic and other (infrasound, hydroacoustic, and radionuclide) sensor stations in 89 nations and Antarctica. To date, 85% of the planned monitoring posts are in operation. The goal of the IMS is to guarantee that the output of any successfully hidden nuclear test explosion would necessarily be so small as to lack any military significance. Thus, the IMS detects and identifies nuclear explosions, with a threshold of 1 kt, in any terrestrial environment in the world. US Department of Energy nonproliferation programs include the 1994 Initiative for Proliferation Prevention and the 1998 Nuclear Cities Initiative, in conjunction with various other government-to-government and lab-to-lab programs.

1.5 SNM DISPOSITION

Almost from the beginning of the nuclear era, SNM spread beyond the early nuclear weapon states, and many nations began their own nuclear programs, mostly under international auspices. The US Atoms for Peace Program transferred 13 t of HEU of US origin to Euratom during the 1950s and 749 kg of Pu to 39 foreign countries between 1959 and 1991. (Less benign dispositions of SNM were to the environment through fallout and the loss of 11 US nuclear bombs in military accidents, not all of which were recovered.)

A modern source of available SNM is the build-down of weapon warheads. Although recent Pu aging studies indicate that nuclear pit lifetimes can be >100 years, a full weapons system will survive reliably for only about 15 years. Historically, the United States has annually monitored its nuclear arsenal through the random inspection and disassembly of 11 weapons of each type in its stockpile. Retired US nuclear weapons are disassembled at the Department of Energy Pantex Plant and separated into categories of HE, fissile SNM, tritium, electronic controls, and miscellaneous parts.

However, the principal modern source of significant SNM is the nuclear power industry. Because of ^{239}Pu proliferation concerns, the US approach to nuclear power is the "once-through" or "open" fuel cycle, which results in the direct disposal of spent fuel. Civilian reactor fuel rods are stored in aboveground holding facilities for several decades until they are permanently buried in a geologic repository. The alternative to the open cycle is to reprocess irradiated reactor elements after their lifetime as efficient power components ends. Although less economical than direct disposal, chemically reprocessing spent U fuel recovers 99.9% of the radiogenic Pu, with separation factors of 10^7–10^8 from the fission products.

Approximately 2 kg of PuO_2 (~25% ^{240}Pu) from a reprocessing facility goes into a welded steel can for subsequent operations. Popular disposal options after that point are vitrification along with mixed fission products into borosilicate glass logs, as well as assimilation into MOX fuel. MOX is a ceramic mixed oxide of Pu in ^{238}U that can be used in existing LWRs. However, because Pu reactor fuels require control rods with enhanced capacity for neutron absorption, an LWR can only burn MOX in about one-third of its core.

Today, nuclear weapons, recycled Pu in civil power programs, spent fuel, and legacy wastes comprise >2200 t of Pu across the world. An estimated 90–100 t of surplus weapons-grade Pu exist in the military programs of both the United States and Russia. However, these quantities are far smaller than the Pu produced in the (continuing) generation of nuclear power, and >200 t of civil Pu have already been separated for MOX utilization. In 1998, the two countries agreed to excess 50 t each of defense Pu. A following summit in Moscow in 2000 designated 34 t of weapons-grade Pu in each nation disposed of by glass immobilization (25%) and in MOX fuel (75%). When the Bush Administration replaced Clinton's in US politics, though, a policy change shifted all 34 t into MOX.

Analogous disposition of HEU also occurs between the superpowers. For example, the 1993 Megatons-to-Megawatts Agreement resulted in Russia selling 500 t of HEU from dismantled warheads to the United States for $12 billion over 20 years. The HEU was down-blended with 1.5% LEU to yield 4.4% LEU for use as reactor fuel. The most successful nonproliferation program in history, Mt-to-Mw, was completed in November–December 2013. On the order of 20,000 nuclear warheads were converted to fuel by this program.

Diplomatic advances for further material dispositions remain necessary for the future, however. In 2010, the Institute for Science and International Security estimated the quantity of fissile materials worldwide at 3800 t, with ~2000 t in weapons-usable forms.

1.6 NUCLEAR PROLIFERATION AND TERRORISM

The United States and USSR both controlled huge arsenals of nuclear weapons during the Cold War, and a widespread fear was nuclear warfare between the two countries. Following the 1991 disintegration of the Soviet Union, however, apprehension over unrestrained nuclear exchanges diminished and was replaced by increasing anxieties over the nuclear ambitions of terrorists and rogue states.

Significant technical, diplomatic, intelligence, and law enforcement assets are committed worldwide to detect and counter the proliferation of illicit nuclear materials and nuclear terrorism. The foremost obstacle to nuclear weapons acquisition is the availability of HEU and Pu, and proliferation prevention largely depends on deterring access to fissionable materials. Production of both HEU and Pu relies on U starting material, the source of which is geologic mines, but the U supply may be further extended in the future by extraction from seawater and fuel recycling by fast breeder reactors.

Extreme but effective examples of nuclear counterproliferation were the Israeli air strikes to destroy Iraq's Osirak reactor at Tuwaitha in 1981 and Syria's clandestine reactor at al-Kibar in 2007. Less drastically, the United States conducted Project Sapphire in 1994 to remove about 600 kg of at-risk HEU from Kazakhstan to the Y-12 facility at Oak Ridge (in cooperation with Kazakhstan).

Technical methods to detect nuclear materials are mixed in their effectiveness. Thus, although Pu can be detected through the abundant 60-keV γ ray of the ^{241}Am decay daughter of ^{241}Pu, detection of HEU is much more difficult because it lacks significant intensity of penetrating radiation. It has been proposed to admix 72-year ^{232}U into HEU to allow surveillance by low-cost radiation monitors. (It is actually ^{208}Tl in equilibrium with ^{232}U that produces [among others] an intense 2615-keV γ ray; this same decay chain also provides intrinsic detectability for ^{233}U produced in the Th fuel cycle, as about 0.1% ^{232}U is also generated in the breeder reactor.) Portal-monitor radiation sensors at site boundaries and border crossings can detect quantities of weapons-grade Pu and HEU, unshielded and in a worst-case geometry (a metal sphere), as well as less than 100 g of Pu within reasonably thick Pb shielding. A related tactic, but not for real-time SNM detection, would be the addition of a specific proportion of long-lived ^{233}U to HEU for source attribution by laboratory nuclear forensic analysis.

Since the initial developments by the five declared nuclear states (United States, USSR/Russia, United Kingdom, France, and China), other countries have joined the nuclear weapons community through concerted national efforts. India pursued the Pu route and detonated a 5–12 kt device underground in 1974, followed by five additional tests in 1998. Pakistan, predominantly emphasizing HEU development, tested six lower-yield devices in 1998 as well. South Africa, perhaps without testing, compiled a clandestine arsenal of six gun-assembled HEU weapons in an undeclared program, but ultimately dismantled them and joined the NPT in 1991. A pseudo-covert nuclear weapons state is Israel, based on a Pu program.

North Korea (DPRK) has tested several Pu-based nuclear weapons, but the forensic evidence to date is marginal and somewhat controversial (including allegations of mimicking underground nuclear tests with detonations of conventional HE). Initial events during 2006 and 2009 were perhaps consistent with explosive yields of ≤1 kt and 1–10 kt, respectively. Another candidate low-yield underground test in 2010 provided no seismic signal, but did release suggestive radioactive Xe. The isotopic ratios of the Xe [6] were unusual, though, perhaps indicative of a mixed source-term, and the possibility of reprocessing activities could not be dismissed by some analysts. However, a recent publication [7] attributes the measurement discrepancies to underground fractionation during melt cooling (see Section 5.4.2). A 2013 test induced a seismic event of magnitude 5.1, consistent with an underground nuclear explosion of several kt. Table 1.2 lists initial weapons tests and SNM production information for the various countries.

Another clandestine nuclear proliferant was Iraq. Using electromagnetic isotope separation (EMIS) and centrifuge technologies, Iraq produced about 10 kg of unirradiated HEU and rather more irradiated 80% ^{235}U. It was estimated that Iraq was perhaps 1–3 years away from producing quantities sufficient for a bomb before being blocked by the 1991 Gulf War. Post-conflict U.N. inspections likewise discovered ^{210}Po material, a radionuclide used with Be as an (α,n) neutron source for initiators of primary fission weapons.

In a neighboring geographic locale, Iran pursues a nuclear power program despite abundant fossil fuel reserves and fails to meet safeguards obligations in violation of the NPT. Dating as far back as the late 1970s, Iran has used and processed undeclared nuclear materials, failed to declare facilities, and has generally refused cooperation with inspections to any significant level. Heavily bunkered underground centrifuge activities for U enrichment have been significantly increasing capacity, and LEU fuel enrichment to 20% ^{235}U began during 2010. By 2012, tens of kg of 20% intermediate-enriched U per month were being stockpiled, thereby dramatically reducing the time and effort that would be needed to increase to >90% weapons-grade HEU. These activities are indicative of the development of a nuclear explosive device and, considered along with the program at Parchin to develop a nuclear payload for missiles, make military motives the natural conclusion.

No treatment of nuclear proliferation would be inclusive without mention of Pakistan's A.Q. Khan, the reigning protagonist for the distribution of weapons technology to rogue states. A metallurgist with the knowledge of classified gas centrifuge technology from the Urenco enrichment plant, Khan established the Kahuta facility in Punjab Province in 1976. By virtue of a broad, covert organization, technology and materials were then acquired for Pakistan's U enrichment program. He also created an extensive clandestine network of >30 companies in 30 countries to distribute enrichment apparatus and expertise to nascent proliferants. In 2004, Khan confessed to having supplied technical and materials assistance since the mid-1980s for nuclear weapons developments in Iran, Libya, and North Korea. North Korea, in turn, served as enabler for Syria and Libya.

Subnational terrorist groups also engender proliferation fears, and acquisition of SNM or weapons-usable materials from civilian power reactors is of concern. In particular, Russian power and CANDU reactors can be unloaded while still operating, thereby making diversion easier. In addition to developing biologic- and chemical-warfare agents, the Japan-based Aum Shinrikyo cult also had programs in fissile material acquisition and weapons production. But despite immense financial resources,

TABLE 1.2
Initial Sources of Fissile Materials and Explosive Device Tests by Weapons States

Country	Material	Location	Facility	Test Date	Test Location
United States	HEU	Oak Ridge, TN	Gaseous diffusion and calutrons	July 16, 1945	Trinity site, Alamogordo NM
	Pu	Hanford, WA	Graphite-moderated LWRs		
USSR	HEU	Tomsk-7, Sverdlovsk-44 and -45, and Krasnoyarsk-45	Gaseous diffusion plants	August 29, 1949	Semipalatinsk, Kazakh SSR
	Pu	Chelyabinsk and Krasnoyarsk-26	Graphite-moderated LWRs		
United Kingdom	Pu	Windscale (Sellafield)	Calder Hall graphite-moderated, air-cooled reactors	October 3, 1952	Montebello Islands (off western Australia)
France	Pu	Chusclan and Codolet communes	Marcoule graphite-moderated, CO_2-cooled reactors	February 13, 1960	Reganne Oasis, Sahara Desert, Algeria
China	HEU	Lanzhou and Heping	Gaseous diffusion plants	October 16, 1964	Lop Nor test site, Xinjiang Uygur Autonomous Region
	Pu	Jiuquan and Guangyuan	Graphite-moderated LWRs		
India	Pu	Trombay	Cirus and Dhruva D_2O research reactors (diversion)	May 18, 1974	Pokhran test range, Jaisalmer, Rajasthan
Pakistan	HEU	Kahuta	Gas centrifuges	May 28, 1998	Ras Koh, Chagai
	Pu	Khushab	D_2O research reactor (diversion)		
North Korea	Pu	North Pyong'an Province	Yongbyon graphite-moderated, light-water research reactors (diversion)	October 9, 2006	Uncertain mountainous region in northeastern area of North Korea
Israel	Pu	Dimona	D_2O research reactor (diversion)	[a]	[a]

[a] No confirmed test by Israel is known. However, on September 22, 1979, a bhangmeter on a Vela satellite registered a double flash, over the Indian Ocean off the coast of South Africa, consistent with a small (2–3 kt) nuclear explosion. It has been widely speculated that this event was an Israeli, or joint Israeli–South African, weapons test. This conjecture remains highly disputed today.

the Aum failed to develop nuclear weapons because they lacked both the expertise and access to the necessary materials. Instead, with 30,000 followers of the sect in Russia at the time of their 1995 sarin attack on the Tokyo subway system, the cult attempted to procure Russian warheads and enlist nuclear weapons specialists.

A less extreme, yet more accessible and probable, nuance of nuclear terrorism is the radiologic dispersal device (RDD; also termed a "dirty bomb"). It consists of highly radioactive material in combination with a conventional explosive to disseminate radiation contamination throughout the near environment without a nuclear explosion. This subject is considered in detail in Chapter 13. Potential RDD source materials have myriad applications and are consequently found throughout the world. An RDD has not been successfully executed as yet, but Iraq is believed to have conducted relevant testing for battlefield deployment.

In 1995, Chechen rebels deployed a ^{137}Cs-dynamite bomb in Izmaylovsky Park in Moscow, but did not detonate it. The Chechens are closely aligned with the al-Qaida terror network, which believes that the acquisition and use of a nuclear device is an Islamic mandate. In Chicago in 2002, the United States arrested US citizen and al-Qaida operative Abdullah al-Muhajir, who, according to Attorney General John Ashcroft, was on a mission to identify targets for an RDD attack. He was incarcerated as an enemy combatant by President Bush. In 2006, Dhiren Barot, a Briton, was sentenced to life in prison for leading a radical cell that schemed to make a terrorist bomb containing 10,000 smoke-detector sources [8]. And in December 2008, containers of H_2O_2, U, Th, Li, thermite, Al powder, Be, B, Mg, and FeO, along with literature for constructing an RDD, were discovered at the Maine residence of James Cummings. He had recently been killed in a domestic dispute and was linked to white supremacist groups.

An RDD is not a weapon of mass destruction but, rather, one of mass disruption. The principal casualties would be individuals proximate to the blast of the HE. For al-Qaida and other terror groups, however, the appeal of RDDs is their shock value. Psychological effect and mass panic, rather than death and destruction, are the operational targets. The chief physical consequence of an RDD would be economic and would result from the widespread decontamination and required demolition in an (likely) urban environment.

Russia and the United States have undertaken a cooperative program to sequester any stray radiation sources across the former Soviet Union. Russia supplied information on the cradle-to-grave control of such medical and industrial materials, while the US Department of Energy funded efforts to locate the sources, with organization and management provided by the IAEA.

1.7 NUCLEAR SMUGGLING

In 1991, the end of the Cold War and dissolution of the USSR resulted in the formation of 12 independent successor states. Of these, Russia, Belarus, Ukraine, and Kazakhstan inherited the strategic nuclear arsenal of the former Soviet Union (FSU). However, many nuclear manufacturing and research facilities had few safeguards to avert theft. Inadequate physical security, as well as poor materials protection, control, and accountability, was common for sites involved in naval fuel production, civilian fuel research and development, weapons fabrication, and warhead storage. By 1996, all FSU nuclear weapons were relocated and centralized in Russia; Russia now shares 20,000 km of border with 14 different countries.

Depressed economic and social conditions in the FSU stimulated illegal smuggling and black-market sales of various nuclear materials. As noted earlier, obtaining the technology to make fissile materials is the major impediment to nuclear proliferation. Even for established national programs, the cost of producing a kilogram of either HEU or weapons-grade Pu is on the order of several tens of thousands of (US) dollars. A successful theft and nuclear smuggling operation, however, could eliminate these technical and economic barriers. In addition to actual warheads and the nuclear materials from dismantlement operations, several hundred tons of legacy SNM that are not incorporated into weapons contribute to the large surplus of materials that constitutes the total threat.

The first identified incidence of illegal trafficking of nuclear material on the black market occurred in 1966. The great majority of similar cases since then has involved hoaxes or sales of non-SNM radioactive materials [9], and such seizures by law enforcement agencies have become familiar throughout Europe. The most popular radioactive contraband involved in these scams are industrial/medical radiation sources, such as ^{60}Co and ^{137}Cs, and miscellaneous nuclear materials, such as D-38 and natural U (both sold as HEU), LEU, ^{241}Am, and ^{252}Cf. Over the last 30 years, a number of recurrent nuclear hoaxes have been investigated at Livermore, and they are grouped into regional and materials categories.

The regional scams comprise four persistent U trafficking variations: Afghanistan (1995–), Southeast Asia (1991–), Philippines (1985–), and Colombia (1996–). An example of the Southeast Asian U hoax is given by the D-38 counterweight described in Chapter 22, and the Afghan U scam is similarly represented in Chapter 27. The Colombian U trafficking has generally attempted to sell innocuous blocks of metal (as U), ^{90}Sr source holders (as U [or Pu]), and various ore materials (as enriched U or HEU).

The materials-focused swindles include sales of ^{187}Os and "red mercury," as well as allegations of the existence of "suitcase nukes." All nuclear scams are characterized by a remarkable naiveté on the part of prospective buyers, and ^{187}Os is an additional curiosity due to an ambiguous aspect of its origin. That is, since 1992, it has been advertised for use in nuclear weapons (of course), as well as in missiles or missile components, and it has been offered for sale for (US) \$1,000–\$80,000/g. But the real ^{187}Os is an otherwise nondescript minor stable isotope of Os, 2% abundant, and one of seven such stable isotopes of the element. So why was this particular isotopic species the choice for the label of the illicit enterprise? Absent any obvious technical relevance of ^{187}Os, the Livermore Nuclear Assessment Program (NAP) speculated that it may have actually originated as ^{137}Cs. With the lower-quality electronic equipment available for nuclear black-market transactions during the early years, as offer sheets, technical specifications, and transport documents were faxed back and forth multiple times, the "3" may have filled in to more resemble an "8," and the "C" similarly to an "O."

Beginning in 1979 and selling for \$200,000–\$500,000/kg, the red-mercury phenomenon has been particularly incredible from a technical standpoint. Alleged to be a unique strategic component of a nuclear weapon, it purportedly allows ignition of thermonuclear fuel without a requisite SNM fission trigger. Red mercury also has myriad other alleged applications, and purported material properties have included densities up to 23 g/cm^3, an isotopic temperature of 160°C, and the presence of Lr as an actinide impurity (the longest-lived known isotope of Lr is ^{262}Lr, with a half-life of only 3.6 h). Additional detail on such questioned specimens analyzed at Livermore National Laboratory (LLNL) is given in Refs. [10,11]. Indeed, entertaining, fringe red-mercury allegations persist to the present time in books, television shows, video games, and other popular culture, including the assertion of counter-rotating cylinders of the substance that emit lethal radiation and allow the manipulation of time (*Die Glocke*, or the "Nazi Bell") [12].

"Suitcase" or "backpack" nukes began their run during the late 1990s, after retired Russian general Alexander Lebed asserted that stolen Russian atomic demolition munitions, ~1-kt devices, were for sale on the black market. However, no known credible source, or credible source of corroborating information, has endorsed this allegation, and the claim was likely politically motivated.

Falsified certificates of authenticity, specification sheets, detailed lab analyses, or documents with official seals and signatures often accompany black-market nuclear materials. The principal individuals implicated in nuclear smuggling are organized crime elements within the FSU, profit-motivated opportunists, and con artists. Trafficking occurs across international borders and along various routes, with main transit channels being through Bulgaria and Turkey, south from Kazakhstan through Iran and Southwest Asia, and south from Russia through Georgia.

Until 1991, all known sales attempts of nuclear materials on the black market were scams. Between 1991 and 1993, most were still hoaxes, although some amounts of SNM emerged. Since 1993, however, although sales remained predominantly fraudulent, kilogram quantities of weapons-usable materials became available. The annual trend in seizures of illicit nuclear materials from 1985 to 2008 is depicted in Figure 1.1. Of these interdictions, few involved either weapons-grade or weapons-usable SNM.

"Counting" the number of illicit radioactivity trafficking incidents worldwide is not an exact science. For example, the IAEA registry listed approximately 1800 such events from 1993 to 2009. However, through 2011, the LLNL NAP followed >2500 cases (nearly all scams that alleged HEU or Pu), of which <800 entailed interdictions of radioactive materials. Part of the discrepancy can be attributed to Livermore's capacity for tracking all such events, including classified investigations, and not restricted solely to open-source information. However, LLNL focuses on thefts, diversions, smuggling, and other illicit activities, while the IAEA additionally includes occurrences that Livermore does not. These include ordinary losses of materials, orphan sources, and situations involving μg-quantity smoke detectors. The IAEA also counts the theft and interdiction of the same sample as two separate incidents.

Figure 1.2 shows seizures of actual weapons-usable U or Pu from 1992 to 2011. As this book is a free-release publication, knowledge of these incidents is unclassified, they number <20, and they are in accord with IAEA information.

During the poster year of 1994, 3 kg of 90% ^{235}U were stolen from a plant near Moscow, but the material was later recovered in St. Petersburg. The largest seizure

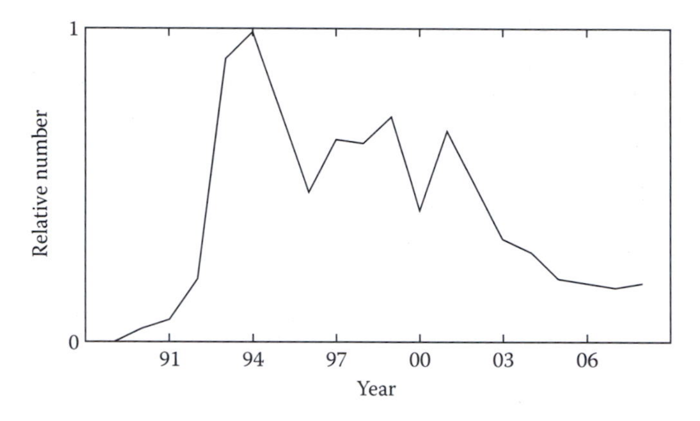

FIGURE 1.1 Radioactive material seizures.

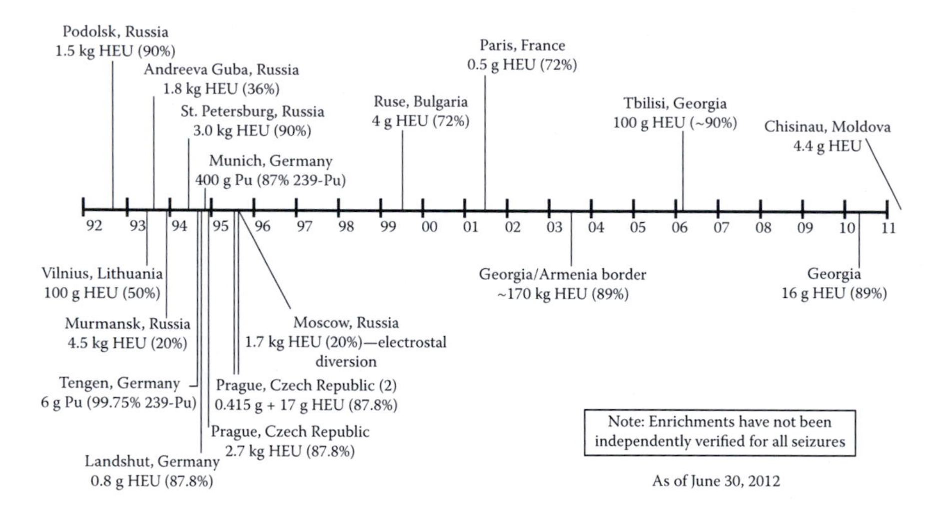

FIGURE 1.2 Seizures of weapons-usable nuclear materials. (Courtesy of Rob Allen and LLNL Nuclear Assessment Program, Livermore, CA.)

to date of bomb-grade, black-market material outside of Russia occurred in Prague in December: two cylinders containing 2.7 kg of 88% ^{235}U as UO_2 were discovered in a parked car. Two Russians and a Czech physicist were arrested. And earlier that year, the German federal intelligence organization (BND) orchestrated a sting that culminated in the seizure of 560 g of a mixed oxide of U and Pu (~400 g of 87% ^{239}Pu mixed with LEU). The contraband also included 210 g of Li metal enriched to 89% in ^{6}Li. A lead-lined suitcase containing the smuggled goods was seized at the Munich airport following a Lufthansa flight from Moscow. The material was of Russian origin, and two Spaniards and a Colombian (with ties to illicit narcotics smuggling) were imprisoned for nuclear trafficking.

A paradigm shift occurred in the smuggling of true SNM between the mid-1990s and the present time. The earliest seizures could be generalized as kg quantities of variable enrichments, occurring three or four times per year. Since then, though, the frequency of such interdictions has decreased to perhaps one every 3 or 4 years. Concurrently, the confiscated amounts of SNM dropped off appreciably, but the sample qualities and HEU enrichments were consistently high. As with quality-assurance specimens in the criminal drug world, the sales pitch of nuclear traffickers now incorporates the availability of representative small samples for technical verification analyses. The claimed further availability of large quantities of identical material at a hidden location is then negotiated with the buyer to complete the transaction. Further detail on recent weapons-grade HEU interdictions may be found in Ref. [13].

Because weapons-usable materials became increasingly accessible on the international black market, global attention has focused on the potential danger of nuclear proliferation by rogue nations, as well as on nuclear extortion and terrorism by subnational groups. To counter these new threats, innovative disposal protocols for dismantlement operations, and better nuclear safeguards in general, have been developed to prevent the diversion of materials from existing national inventories. Further, as the availability of true SNM contraband increased significantly, the various contemporary nuclear smuggling scams warranted increased attention and analysis.

1.8 FORENSIC EFFORT AREAS AND GOALS

Nuclear forensic analysis (NFA) consists of the reliable collection, treatment, analyses, and assessment of evidentiary specimens for elemental, isotopic, chemical, and physical signature species that may provide technical insights into the origins and histories of primary questioned material and associated collateral exhibits. Relationships to military and civil nuclear fuel cycles are particularly important considerations for the attribution of forensic heavy-element profiles of a primary specimen.

NFA was initially the nearly exclusive purview of weapons laboratories and intelligence agencies. However, over the past 20–25 years, NFA has gained international attention beyond those boundaries. An important influence on this increased political and public awareness has been a convergence of heightened transnational terrorism, the indiscriminate application of chemical and biologic WMD against nonmilitary populations, September 11, 2001, and suicide bombing in general, and the wide-ranging regard for the various CSI television series. One result of such interest by personnel of varied and vague credentials was the differentiation of some NFA activities into groupings based on areas of emphasis.

1.8.1 IND versus RDD

Although key NFA focal points are nonproliferation, weapons diagnostics, nuclear smuggling, and related issues, two fearsome potential weapons are particularly worrisome. An improvised nuclear device (IND) is a theoretical bomb, designed or assembled outside an official government agency, which incorporates fissionable and perhaps fusionable fuel, and is intended to produce a nuclear explosion. A nuclear weapon stolen from an existing stockpile and subsequently modified is generally categorized as an IND as well. An RDD is likewise a terrorist device that includes radioactive material(s). However, it is not necessary that the radionuclide(s) produce a nuclear yield. This bomb is meant to locally disseminate radiation-emitting material by means of conventional HE.

The technical complexity of a functional nuclear explosive and safeguards efforts by weapons states, as well as the tremendous difficulties and costs of producing either weapons-grade plutonium or HEU, serve to categorize an IND as a low-probability prospect. However, the aftermath of a successful IND would be cataclysmic. The reverse is true for the case of an RDD. The widespread use of large commercial radiation sources translates to much higher probability for an RDD attack, but one with significantly reduced consequences compared to the IND.

Some countries that concentrate on nuclear terrorism direct most of their attention to the IND danger, having decided to value the potential consequences of such an attack over the relative odds that such an event would occur. The levels of funding support, preparation, and exercises in nuclear counterterrorism, to include NFA, echo that decision. Yet other states are fully focused on the RDD threat, siding with the likelihood statistics for their preparedness profile.

The adopted taxonomy for these disparate efforts has been somewhat forced. Although both terrorist bombs fundamentally depend on radioactive nuclides for their effects, the nonfissionable radioisotopes have been termed "rad" materials, while SNM is set apart as "proper" nuclear matter. Thus, the awkward expression "rad-nuke" has joined the lexicon to arbitrarily separate nuclides based on radical end-use or terror-generated consequences. [Does ^{235}U in an IND transmutate into rad material if there is no nuclear yield, but the HE scatters the HEU into the environment? Allegedly so. Further, to continue the (il)logic toward absurdity, the SNM pit of a functional nuclear weapon is both rad and nuke simultaneously. That is, no nuclear explosion is 100% efficient. So for the Nagasaki bomb, the "nuke" 1-kg fraction of the ^{239}Pu core provided the explosive yield, while the (identical and indistinguishable) 5-kg "rad" component went into fallout.] A relatively "clean," initial all-WMD descriptor, "NBC" (for nuclear, biologic, and chemical), is therefore now "CBRN" or "CBRNE" (explosives). The latter acronyms are currently found in an assortment of national technical units, professional groups, government agencies, and trade magazines.

1.8.2 Pre-Det and Post-Det

Another segregated subject area is predicated on whether a successful explosion was obtained with an IND or RDD (a postdetonation situation), or whether a weapon was intercepted or unsuccessful upon firing (the predetonation condition). These two classifications involve quite disparate consequences for incident response, and they accordingly dictate different tactics for the subsequent NFA as well.

Post-det activities after a successful IND would entail the methodology and measurement systematics created during nuclear testing that focused on radiochemical and other diagnostics. Target analytes would be residual actinides, major and high-value-minor fission products, and activation species. Assessment of device materials, design, operation, near vicinity, and other characteristics is possible with suitable post-shot inorganic and radiochemical data.

A post-det RDD would also be predicated on traditional radiochemistry, complemented by conventional forensic examinations. The latter might assay residual HE, perpetrator-specific exhibits (e.g., latent fingermarks, hair, DNA, or other biometrics), and other potential evidentiary items (paints, fibers, paper, material impressions, etc.). The issue of contaminated evidence (normal forensic processing of radioactively tainted specimens) should be much more important for a post-det RDD, since a functioning IND would presumably eradicate any such articles close to the detonation.

A pre-det analysis of both an IND and an RDD would target a very similar suite of radionuclear analytes as that for a post-det scenario. Materials determinations of IND components would measure actinide elements and their attendant isotopic ratios to gauge threat potential. Fission-product species, perhaps at ultratrace concentrations, might be indicative of analytical separation procedures used to reprocess materials. Neutron-activation species, should a nuclear pit incorporate Pu, could provide chronometry for the time passed since assembly of the weapon.

The pre-det investigation would typically provide many more forensic data. Assessment of IND nuclear source materials would also evaluate device design, metallic apparatus and their trace-element impurities, densities, alloying elements, initiator, possible boost components, and broad-spectrum metallurgy. Sophisticated computer simulations can then predict the performance of a device given empirical forensic data and design parameters as inputs. Information from collateral criminalistics examinations would also be critical in this circumstance for full attribution.

As the provenance of the radioactive source in an RDD would be of considerably less importance than for an IND, collection and analysis of traditional forensic specimens attendant an RDD, in either pre- or post-det situations, are essential. Criminalistics examinations would be the most prone to name and find the perpetrators.

1.8.3 Source and Route

Questions of attribution are conveniently divided into two key forensic areas: source and route (or pathway). Data on the SNM itself provide information on the source of the material (e.g., synthesis methods and specific procedures), whereas analyses of packaging and other collateral items can supply intelligence on the illicit route. Issues related to material source attribution include the identity, quality, and origin of the materials; whether the material was diverted from a legitimate pathway; determination of where legal custody was lost; and potential for the supply of more such material. Route attribution is the domain of criminalistics examinations and geolocation measurements. Relevant issues include the pathway of the contraband after legitimate control was lost, the potential involvement of a black-market network, frequency of shipments, identities of the traffickers, practical capability of the illicit path for large quantities, and likely end-user application.

1.8.4 Casework/Attribution Overview

The primary technical objective of NFA is to determine the attributes of questioned radioactive specimens. Reduced to simple terms, the salient forensic questions for a nuclear sample are as follows: What is it? What was its origin? Is there more of the material? How did it get there? Who was involved? The resulting analyses can be the most multidisciplinary of scientific endeavors. Because law enforcement organizations often drive such investigations, their evidentiary requirements for subsequent prosecution in court are imperative. On-site needs include concerns for public health and safety, timely assessment of the incident, and the collection and preservation of evidence. Subsequent laboratory examinations comprise both nuclear and conventional forensic analyses, the latter perhaps seriously complicated by their conduct on materials contaminated with radioactivity. Samples must be treated and analyzed to conserve any forensic indicators present. These species can range from trace environmental evidence, such as hairs, pollens, explosive residue, DNA, and so forth, to the major and minor elemental isotopes of the principal specimen. As a consequence, meticulous care must be exercised from the time that samples are acquired or interdicted, and similarly continued through transport, storage, analyses, and final destruction. Many law enforcement agencies require that laboratories measuring forensic results for possible introduction at trial have a QA/QC program certified by ISO-17025 accreditation. Interestingly, however, the same analyses performed for intelligence purposes can be considerably less rigid. For example, investigations by the intelligence community may require only appropriate methodology and correct answers (by either certified or uncertified techniques), need not proceed with any presumption of innocence, and may accept a reasonable-doubt threshold of 51% (rather than >99%, e.g., [14]).

The IND post-det scenario is currently overemphasized in the opinion of some analysts. Their argument is that prime counterterrorism efforts have failed profoundly should such a disaster become real. Perhaps more focus should be placed on improved surveillance, better radiation detectors, device standoff sensors, and so forth. Similarly, the worthiness of pre-det source NFA is also somewhat exaggerated at present. While it is indeed critical to determine the identity and isotopic composition of a nuclear threat, further source assays might possibly provide synthesis particulars, the type of production facility, and chronometry data. However, such knowledge would be fundamentally unimportant for relevant attribution should the questioned material or weapon be stolen. In fact, only pre-det analyses in general, as well as RDD post-det route measurements, are potentially able to identify specific perpetrators and pertinent locations. This fact has not been universally appreciated by program managers that administer funded research projects in NFA.

Forensic analyses to determine routes are traditionally encompassed by standard criminalistics techniques, and they have been widely treated in many excellent texts (e.g., Ref. [15]). However, source protocols and geolocation techniques have not been similarly covered, and interesting forensic information can often be obtained by detailed analyses and interpretation of a questioned specimen. For example, radionuclide analyses of enriched U or Pu can provide intelligence on the date of its last chemical purification, sample production technology, chemical tactics for material recovery, reactor design, and ancillary activities ongoing at the plant. The effective attribution of suspect specimens, through measurements of both source and route signature species, is the domain of NFA and the focus of this book. Although selected post-det topics have been added in Chapter 5 of this second edition, the overall emphasis of the work remains pre-det subject matter.

1.9 HISTORICAL PERSPECTIVE

Some believe that NFA is a relatively recent technical endeavor, fostered by the breakup of the FSU during the early 1990s. This opinion has been promoted by the Institute for Transuranium Elements in Karlsruhe, with claims that their start of source analysis of smuggled nuclear materials as a nonweapons laboratory comprised the invention of the discipline in March 1992 [16–18]. Alternatively, some organizations consider that the 1995–1996 formation of the International Technical Working Group on nuclear smuggling (ITWG) established the practice of technical NFA (apparently from a self-centric viewpoint driven by their organizational and operational expediencies after September 11, 2001). However, our opinion is that these assertions are excessively narrow views of NFA, and we encourage a broader scientific perspective. The current signature species of interest in NFA, and their analytic protocols, are principally those of Cold War nuclear weapons programs and intelligence-agency actions. While developments of modern instrumentation, largely mass-spectrometric techniques, have resulted in enhanced forensic abilities today, the measured analytes are generally the same. Some novel applications (e.g., geolocation and chronometry) are particular exceptions to this statement, however, and they are treated later in this book.

The first currently unclassified NFA experiments with tangible results were probably those performed by Enrico Fermi at Trinity, the initial US nuclear weapon test on July 16, 1945. He arrived at the first estimation of the prompt explosive yield of a nuclear device (accurate within ×2) by measuring the dispersal of fragments of paper that he dropped as the blast wave went by. Some hours later, Fermi and Herb Anderson obtained post-shot debris from the ground-zero crater at Trinity by sampling through trap doors in the floors of lead-lined tanks. Subsequent NFA radiochemistry of the specimens resulted in an 18.6-kt yield for Trinity. Three weeks later, Kazuo (Paul) Kuroda determined that the fissile fuel in Little Boy was ^{235}U from his fallout investigations following detonation of a "new kind of bomb" over Hiroshima's Aioi Bridge. These and other examinations from the earliest days of nuclear weapons certainly continue within the efforts of contemporary NFA today.

Moreover, a more expansive definition of NFA would then incorporate perhaps one of the most impressive nuclear forensic studies performed to date. In 1972, scientists noted a 0.003% anomaly in the isotopic composition of uranium ore from the Oklo mine in Gabon, Central Africa. Their resultant investigation into prehistoric geology, enduring Ru and Nd isotopic signatures, and other deliberations led to the startling forensic deduction that a natural nuclear reactor had functioned within the geosphere approximately two billion years ago, at an average energy output of ~100 kW, for several hundred thousand years [19].

Two of us (PG & KM) both developed and utilized NFA methods within the US nuclear test program during the 1980s and early 1990s, but we were only third-generation radiochemistry staff at the Livermore and Los Alamos National Laboratories. The real pioneers of NFA laid the foundations of this discipline many years earlier. The products of their activities were generally classified and unavailable in the open literature, but such constraints surely did not disqualify their accomplishments.

The restrictive and self-focused perspectives of NFA would also reject the LLNL Forensic Science Center (FSC) investigation into the January 1992 fatal explosion of a "cold fusion" electrochemical cell. While cold fusion was (and is) not considered mainstream nuclear science, we regard this accident analysis as our initial NFA inquiry in the public domain. Although we began personal efforts in classified NFA (source) investigations at Livermore during the mid-1980s, the multidisciplinary measurements and ensuing incident assessment of the cold fusion explosion encompassed both source forensic exhibits and associated collateral evidence (see Chapter 24). We are unaware of any prior casework, classified or unclassified, that combined nuclear source examinations with supplemental, nonnuclear forensic analyses for an integrated evaluation of an overall investigation.

1.10 NUANCES OF GRAMMAR

Twenty years ago, the only valid definition of the noun "forensics" referred to formal debate, rhetoric, and argumentation. To specify evidence analysis or connection to courts of law required the adjective, "forensic." Formally trained criminalists and forensic scientists were well aware of this rule of grammar, while others professing to conduct forensic analyses without such preparation were often lacking and at times castigated by virtue of their linguistic errors. Expert referees for forensic journals could be very unkind to authors of submissions that failed to appreciate the grammatical distinction between "forensic" and "forensics."

Today, though, general use of these terms by the general public and popular media has somewhat blurred that division, so that "forensics" is now commonly found to refer to science applied to the law, as well as to rhetoric. Indeed, the Oxford Dictionaries, Merriam-Webster, and the online Free Dictionary all list the noun, as either singular or plural in construct, as acceptable for application to the legal forum. (Interestingly, however, Dictionary.com still limits the use of "forensics" strictly for referring to debating, adhering to the historic imperative.)

However, all of these sources specify that "forensics" may be properly used only as a noun, albeit now with the twofold meanings (i.e., except Dictionary.com). Moreover, they are all unanimous that the only valid adjective is "forensic," also with the dual connotations. So despite arbitrary and subjective combinations of the two words in today's verbiage and literature through popular misuse, as of this writing, the underlying rule remains that "forensics" is the noun and "forensic" is the adjective. Constructs such as "forensics science," "forensics lab," "forensics center," and "forensics applications" remain ungrammatical and indicative, perhaps, of inadequate forensic education (or just bad English).

Thus, the title of this book is **not** *Nuclear Forensics Analysis.*

REFERENCES

1. National Climate Assessment and Development Advisory Committee, Federal advisory committee draft climate assessment report, http://ncadac.globalchange.gov, January 11, 2013.
2. G.H. Marcus and A.E. Levin, New designs for the nuclear renaissance, *Phys. Today*, 55(4), 54, 2002.

3. R. Rhodes, *The Making of the Atomic Bomb*, Simon & Schuster, New York, 1986.
4. L. Hoddeson, P.W. Henriksen, R.A. Meade, and C. Westfall, *Critical Assembly*, Cambridge University Press, Cambridge, U.K., 1993.
5. S.B. McGrayne, *The Theory That Would Not Die*, Yale University Press, New Haven, CT, 2011, pp. 119–128.
6. J.D. Lowrey, S.R. Biegalski, A.G. Osborne, and M.R. Deinert, Subsurface mass transport affects the radioxenon signatures that are used to identify clandestine nuclear tests, *Geophys. Res. Lett.*, 40, 111, 2013.
7. L.-E. De Geer, Reinforced evidence of a low-yield nuclear test in North Korea on 11 May 2010, *J. Radioanal. Nucl. Chem.*, 298, 2075, 2013.
8. J. Barclay, Operation Rhyme—Al-Qaeda's pursuit of a radiological terrorism capability, *IHS Jane's Defense, Risk, & Security Consulting*, IHS Global Limited, Englewood, CO, September 2012, Issue 10, pp. 10–18.
9. P. Williams and P.N. Woessner, The real threat of nuclear smuggling, *Sci. Am.*, 274(1), 40, 1996.
10. P.M. Grant, K.J. Moody, I.D. Hutcheon et al. Forensic analyses of suspect illicit nuclear material, *J. Forensic Sci.*, 43(3), 680, 1998.
11. P.M. Grant, K.J. Moody, I.D. Hutcheon et al. Nuclear forensics in law enforcement applications, *J. Radioanal. Nucl. Chem.*, 235, 129, 1998.
12. H. Stevens, *Hitler's Suppressed and Still-Secret Weapons, Science and Technology*, Adventures Unlimited Press, Kempton, IL, 2007.
13. E.K. Sokova and W.C. Potter, The 2003 and 2006 HEU seizures in Georgia: New questions, some answers, and possible lessons, *Illicit Nuclear Trafficking: Collective Experience and the Way Forward*, STI/PUB/1316, IAEA, Vienna, Austria, 2008.
14. C. Weiss, Expressing scientific uncertainty, *Law Probabil. Risk*, 2, 25, 2003.
15. R. Saferstein, ed., *Forensic Science Handbook*, 2nd edn., 3 vols., Prentice Hall, NJ, 2002, 2005, and 2010.
16. K. Mayer, M. Wallenius, and I. Ray, Nuclear forensics—A methodology providing clues on the origin of illicitly trafficked nuclear materials, *Analyst*, 130, 433, 2005.
17. M. Wallenius, K. Mayer, and I. Ray, Nuclear forensic investigations: Two case studies, *Forensic Sci. Int.*, 156, 55, 2006.
18. K. Mayer, M. Wallenius, and T. Fanghanel, Nuclear forensic science—From cradle to maturity, *J. Alloys Compd.*, 444–445, 50, 2007.
19. G.A. Cowan, A natural fission reactor, *Sci. Am.*, 235(1), 36, 1976.

2 Physical Basis of Nuclear Forensic Science

2.1 BACKGROUND

During the first few years of the twentieth century, the discovery of radioactivity had a profound effect on the chemical sciences. The discipline of radiochemistry was developed to explore the changes in chemical properties that are consequences of transmutations occurring in radioactive decay and to exploit those changes to study the physical properties of the atomic nucleus. The principles of nuclear forensic science are those of radiochemistry.

Chemistry was the first branch of science in which the concept of the atom achieved importance, beginning with Dalton in 1803 [1], even though the concept had been proposed to explain the behavior of gases early in the eighteenth century and was initially postulated by Democritus about 450 BC. Dalton's law of multiple proportions (the weights of elements always combine with each other in small, whole-number ratios) and the law of partial pressures led directly to the atomic hypothesis. The atom (from *atomos*, Greek for indivisible) was defined as the smallest chemical unit unchanged by any chemical or physical processes. Given the success of the application of the atomic hypothesis to many subfields of chemistry (e.g., electrochemistry and the kinetic theory of gases), it is surprising that it was not embraced by the other scientific disciplines until the beginning of the twentieth century.

The concept of the atom led to Avogadro's principle and a series of atomic masses. Taking these masses into account, Mendeleev further organized the elements by imposing an order based on their chemical properties [2]. The resulting "periodic table" (the synthesis of the data of Mendeleev and many of his contemporaries) was used to explain the semiempirical repetition of chemical properties of the elements in 1869. The chemist's most fundamental tool is the periodic table (Figure 2.1). It expresses the fact that elements that are chemically similar have atomic masses that increase in a regular manner, resulting in vertical groupings. It also portrays the observation that the intrinsic valences of the elements required by the law of multiple proportions tend to increase or decrease regularly with atomic mass (e.g., the sequence Na^+, Mg^{2+}, Al^{3+}). Mendeleev used the periodic table to predict the presence and properties of undiscovered elements; his prediction of eka-aluminum in 1870 was borne out in 1875 with the discovery of Ga by de Boisbaudran [3]. Mendeleev also noted that, occasionally, the reported atomic mass of an element needed "correction" to force it to occupy its proper place in the table: the correct ordering of the periodic table is obtained through the atomic

I	II	III	IV	V	VI	VII	VIII	IX	X	XI	XII	XIII	XIV	XV	XVI	XVII	XVIII
H 1																	He 2
Li 3	Be 4											B 5	C 6	N 7	O 8	F 9	Ne 10
Na 11	Mg 12											Al 13	Si 14	P 15	S 16	Cl 17	Sr 18
K 19	Ca 20	Sc 21	Ti 22	V 23	Cr 24	Mn 25	Fe 26	Co 27	Ni 28	Cu 29	Zn 30	Ga 31	Ge 32	As 33	Se 34	Br 35	Kr 36
Rb 37	Sr 38	Y 39	Zr 40	Nb 41	Mo 42	Te 43	Ru 44	Rh 45	Pd 46	Ag 47	Cd 48	In 49	Sn 50	Sb 51	Te 52	I 53	Xe 54
Cs 55	Ba 56		Hf 72	Ta 73	W 74	Re 75	Os 76	Ir 77	Pt 78	Au 79	Hg 80	Tl 81	Pb 82	Bi 83	Po 84	At 85	Rn 86
Fr 87	Ra 88		Rf 104	Db 105	Sg 106	Bh 107	Hs 108	Mt 109	Ds 110	Rg 111	Cn 112	113	Fl 114	115	Lv 116	117	118

Lanthanides	La 57	Ce 58	Pr 59	Nd 60	Pm 61	Sm 62	Eu 63	Gd 64	Tb 65	Dy 66	Ho 67	Er 68	Tm 69	Yb 70	Lu 71
Actinides	Ac 89	Th 90	Pa 91	U 92	Np 93	Pu 94	Am 95	Cm 96	Bk 97	Cf 98	Es 99	Fm 100	Md 101	No 102	Lr 103

FIGURE 2.1 The periodic table of the elements.

number, which only loosely tracks with atomic mass. The atomic numbers and masses of the elements are listed in Table 2.1.

Roentgen discovered x-rays in 1895 [4]. Fluorescence of the glass walls of vacuum tubes was also observed and associated with x-ray production; this discovery led Becquerel to search for the emission of penetrating radiations from phosphorescent substances and resulted (almost by accident) in the discovery of radioactivity in 1896 [5]. Thomson's discovery of the electron in 1897 [6] provided the first evidence that the atom possessed structure. Rutherford's subsequent explanation of the large-angle scattering of α particles by thin foils resulted in the rapid development of the planetary model of the atom, in which most of the atomic mass resided in a "nucleus" of very small volume [7]. Diffraction of x-rays led Moseley to the concept of atomic number and completion of the concept of an element [8]: all atoms with the same atomic number (the same number of protons in the nucleus) belong to the same element, regardless of mass. However, the difference between atomic mass and atomic number required the presence of a neutral particle, the neutron, later discovered by Chadwick in 1932 [9].

The role of radiochemists in the elucidation of atomic structure was key. The Curies observed that the radioactivity content of a sample of uranium ore was much greater than could be accounted for by the U that could be purified from it. This observation was the beginning of the science of radiochemistry, and it led to the chemical isolation of various fractions from the ore, resulting in the discovery of the elements Po and Ra in 1898 [10,11]. The Curies also showed that radioactivity is an intrinsic property of certain atoms, not one stimulated by outside influences. Rutherford and Soddy (1900) isolated short-lived gaseous radioactivities (isotopes of Rn) from purified samples of Th and Ra [12], from which they drew the conclusions that are the basis of all radiochemistry: radioactivity processes result in a change in the chemical properties of the radioactive species, and the intensity of the radiation

TABLE 2.1
Relative Atomic Masses of the Elements (IUPAC 2010)

Atomic Number	Symbol	Name	Atomic Mass
1	H	Hydrogen	1.008
2	He	Helium	4.0026
3	Li	Lithium	6.94
4	Be	Beryllium	9.0122
5	B	Boron	10.81
6	C	Carbon	12.011
7	N	Nitrogen	14.007
8	O	Oxygen	15.999
9	F	Fluorine	18.998
10	Ne	Neon	20.180
11	Na	Sodium	22.990
12	Mg	Magnesium	24.305
13	Al	Aluminum	26.982
14	Si	Silicon	28.085
15	P	Phosphorus	30.974
16	S	Sulfur	32.06
17	Cl	Chlorine	35.45
18	Ar	Argon	39.948
19	K	Potassium	39.098
20	Ca	Calcium	40.078
21	Sc	Scandium	44.956
22	Ti	Titanium	47.867
23	V	Vanadium	50.942
24	Cr	Chromium	51.996
25	Mn	Manganese	54.938
26	Fe	Iron	55.845
27	Co	Cobalt	58.933
28	Ni	Nickel	58.693
29	Cu	Copper	63.546
30	Zn	Zinc	65.38
31	Ga	Gallium	69.723
32	Ge	Germanium	72.63
33	As	Arsenic	74.922
34	Se	Selenium	78.96
35	Br	Bromine	79.904
36	Kr	Krypton	83.798
37	Rb	Rubidium	85.468
38	Sr	Strontium	87.62
39	Y	Yttrium	88.906
40	Zr	Zirconium	91.224
41	Nb	Niobium	92.906
42	Mo	Molybdenum	95.96

(*Continued*)

TABLE 2.1 (*Continued*)
Relative Atomic Masses of the Elements (IUPAC 2010)

Atomic Number	Symbol	Name	Atomic Mass
43	Tc	Technetium	[98]
44	Ru	Ruthenium	101.07
45	Rh	Rhodium	102.91
46	Pd	Palladium	106.42
47	Ag	Silver	107.87
48	Cd	Cadmium	112.41
49	In	Indium	114.82
50	Sn	Tin	118.71
51	Sb	Antimony	121.76
52	Te	Tellurium	127.60
53	I	Iodine	126.90
54	Xe	Xenon	131.29
55	Cs	Cesium	132.91
56	Ba	Barium	137.33
57	La	Lanthanum	138.91
58	Ce	Cerium	140.12
59	Pr	Praseodymium	140.91
60	Nd	Neodymium	144.24
61	Pm	Promethium	[145]
62	Sm	Samarium	150.36
63	Eu	Europium	151.96
64	Gd	Gadolinium	157.25
65	Tb	Terbium	158.93
66	Dy	Dysprosium	162.50
67	Ho	Holmium	164.93
68	Er	Erbium	167.26
69	Tm	Thulium	168.93
70	Yb	Ytterbium	173.05
71	Lu	Lutetium	174.97
72	Hf	Hafnium	178.49
73	Ta	Tantalum	180.95
74	W	Tungsten	183.84
75	Re	Rhenium	186.21
76	Os	Osmium	190.23
77	Ir	Iridium	192.22
78	Pt	Platinum	195.08
79	Au	Gold	196.97
80	Hg	Mercury	200.59
81	Tl	Thallium	204.38
82	Pb	Lead	207.2
83	Bi	Bismuth	208.98
84	Po	Polonium	[209]

(*Continued*)

TABLE 2.1 (*Continued*)
Relative Atomic Masses of the Elements (IUPAC 2010)

Atomic Number	Symbol	Name	Atomic Mass
85	At	Astatine	[210]
86	Rn	Radon	[222]
87	Fr	Francium	[223]
88	Ra	Radium	[226]
89	Ac	Actinium	[227]
90	Th	Thorium	232.04
91	Pa	Protactinium	231.04
92	U	Uranium	238.03
93	Np	Neptunium	237.048
94	Pu	Plutonium	[244]
95	Am	Americium	[243]
96	Cm	Curium	[247]
97	Bk	Berkelium	[247]
98	Cf	Californium	[251]
99	Es	Einsteinium	[252]
100	Fm	Fermium	[257]
101	Md	Mendelevium	[258]
102	No	Nobelium	[259]
103	Lr	Lawrencium	[262]
104	Rf	Rutherfordium	[267]
105	Db	Dubnium	[268]
106	Sg	Seaborgium	[269]
107	Bh	Bohrium	[274]
108	Hs	Hassium	[270]
109	Mt	Meitnerium	[278]
110	Ds	Darmstadtium	[281]
111	Rg	Roentgenium	[281]
112	Cn	Copernicium	[285]
113			[286]
114	Fl	Flerovium	[289]
115			[289]
116	Lv	Livermorium	[293]
117			[294]
118			[294]

associated with a given preparation changes with time in a way that is characteristic of the radioactive species that make up that preparation. Rutherford and Soddy also observed that, in those instances in which a short-lived radioactive species is separated from a longer-lived species that acts as its progenitor, if the "daughter" species decays appreciably over a given time interval, it is resupplied within the "parent" preparation at a similar level.

Thus, by 1905, the first radiochemists had demonstrated that radioactive decay involved the spontaneous transmutation of an atom of one chemical element into the atom of another, accompanied by the observed radioactive emissions. The intensity of the radioactivity, and therefore the rate of transmutation, was known to be proportional to the number of atoms in a given chemical preparation. The speed with which the change took place was shown to be dependent on the identity of the radioactive species. These observations gave rise to the statistical theory of radioactive decay, which will be discussed later. The atom was no longer seen as structureless and indivisible.

The different radioactive decay modes and decay constants associated with a given element led to the concept of isotopes [13], in which a change in the nuclear mass (attending a change in the number of neutrons) changes the radioactive properties of an atom but not its chemical behavior. This idea was necessary to reconcile the growing number of observed radioactive species within the limited number of available slots in the periodic table. These radiochemical conclusions predated the postulation of the atomic nucleus and the neutron by a decade.

2.2 TYPES OF RADIOACTIVE DECAY

Becquerel likened the penetrating radiations emitted by salts of uranium to x-rays. However, by the next year (1897), Rutherford found that the radiations were of more than one type, with various degrees of penetrating power. He termed the less penetrating emissions α rays and the more penetrating discharges β rays. It was quickly determined that the β rays could be deflected by a magnetic field and that the particles had the same charge-to-mass ratio (5.273×10^{17} esu/g) as the electrons that had recently been discovered by Thomson in the operation of a cathode-ray tube. From the absorption properties of α rays, Marie Curie deduced that they were much heavier particles, following which discovery Rutherford was successful in deflecting them with a stronger magnetic field [14]. The charges of α and β particles were determined to be of opposite signs. Rutherford and Royds proved that α particles were He nuclei in 1909, when they successfully collected the radioactive emissions from a sample of radium in an evacuated tube and made diagnostic spectroscopic measurements [15].

In 1900, Villard identified a third kind of radiation emitted by radioactive materials, and he naturally dubbed it gamma radiation [16]. Gamma rays are not deflected by electric or magnetic fields, and must therefore carry no charge. The connection between radioactivity and x-rays was thereby realized: like x-rays, γ-rays are short-wavelength photons. X-rays result from the rearrangement of electrons around the nucleus, whereas γ-rays arise from transitions within the nucleus itself. Although electrons are bound in atoms with energies of eV–keV, nucleons are bound in atomic nuclei with MeV energies. Although there is some overlap, particularly in the heavy elements, γ-rays tend to be higher in energy than x-rays.

Alpha decay is the spontaneous emission of a ^{4}He nucleus, an agglomeration of two protons and two neutrons. Even though the kinetic energies of emission of these particles are quite high, the α-particle mass is large enough that its velocity is only a few percent of the speed of light; the kinematic behavior of α particles can therefore be treated nonrelativistically. Alpha particles slow down and stop with a definite range in matter (the distance traversed by a particle passing through an intervening substance), dependent on the initial energy of the particle and the nature of the stopping material. The range of α radiation is quite short in comparison to that of the other types of radioactive emissions. Alpha particles are ejected with well-defined energies that are characteristic of the emitting nuclei, thereby permitting spectroscopic measurements (measurements in which the energy of the particle is determined). However, careful protocols must be exercised because of α-particle energy loss in even the most tenuous of matter. Alpha decay is a prominent decay mode of the heavy elements, and its probability increases with decay energy in a regular way for the isotopes of a given element [17,18].

Beta decay (i.e., β^- decay) entails the emission of an electron at high velocity, near the speed of light. This mode is a consequence of the conversion of a neutron into a proton in the emitting nucleus and results in no change in the number of nucleons and, consequently, the atomic mass number. Beta particles have a high charge-to-mass ratio and are thus easy to deflect with an electric or magnetic field. The interaction of β particles with matter is complex; the probability of absorption is approximately exponential until a certain thickness of absorber is achieved, whereupon absorption then becomes complete. The deceleration of the electrons in the scattering of β particles off atomic electrons results in a shower of low-energy electrons and photons called Bremsstrahlung, a German word meaning "braking radiation." The ranges of β particles are generally much longer than those of α particles. The emission of electrons by radioactive nuclei initially supported a view that the nuclear mass in excess of the number of protons was attributed to closely paired protons and electrons. However, it was observed that an electron, to be confined within a nucleus, was required to have a wavelength on the order of nuclear dimensions, which would necessitate a much larger electron kinetic energy than is credible.

Although the complete theory is too detailed for treatment here, electrons are emitted in β^- decay with energies spanning the range from all that is available (the Q-value) to zero. This property is a consequence of the rule of conservation of lepton number, which requires the simultaneous emission of the electron with an antineutrino for β^- decay. The antineutrino is a nearly massless particle, undetectable by ordinary methods, that carries off some of the decay energy. Neutrinos undergo such weak interactions with matter that they were not detected [19,20] until more than 20 years after they were first postulated by Pauli to explain the apparent violation of the Law of Conservation of Energy in β decay [21]. The electron is a lepton and the antineutrino an antilepton, making the number of leptons emitted in the β-decay process conserved (a net of zero). As a result of the energy division between the electron and antineutrino, there is little reason to perform spectroscopic measurements on the charged particles emitted in β decay. However, β-radiation counters can be quite sensitive to the energy distribution of the electrons and must therefore be calibrated against standard sources with radionuclides of various decay energies. For a compete discussion of the β-decay process, the classic reference is Fermi [22,23].

Although not very important for nuclear forensic applications, two other types of radioactive decay are classified as β decay. Both result in conversion of a proton into a neutron in the emitting nucleus without a change in atomic mass number. The first involves emission of a positron (β^+ decay), which is mechanistically identical to the emission of an electron. The second is electron capture (EC decay), discovered by Alvarez in 1938 [24], in which an atomic electron is captured by the nucleus. Both processes (one involving the emission of an antilepton and the other the absorption of a lepton) are accompanied by emission of a neutrino. These modes of β decay will be discussed further later.

Gamma decay involves the emission of photons and is generally observed following either β- or α-decay processes that leave daughter nuclides in excited nuclear states. These states, or nuclear energy levels, are analogous to the energy levels of atomic and molecular spectroscopy. Nuclear levels occur at discrete energies, and γ-rays from transitions between these states are emitted with well-defined energies. The lowest energy level of any nucleus is its ground state. Emission of γ-rays is usually prompt after the initial α or β emission. The uncertainty principle [25] requires that the product of the lifetime (τ) of a nuclear state and the uncertainty in its energy (ΔE) be greater than or equal to Planck's constant of motion ($\tau\Delta E \geq h/2\pi$). There is thus an intrinsic accuracy with which the energies of the emitted photons can be known that is dependent on the lifetimes of the levels involved in the transitions. For the γ-rays of interest in nuclear forensic science, the lifetimes of the nuclear states are long enough that the uncertainty in the energies of the emitted photons is smaller than the resolution of even the best detector.

Emission of a γ-ray does not change the neutron or proton number of the emitting nucleus, but it does cause a decrease in its energy state. The spectroscopy of γ-rays arising from the decay of mixtures of radionuclides is a well-developed laboratory technique and is one of the most valuable tools available to the radiochemist. It is reiterated that γ-rays subsequent to a primary β or α decay arise from the states of the daughter nucleus.

The process of internal conversion (IC) competes with γ-ray emission in the depopulation of excited nuclear states. The external electromagnetic field of a nucleus can interact with atomic electrons such that energy and angular momentum are transferred to an electron, causing it to be ejected from the atom. In this process, the electron is not created but is merely dislodged from its atomic level; IC is therefore not accompanied by emission of an antineutrino. IC is, however, often accompanied by x-ray production induced by the rearrangement of the atomic electrons to eliminate the orbital vacancy. The ejected electron will have a discrete kinetic energy, which is equal to the nuclear transition energy less the atomic electron binding energy. A γ-ray photon is never produced in this process, even as an intermediate, and the nucleus interacts directly with the atomic electrons. A consequence of the competition between IC and γ-ray emission is that, even if the population of a given nuclear state in a decay process is known (as in simple low-energy decays), the absolute intensities of the γ-rays depopulating the state are not known, but only the sum of the intensities of γ-ray emission and IC decay. As a consequence, the absolute intensities of γ-rays emitted in radioactive decay (photons per decay) should be obtained from tables [26]. As an example, in the β decay of ^{203}Hg, essentially 100% of the decays populate the state in ^{203}Tl at an energy of 279 keV above its ground state. In the deexcitation of the 279-keV state, γ-ray photons are emitted only 81.5% of the time, with the balance of the decays proceeding by IC. The competition between γ-ray emission and IC decay is explored further later.

Absorption of γ-rays in matter occurs via mechanisms different from those governing the interactions of charged particles. Gamma rays are absorbed according to an exponential law characterized by an absorption coefficient (often expressed as a half-thickness). The energies of the photons escaping the absorber are unchanged, but they are reduced in intensity. Full-energy photons are accompanied by a continuum of lower-energy photons and electrons (Bremsstrahlung) that arise in the photoelectric or Compton interactions with atomic electrons. These result in the attenuation of the primary γ-rays. High-energy photons of greater than 1.02-MeV energies can also interact with matter through the production of positron–electron pairs, resulting in annihilation radiation in the γ-ray spectrum at 511 keV [27,28].

There is a type of radioactive decay that is limited to the heaviest elements, occurring with low probability even in the U and Pu isotopes. Spontaneous-fission (SF) decay (discovered by Flerov in 1940 [29]) involves the splitting of heavy nuclei into two fragments of comparable size, accompanied by a large release of energy. Although it is a rare decay mode in nuclides of interest to nuclear forensic science, the forensic radiochemist can gain productive information from the decay products, particularly in the analysis of mixtures of Pu isotopes. Occurring infrequently in comparison to α decay, the SF process emits neutrons that impose limitations on the designs of nuclear explosive devices. The range of the decay products of SF in matter is quite small—so low that the thickness of an otherwise invisible deposit in a source preparation can result in significant energy loss and make spectroscopic measurements difficult.

The division of charge and mass between the two fission fragments in a given fission event is undefined, but isotopes between Zn and Tb are produced with well-defined probabilities. The analyst can isolate these products to realize inferences and conclusions about a sample. The fission process will be discussed in more detail later.

2.3 RATE LAWS IN RADIOACTIVE DECAY

The early radiochemists determined that the intensity of the radioactivity of a given preparation containing only one radioactive species was proportional to the number of radioactive atoms contained in the preparation. Rutherford and Soddy [12] showed that the radioactivity of a sample of gaseous radionuclides decreased exponentially with time, with a decay constant that was characteristic of the radionuclide. In 1905, von Schweidler developed the statistical theory of radioactive decay, which showed that these two statements were equivalent [30].

Consider a radioactive species, A. If a preparation contains N_A atoms of A, the probability of a given atom of A disintegrating in a particular time interval (Δt) is independent of the past history of the atom and depends only on the duration of Δt for short periods. The radioactive decay rate of A is given by the equation

$$-\mathrm{d}N_A/\mathrm{d}t = \lambda_A N_A. \tag{2.1}$$

The rate of decay of A (given a negative sign since A is being depleted) is proportional to the number of atoms of A. The constant of proportionality, the decay constant λ_A, is characteristic of the identity of A. Rearranging,

$$dN_A/N_A = -\lambda_A dt.$$

Solving the differential equation and imposing the boundary condition that at reference time t_0 (e.g., the time when the sample was prepared or counted for the first time) there was an established quantity of A present in the sample, ${}^0N_A$, result in

$$N_A = {}^0N_A \exp\left[-\lambda_A\left(t-t_0\right)\right]. \tag{2.2}$$

This equation is the exponential dependence on time noted by Rutherford. The activity or intensity of the radioactivity of N_A atoms of A is equal to $\lambda_A N_A$.

For the special case in which sufficient time elapsed such that only one-half of the original material remains, Equation 2.2 becomes, at time t_1,

$$N_A/{}^0N_A = \exp\left[-\lambda_A\left(t_1-t_0\right)\right] = 1/2.$$

Taking the natural logarithm of both sides of the equation, the elapsed time, $t_1 - t_0$, necessary for one-half of the atoms of A to decay is

$$t_1 - t_0 = \ln\left(2\right)/\lambda_A$$

and is termed the radioactivity half-life, $t_{1/2}$. This reciprocal relationship between decay constant and half-life is valid from any time of reference. As a consequence, if one-half of the radioactivity in a given sample decays over a time interval of $t_{1/2}$, then an additional $t_{1/2}$ interval results in one-fourth of the initial activity remaining, one-eighth after another interval, and so on.

A given nuclide may undergo transmutation by more than one decay mode, and this feature is common for heavy elements of forensic interest. In these cases, the decay constant is the sum of the partial constants for each decay process. For instance, the nuclide ^{212}Bi occurs in the Th decay chain. The half-life of ^{212}Bi is 60.6 min, decaying 36% of the time by α emission and 64% of the time by β emission. Therefore, the decay constant for α decay is (ln 2/60.6 min) × 0.36 = 0.00412 min^{-1}, and (ln 2/60.6 min) × 0.64 = 0.00732 min^{-1} for β decay. This leads directly to the concept of a partial half-life, which is ln 2 divided by the decay constant for the specific process; thus, for ^{212}Bi, the partial half-life for α decay is 168.2 min, whereas for β decay it is 94.7 min. The fraction of decays by rare decay processes is often expressed as a partial half-life. As an example, ^{240}Pu is an α emitter with an overall half-life of 6563 years. Very infrequently, it decays by SF with a partial half-life of 1.14×10^{11} years; therefore, $\lambda_{SF}/(\lambda_{SF} + \lambda_\alpha) \simeq t_{1/2}(\alpha)/t_{1/2}(SF) = 5.75 \times 10^{-8}$ (5.75×10^{-6}%) of the decays are via SF.

Only total decay constants should be used in the equations that follow. Consider the case in which A decays to B, which in turn decays with a decay constant λ_B. If a decay equation analogous to Equation 2.1 is constructed,

$$-dN_B/dt = \lambda_B N_B - \lambda_A N_A. \tag{2.3}$$

As before, the rate of decay of B is proportional to the number of atoms of B, N_B. In this case, the rate of ingrowth of B is equal to the instantaneous rate at which A is decaying, $\lambda_A N_A$. If Equation 2.2 is substituted into Equation 2.3 to eliminate N_A, and the differential equation is solved,

$$N_B = \left[\lambda_A/(\lambda_A - \lambda_B)\right]{}^0N_A\left\{\exp[-\lambda_A(t-t_0)] - \exp[\ \lambda_B(t-t_0)]\right\}$$

$$+\,{}^0N_B \exp[-\lambda_B(t-t_0)], \tag{2.4}$$

where ${}^0N_B$ is the value of N_B at reference time t_0. For the case in which ${}^0N_B$ is 0 (as in a sample of A that is freshly purified at the reference time), the second term becomes zero.

A discussion of some of the consequences of Equation 2.4 follows. First, consider the special case in which the decay product B is a stable nuclide ($\lambda_B = 0$). Simplifying the equation yields

$$N_B(\text{stable}) = {}^0N_B + {}^0N_A\left\{1 - \exp[-\lambda_A(t-t_0)]\right\}.$$

At the limit of elapsed times that are long compared to the half-life of A, the equation further reduces to

$$N_B(\text{stable}) = {}^0N_B + {}^0N_A$$

as expected; that is, the final number of B atoms is equal to the initial number plus all of the initial atoms of A (which have decayed to B).

For those cases in which less than 100% of the decay of A results in the production of B, Equation 2.3 is modified to

$$-dN_B/dt = \lambda_B N_B - \lambda_A N_A b,$$

where b is the fraction of the decays of A that produces B. Algebra results in an equation identical to Equation 2.4, except that ${}^0N_A b$ replaces ${}^0N_A$. The same is true for both subsequent equations, where N_B(stable) is calculated.

In the situation in which the half-life of A is much longer than that of B, $\lambda_A \ll \lambda_B$ and the parent activity does not significantly decrease over the interval of several daughter half-lives. In this case, Equation 2.4 can be approximated by

$$N_B = \lambda_A N_A/\lambda_B + \left({}^0N_B - {}^0N_A\right)\exp\left[-\lambda_B(t-t_0)\right].$$

If the times $(t - t_0)$ are restricted to those that are long relative to the half-life of B, the equation further reduces to

$$\lambda_B N_B = \lambda_A N_A, \tag{2.5}$$

and the activities of parent and daughter are equal at all long time intervals (a condition known as secular equilibrium, which is very useful to the nuclear forensic analyst). As an example, ^{237}Np is an α emitter with a half-life of 2,140,000 years. Its decay daughter, ^{233}Pa, has a half-life of 27 days. A sample of ^{237}Np is chemically processed so that some (not necessarily all) of the ^{233}Pa is removed. The sample is then set aside for several months, during which time the ^{233}Pa activity grows back into the sample according to Equation 2.4. After a year, it does not matter how completely the ^{233}Pa was removed from the sample, as it has grown back to the extent that its activity is equal to that of the ^{237}Np (according to Equation 2.5), within the accuracy of our ability to measure it. The γ-rays from the decay of ^{233}Pa can be more readily measured than those from the decay of ^{237}Np. As a consequence, when an aged ^{237}Np sample is counted, the decay rate of the parent is identical to that obtained for the ^{233}Pa progeny in the specimen.

Secular equilibrium is a special case of transient equilibrium; in transient equilibrium, the parent is longer-lived than the daughter, $\lambda_A < \lambda_B$, but not overwhelmingly so. In this case, when $(t - t_0)$ becomes sufficiently large, Equation 2.4 reduces to

$$(\lambda_B - \lambda_A) N_B = \lambda_A N_A. \tag{2.6}$$

If the parent is shorter-lived than the daughter (a common situation among actinide nuclides and fission products), such that $\lambda_A > \lambda_B$, an equilibrium condition is never attained, and Equation 2.4 must be applied in all cases.

In the determination of the age of a sample of U or Pu, it is often desirable to use the concentrations of ingrown radionuclides produced by the radioactive decay to daughter products (species "B" in the equations given earlier), or species even more remote. For instance, ^{234}U in a sample of high-enriched uranium decays to ^{230}Th, which in turn decays to ^{226}Ra; any combination of two of the three radionuclide concentrations can be used for chronometry. The general solution for the concentration of the nth species generated in a chain of radioactive decays as a function of time was first developed by Bateman in 1910 [31]. For a sample that originally contains only the parent activity (species 1) at t_0,

$$N_n = C_1 \exp[-\lambda_1(t - t_0)] + C_2 \exp[-\lambda_2(t - t_0)] + \cdots + C_n \exp[-\lambda_n(t - t_0)], \tag{2.7}$$

where

$$C_1 = [\lambda_1 \lambda_2 \ldots \lambda_{(n-1)}\, {}^0N_1] / [(\lambda_2 - \lambda_1)(\lambda_3 - \lambda_1)\ldots(\lambda_n - \lambda_1)];$$

$$C_2 = [\lambda_1 \lambda_2 \ldots \lambda_{(n-1)}\, {}^0N_1] / [(\lambda_1 - \lambda_2)(\lambda_3 - \lambda_2)\ldots(\lambda_n - \lambda_2)],$$

and so forth.

The first term of Equation 2.4 is obtained when Equation 2.7 is solved for the second member of a decay chain; the second term is necessary when the initial concentration of species B is not zero at t_0. The forensic application of the decay equations will be treated in more detail in the later section on chronometry (Chapter 6).

2.4 ATOMS, BINDING ENERGY, AND CHART OF THE NUCLIDES

An atom consists of a positively charged nucleus surrounded by a cloud of negative electrons. The charge of the nucleus is determined by the number of protons it contains, and this charge defines both the number and arrangement of electrons around the atom and, as a result, its chemical properties. Though vastly different in mass, the intrinsic charges of the electron and the proton are equal but opposite (the mass of an electron is only 1/1836 of the mass of the proton). The balance of mass in the nucleus is made up of neutrons, which exert only an infinitesimal effect on the chemical properties of an atom, principally through the small change in its inertia. The mass of the neutron is 0.14% larger than that of the proton and 0.08% larger than the mass of a hydrogen atom, an important difference that defines the radioactive properties of the free neutron [32].

The nuclear properties of a given atom are determined by the number of neutrons and protons in its nucleus. In radiochemistry, an atom with a defined number of neutrons and protons is referred to as a nuclide. Collectively, neutrons and protons are referred to as nucleons. The abbreviated designation for a given nuclide is the element symbol from the periodic table (Figure 2.1), and the atomic mass (the total number of nucleons in the nucleus) is used as a preceding superscript. In some publications, the number of protons has been given as a preceding subscript. This convention is redundant and will not be followed here. For example, a carbon nucleus containing seven neutrons is designated as ^{13}C, where 13 is the number of nucleons in the nucleus (seven neutrons and six protons). Two or more nuclides of a given element are called isotopes of that element. The term "isotope" is often used incorrectly in place of the word "nuclide." Formally, "isotope" should only be used when discussing or comparing a collection of two or more nuclides comprising the same element. Thus, ^{235}U is a nuclide and ^{238}U is a nuclide, and both are isotopes of U (element 92, $Z = 92$).

The concept of isotopes was developed during the early part of the twentieth century to explain the fact that two samples of a given chemical element, arising from the radioactive decay of two different parent species, can display different radioactive properties and decay constants. The first direct proof of the existence of different isotopes arising from radioactive decay came from atomic weight measurements of Pb in 1914 [33,34]. It was found that the atomic weight of Pb isolated from the Th mineral thorite was significantly different from that of the Pb isotope isolated from the U mineral clevite, nominally 208 *vs* 206, respectively. The first observation of stable isotopes not involving radioactive phenomena was made by Thomson in 1912 [35]. Using a primitive mass spectrograph, he demonstrated that the inert gas Ne was composed mainly of two isotopes, with masses of 20 and (less abundantly) 22. Observations such as these were used to explain the fact that the atomic masses of some elements are not even multiples of the mass of a nucleon but, rather, have some fractional value (see Table 2.1). If the naturally occurring element consists of two or more isotopes, its apparent atomic mass will be some (noninteger) weighted average of the masses of the individual isotopes.

Part of the initial resistance to the planetary model of the atom and the existence of the nucleus involved the confinement of the positively charged protons in a very small volume, with dimensions on the order of 10^{-12} cm. Even though the charge on an individual proton is very small (4.8×10^{-10} esu), the electrostatic repulsion between two protons held at nuclear distances is enormous on that scale. The existence of the nucleus requires a strong, short-range, attractive force. It was suspected that this force was mediated by the neutron, which turned out to be too simple an explanation. The special theory of relativity provides the means by which the electrostatic repulsion is overcome: some of the mass of the individual nucleons is converted to binding energy, holding the nucleus together.

The binding energy of an atom is defined as the energy that would be released if the atom were to be synthesized from the appropriate numbers of neutrons and hydrogen atoms. For any atom of interest to the forensic analyst, synthesis from fundamental components leads to a release of energy, making binding energy always positive for nuclei that live beyond nuclear rotation times (~10^{-19} s). The most tightly bound nuclei are the most stable. In principle, any spontaneous transformation that involves a net decrease in mass can occur. However, in practice, only those transformations involving simple nuclear rearrangements, such as those occurring in the traditional radioactive decay modes, are probable unless the energy release is large (as in fission).

It is beyond the scope of this book to provide a comprehensive explanation of the way in which binding energy arises, rooted in the nucleon–nucleon exchange of virtual pions that acts as the short-range "glue" that holds the positively charged nucleus together [36]. In brief, the nucleons in a nucleus occupy definite quantum states very similar to the quantum states of the atomic electrons in atoms. The major difference is that there are two types of nucleons, and only a limited number of nucleons are permitted in each quantum state by the exclusion principle [37]. However, a quantum state that is filled with neutrons can still accommodate protons. These states can be thought of as occupying a well that is defined by the nuclear potential, filling with particles from bottom to top. Even though the structure of the proton states is similar to that for the neutrons states, the proton states are separated farther in energy from one another as a consequence of the mutual Coulomb repulsion of the protons. If the well is filled with "nuclear fluid" to a certain level (corresponding to the atomic mass), the most stable configuration is that in which the highest-filled proton and neutron levels are at the same energy. For light nuclei, for which Coulomb effects are minimal, the most stable nuclei are those that contain equal numbers of protons and neutrons. For heavier nuclei, it takes more neutrons to fill the neutron energy levels to the same depth to which the proton levels are filled, resulting in a shift of stability for a given element to greater neutron excess as one proceeds up the periodic table to heavier elements. As the nuclear configuration moves away from this balance (conversion of protons to neutrons or vice versa), the system becomes less stable and, therefore, more susceptible to radioactive decay processes.

The mass of a given nuclide can be expressed as the sum of the masses of the nucleons from which it is assembled, plus the nuclear binding energy; however, it is more conveniently given in terms of the mass excess. The mass excess is defined as the difference between the exact atomic mass M_A and the mass associated with A nucleons on the ^{12}C scale, m_A. The atomic mass is defined by assignment of Avogadro's number, N_A, such that the atomic mass of ^{12}C is exactly 12.000 atomic mass units (amu) and its mass excess is exactly zero ($N_A = 6.0221 \times 10^{23}$ atoms per mole). The mass excess in energy units is obtained by multiplying the mass excess in mass units by 931.494 MeV/amu. The mass excess of a bound nucleus need not be of any particular sign, as the binding energy incorporated in the ^{12}C reference is included in the basis of the mass. However, the most tightly bound nuclides are assigned a negative mass excess.

The energy released in radioactive decay, the Q-value, is defined as the mass of the parent species less the mass of its daughters, expressed in energy units. In defining it this way, a net release of energy results in a positive Q-value, the opposite convention from chemistry, where ΔH is negative for an exothermic reaction. The mass of an atom of nuclide A is given by $M_A = m_A + \Delta A$, where ΔA is the mass excess. For a reaction in which A decays to B and C, the Q-value is given as

$$Q = M_A - (M_B + M_C) = m_A + \Delta A - m_B - \Delta B - m_C - \Delta C$$
$$= (m_A - m_B - m_C) + (\Delta A - \Delta B - \Delta C) = \Delta A - \Delta B - \Delta C$$

As with leptons, the number of nucleons must be conserved in radioactive decay, and nuclear reactions that destroy baryons are the province of high-energy physics. As a consequence, the term m_A-m_B-m_C equals zero, and only mass excesses remain. Mass excesses are calculated from neutral atoms (which incorporate the binding energies of the atomic electrons) and are tabulated. Those mass excesses relevant to nuclear forensic science are listed in Table 2.2, and others can be found in the Table of Isotopes [26].

Figure 2.2 shows the mass excesses of the isobaric (having the same mass number) nuclides with 143 nucleons. The most tightly bound of these, and consequently the most stable, is ^{143}Nd, having 60 protons and 83 neutrons in its nucleus. For those species to the left side of the figure, which have excess neutrons relative to ^{143}Nd, it is energetically favorable for the nucleus to eject an electron and undergo negative beta (β^-) decay—which results in the conversion of a neutron to a proton—and become more stable. For example,

$$^{143}\text{La} = {}^{143}\text{Ce} + Q_{\beta^-}.$$

The mass excesses of ^{143}La and ^{143}Ce are −78.191 and −81.616 MeV, respectively, so the energy released in the emission of an electron, $Q_{\beta-} = 3.42$ MeV. In this decay, the ^{143}La atom, containing 57 protons, 86 neutrons, and 57 electrons, converts a neutron into a proton, resulting in 58 protons, 85 neutrons, and 58 electrons. This is nominally the configuration of a neutral ^{143}Ce atom. Even though the decay electron was likely ejected from the atom to result in ionized ^{143}Ce and one or more free electrons, the difference in binding energies of the electrons, caused by their disarray in the

TABLE 2.2
Mass Excesses Relevant to Nuclear Forensic Analysis; Ground States

Nuclide	Mass Excess (MeV)	Nuclide	Mass Excess (MeV)
^{1}n	8.071	^{234}Pa	40.337
^{1}H	7.289	^{232}U	34.601
^{2}H	13.136	^{233}U	36.912
^{3}H	14.950	^{234}U	38.140
^{3}He	14.931	^{235}U	40.913
^{4}He	2.425	^{236}U	42.440
^{6}Li	14.086	^{237}U	45.385
^{7}Li	14.908	^{238}U	47.304
^{9}Be	11.348	^{239}U	50.570
^{12}C	0.000 (defined)	^{240}U	52.708
^{223}Ra	17.230	^{236}Np	43.380
^{224}Ra	18.818	^{237}Np	44.867
^{225}Ra	21.986	^{238}Np	47.450
^{226}Ra	23.661	^{239}Np	49.304
^{228}Ra	28.935	^{236}Pu	42.893
^{225}Ac	21.629	^{238}Pu	46.158
^{227}Ac	25.846	^{239}Pu	48.583
^{228}Ac	28.889	^{240}Pu	50.120
^{227}Th	25.801	^{241}Pu	52.950
^{228}Th	26.763	^{242}Pu	54.712
^{229}Th	29.579	^{244}Pu	59.799
^{230}Th	30.856	^{241}Am	52.929
^{231}Th	33.810	^{242}Am	55.463
^{232}Th	35.443	^{243}Am	57.167
^{234}Th	40.610	^{242}Cm	54.798
^{231}Pa	33.420	^{244}Cm	58.447
^{233}Pa	37.483		

decay process, is small compared to the energy released in the nuclear transformation. Thus, the error associated with ignoring this effect is small.

Those nuclides on the right side of Figure 2.2 are also unstable relative to ^{143}Nd. Decay can occur through capture of an electron by the atomic nucleus or by emission of a positron. For example, consider the decay of ^{143}Eu by EC:

$$^{143}\text{Eu} = {}^{143}\text{Sm} + Q_{\text{EC}}.$$

The ^{143}Eu nuclide, with 63 protons, 80 neutrons, and 63 electrons and a mass excess of −74.360 MeV, loses an electron from atomic space, converting a proton into a neutron and resulting in a ^{143}Sm nuclide containing 62 protons, 81 neutrons, and 62 electrons (still effectively a neutral atom). The mass excess is −79.527 MeV, and the Q-value for this process is 5.17 MeV. Positron (β^{+}) emission competes with EC in

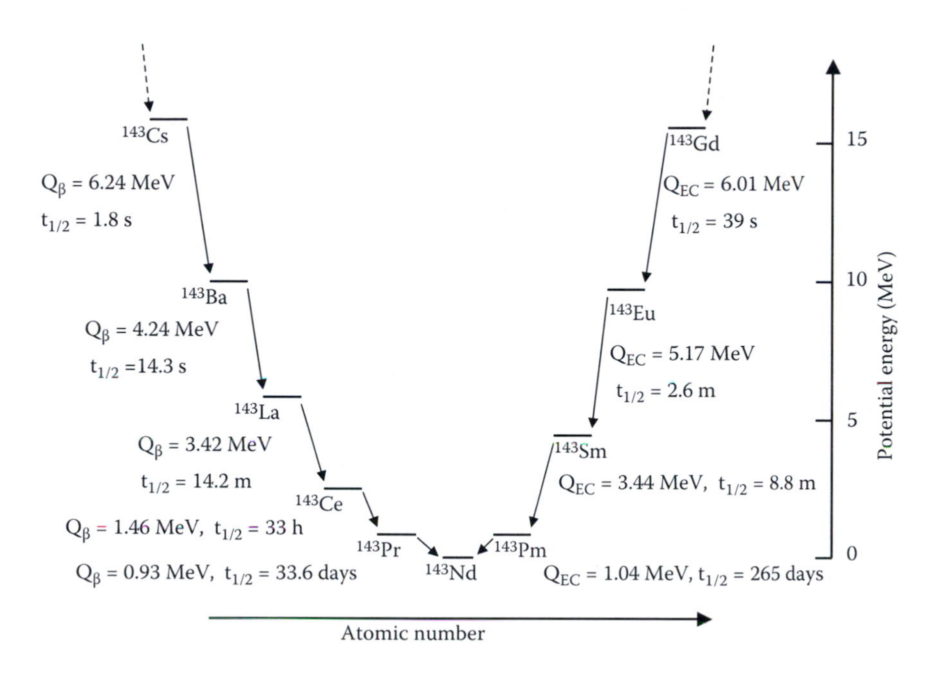

FIGURE 2.2 The mass excesses of the isobaric nuclides with A = 143 as a function of atomic number. The most stable nuclide at this mass is ^{143}Nd.

the decay of ^{143}Eu. In that case, however, the 63-proton, 80-neutron, and 63-electron assemblage decays to 62 protons, 81 neutrons, 63 electrons, and a positron. The use of neutral-atom mass excesses no longer holds, and the proper equation is now

$$^{143}\mathrm{Eu} = {}^{143}\mathrm{Sm} + \mathrm{e}^- + \mathrm{e}^+ + Q_{\beta^+}.$$

Because the extra electron and the β^+ both have masses of 0.511 MeV (energy equivalent), the Q-value for β^+ decay is 1.022 MeV lower than that for EC. A consequence of this is that β^+ emission is energetically forbidden if Q_{EC} is less than 1.022 MeV, and only EC can occur.

Figure 2.3 shows the masses of the isobaric nuclides with A = 144. Unlike atomic electrons, an energy advantage arises from all nucleons being paired—a consequence of the strong nuclear force. This behavior results in nuclides with even numbers of both protons and neutrons possessing an energy advantage over those with unpaired nucleons. For even-mass isobars, this results in the possibility of two, or even three, nuclides being stable to β-decay processes (in this case, ^{144}Nd and ^{144}Sm), as well as nuclides like ^{144}Pm that are unstable to both β decay and EC decay. The decay of ^{144}Sm into ^{144}Nd is energetically possible via the simultaneous absorption of two atomic electrons into the nucleus (along with the emission of two neutrinos). However, this "double-EC" process is extremely unlikely and has never been observed. Although double-β decay has been measured [38], the lifetime of the process is so long that it has no applicability to nuclear forensic analysis.

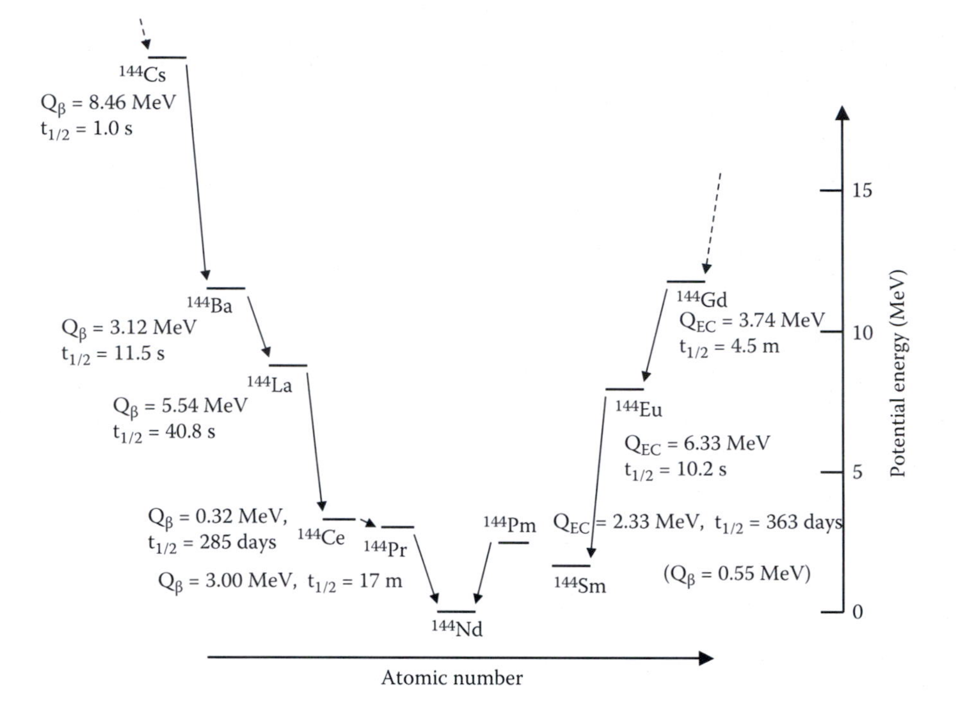

FIGURE 2.3 The mass excesses of the isobaric nuclides with A = 144 as a function of atomic number. There are two nuclides that are stable to β-decay processes, a consequence of the odd–even pairing effect.

At each mass number, there are one or more nuclides that are stable with regard to decay processes involving the emission or absorption of electrons; this general class of decay, consisting of β^- or β^+ emission or EC, is termed "β-decay processes." If mass-excess curves, such as those shown in Figures 2.2 and 2.3, are arranged in the order of increasing mass, the proton and neutron numbers of the stable species occupy the minimum of a pseudo-parabolic valley. This "valley of beta stability" begins in the light elements, where it is occupied by nuclides with nearly equal numbers of neutrons and protons. As mass number increases, the valley bends toward neutron excess because extra neutrons are required to dilute the increasing Coulomb repulsion between the nuclear protons. As a consequence of this bend, there are more than two β-stable isotopes per element on average. However, the alternating of smooth and stepped mass-excess curves associated with odd and even mass numbers, respectively, results in fewer β-stable nuclides with odd Z than with even Z; further, there are no stable nuclides with $Z = 43$ or $Z = 61$. Thus, the elements Tc [39] and Pm [40] do not exist in nature.

The relative stability of nuclides across the entire periodic table can be examined by calculating the packing fractions at each mass number. The packing fraction of a particular nuclide is defined as its mass excess divided by its mass number, $\Delta A/A$. It is a measure of how tightly bound the nuclear system is on a per-nucleon basis. In Figure 2.4, the packing fraction for the most stable nuclide

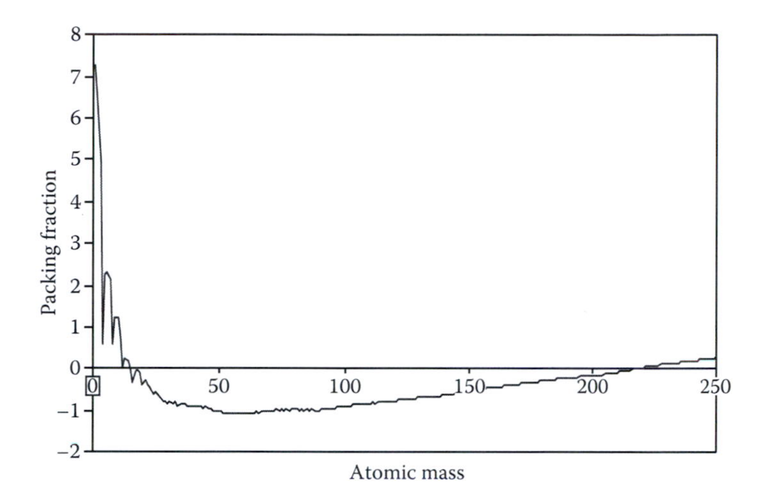

FIGURE 2.4 The nuclear packing fraction as a function of mass for the most stable nuclei at each mass number. The increase in packing fraction with increasing mass beyond the broad minimum at A = 60 is responsible for the onset of α decay for nuclei with A ≥ 150, and SF for A ≥ 240.

at each mass number is plotted as a function of mass. Clearly, the most tightly bound nuclei are those that occupy the broad minimum in the region of A = 60. At higher mass numbers, the packing fraction rises slowly as the nucleons in the nuclides become less tightly bound, induced by the increased Coulomb repulsion between the nuclear protons. As a consequence, decay modes in which the mass of the nuclide can be reduced become energetically possible, and then probable, with the onset of α decay near A = 150 and then SF near A = 240. The heaviest nuclide considered stable is ^{209}Bi. Although it is energetically unstable to α decay, the rate is so slow that it has never been observed. The slope of this packing-fraction curve is the origin of the large energy released in the fission process. Similarly, Figure 2.4 can also explain the energy released in the fusion of the lightest elements, particularly the hydrogen isotopes, as will be discussed in more detail later. The structure at mass numbers 4, 8, 20, and so forth is caused by closed configurations of nucleons, analogous to the electron configurations that result in the noble gases.

Figure 2.5 depicts two isobaric mass chains that are linked by α decay. In the figure, it is evident that ^{242}Pu and ^{242}Cm are stable to both β and EC decay processes, but are unstable to α decay (emission of a ^{4}He nucleus). For ^{242}Cm, the Q-value for the decay process $^{242}\text{Cm} = {}^{238}\text{Pu} + {}^{4}\text{He} + Q_\alpha$ is given by $\Delta(^{242}\text{Cm}) - \Delta(^{238}\text{Pu}) - \Delta(^{4}\text{He})$ = 6.216 MeV. For ^{242}Cf, the α-decay energy is so large that the time associated with the process is short compared with that expected for the low-energy EC decay to ^{242}Bk. In this case, α decay competes so favorably with EC that the latter is not observed. Similarly, even though ^{242}Np is unstable to α decay (to ^{238}Pa), α emission does not compete successfully with the much faster β decay to ^{242}Pu.

Much of the information contained in Figures 2.2, 2.3, and 2.5 can be extracted from the chart of the nuclides (Figure 2.6). This chart (sometimes called a Segre

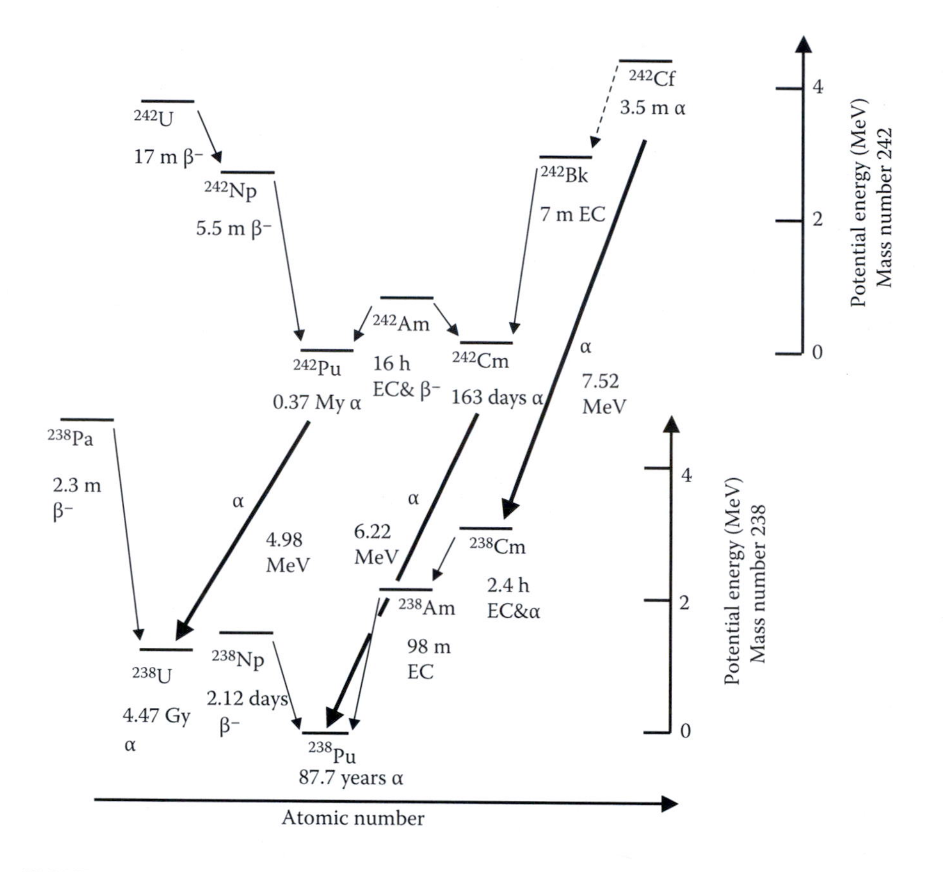

FIGURE 2.5 The mass excesses of nuclei with A = 238 and A = 242 as a function of atomic number. The nuclei in these two isobars are linked via α decay.

chart) is the most fundamental tool of the radiochemist, allowing the tracing of genetic lines linking radionuclides with their decay daughters. All known nuclides are plotted on the chart, with the atomic numbers increasing from bottom to top and the neutron numbers increasing from left to right. Thus, in any horizontal row, all nuclides possess the same nuclear charge and are isotopes of the same element. Each isobar (e.g., Figures 2.2 and 2.3) occupies a "northwest-to-southeast" diagonal slice through the chart, with nuclei unstable to β^- decay positioned below the stable nuclides and those unstable to EC and β^+ decay positioned above. This behavior is observed throughout the chart, but is masked at the heavy end by the onset of α decay. The valley of β stability is clearly visible, marked by stable nuclides veering toward the neutron-excess region.

Radioactive decay can be tracked across the chart of the nuclides, analogous to the movements of chess pieces on a chess board. If a particular nuclide undergoes β^- decay, the daughter nuclide of the decay process is one space diagonally up and to the left of the parent nuclide. Similarly, EC or β^+ decay causes a movement diagonally down and to the right. Emission of an α particle results in a nucleus that is two

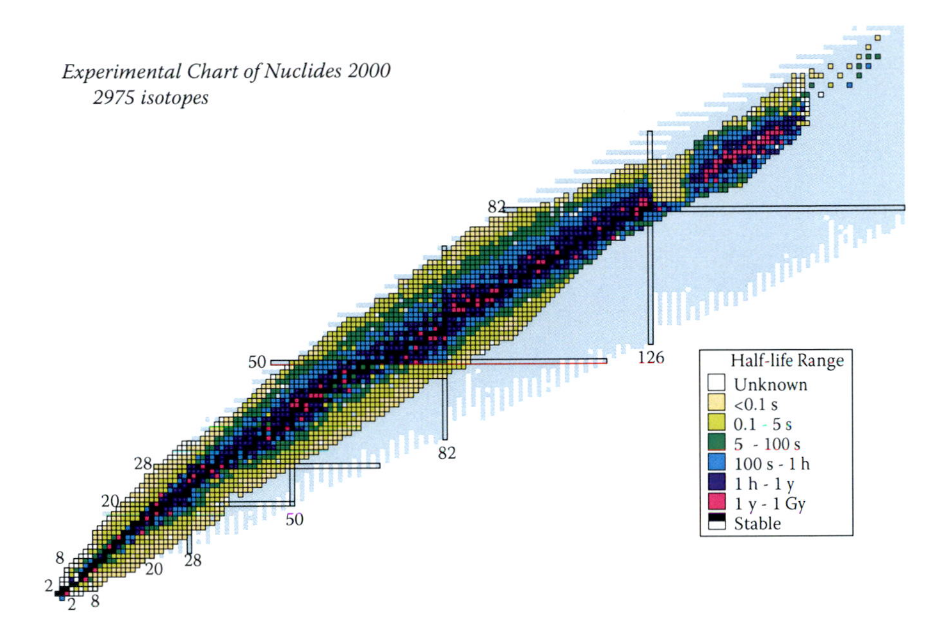

FIGURE 2.6 The chart of the nuclides, with the neutron number increasing from left to right and the proton number increasing from bottom to top. Radioactive decay can be tracked on the chart according to simple rules.

units down and two units to the left from the emitting nuclide. Gamma decay and IC involve transitions that do not result in a change in the position of the emitting nuclide on the chart, although an isomeric state may exist. Only SF decay cannot be tracked on the chart of the nuclides, as the final products of this decay process are undefined.

Each nuclide is represented by a square sufficiently large to contain a brief summary of some of its physical properties. When isomeric states exist (this topic is covered later in this book), the square is subdivided to present the additional information. Sometimes this requires that the chart be broken into smaller sections to avoid becoming inconveniently large. Consider, for example, the low end of the chart, depicted in Figure 2.7. The valley of β stability begins with normal hydrogen, also known as protium (^{1}H), shown as a dark box in the lower left corner. The neutron, whose mass excess is somewhat greater than that of the proton and is consequently unstable to β decay, is denoted by the nearby shaded box. For most of the chart, the edge is determined by the limits of the explored nuclei. At the light end of the chart, however, it also depicts the limit between bound (those with positive binding energy) and unbound nuclei. Deuterium (^{2}H, D) and tritium (^{3}H, T) are the only bound isotopes of hydrogen heavier than protium. Deuterium is stable, whereas T is unstable to β decay to ^{3}He. Beyond He, the nuclides in Figure 2.7 that are most important to nuclear forensic applications are the stable ones (dark boxes), as will be discussed later.

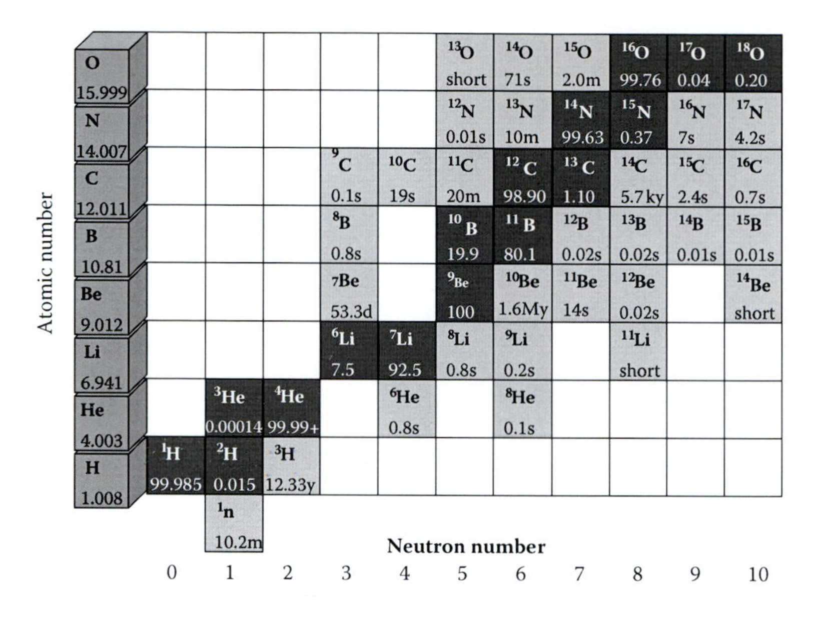

FIGURE 2.7 Low-mass end of the chart of the nuclides.

The decay modes, Q-values, and half-lives for the nuclides of importance to nuclear forensic science are given in Table 2.3; others can be found in the Table of Isotopes [26].

2.5 NUCLEAR STRUCTURE, ISOMERISM, AND SELECTION RULES

The theory of nuclear structure is markedly different from the theory of the structure of atomic electrons around the nucleus in two fundamental regards: first, the forces between atomic electrons are purely electrostatic, whereas the forces between the nucleons are both complicated and incompletely understood; second, the interaction between electrons can be treated as a perturbation of the central attraction between the electrons and the atomic nucleus, whereas the interaction between nucleons is short-ranged, saturated, and not oriented toward a center. In one parallel with atomic theory, however, even if an analytic expression for the interaction between two nucleons existed, an accurate description of most nuclei would be intractable because of the many-body nature of the problem.

Nuclear structure theory relies on a set of models in which the behavior of the nucleus is approximated under a given set of rules. The most successful of these models treat the nucleus as a whole, creating an attractive nuclear potential well in which the nucleons are confined. Like particles in the quantum-mechanical "particle-in-a-box" problem, the shape of the well and the nucleon–nucleon force define a set of energy levels, which the nucleons fill from the bottom. Each level is characterized by a binding energy, an angular momentum, and parity. A primitive nuclear model to explain the bend of the valley of β stability toward neutron excess will be presented later, and it will prove useful in the demonstration of other issues of nuclear structure (see Figure 2.8). For more information on nuclear models, references [41,42] are good sources (the model under discussion here is similar to the Fermi gas model).

TABLE 2.3
Radioactive Decay Properties Relevant to Nuclear Forensic Science

Nuclide	Decay Mode	Decay Q-Value (MeV)	Partial Half-Life	Total Half-Life
^{1}n	β^-	0.782	615 s	
^{3}H	β^-	0.019	12.33 years	
^{207}Tl	β^-	1.423	4.77 min	
^{208}Tl	β^-	5.001	3.053 min	
^{209}Tl	β^-	3.980	2.20 min	
^{210}Tl	β^-	5.484	1.30 min	
^{209}Pb	β^-	0.644	3.253 h	
^{210}Pb	β^-	0.064	22.3 years	22.3 years
	A	3.792	1.17×10^9 years	
^{211}Pb	β^-	1.373	36.1 min	
^{212}Pb	β^-	0.574	10.64 h	
^{214}Pb	β^-	1.023	26.8 min	
^{210}Bi	β^-	1.163	5.013 days	5.013 days
	A	5.037	1.04×10^4 years	
^{211}Bi	β^-	0.579	12.9 h	2.14 min
	α	6.750	2.15 min	
^{212}Bi	β^-	2.254	1.575 h	60.6 min
	α	6.207	2.808 h	
^{213}Bi	β^-	1.426	46.56 min	45.6 min
	α	5.982	1.515 days	
^{214}Bi	β^-	3.272	19.9 min	19.9 min
	α	5.621	65.8 days	
^{210}Po	α	5.407	138.4 days	
^{211}Po	α	7.594	0.516 s	
^{212}Po	α	8.954	0.299 μs	
^{213}Po	α	8.537	4.2 μs	
^{214}Po	α	7.833	164.3 μs	
^{215}Po	α	7.526	1.781 ms	
^{216}Po	α	6.906	0.145 s	
^{218}Po	β^-	0.264	10.8 days	3.10 min
	α	6.115	3.10 min	
^{215}At	α	8.178	0.10 ms	
^{217}At	β^-	0.740	4.49 min	32.3 ms
	α	7.202	32.3 ms	
^{218}At	α	6.874	1.6 s	
^{219}At	β^-	1.700	31 min	56 s
	α	6.390	58 s	
^{219}Rn	α	6.946	3.96 s	
^{220}Rn	α	6.405	55.6 s	
^{222}Rn	α	5.590	3.824 days	
^{221}Fr	α	6.458	4.9 min	
^{223}Fr	β^-	1.149	21.8 min	21.8 min
	α	5.430	250 days	

(Continued)

TABLE 2.3 (*Continued*)
Radioactive Decay Properties Relevant to Nuclear Forensic Science

Nuclide	Decay Mode	Decay Q-Value (MeV)	Partial Half-Life	Total Half-Life
^{223}Ra	α	5.979	11.435 days	
^{224}Ra	α	5.789	3.66 days	
^{225}Ra	β⁻	0.357	14.9 days	
^{226}Ra	α	4.871	1,600 years	
^{228}Ra	β⁻	0.046	5.75 years	
^{225}Ac	α	5.935	10.0 days	
^{227}Ac	β⁻	0.045	22.08 years	21.77 years
	α	5.042	1,578 years	
^{228}Ac	β⁻	2.127	6.15 h	
^{227}Th	α	6.146	18.72 days	
^{228}Th	α	5.520	1.9131 years	
^{229}Th	α	5.168	7,340 years	
^{230}Th	α	4.770	75,380 years	
^{231}Th	β⁻	0.390	25.52 h	
^{232}Th	α	4.083	1.405×10^{10} years	
^{233}Th	β⁻	1.245	22.3 min	
^{234}Th	β⁻	0.273	24.10 days	
^{231}Pa	α	5.149	32,760 years	
^{233}Pa	β⁻	0.570	26.967 days	
^{234m}Pa	β⁻	2.277	1.17 min	1.17 min
	IT	0.080	12.2 h	
^{234g}Pa	β⁻	2.197	6.70 h	
^{232}U	α	5.414	68.9 years	
^{233}U	α	4.908	1.592×10^{5} years	
^{234}U	α	4.858	2.455×10^{5} years	
^{235}U	α	4.679	7.038×10^{8} years	
^{235m}U	IT	0.00008	25 min	
^{236}U	α	4.572	2.342×10^{7} years	
^{237}U	β⁻	0.519	6.75 days	
^{238}U	α	4.270	4.468×10^{9} years	
^{239}U	β⁻	1.265	23.45 min	
^{240}U	β⁻	0.388	14.1 h	
^{236m}Np	β⁻	0.550	1.953 days	22.5 h
	EC	1.000	1.803 days	
^{236g}Np	β⁻	0.490	1.232×10^{6} years	1.54×10^{5} years
	EC	0.940	1.764×10^{5} years	
	α	5.020	9.6×10^{7} years	
^{237}Np	α	4.959	2.14×10^{6} years	
^{238}Np	β⁻	1.292	2.117 days	
^{239}Np	β⁻	0.722	2.3565 days	
^{240m}Np	β⁻	2.200	7.22 min	
^{236}Pu	α	5.867	2.858 years	
^{238}Pu	α	5.593	87.74 years	

(*Continued*)

TABLE 2.3 (*Continued*)
Radioactive Decay Properties Relevant to Nuclear Forensic Science

Nuclide	Decay Mode	Decay Q-Value (MeV)	Partial Half-Life	Total Half-Life
^{239}Pu	α	5.244	24,110 years	
^{240}Pu	α	5.256	6,563 years	
^{241}Pu	β^-	0.021	14.35 years	14.35 years
	α	5.140	5.86×10^5 years	
^{242}Pu	α	4.983	3.733×10^5 years	
^{243}Pu	β^-	0.582	4.956 h	
^{244}Pu	α	4.666	8.08×10^7 years	
^{245}Pu	β^-	1.205	10.5 h	
^{241}Am	α	5.638	432.2 years	
^{242m}Am	α	5.588	3.07×10^4 years	141 years
	IT	0.049	141 years	
^{242g}Am	β^-	0.665	19.37 h	16.02 h
	EC	0.751	3.86 days	
^{243}Am	α	5.438	7,370 years	
^{244m}Am	β^-	1.516	26 min	26 min
	EC	0.164	50.2 days	
^{244g}Am	β^-	1.428	10.1 h	
^{245}Am	β^-	0.894	2.05 h	
^{242}Cm	α	6.216	162.8 days	
^{244}Cm	α	5.902	18.10 years	

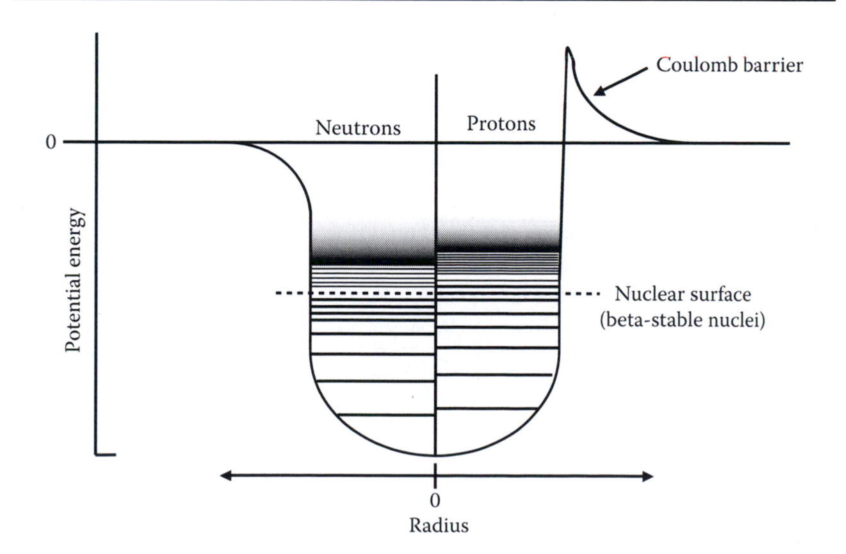

FIGURE 2.8 A simplistic nuclear-level model. Proton levels are inclined to larger separations than neutron levels because of the Coulomb interaction between protons, which also results in a barrier to charged-particle interactions. Beta-stable nuclides, with protons and neutrons filling the well to the same potential energy, tend to have more neutrons than protons.

In Figure 2.8, the nucleus defines two separate potential wells, one for neutrons and one for protons. The Pauli exclusion principle [37] applies to nucleons in the nucleus (which obey Fermi–Dirac statistics), just as it does to electrons in the atom. In atoms, no two electrons may be described by the same set of quantum numbers; in nuclei, no two protons may occupy the same state, and no two neutrons may do so, but as neutrons and protons are dissimilar particles, they may have identical quantum states. The quantum states associated with neutrons tend to be closer together in energy than those associated with protons because of the Coulomb force. Similarly, the nucleus as a whole has a Coulomb barrier to charged particles that does not apply to neutrons. This is an important distinction that will be explored in more detail in the section on nuclear reaction theory.

The twofold well and energy levels shown in Figure 2.8 do not exist without the presence of nucleons. This is the fundamental disconnect of the model. The well is constructed and nucleons added, either in the form of a "nuclear fluid" or in the form of nuclear particles. The most stable nucleus at a given mass number is the one where the "fluid level" of protons and neutrons is approximately equal. With an imbalance, spontaneous decay processes can occur in which protons become neutrons or vice versa, depending on the energy associated with the conversion. As remarked earlier, there is an additional odd–even effect associated with the extra binding energy attained by nuclei with completely paired nucleons. It is accommodated in the same fashion employed by atomic theorists, invoking a spin quantum number that allows each state to be occupied by two similar nucleons with the same major angular momentum quanta, but with spins of opposite sign.

In atomic spectroscopy, there are configurations of electrons associated with high ionization potential and, consequently, relative chemical inertness. There are analogous configurations of nucleons associated with extra nuclear stability (i.e., local maxima in binding energy). These "magic numbers" of nucleons are caused by gaps in the energy-level spacings in a nucleus, which arise from the motions of single particles in an average nuclear potential and are explained by the nuclear shell model [43]. Nuclei with magic numbers (2, 8, 20, 28, 50, 82, 126, and 184) of protons or neutrons tend to be spherical. Nuclear levels in these species are far apart and are associated with the promotion of single particles (or the resulting "holes" in single-particle states) from one nuclear level to another. Nuclei that have nucleon numbers far from the magic numbers are often better described by collective models [44,45], in which groups of nucleons move collectively and are characterized by nuclear rotational and vibrational levels.

Collective models give rise to nonspherical nuclear shapes, and collective states arising from the motion of the nucleus as a whole are more closely spaced than are the single-particle states. The nuclei of the U and Pu isotopes of interest to the nuclear forensic analyst are strongly nonspherical. Their nuclear spectroscopy is dominated by ladders of prolate rotational states superimposed on the single-particle states [26].

As noted previously, there is a binding energy associated with the pairing of nucleons of opposite spin. It results in nuclei containing even numbers of both protons and neutrons (the "even–even" nuclei), with ground states that are $J^{\pi} = 0^{+}$. The lowest single-particle excited states have much higher energies than do those nuclei with an odd number of protons or neutrons (or both). In the ground state of even–even nuclei, all angular momenta are canceled by the pairing of nucleons with the same principal quantum numbers and opposite spin. In the heavy elements, the first single-particle states in even–even nuclei are approximately 500–600 keV above the ground state, with only rotational collective states occurring at lower excitation energies (in the pairing gap). The odd-nucleon actinides have single-particle states that can occur at excitation energies as low as a few electron volts above the ground state.

Excited states in nuclei arise by promoting nucleons (singly or collectively) from the most bound states to higher states through the absorption of energy (as in nuclear reactions), or through decay processes that may not proceed in a single step and may leave the daughter nucleus in a transient state higher in energy than the ground state. As in atomic electron spectroscopy, selection rules govern which states tend to be populated in decay processes, even when energetically achievable. These nuclear selection rules arise in various nuclear models in which the quantum numbers occur. Decays involving small changes in quantum numbers tend to be favored over those involving large changes; decay processes involving large changes in energy tend to be favored over those involving small changes in energy; and the probability of populating a given state in the daughter nucleus is determined by a combination of these two effects.

The definition of the models and the resultant selection rules are beyond the scope of this book, which is focused on practical matters of nuclear forensic science. The salient points of this section are that nuclei exist in discrete energy states, similar to those occupied by the atomic electrons. Each state is characterized by an excitation energy, an angular momentum, and a parity. Decay processes can result in excited states being occupied, subject to the limitations of the law of conservation of energy; however, selection rules, based on the quantum-mechanical properties of these states, exist that make some decay processes improbable.

The number of states in a given energy interval (the level density) increases with an increase in excitation energy (see Figure 2.8). As mentioned earlier, the energy width of each state is nonzero, depending on the lifetime of that state. In the heavy elements, at excitation energies of a few million electron volts, the states become so closely packed that they "overlap" and form a continuum. At these energies, nuclear spectroscopy becomes a statistical science rather than the study of discrete states. The transition between discrete states and continuum states occurs at excitation energies of concern in nuclear reactions, but radioactive decay processes almost always involve transitions between states well below the continuum threshold.

The effect of nuclear structure on the decay properties of radionuclides is likely best illustrated through a set of examples, taken mainly from the Table of Isotopes [26]. The excited states in light nuclei tend to be far apart in energy. In addition, collective effects are minimized in nuclei containing a small number of nucleons. Thus, decay processes in light nuclei are usually simple and straightforward to understand. Consider the EC decay of ^{7}Be, depicted in Figure 2.9. The half-life for the decay is 53.3 days, and the total energy available from the conversion of a proton in ^{7}Be to a neutron in ^{7}Li (the difference in their mass excesses) is 862 keV. Because emission of a β^+ requires 1022 keV more energy than does EC, β^+ emission does not compete with EC in the decay of ^{7}Be. Only two states in ^{7}Li are energetically available to receive the decay: the $J^\pi = 3/2^-$ ground state and a $J^\pi = 1/2^-$ excited state at 478 keV. Because

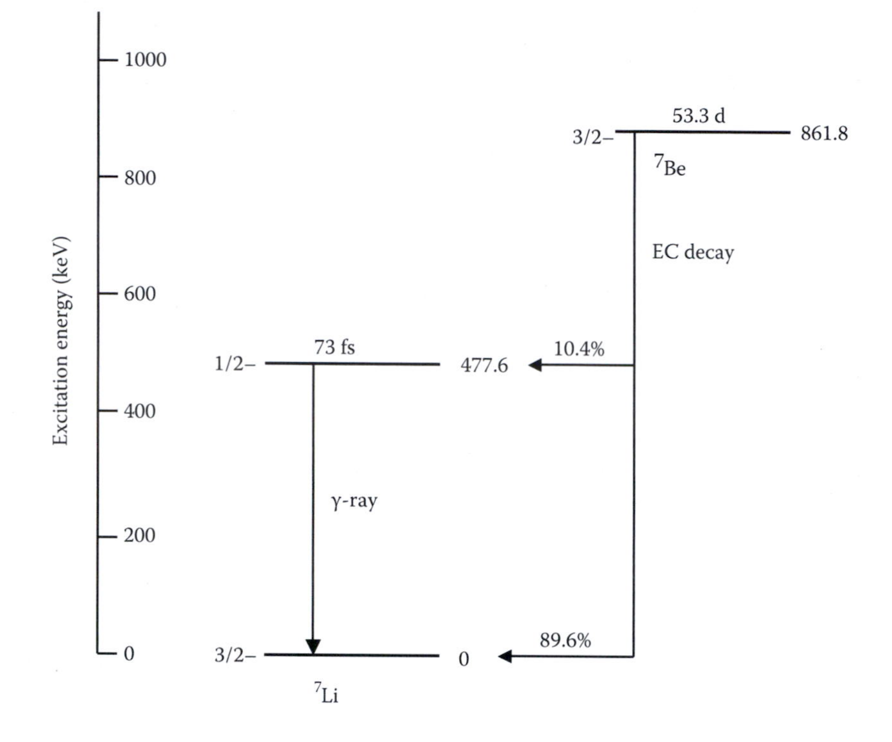

FIGURE 2.9 The EC decay of ^{7}Be to levels in ^{7}Li.

both ^{7}Li and ^{7}Be have an odd number of nucleons, there is always an unpaired nucleon whose intrinsic spin (±1/2) results in all states in both nuclei being half-integer.

^{7}Li does have higher-energy states, discovered in nuclear reaction experiments, but the lowest of these states has an excitation energy of 4630 keV and is prohibited from participating in the decay by the law of energy conservation. The quantum numbers of both accessible daughter states are close to those of the ^{7}Be ground state, so the energy of the decay is more important than the selection rules for deciding the relative probability with which the states in ^{7}Li are populated. Ninety percent of the decays of ^{7}Be directly populate the ground state of ^{7}Li, and 10% populate the first excited state. Decays populating the first excited state proceed promptly to the ground state through the emission of a 478-keV γ-ray, with a half-life of 7.3×10^{-14} s. From the uncertainty principle, the uncertainty in excitation energy of this state is on the order of 0.01 eV, which is insignificant compared with the measurement accuracy of ΔE. The decay energy of the first excited state is very large relative to the binding energies of the electrons in ^{7}Li, so IC does not significantly compete with γ emission for depopulation of the excited state [46].

The decay of ^{22}Na is shown in Figure 2.10. The half-life for the decay is 2.6 years, and the total decay energy available in the conversion of a proton in ^{22}Na to a neutron in ^{22}Ne is 4790 keV. Because the decay energy of ^{22}Na is significantly larger than that for ^{7}Be, the selection rules involved in ^{22}Na decay must be unfavorable to result in its half-life being significantly longer than that of ^{7}Be. As the decay energy is much

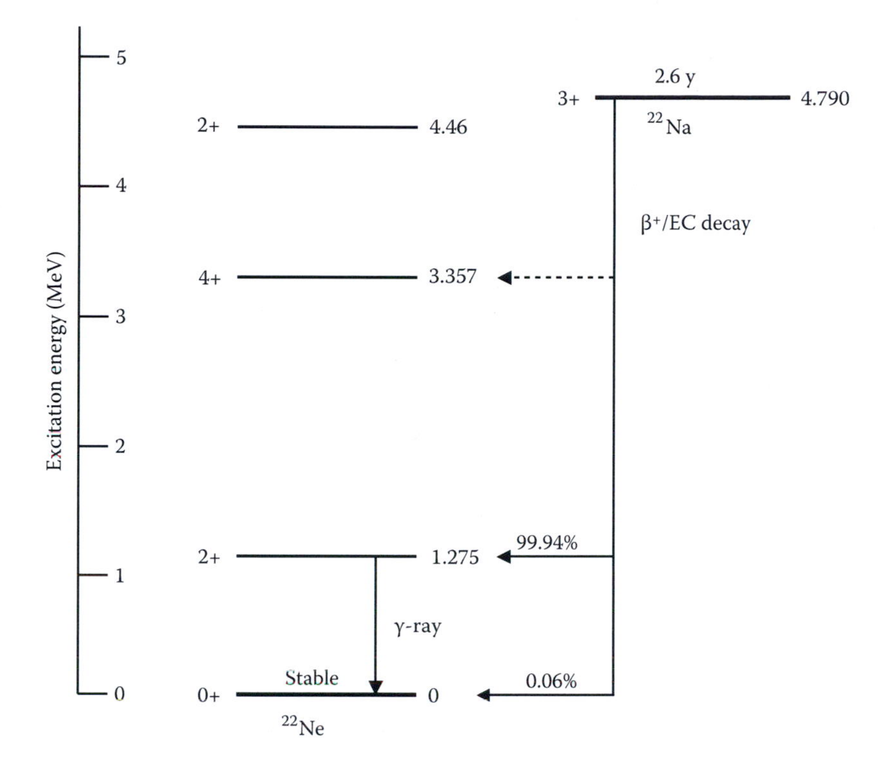

FIGURE 2.10 The EC/β^+ decay of ^{22}Na to levels in ^{22}Ne.

larger than 1022 keV, β^+ emission can compete with EC (in fact, 90.5% of the decays populating the first excited state involve emission of a β^+). ^{22}Ne is an even–even nucleus: in its ground state, all nucleons are paired and result in a $J^\pi = 0^+$ configuration. Because an even number of nucleons exist in both ^{22}Ne and ^{22}Na, any nuclear state in either of them must involve an even number of unpaired nucleons, each with spin 1/2. Thus, all states in both nuclei must be integer. Only four states in ^{22}Ne are energetically available to receive the decay, and of these, nearly all of the decay populates the $J^\pi = 2^+$ first excited state at 1275 keV.

The relative lack of intensity for transitions to other states is a result of a combination of decay energetics and selection rules. The energies released in decays to the levels at 3357 and 4460 keV are so much less than that for decay to the 1275-keV state that the probabilities for the former are very small. Decay to the ground state, although energetically favored, is hindered by the requirement that it remove three units of angular momentum, a transition from $J^\pi = 3^+$ ^{22}Na to $J^\pi = 0^+$ ^{22}Ne (ground state). In contrast, decay to the first excited state involves only a single unit of angular momentum (from $J^\pi = 3^+$ ^{22}Na to $J^\pi = 2^+$ ^{22}Ne at 1275 keV). Decay populating the first excited state proceeds promptly to the ground state through the emission of a 1275-keV γ-ray. As for ^{7}Be, IC does not compete with γ emission for depopulation of the excited state.

In heavier nuclei, particularly those involving unpaired nucleons, there are many low-lying excited states at energies accessible to decay processes. This allows excited states with long half-lives, a consequence of the selection rules for γ decay. These long-lived states, known as nuclear isomers, may be so hindered to γ-decay processes by the necessity of large changes in quantum number that other decay processes can successfully compete with γ emission and IC. Soddy proposed the existence of isomers in 1917 [47,48], and Hahn subsequently observed the phenomenon, assigning the classical uranium chain decay products, UX_2 and UZ, to different states in ^{234}Pa [49,50]. For many years [51], these were the only known isomeric pair: nuclei that possess the same number of protons and neutrons, but with quite different radioactive properties. Although dozens of such states are now known, the ^{234}Pa isomer remains the most important isomeric state for the nuclear forensic scientist.

A portion of the A = 234 isobaric decay scheme is shown in Figure 2.11. The even–even nuclide, ^{234}Th, β-decays to ^{234}Pa with a half-life of 24.1 days. The total decay energy available in the conversion of a neutron in ^{234}Th to a proton in ^{234}Pa is 273 keV. The ground state of ^{234}Pa is $J^{\pi} = 4^{+}$, and it is not surprising that an immeasurably small fraction of the β^- decay of $J^{\pi} = 0^{+}$ ^{234}Th proceeds directly to the ground state of ^{234}Pa. Of the ^{234}Th decay, 70.3% populates a $J^{\pi} = 0^{-}$ state in ^{234}Pa at an excitation energy of approximately 80 keV, while 29.7% populates a cluster of $J = 1$ states at excitation energies between 173 and 193 keV. Gamma transitions involving small changes in the quantum numbers of the nucleus are favored by the γ-decay selection rules. The $J = 1$ states decay by a combination of photon emission and IC to the 80-keV state. Essentially all of the decay intensity of ^{234}Th collects in this isomeric state, which we denote ^{234m}Pa. Only two states are energetically accessible to receive the γ/IC decay of ^{234m}Pa, and the combination of small decay energies and large change in quantum numbers makes the process very slow, proceeding with a partial half-life of approximately 15 h.

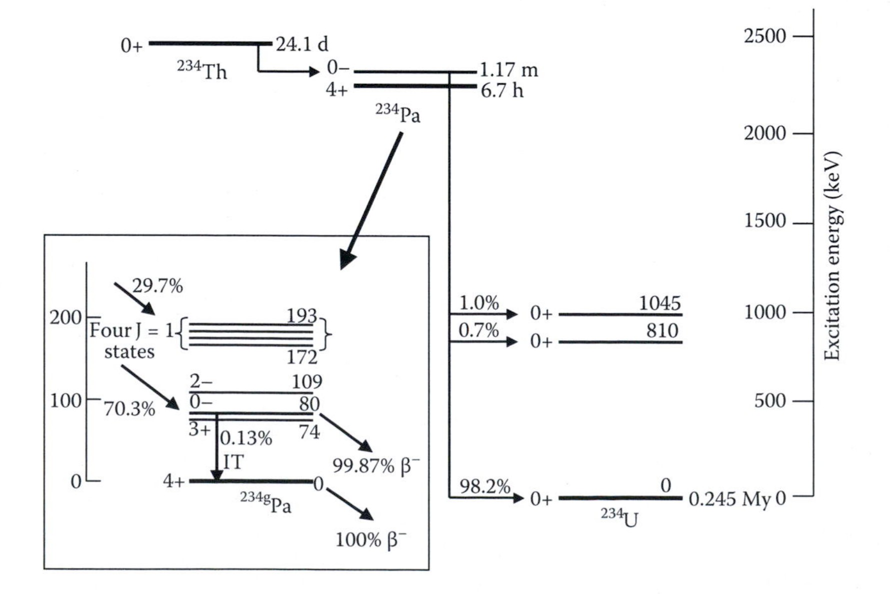

FIGURE 2.11 The β^- decays of ^{234}Th and ^{234}Pa. The decay of ^{234}Th populates the isomeric state in ^{234}Pa; β^- decay and IT compete in the decay of ^{234m}Pa.

As Figure 2.11 shows, the Q-value for β^- decay from ^{234m}Pa is quite large (2277 keV). As a consequence, β decay competes successfully with decay processes internal to ^{234}Pa (referred to collectively as isomeric transition, or IT). ^{234m}Pa decays with a half-life of 1.17 min, 99.87% of the time through β^- decay to low-J states in ^{234}U, and 0.13% of the time by IT to the $J^\pi = 4^+$ ^{234}Pa ground state (sometimes written as ^{234g}Pa). ^{234g}Pa, in turn, decays with a half-life of 6.75 h to completely different states in ^{234}U. The γ transition from ^{234m}Pa to ^{234g}Pa is so highly converted that no 80-keV photons are observed, and the decay occurs by 100% IC.

Anticipating our impending treatment of nuclear chronometry, the decay equations given previously can be applied to the A = 234 isobaric nuclides. After the chemical isolation of a sample of ^{234}Th, secular equilibrium will be established between ^{234m}Pa and ^{234}Th within a few minutes; their activities are related through Equation 2.5. At some later time (on the order of a day or more), a transient equilibrium is established between ^{234g}Pa and ^{234}Th, whose half-lives are more comparable with one another than are those of ^{234m}Pa and ^{234}Th. The activities of ^{234}Th and ^{234g}Pa are related by Equation 2.6, incorporating the IT branch in ^{234m}Pa:

$$[\lambda(^{234g}Pa) - \lambda(^{234}Th)]\, N(^{234g}Pa) = 0.0013\, \lambda(^{234}Th)\, N(^{234}Th).$$

The α decay of 4.47×10^9-year ^{238}U produces ^{234}Th; if a ^{238}U source is chemically isolated, after several months, secular equilibrium is established among ^{238}U, ^{234}Th, ^{234m}Pa, and (fractionally) ^{234g}Pa.

The α decay of ^{234}U to ^{230}Th is shown in Figure 2.12, and this decay process is the basis of one of the most important chronometers in nuclear forensic analysis. Similar to the other modes of radioactive decay, α-particle emission is also subject to selection rules and is very sensitive to the magnitude of the decay energy. Because ^{230}Th is an even–even nuclide, there is a pairing gap (a consequence of the extra binding energy obtained when all nucleons are paired) such that the only levels at excitation energies below 500 keV are the collective states built on the ground-state configuration. The 2^+ and 4^+ states in ^{230}Th, shown in Figure 2.12, are caused by rotation of the prolate nucleus as a whole, involving no breaking of paired nucleons. The fact that only J = even states are present in the rotational band is a quantum-mechanical effect specific to rotational states built on $J = 0$ intrinsic states [37,41]. (Later examples will demonstrate that the rotational states built on J = nonzero intrinsic states are spaced by single units of angular momentum.)

Because the lowest three states in ^{230}Th have the same single-particle configurations, α decay to the 2^+ and 4^+ states is only weakly hindered by the selection rules, with reduced decay energy responsible for most of the decreased probability of their population in the decay process. Only an insignificant fraction of ^{234}U decay populates states above the pairing gap in ^{230}Th.

Inset in Figure 2.12 is the diagram of a spectrum that a nuclear forensic analyst might interpret following the count of a ^{234}U sample with an α counter. Note that even though the α-decay Q-value is 4.858 MeV, the energy of the highest observed α-particle group in the spectrum is 4.775 MeV. This fact is a consequence of the mass of the

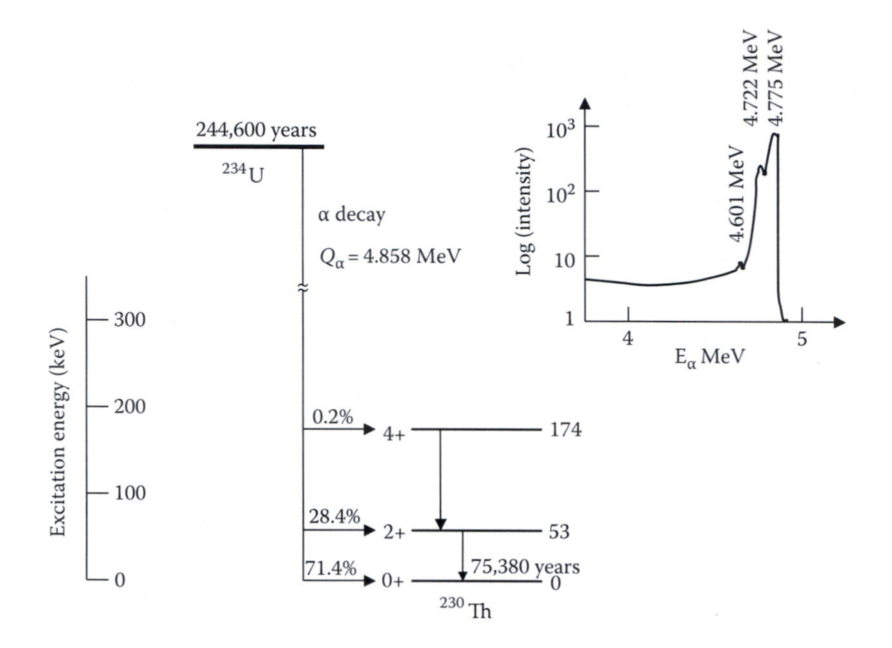

FIGURE 2.12 The α decay of ^{234}U, populating states in ^{230}Th. Inset is a schematic diagram of the α spectrum, showing the decay intensity to the different members of the ground-state rotational band.

α particle and the law of conservation of momentum. In α decay, the energy released in the emission of a ^{4}He nucleus is divided between the outgoing decay products:

$$'Q_\alpha = E(^4\text{He}) + E(\text{daughter}),$$

where

$'Q_\alpha$ is the energy released in the α decay (the Q-value less the excitation energy left in the daughter nucleus, if any)

$E(^4\text{He})$ is the kinetic energy of the emitted α particle

$E(\text{daughter})$ is the recoil energy of the decay daughter

The α particle and the daughter product must be created with equal, but opposite, momenta:

$$p(^4\text{He}) = p(\text{daughter}).$$

Because α particles are emitted at subrelativistic velocities, classical mechanics applies and $E = p^2/2m$. Substituting and rearranging results in the following formula:

$$E(^4\text{He}) = {'Q_\alpha} \left\{ m(\text{daughter}) / [m(^4\text{He}) + m(\text{daughter})] \right\}. \qquad (2.8)$$

Therefore, the energy of the α particle emitted in the direct decay of ^{234}U to the ground state of ^{230}Th is (4.858 MeV) × 230/(230 + 4) = 4.775 MeV. Equation 2.8

also applies to the other modes of radioactive decay. However, except for the fission process, the mass of the outgoing particle is so small that the equation reduces to E(particle) = $'Q$, and essentially no recoil energy is imparted to the residual daughter nucleus.

The α decay of ^{241}Am to ^{237}Np is shown in Figure 2.13. Again, this is also a very important chronometric decay scheme. Because ^{237}Np is an odd-mass nuclide, all states have half-integer angular momenta (J-values), and there is no pairing gap between the $5/2^+$ ground state and the first intrinsic (noncollective) excited state, the $J^\pi = 5/2^-$ state at 60 keV. The $7/2^+$ state at 33 keV and the $9/2^+$ state at 76 keV are the first members of a rotational band built on the ground state. Similarly, the $7/2^-$ state at 103 keV is a rotational state based on the intrinsic $5/2^-$ state at 60 keV. As a result, they have very similar quantum numbers. Most of the ^{241}Am α-decay intensity populates the $5/2^-$ rotational band starting at 60 keV, even though the ground-state band is populated by transitions that are of higher energy and, hence, are potentially more probable. This result is a consequence of the 60-keV state in ^{237}Np being of the same quantum-mechanical configuration as the ^{241}Am ground state, and the population of this "analog state" is favored by the α-decay selection rules.

The 60-keV state is not isomeric, however: the gamma selection rules provide for its rapid depopulation ($t_{1/2}$ = 68 ns) by γ transitions to the ground state. This transition is substantially converted, with only about one-third of the decays involving

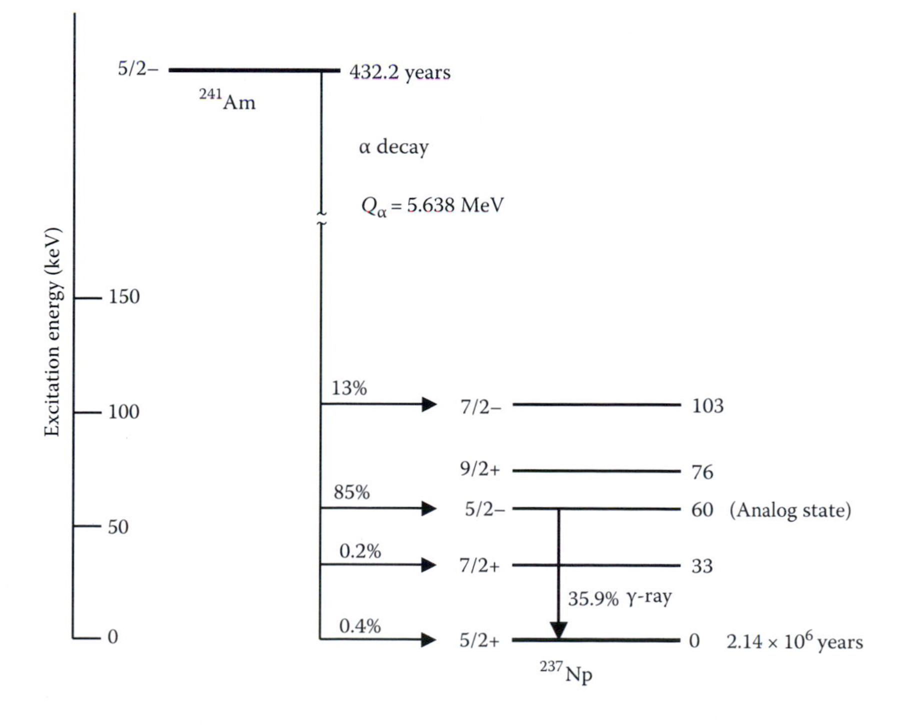

FIGURE 2.13 The α decay of ^{241}Am, populating states in ^{237}Np. Decays to the ground state are hindered by the change in parity. The analog state at 60 keV in ^{237}Np has the same quantum numbers as the ^{241}Am ground state.

emission of a γ photon. The energy of the α particle most probably emitted in the decay of ^{241}Am is (5.638 – 0.060) × 237/(237 + 4) = 5.486 MeV (Equation 2.8).

An analog state that is also isomeric is populated in the α decay of ^{239}Pu. The 23-min ^{235m}U isomer, which has the same quantum numbers as the $J^{\pi} = 1/2^{+}$ ^{239}Pu ground state, has an associated excitation energy of only 0.076 keV. It is one of the lowest-energy nuclear excited states known, so low that changing the electronegativity of the chemical compound incorporating the atom significantly changes the half-life of its decay by internal transition [52–54]. The IT decay is completely converted, with electrons emitted from the atomic P- and Q-shells. In the discovery experiments, α decay in a thin source of ^{239}Pu produced ^{235m}U atoms that were propelled from the source by nuclear recoil and stopped in a volume of gas, a common radiochemical technique.

The final example involves the decay of ^{242m}Am, depicted in Figure 2.14. The odd–odd ^{242}Am nuclide is an exemplary workshop in nuclear structure and decay. The ground state, with $J^{\pi} = 1^{-}$, is unstable to both β^{-} decay to ^{242}Cm and EC decay to ^{242}Pu, both of which are even–even nuclides. Decays to the 0^{+} and 2^{+} levels in the ground-state rotational bands of both daughters are favored by both decay energy and the selection rules, so both processes occur. The β^{-} decay to ^{242}Cm proceeds with a partial half-life of 19.37 h, whereas EC decay to ^{242}Pu proceeds with a partial half-life of 3.86 days. The decays combine for an overall half-life of 16.02 h. The α decay of ^{242g}Am is expected to proceed with a partial half-life of thousands of years, and it is consequently not observed. The isomeric state in ^{242}Am, at an excitation energy of 49 keV, has $J^{\pi} = 5^{-}$. The selection rules for β^{-} and EC decays to the ground-state ($J^{\pi} = 0^{+}$) bands of ^{242}Cm and ^{242}Pu preclude these processes from occurring with any significant probability; only decays to states above the pairing gaps in the daughters are allowed.

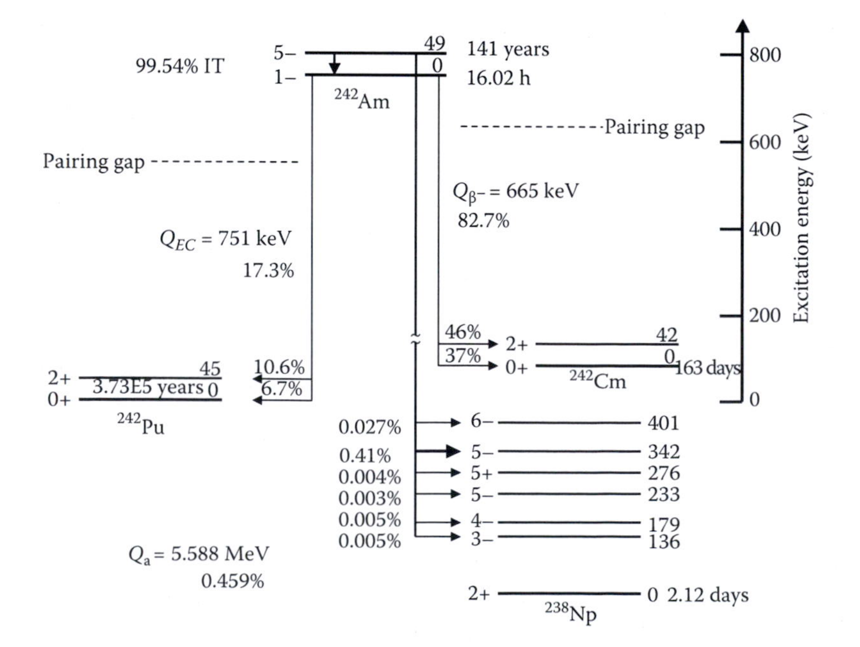

FIGURE 2.14 The complicated decay scheme of ^{242m}Am.

Because the excitation energies of the states that would receive the decay intensities are comparable to the decay energies themselves, ^{242m}Am does not decay by either β^- or EC decay. Therefore, even though the change in quantum numbers is substantial in going from the $J^\pi = 5^-$ state to the $J^\pi = 1^-$ state in ^{242}Am, a highly converted IT is the dominant mode in the decay of ^{242m}Am. The half-life for this process is 141 years, long enough that α decay to odd–odd ^{238}Np (with a partial half-life of 3.07×10^4 years) proceeds with a significant intensity (0.46%). Several of the low-lying states in ^{238}Np are not shown in the figure; only the ground state and states receiving significant intensity from the α decay of ^{242m}Am are presented. Clearly, the $J^\pi = 5^-$ state in ^{238}Np at 342 keV is the level with quantum numbers analogous to those of ^{242m}Am. The decay equations are instructive: if an aged sample of ^{242m}Am is chemically purified, it will immediately be in secular equilibrium with ^{242g}Am (they cannot be chemically separated from one another). After 2 weeks, ^{238}Np will be in secular equilibrium such that Equation 2.5 holds, and $\lambda(^{238}\text{Np})N(^{238}\text{Np}) = 0.0046\ \lambda(^{242m}\text{Am})N(^{242m}\text{Am})$. After a few years, transient equilibrium is established with ^{242}Cm, as defined by Equation 2.6, so that $0.9954 \times 0.827\ \lambda(^{242m}\text{Am})N(^{242m}\text{Am}) = [\lambda(^{242}\text{Cm}) - \lambda(^{242m}\text{Am})]N(^{242}\text{Cm})$. Equilibrium is never established between ^{242m}Am and ^{242}Pu, whose half-life is the larger of the two.

Through the discussion and examples depicted in Figures 2.9 through 2.14, adequate background knowledge of nuclear structure should be established to understand its influence on the following subject areas: nuclear reactions, the fission process, and forensic chronometry.

2.6 NUCLEAR REACTIONS

A nuclear reaction entails the interaction of a nucleus with another nucleus or a fundamental particle, like a nucleon, in such a way as to induce a change in the reactants. The first nuclear reactions were conducted using α particles arising from the decay of natural sources. In 1919, Rutherford [55] observed the production of protons in the irradiation of nitrogen with α particles:

$$^{14}\text{N} + {}^{4}\text{He} \rightarrow {}^{17}\text{O} + {}^{1}\text{H}.$$

Simple binary reactions such as this can also be expressed in the shorthand notation ^{14}N(α,p)^{17}O, where the nuclei inside the parentheses are the incoming and outgoing light particles, respectively. As in decay equations, the numbers of protons and neutrons on both sides of the equation balance. Rutherford took advantage of the fact that some α emitters that occur in nature produce He nuclei of sufficient energy to overcome the Coulomb barriers (the mutual repulsion of two positively charged nuclei as they approach one another) of light elements. Thus, for instance, Chadwick discovered the neutron by irradiating a Be foil ($Z = 4$) with α particles emitted by a ^{210}Po source [9]:

$$5.304\text{-MeV}\ {}^{4}\text{He} + {}^{9}\text{Be} \rightarrow {}^{12}\text{C} + {}^{1}\text{n}$$

or

$$^{9}\text{Be}(\alpha,\text{n})^{12}\text{C}.$$

In this reaction [an (α,n) reaction], the α particle and the Be nucleus fuse, resulting in a ^{13}C nucleus with sufficient excitation energy that it emits a neutron to deexcite. In 1934, Joliot and Curie observed β^+ radioactivities in the irradiation of light elements with α particles from a ^{210}Po source [56]. This first production of artificial radioactivity was the beginning of the science of nuclear chemistry and was followed rapidly by the development of particle accelerators.

In many ways, nuclear reactions can be considered similar to chemical reactions, with some differences: first, in chemical reactions, the elements involved in the process remain unchanged, but they enter into new combinations with one another through the agency of the reaction; however, a nuclear reaction can result in the formation of new nuclides that do not necessarily constitute the same elements as the reactants. Second, the energy released in chemical reactions is usually much less than that released from nuclear reactions; for example, the complete fission of 1 lb of U liberates energy equivalent to that obtained by burning 1400 t of coal. Third, the traditional chemist thinks in terms of chemical reactions involving macroscopic quantities (e.g., moles of reactants and products); in nuclear reactions, the radiochemist considers changes involving individual atoms and ultratrace quantities.

Following the formalism of Weisskopf [57], a low-energy nuclear reaction can be considered as occurring in three successive steps: the initial, independent-particle stage; the compound-system stage, during which nuclear matter is in contact; and the final outgoing-particle stage. In the independent-particle stage, the incident projectile interacts with the nucleus (and vice versa) through their potentials. The projectile is either scattered away or absorbed, depending on the incident kinetic energy and whether the collision is central or peripheral. Absorption is any process in which the interaction between projectile and target nuclei cannot be described by a potential alone.

Figure 2.15 shows an idealized nuclear reaction (Weisskopf stage 1), introducing the concept of an impact parameter. The incoming charged particle is characterized by a nuclear radius r_1, and the target nucleus by a radius r_2. At infinite distance, the trajectory of the incoming projectile is parallel to the reaction axis, shown by the dashed

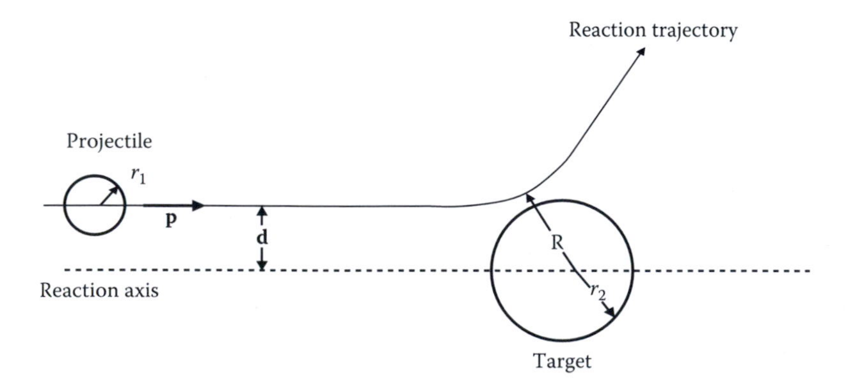

FIGURE 2.15 Schematic diagram of a charged-particle nuclear reaction. The angular momentum of the system is $\mathbf{p} \times \mathbf{d}$, the vector product of the linear momentum and the impact parameter.

line, which is offset by a distance d, the impact parameter. If d is zero and the energy of the projectile is sufficient to overcome the electrostatic repulsion between the nuclei, a central collision occurs in which there is no off-axis momentum and all motion of the reactants and the products is along the reaction coordinate. If d is nonzero, there will be an electrostatic deflection of the incoming ion. Should the distance of closest approach, R, be greater than $r_1 + r_2$, no nuclear reaction will ensue. At intermediate impact parameters, the angular momentum of the collision is given by the vector product $\mathbf{p} \times \mathbf{d}$. The total angular momentum of the colliding nuclear system will be the vector sum of the spins of the incident particles and the collisional angular momentum and will be divided among the outgoing particles. As a result of the collision, particles and γ-rays will usually be emitted from the outgoing particles, which will also carry away angular momentum as well as excitation energy.

In central and near-central collisions involving low-mass projectiles and low nuclear excitation, formation of a compound nucleus is favored. The projectile and target fuse, and nucleons from the projectile become indistinguishable from those of the target. A compound nucleus is characterized by an excitation energy, defined by the reaction Q-value and the relative kinetic energies of the incoming particles. This excitation energy is removed by the emission of particles (mostly p, n, and α) and γ-rays. Other absorptive processes, usually attributed to more peripheral collisions, are twofold: direct reactions, in which an incident, low-mass projectile can collide with a single nucleon or a small cluster of nucleons in the target, ejecting it from the nucleus; and transfer reactions, in which nucleons are exchanged between the projectile and target without equilibrium being established between them. Both of these peripheral mechanisms usually result in residual nuclei with excitation energy, but these energies are not completely defined by the kinematics of the reaction. The outgoing-particle stage of a compound nucleus reaction consists of "evaporated" particles (which have a characteristic spectrum defined by the nuclear temperature) and the residual nucleus (which can also be fission fragments for targets of heavy nuclei). For direct reactions, the outgoing particles include the "knocked-out" nucleons, and for transfer reactions, the outgoing particles include both target-like and projectile-like products from the nonequilibrium nucleon exchange.

Independent of mechanism, if nuclear species A and B react to form species C and D, the reaction can be expressed as

$$\mathrm{A} + \mathrm{B} \leftrightarrows \mathrm{C} + \mathrm{D} + Q,$$

where, as before, Q is the energy release in the reaction (and is defined as the difference between the sum of masses of the reactants and the sum of masses of the products):

$$Q = m_\mathrm{A} + m_\mathrm{B} - m_\mathrm{C} - m_\mathrm{D} \quad \text{(in energy units, 1 amu = 931.5 MeV).}$$

Also as before, mass excesses can be used in the equation. If Q is negative, then the missing energy for the reaction must be supplied via the kinetic energy of the incident particle. However, not all of the kinetic energy of the incident particle is available as reaction energy; a portion must appear as the recoil energy of the composite system, determined from the law of conservation of momentum. The energy released in nuclear reactions is manifest in the kinetic energies of the products, distributed approximately inversely proportional to the masses in a binary reaction.

In general, as the kinetic energy of A increases relative to B, the possibility of creating different reaction products increases, particularly for complex nuclear reactants. At higher energies, $A + B \leftrightarrows E + F$ may become possible, competing with the production of C and D. The probability of such a given reaction "channel" is stated as a cross-sectional area: a highly probable reaction will have a larger cross section than will a less probable reaction. The cross section represents the classical area that a target nucleus would present to an incoming projectile if the outcome of the interaction were predetermined. Cross sections are usually denoted by the symbol σ and are often expressed in units of barns, where 1 barn (b) = 1×10^{-24} cm^2 and is a unit comparable to nuclear dimensions.

The rate at which C is produced in a reaction in which a target of A is irradiated with B particles is dependent on the areal density of target atoms ($\mathcal{N}_A$), the flux of particles through the target (ϕ_B), and the cross section for the production of C (σ_C):

$$dN_C/dt = \sigma_C \phi_B \mathcal{N}_A. \tag{2.9}$$

If C is a stable species, like ^{17}O in the Rutherford reaction provided earlier, at the end of an irradiation that is Δt seconds long, the number of atoms of C produced is

$$N_C = \sigma_C \, \phi_B \mathcal{N}_A \Delta t. \tag{2.10}$$

If C is a radioactive species, Equation 2.9 must be modified to allow for the atoms of C that are produced near the beginning of the irradiation, but which decay before the end. The instantaneous rate of production of C is

$$dN_C/dt = \sigma_C \phi_B \mathcal{N}_A - \lambda_C N_C, \tag{2.11}$$

where λ_C is the decay constant of C (or ln $2/t_{1/2}$). The positive first term of the equation is the rate of production, and the negative second term is the rate of decay. Solving the differential Equation 2.11 for the number of atoms of C produced in an irradiation of duration Δt yields

$$N_C = \sigma_C \phi_B \mathcal{N}_A / \lambda_C \times [1 - \exp(-\lambda_C \Delta t)] = A_C / \lambda_C, \tag{2.12}$$

where A_C is the activity of species C. For isotope production irradiations, Equation 2.12 is referred to as the Activation Equation and is solved for A_C for the various irradiation and target parameters.

A typical application of Equation 2.12 is shown in the following example. A Mn foil is exposed to a flux of 10^{10} thermal neutrons/s in a reactor pool. The cross section for producing 2.6 h ^{56}Mn by the reaction of thermal neutrons with ^{55}Mn is 13.3 b. The areal density of the metal target is 20 mg/cm^2. At the end of a 2 h irradiation, calculation of the quantity of ^{56}Mn produced in the target is as follows. In this case, $\phi = 10^{10}$/s; N = 20×10^{-3} g/cm^2/55 g/mol $\times 6.02 \times 10^{23} = 2.23 \times 10^{20}$ atoms/cm^2;

$\sigma = 13.3 \times 10^{-24}$ cm^2; $\Delta t = 7200$ s; and $\lambda = \ln 2/2.6$ h/3600 s/h $= 7.4 \times 10^{-5}$/s. At the end of the irradiation, therefore, 1.66×10^{11} atoms of ^{56}Mn will be present in the target.

If the beam interaction results in the destruction of the product C (a common occurrence in high-flux reactor irradiations), a third term must be added to the right side of Equation 2.11: destruction $= -\sigma_d \phi_B N_C/\mathcal{A}$, where σ_d is the cross section for destruction of C and $\mathcal{A}$ is the area of the target being struck by the beam. Solving the equation gives a result very similar to that of Equation 2.12, but with $(\lambda_C + \sigma_d \phi_B/\mathcal{A})$ substituted for λ_C everywhere it occurs.

In nuclear reactors, the case often arises in which the neutron flux is so high that the number of target atoms changes significantly during irradiation, such that $\mathcal{N}_A$ can no longer be treated as constant. In this case, the number of target atoms at time t after the start of the irradiation is $\mathcal{N}_A = {}^0\mathcal{N}_A \exp(-\phi_B \sigma_X t/\mathcal{A})$, where ${}^0\mathcal{N}_A$ is the initial areal target density, $\mathcal{A}$ is the area of the target struck by the beam, and σ_X is the cross section for reactions that destroy A. Equation 2.11 becomes

$$dN_C/dt = \sigma_C \phi_B {}^0\mathcal{N}_A \exp\left(-\phi_B \sigma_X t/\mathcal{A}\right) - \lambda_C N_C \quad (2.13)$$

Solving Equation 2.13 results in

$$N_C = \sigma_C \phi_B {}^0\mathcal{N}_A \mathcal{A} / \left(\lambda_C \mathcal{A} - \phi_B \sigma_X\right) \times \left[\exp\left(-\phi_B \sigma_X \Delta t/\mathcal{A}\right) - \exp\left(-\lambda_C \Delta t\right)\right]. \quad (2.14)$$

In the case in which σ_X is very small (i.e., the target is not progressively destroyed), Equation 2.14 reduces to Equation 2.12 as expected.

The experienced nuclear chemist knows that Equation 2.12 may sometimes be of limited utility if irradiations of any significant duration are conducted. Particularly at particle accelerators, the particle flux ϕ_B is rarely constant. Assume that a long irradiation of length Δt can be divided into a series of n irradiation intervals, Δt_i, during which the flux, ϕ_{Bi}, is approximately constant. In any particular interval, ϕ_{Bi} can be zero. The total number of atoms of C present at the end of the irradiation is

$$N_C = \sum_{i=1,n} \left\{\sigma_C \phi_{Bi} \mathcal{N}_A/\lambda_C \times [1 - \exp(-\lambda_C \Delta t_i)] \exp(\lambda_C t_i)\right\}, \quad (2.15)$$

where t_i is the time between the end of the ith irradiation interval and the end of the irradiation.

Finally, consider the case of competing reactions, in which the interaction of A and B not only produces C and D, but can also produce E and F. The presence of other reaction channels does not affect the production of C, the probability of which is defined by σ_C. Production of C is independent of σ_E, or of the cross sections for any other reaction channels, unless one of the products decays to C during the irradiation. If E is also produced in the target during irradiation, with cross section σ_E, and E decays to C with decay constant λ_E, then Equation 2.11 must be modified by the addition of a (positive) term for the creation of C through decay of E (i.e., $\lambda_E N_E$). If E has no parent activities of its own, then N_E is given by Equation 2.12. Substituting and solving for N_C gives

$$N_C = \phi_B \mathcal{N}_A/\lambda_C \times \big(\sigma_C[1-\exp(-\lambda_C\Delta t) + \sigma_E/(\lambda_E - \lambda_C) \times \{\lambda_E[1-\exp(-\lambda_C\Delta t)] - \lambda_C[1-\exp(-\lambda_E\Delta t)]\}\big). \quad (2.16)$$

If σ_E is very small compared with σ_C, or if the half-life of E is very long compared to the length of the irradiation, Equation 2.16 also reduces to Equation 2.12.

From the earlier examples, the analyst can construct and solve the necessary differential equations for any radionuclide production scenario that arises in a particular forensic application.

Nuclear reactions can be divided into two distinct categories—those involving irradiations with charged particles and those involving irradiations with neutrons. There are two fundamental differences between the two: first, a Coulomb barrier to charged-particle reactions exists—an electrostatic repulsion between positively charged nuclei that must be overcome before the nuclear matter of the two reactants can come into contact. There is no electrostatic barrier to interactions between neutrons and nuclei, however, and the incoming trajectory shown in Figure 2.15 becomes a straight line. Second, at subrelativistic velocities, the range of charged particles in matter is relatively small compared to that of neutrons; this limits the effective quantity of target material that can be irradiated by charged particles in a given experiment. As a consequence, extremely large quantities of nuclear materials cannot be made by charged-particle reactions. However, some useful nuclear materials cannot be produced in low-energy neutron reactions (e.g., β^+-emitting ^{22}Na; Figure 2.10). Even though they are of less interest in nuclear forensic investigations than nuclear reactions with neutrons, some treatment will be afforded reactions induced by charged particles.

The height of the Coulomb barrier, B_C, depends on the atomic numbers of the target nucleus and the projectile nucleus, Z_T and Z_P, respectively. The electrostatic repulsion between the two charged particles at some distance r is given by

$$E_C = 1/(4\pi\varepsilon_0) \times Z_T e \times Z_P e/r,$$

where $Z_T e$ and $Z_P e$ are the charges on the two nuclei and the permittivity constant $\varepsilon_0 = 8.854 \times 10^{-12}$ coul^2/nt/m^2. When E_C has units of million electron volts (MeV) and r is in centimeters, this equation reduces to

$$E_C(\text{in MeV}) = 1.44 \times 10^{-13}\, Z_T Z_P / r\,(\text{in cm}).$$

As the two nuclei approach each other, the electrostatic repulsion between them increases as the inverse of the separation. For large relative velocities, this repulsion continues until the nuclei come within range of the attractive nuclear force, which is specified as the "touching radius." As mentioned earlier, the nuclear force is short-ranged and saturated. Therefore, the number of nucleons in a given volume of nuclear matter is reasonably constant. Treating nuclei as roughly spherical, the result is that the nuclear radius is proportional to the cube root of the atomic mass A: $r = r_0 A^{1/3}$. Experimentally, for reactions with light ions, the proportionality constant r_0 has been determined to be ~1.4×10^{-13} cm. The Coulomb barrier is equal to E_C evaluated at the touching radius. Substituting into the preceding equation,

$$B_C\left(\text{in MeV}\right) = 1.03\, Z_T Z_P \big/ \left(A_T^{1/3} + A_P^{1/3}\right). \tag{2.17}$$

It was mentioned previously that the kinetic energy of the helium nucleus emitted in α decay is less than the *Q*-value for the decay because of the recoil of the residual nucleus required by conservation of momentum. Here the problem is the same: when the target and projectile nuclei encounter each other, the composite nuclear system must move with the same momentum as the original projectile. This requires that the lab-frame energy of the projectile be higher than the center-of-mass Coulomb barrier by a factor of $(A_T + A_P)/A_T$.

For Rutherford's $^{14}N(\alpha,p)^{17}O$ reaction, the Coulomb barrier is 3.6 MeV, requiring an α-particle kinetic energy of 4.6 MeV to overcome it. The radioactive decay of several natural α emitters produce He nuclei with sufficient kinetic energy to induce the reaction. However, the Coulomb barrier to the interaction of α particles with uranium nuclei is 24 MeV, well beyond the decay energy of any known α emitter.

Charged particles can be accelerated by technical means. Thus, in Van de Graaf or Cockroft–Walton electrostatic accelerators, the terminal is charged to a high voltage, and ions that are produced in a centrally located ion source are accelerated when they leave the boundary of the terminal. Unfortunately, the use of voltages above ~10 MV is difficult because of the breakdown of insulators and the appearance of corona discharge. Nevertheless, electrostatic accelerators are quite useful for the study of low-energy nuclear physics. To produce particles with higher energies, the accelerator must make use of carefully synchronized accelerating potentials of smaller magnitude, which are accomplished in machines such as cyclotrons or linear accelerators. Some can accelerate any stable nuclide in the chart of the nuclides to nuclear reaction energies.

Nuclei created in charged-particle bombardments almost always have excitation energies in large excess of the binding energies of protons, neutrons, and α particles. The emission of γ-rays is a slow process compared to particle emission, so emission of particles is almost always preferred until the nucleus deexcites or "cools" to excitation energies comparable to particle-binding energies. Because there is no Coulomb barrier to the process, the emission of neutrons nearly always takes place in preference to the emission of charged particles in reactions with low-mass projectiles.

In a highly excited nucleus, the excitation energy is shared randomly among the nucleons, and any given nucleon has a constantly varying share of the total energy. This situation closely matches the chemical description of the kinetic motion of a system characterized by a temperature. "Evaporation" of a neutron occurs only when a quantity of energy greater than its binding energy is concentrated on it.

The difficulty of concentrating large amounts of energy in a single neutron implies that the evaporated nucleons will be emitted with low kinetic energies. The energy distribution of emitted neutrons is a function of the temperature of the excited nucleus and is of the form

$$P(E) = E\exp(-E/T), \tag{2.18}$$

where $P(E)$ is the probability of emission of a neutron of energy E, leaving the residual nucleus with a temperature T. The most probable neutron kinetic energy is $E \simeq T$. The expected energy spectrum of neutrons evaporated from nuclei with a temperature of 2 MeV is shown in Figure 2.16.

Experimentally, the most common sources of neutrons are nuclear reactions, usually taking place at an accelerator or in a nuclear fission reactor. In charged-particle irradiations involving light nuclei, mono-energetic neutrons can be produced. For instance, the $^{3}H(d,n)^{4}He$ reaction has a Q-value of 17.6 MeV; this reaction system is kinematically well defined for low irradiation energies, and conservation of momentum results in a neutron energy of ~14 MeV in the forward direction in the lab frame. This reaction, responsible for boosting the performance of the fission components of nuclear weapons, will be discussed later. Strong sources of neutrons are produced by inserting a thin Be foil into an intense beam of deuterons delivered by a particle accelerator. In this case, neutrons with a distribution of energies up to ~3.5 MeV larger than that of the incident deuteron energy are obtained, arising from a combination of the reaction kinematics and evaporation processes from the compound nucleus.

By far, the most prolific known sources of neutrons are nuclear chain reactions. Neutrons emitted during the fission of U reactor fuel have kinetic energies ranging

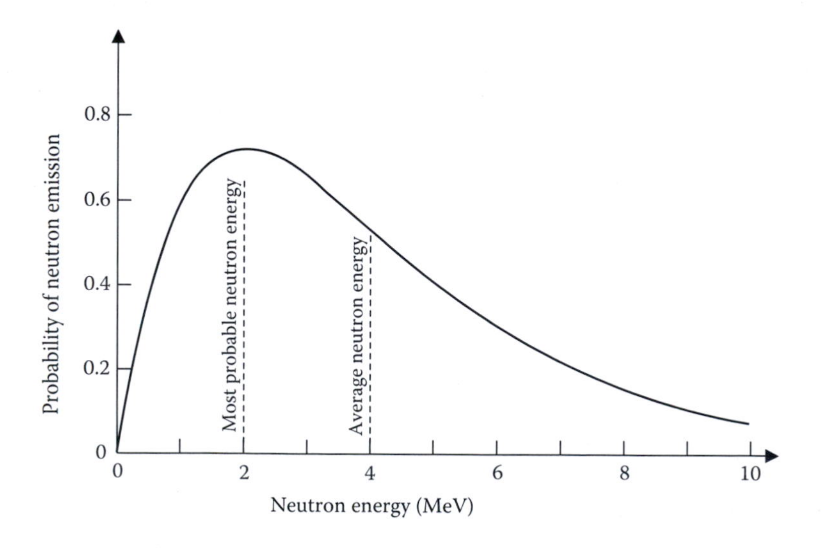

FIGURE 2.16 Evaporation spectrum of neutrons emitted by $T = 2$ MeV nuclei.

from zero to a few MeV. Neutrons escaping from the fuel assembly to irradiate an external target are moderated to lower energies through elastic scattering from low-Z materials (see the following text). It is possible to irradiate targets with very large fluxes of neutrons in a reactor environment, and nuclear fission reactors will be discussed in Chapter 3.

The first neutron irradiations were carried out with (α,n) sources. Fermi used a small glass bulb containing Be powder and a large quantity (800 mCi) of radon gas [58], a source that emitted approximately 10^7 neutrons/s. The Q-value for the ^{9}Be(α,n)^{12}C reaction is 5.6 MeV, so that neutrons with energies of up to 12.3 MeV were emitted by the source. Fermi noted that there were three types of nuclear reactions giving rise to artificial radioactivities: ^{m}Z(n,γ) $^{m+1}$Z, ^{m}Z(n,^{1}H)^{m}(Z – 1), and ^{m}Z(n,^{4}He) $^{m-3}$(Z – 2), where ^{m}Z is the symbol for a nuclide with atomic number Z and mass number m, and n (or ^{1}n) is the symbol for the neutron. Shortly afterward, Heyn observed ^{m}Z(n,2n)$^{m-1}$Z reactions induced by a similar source [59,60]. Following the observation of the elastic scattering of neutrons in a ^{14}N cloud chamber by Feather [61], Fermi discovered that surrounding both the neutron source and the object to be irradiated with large blocks of paraffin caused activations to increase by factors of as much as 100. He also observed neutron absorption maxima that occurred for "bands" of neutron energies below 100 eV.

The elastic scattering of neutrons from certain materials, particularly those containing H and C, provides the means by which the high-energy neutrons emitted in charged-particle reactions or fission can be moderated to very low velocities, with speeds comparable to motions arising from the random thermal energy of their environment. An elastic-scattering interaction is comparable to the interaction between billiard balls: no excitation energy is passed to the target nucleus, only kinetic energy. In charged-particle reactions, elastic scattering can be explained as resulting from electrostatic repulsion. It is more probable at low bombarding energies and high impact parameters. In neutron-induced reactions, neutrons must elastically scatter from the nuclear potential of the target nucleus, which is an entirely different process. Energy transfer is greatest when the masses of the collision partners are identical. The energy transferred from the bombarding particle to the target is also a function of impact parameter (see Figure 2.15) and is greatest for central collisions in which the target recedes in the original direction of the incident particle. The average decrease (per collision) in the natural log of the neutron energy is

$$\xi = \left(\ln E_{\text{initial}} - \ln E_{\text{final}}\right)_{\text{ave}} \simeq 1 + (A-1)^2/(2A) \times \ln\left[(A-1)/(A+1)\right], \quad (2.19)$$

where A is the nucleon number of the target nucleus. For A = 1 (a H nucleus), ξ is 1. It takes about 20 collisions with hydrogen to slow a 2-MeV neutron to thermal energy, E_{th} (0.025 eV, or 2200 m/s).

From Equation 2.19, hydrogen is deduced to be the most effective moderator for reducing the energy of an incident beam of neutrons. This conclusion is correct in the strictest sense if one factors in differences caused by the densities of the scattering media. In engineering applications, however, it is somewhat simplistic, as the probability of losing neutrons through inelastic processes (including capture) must be considered. Thus, it might be preferable to use a less effective scattering medium if the neutron losses were smaller. This issue will be discussed in greater detail in the section on moderation in nuclear reactors.

Because the neutron carries no charge, it can easily approach a target nucleus without being repulsed by the Coulomb field, even at low energies. This means that exothermic nuclear reactions can be initiated by neutrons moving with very low kinetic energies. The slower a neutron is moving, the more time it spends in the vicinity of a particular target nucleus. As a consequence, the reaction cross section for capture of a neutron (σ_C) is inversely proportional to the neutron's velocity, $\sigma_C \sim 1/v \sim E^{-1/2}$. At low reaction energies (neutrons with energies <1 eV), this relationship holds true for most nuclei. The most important reaction involving thermal neutrons is capture, resulting in the formation of a compound nucleus, as described earlier for charged-particle reactions. The excitation energy of the compound nucleus is determined by the reaction Q-value, and it narrowly exceeds the neutron-binding energy in reactions with slow neutrons. Usually, the evaporation of particles from these nuclei is not energetically favorable, and the compound nucleus cools by the emission of electromagnetic radiation (hence the abbreviation (n,γ) to describe these reactions).

The probability of a nucleus capturing a thermal neutron is strongly dependent on the nuclear structure of that nucleus and is determined by the nuclide that the nucleus will form. Capture of thermal neutrons by ^{135}Xe proceeds with a capture cross section of $\sigma_{C,th} = 2.6 \times 10^6$ b, by ^{157}Gd with $\sigma_{C,th} = 2.5 \times 10^5$ b; conversely, for ^{15}N, $\sigma_{C,th} = 2 \times 10^{-5}$ b, for ^{14}C, $\sigma_{C,th} < 1 \times 10^{-6}$ b, and for ^{4}He, $\sigma_{C,th} = 0$. For most nuclei, $\sigma_{C,th}$ is on the order of barns to tens of barns.

A special class of reactions with thermal neutrons involves targets with low atomic numbers, in which charged-particle emission from the excited compound nucleus competes favorably with photon emission. Such threshold reactions proceed because the height of the Coulomb barrier to the emission of the charged particle is comparable to, or smaller than, the excitation energy of the compound nucleus. Examples of these reactions of interest to the nuclear forensic analyst are ^{3}He(n,p)^{3}H, with $\sigma_{p,th} = 5300$ b; ^{14}N(n,p)^{14}C, with $\sigma_{p,th} = 1.8$ b; and ^{6}Li(n,α)^{3}H, with $\sigma_{\alpha,th} = 940$ b.

Reactions with neutrons of somewhat higher energy, from 1 to 100 eV, have cross sections that no longer follow the $1/v$ relationship. In this, the resonance region—first observed by Fermi—the capture cross section fluctuates wildly with small changes in the energy of the incident neutron. Each of the peaks in the cross section corresponds to the formation of a definite energy level in the compound nucleus, most probably only ±1/2 unit of angular momentum away from the ground state of the target. Slow-moving neutrons have low momenta, and consequently low angular momenta, even at large impact parameters (Figure 2.15). These states are at far greater excitation energies than those populated by decay processes. For heavy nuclei, the binding energy of the neutron, and consequently the excitation energy of the compound nucleus, is several MeV, allowing the number of peaks in the resonance spectrum to be quite large.

Each state has a mean lifetime to decay processes, mainly deexcitation via γ-ray emission, and therefore has a calculable energy width. The shape of each peak in the neutron cross-section spectrum is defined by its decay width through the Breit–Wigner relationship [62]. Eventually, for reactions involving neutrons of energies >100 eV, the widths of the cross-section peaks become comparable to their separation, and the resonant behavior ends. At these energies, the energy levels may be treated statistically, and the properties of the excited compound nucleus can be best described by means of a level density and the nuclear temperature. The variation of the capture-reaction cross section with neutron energy becomes more complicated. In general, the reaction cross section decreases with increasing neutron energy beyond the resonance region. The products observed following the reaction are most probably those observed by Fermi and Heyn (see the preceding text), with (n,γ) dominating at low energies and the endothermic (n,p), (n,α), and (n,2n) reactions becoming important for neutron energies beyond their respective thresholds. In the heavy elements, (n,fission) also competes.

Neutron reactions resulting in fission are very important to the nuclear forensic scientist, as they drive both reactor and weapon technologies. Neutron-induced fission will be discussed later.

2.7 NATURAL RADIOACTIVITY

Detailed examination of the chart of the nuclides shows that only 81 elements have isotopes that are considered stable against all modes of radioactive decay. However, approximately 90 elements can be considered to be naturally occurring. In fact, many radioactive species occur in nature (at last count, there were 60–80 naturally occurring radionuclides). These nuclides are generally one of three types: unstable nuclides having lifetimes comparable to the age of the earth, short-lived nuclides that are replenished through secular equilibrium with long-lived species of the first type, and nuclides that are produced in natural nuclear reactions.

Nuclides of the first type, those with half-lives long enough that some fraction of them remain from the time the elements were formed, are listed in Table 2.4. The materials most familiar are the isotopes of U and Th, which occur naturally in ores. The nuclide ^{40}K, which makes up 0.0117% of natural elemental K, is responsible for the detectable radioactivity in K-rich foods, such as bananas. With the exception of ^{40}K and the actinides, the nuclides listed in Table 2.4 have such long half-lives that they are difficult to detect without special preparation, and they do not significantly add to the natural radiation environment. The inclusion of nuclides in Table 2.4 is somewhat arbitrary. For instance, ^{82}Se, ^{116}Cd, ^{130}Te, and ^{132}Te have been observed to decay by double-β emission, with half-lives of $>10^{19}$ years [38]. Although ^{48}Ca and ^{180}Ta have not been observed to decay, they clearly must, as they are unstable to β-decay processes. In fact, naturally occurring ^{180}Ta is actually an isomeric state [26] and is therefore included in Table 2.4.

The limit on the half-life of ^{48}Ca is quite high ($>10^{20}$ years) because of unfavorable selection rules, so it is excluded from the table. In fact, the double-β decay half-life of ^{48}Ca may actually be as short as the single-β process [63]. Most otherwise-stable nuclides with $A > 160$ are unstable to α decay, but few of those with $A < 209$ have half-lives short enough to measure. For instance, the limit on the α-decay half-life of ^{204}Pb is greater than 1.4×10^{17} years [26].

Nuclides of the second type are formed in the decays of ^{232}Th, ^{235}U, and ^{238}U. They occur in long decay chains—the radioactive series. Because the progenitors of the series were created with the earth, the members of the decay chains exist in

TABLE 2.4
Long-Lived, Naturally Occurring Radionuclides with Half-Lives Less Than 10^{18} Years

Nuclide	Decay Mode	Half-Life (Years)
^{40}K	β^-, β^+	1.27×10^9
^{50}V	β^-, EC	1.4×10^{17}
^{87}Rb	β^-	4.88×10^{10}
^{113}Cd	β^-	9×10^{15}
^{115}In	β^-	4.4×10^{14}
^{123}Te	EC	1.3×10^{13}
^{138}La	β^-, EC	1.05×10^{11}
^{144}Nd	α	2.1×10^{15}
^{147}Sm	α	1.06×10^{11}
^{148}Sm	α	7×10^{15}
^{152}Gd	α	1.1×10^{14}
^{176}Lu	β^-	3.78×10^{10}
^{174}Hf	α	2.0×10^{15}
^{180}Ta	β^-, EC	$>1.2 \times 10^{15}$
^{187}Re	β^-	4.3×10^{10}
^{186}Os	α	2×10^{15}
^{190}Pt	α	6.5×10^{11}
^{232}Th	α	1.405×10^{10}
^{235}U	α	7.038×10^8
^{238}U	α	4.468×10^9

concentrations defined by their decay properties and the law of secular equilibrium (Equation 2.5). The three series are named the uranium, actinium, and thorium series, and they are shown in Figures 2.17 through 2.19, respectively. The decay modes of the individual nuclides making up the decay series are given, and their detailed decay properties are provided, in Table 2.3. The uranium series (Figure 2.17), which begins at ^{238}U and ends at stable ^{206}Pb, is also known as the (4n + 2) series because the mass numbers of each member of the decay chain can be described by this formula, with n = 51–59. It is noted that ^{234}U ($t_{1/2} = 2.46 \times 10^5$ years) and ^{214}Po ($t_{1/2} = 164$ μs) are both members of the U series. Despite the disparity in half-life, they share the property that neither would exist in nature without replenishment through radioactive decay, as both half-lives are very short compared to the age of the earth. Both natural and anthropogenic processes can cause disequilibrium between ^{234}U and ^{238}U that can be measured and interpreted by the nuclear forensic analyst, as the ^{234}U half-life is long compared with the span of history of the chemical sciences. Conversely, for all practical purposes, ^{214}Po will always be in equilibrium with ^{214}Bi.

The actinium series (Figure 2.18), starting at ^{235}U and ending at stable ^{207}Pb, is also known as the (4n + 3) series. The thorium series (Figure 2.19), beginning at ^{232}Th and ending at stable ^{208}Pb, is also known as the (4n) series. Both of these series also contain long-lived radionuclides whose disequlibria can be interpreted by the analyst. These

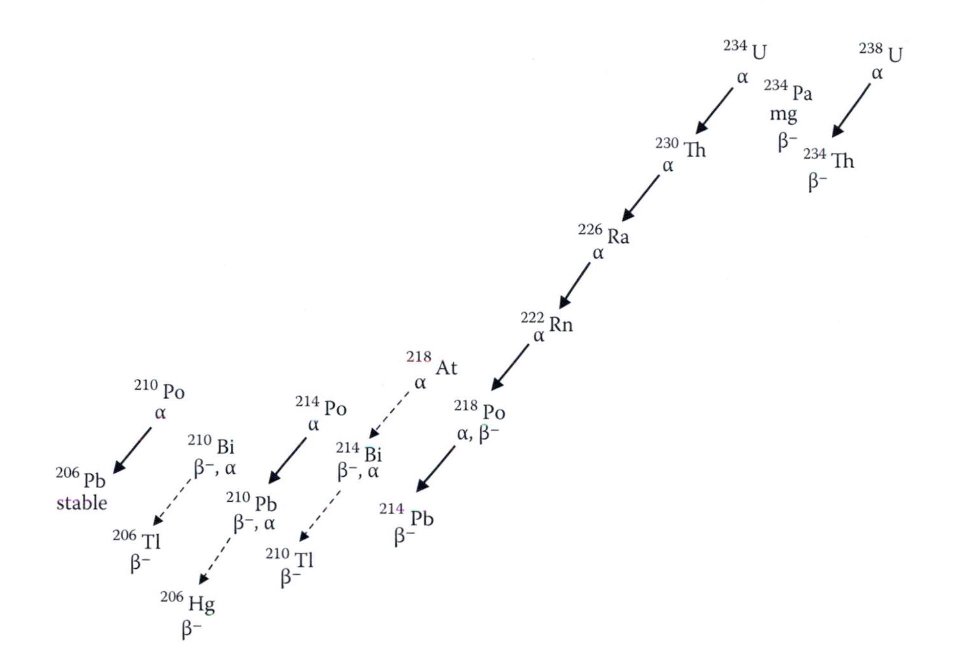

FIGURE 2.17 The uranium (4n + 2) series of naturally occurring radionuclides. Descendants of primordial ^{238}U, these activities are in secular equilibrium in nature. Arrows denote α decay, with a dashed arrow designating a rare process; β^- decay results in a transition upward and to the left.

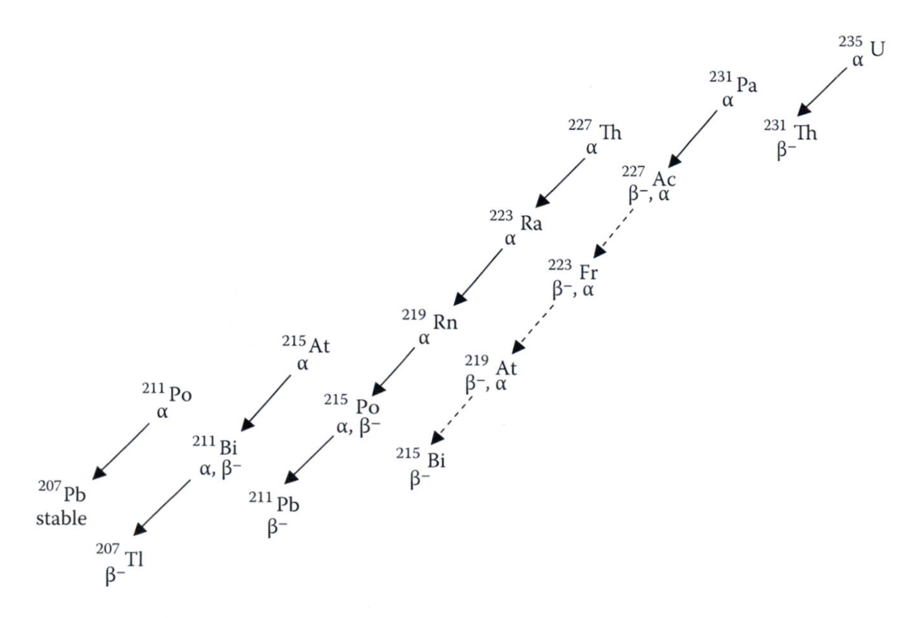

FIGURE 2.18 The actinium (4n + 3) series. Descendants of primordial ^{235}U, these activities are in secular equilibrium in nature. Arrows denote α decay, with a dashed arrow designating a rare process; β^- decay results in a transition upward and to the left.

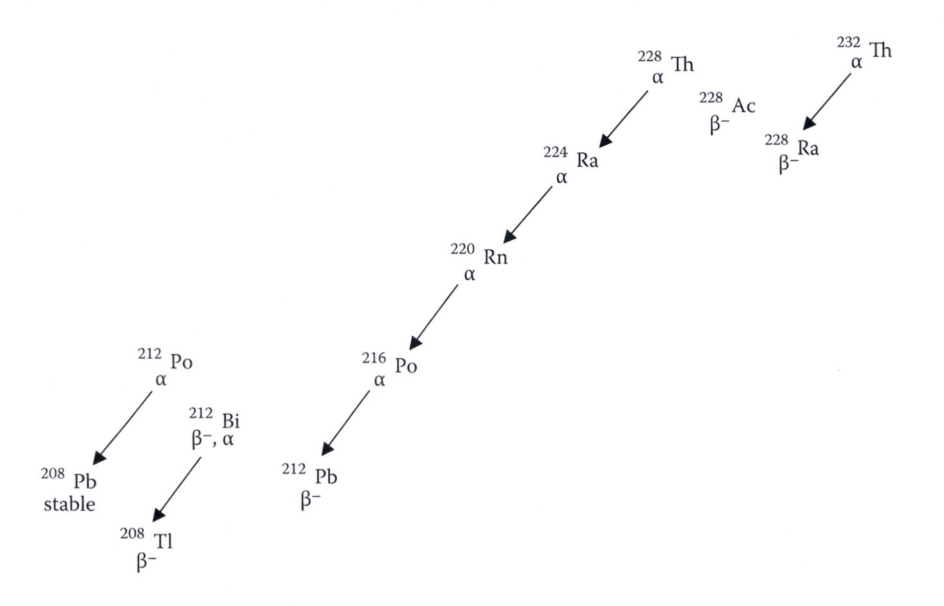

FIGURE 2.19 The thorium (4n) series. Descendants of primordial ^{232}Th, these activities are in secular equilibrium in nature. Arrows denote α decay, with a dashed arrow designating a rare process; β^- decay results in a transition upward and to the left.

three natural radioactive series provide the means by which at least one nuclide of each element between Bi ($Z = 83$) and Th ($Z = 90$) exists in nature, and they supplied the clues that led early radiochemists to deduce the concept of isotopes and the radioactive decay laws before the atomic nucleus was even postulated. The absence of a (4n + 1) series in nature is caused strictly because the half-life of the progenitor, ^{237}Np, is short compared to the age of the earth. The neptunium series, even though it does not exist in nature, is relevant to synthetic materials, particularly those containing the isotopes of Pu. The neptunium series, starting with ^{237}Np and ending at stable ^{209}Bi, is shown in Figure 2.20. As before, disequilibrium between its members often provides the basis for a forensic interpretation. As for the three natural series, the decay properties of the individual members of the (4n + 1) series can be found in Table 2.3.

Nuclides of the third type are those formed in natural nuclear reactions. Low-mass radionuclides such as ^{3}H, ^{7}Be, ^{10}Be, ^{14}C, and ^{36}Cl are produced in the upper atmosphere by the interaction of cosmic rays with the components of air. Air samples obtained for the purposes of environmental monitoring often contain measurable amounts of 53-day ^{7}Be (Figure 2.9), observable by its 478-keV γ-ray emission. The 5700-year radionuclide ^{14}C has been used to determine the ages of objects that contain it. However, this technique, which depends on a constant rate of production in the atmosphere, is more complicated when applied to natural materials created since the dawn of the atomic age. Neutrons generated in atmospheric nuclear explosions reacted with ^{14}N in the air to produce ^{14}C via the (n,p) reaction, far outweighing the amount produced naturally in the upper atmosphere during the 1945–1962 time frame.

Other naturally occurring radionuclides of the third type are found in the vicinity of deposits of U and Th. Alpha particles emitted in the decay of the short-lived members of

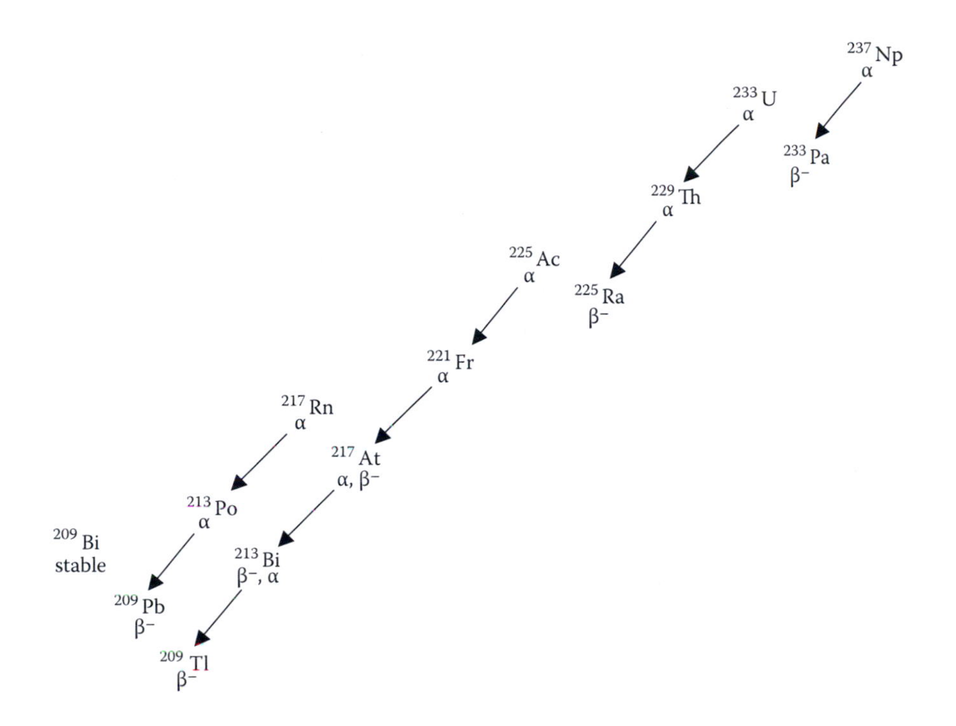

FIGURE 2.20 The neptunium (4n + 1) series. The longest-lived member of this series is ^{237}Np, with $t_{1/2} = 2.14 \times 10^6$ years; consequently, this series does not exist in nature.

the radioactive series can react with light elements present in the ore. SF of ^{238}U releases neutrons that are captured by the other components of the deposit. Neutron capture by ^{235}U results in the formation of fission products, and neutron capture by ^{238}U, followed by two β decays, results in the production of ^{239}Pu. All of these radionuclides are present at very low levels in ore bodies, well below the levels introduced into the environment by the aboveground nuclear explosions. However, this was not always the case. Kuroda proposed that, during the early history of the earth, the ratio of ^{235}U to ^{238}U in natural U would have been high enough that an assembly of U and water could have supported a chain reaction and formed a natural nuclear reactor [64]. Subsequently, the discovery of isotopic anomalies in the 1.7-billion-year-old U deposits at Oklo [65] proved that, at one point in the earth's past, radionuclides produced by neutron-induced reactions were considerably more prevalent. Reference [65] is highly recommended as one of the first nuclear forensic investigations reported in the open literature.

2.8 FISSION, BARRIER PENETRATION, AND ENERGY PRODUCTION

Fission was discovered by radiochemists in 1938. Six years later, large-scale reactors based on the fission process were running to produce a nuclear fuel for use in a new class of explosive ordinance—the atomic bomb. The speed with which the laboratory experiment was explained by theory and subsequently scaled up to industrial-size processes is one of the more amazing technical feats in history. However, though more than 60 years have passed since the original observation of the phenomenon, nuclear fission remains a challenging field of research. In fact, it is such a complicated process that it is still poorly understood.

During the investigation of neutron-induced reactions in various elements, Fermi discovered several different β^- radioactivities when U was used as a target [58]. Because the β^- decay process results in the conversion of the capture-product nuclide into an isotope of the next higher chemical element, Fermi assumed that a transuranium element had been produced. The wide variety of β^- half-lives resulting from these experiments led him to believe that he had produced several of these elements by successive neutron captures (with enormous cross sections) following β^- decay during irradiation. The experiments were repeated by Hahn and Strassmann in 1938 [66,67]: they discovered that at least part of the radioactivity could not be chemically isolated from Ba (atomic number $Z = 56$) and must therefore be caused by decays of an isotope of that element. Meitner and Frisch [68] suggested that, following capture of a neutron, the U nucleus had split into two roughly equal parts, with the complementary nucleus being an isotope of krypton ($Z = 36$), so that the number of protons in the product nuclei summed to the number in the original U nucleus ($Z = 92$). These researchers were the first to call the reaction "fission"—a term previously reserved by the biologists to describe cell division.

Meitner and Frisch recognized that the fission process must be accompanied by the release of considerable energy, and Frisch [69] demonstrated that most of the energy was manifest in the form of the kinetic energies of the fission fragments receding from one another through the mutual Coulomb repulsion of their nuclei. Jentschke and Prankl [70] demonstrated the presence of a low-energy group and a high-energy group in the kinetic-energy spectrum, indicating that the favored mass division in the neutron-induced fission of the actinides is asymmetric.

Nuclear fission is of great interest to the forensic analyst, not only because it is the underlying process in nuclear reactors that gives rise to some of the nuclear fuels and changes the isotopic content of the starting materials, but also because of its use in radionuclide chronometry and in determining reprocessing technologies. Isotopes of approximately 40 elements are formed in the fission of actinides. Fission is not only a physical change, but also a profound chemical change. This is the reason that radiochemists discovered fission and made the majority of the important contributions to its experimental study, dominating the field until the 1960s [71,72]. Even with the advent of physical techniques that could determine fission properties over a wide range of possible fission-product yields, radiochemistry is still used to great advantage in situations where rare processes must be studied in the presence of overwhelming backgrounds (e.g., those situations encountered in nuclear forensic analysis).

Shortly after the discovery of fission, Bohr and Wheeler [73] proposed a theoretical explanation, based on a charged liquid drop, with great success. In this nuclear model, the stability of the nucleus against fission is seen as resulting from a competition between the short-range forces that hold the nucleus together and the repulsive electrostatic forces between the protons, which try to disrupt the nucleus. Electrostatically, the worst-case nuclear configuration is the sphere: a deformed nucleus is subjected to less Coulomb repulsion, whereas the saturated strong force is little changed at small deformations. In the heavy elements, this phenomenon results in the most stable nuclear configuration being nonspherical, with low-energy rotational structures in their level schemes.

As a charged liquid drop deforms, it performs work against its surface tension, which results in an initial increase in nuclear potential energy that is more rapid than the corresponding decrease in Coulomb energy. If the nucleus continues to deform along an axis, it will reach a critical deformation at which the rate of change of the liquid-drop surface energy becomes equal to the rate of change of the electrostatic energy. Beyond this deformation, fission is unavoidable. Figure 2.21 shows a schematic representation of the potential energy of a charged liquid drop as a function of its deformation. The hump in the potential energy along the deformation axis is the fission barrier (similar to the activation energy for reaction in chemistry), and it arises in both decay studies and reaction studies.

SF, a rare decay mode occurring only in the heavy elements, can be treated as a barrier-penetration problem. The nucleus in its ground-state configuration must tunnel through the fission barrier for the process to occur. In nuclear reactions, sufficient energy can be supplied by the interaction that a heavy product nucleus can possess excitation energy comparable to, or more than, the height of the potential barrier, making fission a much more probable process. The liquid-drop model predicts that the height of the fission barrier is proportional to Z^2/A for a given nucleus. Thus, fission processes are not important in the low-energy nuclear physics of light elements. The height of the liquid-drop fission barrier decreases rapidly with increasing nuclear

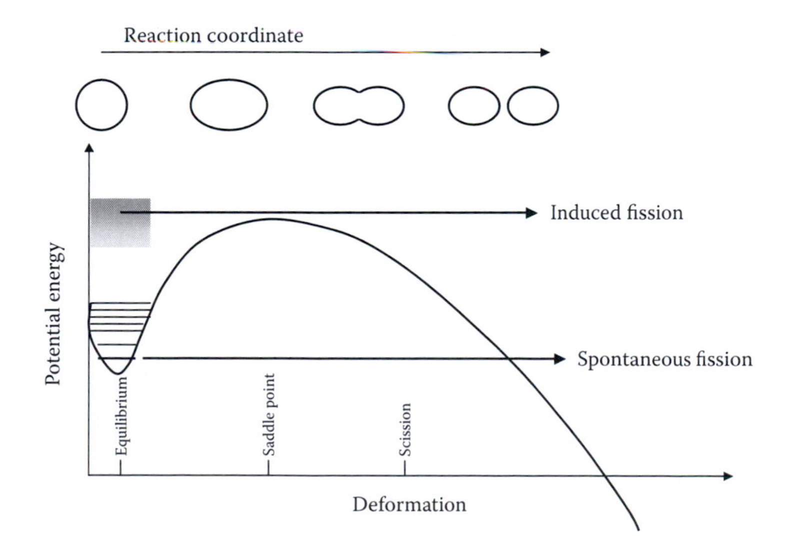

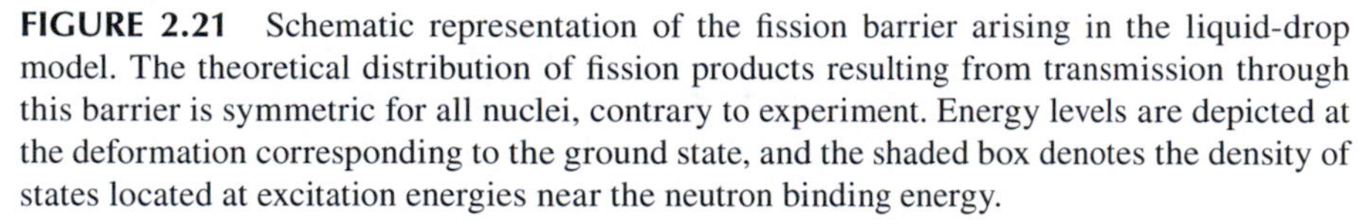
FIGURE 2.21 Schematic representation of the fission barrier arising in the liquid-drop model. The theoretical distribution of fission products resulting from transmission through this barrier is symmetric for all nuclei, contrary to experiment. Energy levels are depicted at the deformation corresponding to the ground state, and the shaded box denotes the density of states located at excitation energies near the neutron binding energy.

charge, initially decreasing below 5 MeV at the U isotopes. Although the division of ^{208}Pb into two nuclei of ^{104}Nb is energetically advantageous (it would release 123 MeV; see Figure 2.4), this decay process does not occur spontaneously and is observed only in nuclear reactions that supply enough energy to overcome the barrier. The partial half-lives for the SF of ^{238}U, ^{242}Pu, ^{246}Cm, and ^{250}Cf (nuclides related to one another through a series of α decays) are 8.2×10^{15}, 6.7×10^{10}, 1.8×10^{7}, and 1.7×10^{4} years, respectively [26], demonstrating the rapid increase in the penetrability of the fission barrier and the corresponding increase in the SF rate with increasing Z of the fissioning nucleus.

Beyond the critical deformation (a saddle point at which Coulomb forces and nuclear forces are in a precarious balance), the deformation of the nucleus increases rapidly, and a neck forms and then ruptures, leaving "prompt" fission fragments. On occasion, small particles can be emitted in the splitting process. The nuclear geometry just before neck rupture is called the scission configuration, and the two nascent fission fragments are highly deformed because of their mutual Coulomb repulsion. The resulting fission fragments recede with less kinetic energy than if the postscission configuration were two touching spheres (about 175 MeV, as opposed to 250 MeV, divided between the fragments according to the law of conservation of momentum). The primary fragments reach 90% of their final kinetic energies within 10^{-20} s of scission. The deformation in each fragment is transformed into the internal excitation energy of the nuclei as they relax into more spherical configurations. This excitation energy results in the prompt evaporation of neutrons and the subsequent emission of γ-rays. Following these processes, the residual fission fragments are referred to as primary fission products. These products are formed in either their ground states or in isomeric states.

However, the liquid-drop model of the nucleus ignores nuclear structure effects and results in a simplified fission barrier. From the previous discussion of nuclear levels, it is apparent that nuclei are much more than homogeneous balls of charged, incompressible fluid. For example, in SF, the quantum numbers of the nucleus just before scission must be traceable to the initial ground-state configuration. For even–even nuclei, the lowest potential-energy state at any deformation is expected to be $J^{\pi} = 0^{+}$. For nuclei with odd or unpaired nucleons in the ground state, preservation of quantum numbers results in the nucleus passing through the saddle-point configuration in states that may not be the lowest possible potential energy, effectively increasing the height of the fission barrier. This hinders the decay process such that SF decay of an odd-nucleon actinide proceeds much more slowly than does that of its even–even neighbors. Another limitation of the liquid-drop model is the simple Z^2/A dependence of the fission barrier. Were it the situation, this property would result in the theoretical end of the periodic table near the end of the actinide series, at $Z \simeq 104$, where the liquid-drop fission barrier vanishes. This is assuredly not the case [26].

The shell model of the nucleus, in which each nucleon is treated as moving independent of the other nucleons in an average potential, was first applied to deformed nuclei by Nilsson [74]. However, theorists were unsuccessful in extending Nilsson's model to strongly deformed nuclei, such as those occurring intermediate in the fission process. The classical shell model alone does not predict a fission barrier as a function of deformation, and it is also unsuccessful in explaining the fission isomers. The first fission isomer, discovered by Polikanov [75], was eventually attributed to the SF of an excited state in ^{242}Am with a half-life of 14 ms. This half-life is 20 orders of magnitude smaller than that expected for the ground state on the basis of systematics. Fission theory would require that this state have an excitation energy of more than 2.5 MeV for tunneling through the ^{242}Am fission barrier to proceed at this rate. However, such an excited state should decay back to the ground state by γ-ray emission with a much shorter half-life than that observed. At present, more than two dozen fission isomers are known [76], all with properties that seem inconsistent with their assignments of A and Z.

The existence of fission isomers was explained by Strutinsky [77,78], who created a formalism by which the shell model could be incorporated into the liquid-drop model to elucidate the properties of strongly deformed nuclei. This "macroscopic–microscopic" model results in a double-humped fission barrier when applied to the actinide nuclei (see Figure 2.22). The potential barrier to fission has two maxima, separated by a minimum with an excitation energy of only 2–3 MeV, relative to the ground state. The fission isomers are low-lying states in the second well of the fission barrier. Nuclei in this well are more strongly deformed than are those in the first well and are sometimes referred to as shape isomers. The enhanced SF rate for states in the second potential well is explained by the much smaller barrier through which the nucleus must tunnel to reach the scission point. The fission isomers' reduced rate of decay to states in the first potential well arises because of the inner potential barrier: decay to states in the first well would require a change in shape of the nucleus as a whole, which is a relatively slow process. To spontaneously fission, a nucleus occupying a state in the first potential well must tunnel through the entire barrier. While occupying a state in the second well, however, it must only tunnel through the second hump of the barrier.

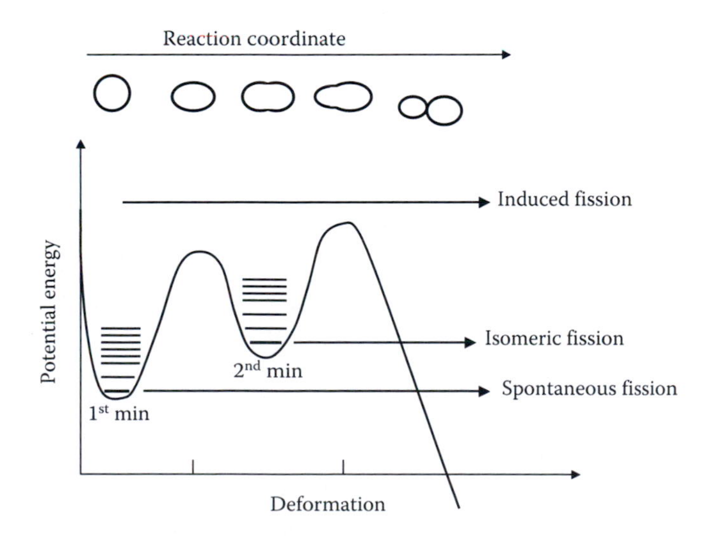

FIGURE 2.22 Schematic diagram of the double-humped fission barrier arising in the macroscopic–microscopic model. States in the first and second wells are shown, in addition to the double barrier. SF of nuclei in their ground states requires tunneling through the entire barrier, while fission isomers must only pass through the outer barrier to decay. The outer barrier results in the asymmetric mass division in the fission of the actinides.

For excited states in the first potential well, the fission process competes successfully with γ emission when the excitation energy is sufficiently high that the barrier penetration can occur in 10^{-15} s (the timescale on which γ emission takes place). Gamma deexcitation occurs when the states are just below the top of the barrier. For larger excitation energies, the fission cross section quickly rises to $\gg$1 b. This is the reason that fission of odd-mass actinide nuclei can be induced by the capture of slow neutrons, while even–even actinides require higher-energy neutrons to fission with significant probability. An extra stability is associated with the ground states of nuclei in which all of the nucleons are paired. For example, capture of a slow neutron by ^{235}U results in an excitation energy in the even–even ^{236}U compound nucleus of 6.54 MeV, whereas a similar capture by ^{238}U produces odd-mass ^{239}U with an excitation energy of just 4.80 MeV (Table 2.2). Because the fission barrier in the U isotopes is ~5.5 MeV high, ^{235}U is induced to fission by irradiation with slow neutrons, but ^{238}U is not. The cross section for neutron-induced fission of ^{238}U increases by nearly two orders of magnitude if the incident neutron energy is increased from 0.5 to 1.0 MeV [79].

The three common nuclides that are induced to fission with high probability by thermal neutrons are ^{233}U, ^{235}U, and ^{239}Pu. Their importance in nuclear weapons applications and the nuclear fuel cycle stems not only from the fact that they undergo fission in reactions with neutrons of all energies (including thermal), but also from the fact that they can be produced and isolated in large quantities. The cross sections for the thermal-neutron-induced fission of ^{233}U, ^{235}U, and ^{239}Pu (i.e., the (n_{th},f) reaction) are 520, 580, and 750 b, respectively, whereas the corresponding (n_{th},γ) cross sections are 50, 100, and 250 b. For comparison, the (n_{th},f) and (n_{th},γ) reactions on ^{238}U proceed with cross sections of $<5 \times 10^{-4}$ and 2.8 b, respectively. The capture cross section is reduced relative to those of the thermal-neutron-induced fissioning nuclides because of the lower-level density arising from the pairing gap in even–even ^{238}U.

Another shortcoming of the liquid-drop model is its prediction of a symmetric mass division in the scission process of all fissioning nuclei (see Figure 2.21). The observed asymmetric division of the actinides can only be explained within the context of the macroscopic–microscopic model of fission. It can be shown [80,81] that the potential energy in the vicinities of the first maximum, and the first and second minima, of the fission barrier is lowest for deformations that are reflection-symmetric along the elongation axis. However, the outer barrier is lowered if the nucleus deforms asymmetrically, leading to pear-shaped nuclei as the scission point is approached (see Figure 2.22). This is at least partially driven by the presence of closed nuclear shells in the nascent fission fragments.

In reality, the potential-energy surface that defines the fission process is considerably more complicated than that slice along the reaction coordinate shown in Figure 2.22. There is no one, single outcome of the mass division in the scission of a particular nucleus. Several mass divisions are allowed and are best given as a probability distribution. We show this probability in Figure 2.23 for the fission of ^{235}U by thermal neutrons (solid curve) and by 14-MeV neutrons (dashed curve). Maximum fission-product yields occur for divisions resulting in fragment masses near A = 95 and A = 140. Although symmetric fission of the ^{236}U compound nucleus into two

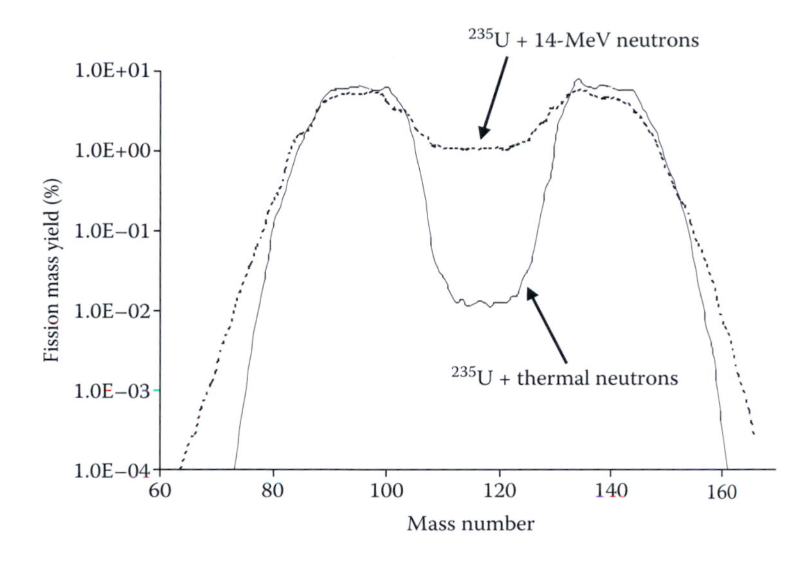

FIGURE 2.23 Yields of fission-product mass chains as a function of mass number for thermal-neutron fission of ^{235}U (solid curve) and 14-MeV neutron fission of ^{235}U (dashed curve). Nuclear structure in the scission configuration becomes less important with increasing excitation energy of the compound nucleus.

equal fragments, induced by the capture of a thermal neutron, is unfavorable from the standpoint of the fission barrier, it is not absolutely forbidden, just less probable by more than two orders of magnitude. In the interaction of ^{235}U with 14-MeV neutrons, however, the excess energy of the compound nucleus gives it more freedom in selecting trajectories through the barrier other than that of minimum potential energy. This results in greater probabilities for both symmetric mass division and more extreme asymmetric divisions: the valley between the two yield maxima becomes more shallow, and the yields of "wing" products at the extreme ends of the distribution increase at the expense of some of the intensity at the yield maxima. In the fission of ^{239}Pu, the structure of the mass distribution washes out and becomes a single, broad peak for excitation energies >40 MeV. Conversely, the narrowest distributions with the highest peak-to-valley ratios of fission probability are obtained in SF. Barring exotic decay modes, two fragments are generated by fission. Therefore, the sum over all masses in a given mass–yield curve (e.g., Figure 2.23) results in approximately 200% of total fission yield.

Fission almost always results in two primary fragments. However, occasionally fission into three fragments is observed, with the third fragment being a light particle such as ^{6}He, ^{8}He, or ^{6}Li, with ^{3}H and ^{4}He being the most common. The origin of these light particles is thought to be the neck rupture at scission. For the actinide nuclei of interest in nuclear forensic science, these "ternary-fission" phenomena constitute about 0.2% of the total fission intensity.

Frisch noted that, by far, the largest component of the energy released in fission appears as kinetic energy of the fragments. The primary fragments separate from one another with equal but opposite momenta. Evaporation of neutrons and γ-rays occurs long before the fragments have receded a sufficient distance from one another to be observed, so the law of conservation of momentum does not strictly apply to the motions of the primary fission products. However, the energies and momenta associated with the neutrons and photons are relatively small, so momentum conservation can be assumed to first order. The kinetic energies of the light and heavy fission products (E_L and E_H, respectively) relate to their masses (M_L and M_H, respectively) according to the relationship

$$E_{\mathrm{L}}/E_{\mathrm{H}} = M_{\mathrm{H}}/M_{\mathrm{L}}\,.$$

Thus, the kinetic energy and velocity of the light fragment in an asymmetric fission are greater than those of the heavy fragment.

Figure 2.23 is not reflection-symmetric around A = 236/2 = 118 because of the neutrons that are emitted from the prompt fission fragments as they deexcite to form the primary fission products. The internal excitation energies of the fission fragments arise as they become more spherical immediately following scission. Just as for excited nuclei formed in nuclear reactions, neutrons are evaporated from the fragments until they reach excitation energies comparable to the neutron-binding energy, after which γ-rays are emitted ($\sim 10^{-12}$ s). The total kinetic energy of the neutrons is approximately 5 MeV. Although this nuclear system is more complicated than can be described by a simple temperature (Equation 2.18), in the lab frame, the neutron spectrum generated by fission is very similar to that shown in Figure 2.16, but with an effective temperature of 1.3 MeV. The total energy removed by the emission of prompt γ-rays is approximately 7 MeV.

A very striking feature of the mass-yield distribution for ^{235}U(n$_{th}$,f), shown in Figure 2.23, is the two spikes that occur at complementary masses close to the valley side of the yield maxima. These sharp peaks are thought to be caused by the primary fission process favoring the formation of products near the doubly closed-shell nucleus, ^{132}Sn (N = 82 and Z = 50). Preferential formation of these products forces an increase in the probability of formation of their complementary products on the other side of the fission valley. The effect is washed out with increasing excitation energy in the compound nucleus and is absent in the resultant mass-yield distribution from incident 14-MeV neutrons.

For the thermal-neutron-induced fission of ^{235}U, the average number of neutrons emitted per fission, the neutron multiplicity (ν) = 2.4, results in reflection symmetry of the mass distribution around A = (236 − 2.4)/2 = 116.8. For the reaction of 14-MeV neutrons with ^{235}U, ν = 4.1. There is a general upward trend in multiplicity with incident neutron energy, but the resonance region often exhibits abrupt changes in multiplicity, fission probability, and product mass asymmetry from one resonance to the next. Similarly, there is a trend toward greater ν with increasing Z of the fissioning system (e.g., ^{233}U(n$_{th}$,f) yields ν = 2.5, ^{239}Pu(n$_{th}$,f) proceeds with ν = 2.9, and the SF of ^{252}Cf gives ν = 3.8). Of course, these ν values are averaged data, as a probability distribution of the number of prompt neutrons exists for a given fissioning system. For example, when ^{235}U is irradiated with 1.25-MeV neutrons, ν = 2.64; the distribution is such that 2% of the fissions emit no neutrons, 11% emit one neutron, 30% emit two neutrons, 41% emit three neutrons, 10% emit four neutrons, and 6% emit five or more neutrons.

The multiplicity of neutrons in the fission process provides the means by which a chain reaction can be sustained—a necessary condition for the operation of both nuclear reactors and explosives. For a chain reaction to proceed, on average, at least one neutron created in the fission process must be absorbed by a nearby fissile nucleus and result in a fission. Some neutrons escape from the fuel region, making a fraction of ν unavailable for subsequent reaction. In nuclear reactors, there are reflection techniques for returning neutrons to the fuel. Other neutrons are lost to (n,γ) reactions on the components of the fuel and is the reason that quantities of nonfissile species in fission fuels are usually reduced to some extent (including the even–even isotopes of U and Pu).

Just as mass is distributed among the fission products, nuclear charge (in the form of protons) must also be distributed. As before, the potential-energy surface does not dictate a unique number of protons in a given fission fragment with a mass number A. Rather, there is a distribution of charge probabilities centered around a most probable proton number, $Z_p(A)$. Protons are rarely evaporated from the fission products (the neutron-binding energies of the neutron-rich primary fission fragments are much lower than the binding energies of p and α particles), so the sum of the number of protons in the primary fission products is equal to the original Z of the fissioning system. Fission products of mass = A, with Z greater or less than $Z_p(A)$, are formed with lower yields. The value of Z_p for any given A can be calculated to good accuracy from the postulate of equal charge displacement: the most probable charges (Z_p and Z'_p) for a given mass number A and its complementary product A′ are displaced equally from the most β-stable charges for those masses. The distribution of protons among a group of primary fission products of given mass is roughly Gaussian about Z_p, although significant deviations from this behavior are caused by shell and nucleon pairing effects. As observed in the distribution of mass, the charge distribution is broadened by increasing the excitation energy of the fissioning system [82].

From Figure 2.23, the probability that a given mass number A arises in the thermal-neutron fission of ^{235}U can be obtained. This probability is divided among all the possible fission products at that mass, with the most going to the product whose proton number is closest to $Z_p(A)$. The probability for the formation of a given primary fission product is known as its independent yield.

As noted earlier, the valley of β stability bends toward neutron excess in the heavy elements. The neutron/proton ratio of ^{236}U is 1.565. Assuming that 2.4 neutrons are evaporated from the prompt fission fragments, the average neutron/proton ratio of the primary fission products must therefore be $(144 - 2.4)/92 = 1.539$. The stable isotopes of the fission-product elements have neutron/proton ratios that vary from 1.25 to 1.45. Thus, the most probable fission products are neutron-rich and fall on the β^--unstable side of the valley of stability. The widths of the charge distributions are narrow enough that very few nuclides that are neutron-poor relative to the stable nuclides have significant independent yields.

Most data on the distribution of charge at a given mass were obtained through radiochemical measurements of independent yields. Each such measurement requires careful and, frequently, rapid separation of the relevant fission product so that the impact of β decays on the independent yields is negligible. Only a few "shielded" nuclides (e.g., ^{126}Sb and ^{136}Cs), which are blocked from receiving decay intensity by a stable nuclide or a very long-lived precursor, can have their independent yields measured directly. It is often not possible to measure the independent yield of a given nuclide before essentially all of its short-lived β^- decay precursors have decayed. This results in the concentration of that nuclide in a mixture of fission products being cumulative, rather than independent. The cumulative yield is the sum of the independent yields of the nuclide itself and all of its precursors. The extraction of independent yields from cumulative-yield measurements is greatly facilitated by the calculation of $Z_p(A)$ and the assumption of a Gaussian charge distribution that changes slowly with fission-product mass number.

Figure 2.24 shows the independent yields (or charge dispersion) for products with A = 141 arising from the thermal-neutron fission of ^{235}U [83]. In this case, $Z_p(141) = 54.97$, and 100% of the fission yield at A = 141 is on the neutron-rich side of the beta-stable nuclide, 59-proton ^{141}Pr. If a mixture of fission products is allowed to decay for a day before processing, all of the mass-141 nuclides will transmute to the product with 58 protons, 32.5-day ^{141}Ce. Even though the independent yield of ^{141}Ce is immeasurably small, its cumulative yield is approximately equal to the sum of the independent yields of all other members of the mass chain and is equal to the mass yield at A = 141 (plotted in Figure 2.23).

To complete the discussion of the fission process, a complication that arises in folding the independent fission yields into a mass-yield curve (such as that shown in Figure 2.23) is the presence of β^- delayed neutron emitters. These nuclides β-decay

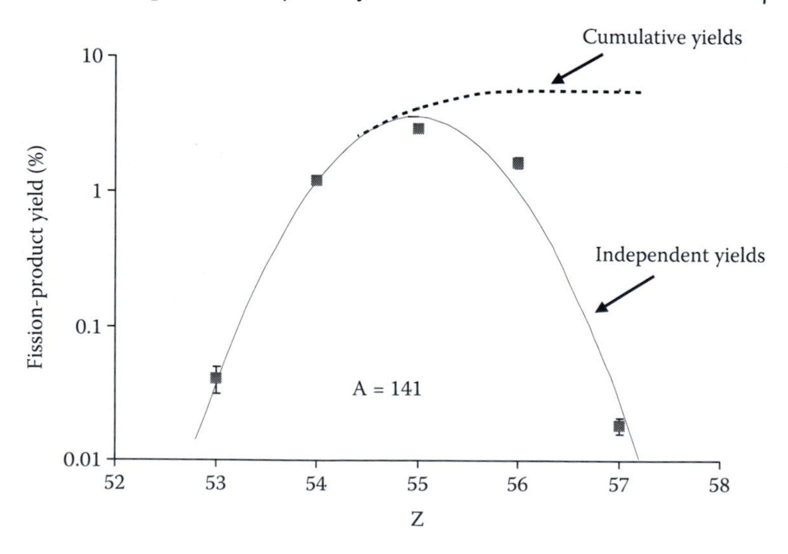

FIGURE 2.24 The charge dispersion for fission products with A = 141 from the thermal neutron fission of ^{235}U. Experimentally determined independent yields are shown as data points with a fitted Gaussian curve. These are the yields of the primary fission products as formed, before β^--decay processes have time to proceed. The dashed curve illustrates the cumulative yields, which are measured after all of the precursors of a particular nuclide have β^--decayed.

with some probability to states that have excitation energies greater than the neutron-binding energy of the decay daughter. In these decays, neutron emission successfully competes with γ decay as a deexcitation process. This process was first predicted by Hahn and Strassmann almost immediately after their discovery of fission [84] and was soon observed experimentally [85,86]. While over 60 delayed-neutron emitters have been identified, five contribute the majority of the delayed-neutron intensity in a mixture of fission products. They are 55.6-s ^{87}Br, 24.5-s ^{137}I, 4.4-s ^{89}Br, 2.8-s ^{94}Rb, and 1.7-s ^{135}Sb. There is also significant intensity from a mixed group of shorter-lived products with half-lives between 0.1 and 0.6 s. In the thermal neutron fission of ^{235}U, β-delayed neutron emission accompanies 1.6% of the fissions.

It is these delayed neutrons that provide the means by which the chain reaction in a nuclear reactor can be easily controlled. Delayed neutrons increase the average time between neutron generations substantially. For chain-reacting systems operating at a steady state ($k = 1$), this delay provides the opportunity for control.

REFERENCES

1. F. Greenaway, *John Dalton and the Atom*, Cornell University Press, Ithaca, NY, 1966.
2. W.V. Farrar, Nineteenth-century speculations on the complexity of the chemical elements, *J. Br. Soc. Hist. Sci.*, 2, 297, 1965.
3. P.E.L. de Boisbaudran, Extraction du gallium de ses minerals, *Acad. Sci. C. R.*, 82, 1098, 1876.
4. W.K. Roentgen, Ueber eine neue Art von Strahlen, *Sitzungberichte der Wuerzburger Physikalischen-Medicinischen Gesellschaft*, December 1895.
5. H. Becquerel, Sur les radiations invisibles emises par les corps phosporescents, *Acad. Sci. C. R.*, 122, 422, 1896.
6. J.J. Thomson, Cathode rays, *Philos. Mag.*, 44, 293, 1897.
7. E. Rutherford, The scattering of α and β particles by matter and the structure of the atom, *Philos. Mag.*, 21, 669, 1911.
8. H.G.J. Moseley, The high-frequency spectra of the elements, *Philos. Mag.*, 26, 1024, 1913.
9. J. Chadwick, The existence of a neutron, *Proc. R. Soc. (London)*, A136, 692, 1932.
10. P. Curie and M.S. Curie, Sur une substance nouvelle radio-active, contenue dans la pechblende, *Acad. Sci. C. R.*, 127, 175, 1898.
11. P. Curie, M.S. Curie, and G. Bemont, Sur une nouvelle substance fortement radio-active, contenue dans le pechblende, *Acad. Sci. C. R.*, 127, 1215, 1898.
12. E. Rutherford and F. Soddy, Radioactive change, *Philos. Mag.*, 5, 576, 1903.
13. F. Soddy, Intra-atomic charge, *Nature*, 92, 399, 1913.
14. E. Rutherford, J. Chadwick, and C.D. Ellis, *Radiations from Radioactive Substances*, Macmillan, New York, 1930.
15. E. Rutherford and T. Royds, The nature of the α particle from radioactive substances, *Philos. Mag.*, 17, 281, 1909.
16. P. Villard, Sur le rayonnement du radium, *Acad. Sci. C. R.*, 130, 1178, 1900.
17. H. Geiger and J.M. Nuttal, The ranges of the α-particles from various radioactive substances and a relationship between range and period of transformation, *Philos. Mag.*, 23, 439, 1912.
18. G. Gamov, Zur Quantentheorie des Atomkernes, *Z. Phys.*, 51, 204, 1928.
19. C.L. Cowan Jr., F. Reines, F.B. Harrison, H.W. Kruse, and A.D. McGuire, Detection of the free neutrino: A confirmation, *Science*, 124, 103, 1956.
20. F. Reines, Neutrino interactions, *Ann. Rev. Nucl. Sci.*, 10, 1, 1960.

21. W. Pauli, *Rapports du Septieme Conseil de Physique Solvay, Brussels, 1933*, Gauthier-Villars and Cie, Paris, France, 1934, Chap. 17.
22. E. Fermi, Tentativo di una teoria dell'emissione dei raggi β, *Ric. Sci.*, 4, 491, 1933.
23. E. Fermi, Versuch einer Theorie der β-Strahlen, *Z. Phys.*, 88, 161, 1934.
24. L.W. Alvarez, Electron capture and internal conversion in gallium 67, *Phys. Rev.*, 53, 606, 1938.
25. W. Heisenberg, Ueber den anschlaulichen Inhalt der quantentheoretischen Kinematik und Mechanik, *Z. Phys.*, 43, 172, 1927.
26. R.B. Firestone and V.S. Shirley, eds., *Table of Isotopes*, 8th edn., Wiley, New York, 1996.
27. P.A.M. Dirac, A theory of electrons and protons, *Proc. R. Soc. (London)*, 126, 360, 1930.
28. C.D. Anderson, The positive electron, *Phys. Rev.*, 43, 491, 1933.
29. G.N. Flerov and K.A. Petrzhak, Spontaneous fission of Uranium, *Phys. Rev.*, 58, 89, 1940.
30. S. Meyer and E.R. von Schweidler, Untersuchungen ueber radio-active Substanten (VI. Nitteilung) Ueber Radium F (Polonium), *Sitzber. Akad. Wiss. Wien [IIa]*, 115, 63, 1906, and references therein.
31. H. Bateman, Solution of a system of differential equations occurring in the theory of radio-active transformations, *Proc. Camb. Philos. Soc.*, 15, 423, 1910.
32. J.M. Robson, Radioactive decay of the neutron, *Phys. Rev.*, 83, 349, 1951.
33. F. Soddy, *Chemistry of the Radio-Elements, Part II*, Longmans, Green, London, U.K., 1914, Chaps. 1–3.
34. F.A. Paneth and G. Hevesy, Uberversuche zur Trennung des Radium Duon Blei, *Sitzber. Akad. Wiss. Wien, Math.-naturw. Kl.*, 122(IIa), 993, 1913.
35. G.P. Harnwell and J.J. Livingood, *Experimental Atomic Physics*, McGraw-Hill, New York, 1933, Chap. 3.
36. H. Yukawa, On the interaction of elementary particles, *Proc. Phys. Math. Soc. Jpn.*, 17, 48, 1935.
37. D. Bohm, *Quantum Theory*, Prentice Hall, New York, 1951, Chap. 4.
38. M. Moe and P. Vogel, Double beta decay, *Ann. Rev. Nucl. Part. Sci.*, 44, 247, 1994.
39. C. Perrier and E. Segre, Some chemical properties of element 43, *J. Chem. Phys.*, 5, 712, 1937.
40. J.A. Marinsky, L.E. Glendenin, and C. Coryell, The chemical identification of radioisotopes of neodymium and of element 61, *J. Am. Chem. Soc.*, 69, 2781, 1947.
41. B.L. Cohen, *Concepts of Nuclear Physics*, McGraw-Hill, New York, 1971.
42. E. Segre, *Nuclei and Particles*, 2nd edn., Benjamin, Reading, MA, 1977.
43. M. Goeppert Mayer, Nuclear configurations in the spin-orbit coupling model, I. Empirical evidence, *Phys. Rev.*, 78, 16, 1950.
44. A. Bohr and B.R. Mottelson, Collective and individual-particle aspects of nuclear structure, *Kgl. Dan. Vidensk. Selsk. Math.-Fys. Medd.*, 27, 1, 1953.
45. A. Bohr, On the structure of atomic nuclei, *Am. J. Phys.*, 25, 547, 1957.
46. M.E. Rose, Theory of internal conversion, in *Alpha-, Beta- and Gamma-ray Spectroscopy*, Vol. 2, K. Siegbahn, ed., North-Holland, Amsterdam, the Netherlands, 1965, Chap. 16.
47. F. Soddy, The complexity of chemical elements, *Proc. R. Inst. Gr. Brit.*, 22, 117, 1917.
48. C.F. von Weizsaecker, Metastable states of atomic nuclei, *Naturwissenschaften*, 24, 813, 1936.
49. O. Hahn, Ueber eine neue radioactive Substanz im Uran, *Ber. Deut. Chem. Ges.*, 54, 1131, 1921.
50. O. Hahn, Ueber ein neues radioaktives Zerfallsproduct im Uran, *Naturwissenschaften*, 9, 84, 1921.
51. B. Pontecorvo, Nuclear isomerism and internal conversion, *Phys. Rev.*, 54, 542, 1938.
52. J.R. Huizenga, C.L. Rao, and D.W. Engelkemeir, 27-minute isomer of U^{235}, *Phys. Rev.*, 107, 319, 1957.

53. H. Mazaki and S. Shimizu, Effect of chemical state on the decay constant of U^{235m}, *Phys. Rev.*, 148, 1161, 1966.
54. M.N. de Mevergnies, Chemical effect on the half-life of U^{235m}, *Phys. Rev. Lett.*, 23, 422, 1969.
55. E. Rutherford, Collision of particles with light atoms IV. An anomalous effect in nitrogen, *Philos. Mag.*, 37, 581, 1919.
56. F. Joliot and I. Curie, Artificial production of a new kind of radio-element, *Nature*, 133, 201, 1934.
57. V.F. Weisskopf, Nuclear physics, *Rev. Mod. Phys.*, 29, 174, 1957.
58. E. Fermi, Artificial radioactivity produced by neutron bombardment, in *Nobel Lectures: Physics, 1922–1941*, Elsevier, Amsterdam, the Netherlands, 1965, p. 414.
59. F.A. Heyn, Evidence for the expulsion of two neutrons from Cu and Zn by one fast neutron, *Nature*, 138, 723, 1936.
60. F.A. Heyn, Radioactivity induced by fast neutrons according to the (n,2n) reaction, *Nature*, 139, 842, 1937.
61. N. Feather, Collisions of neutrons with light nuclei, *Proc. R. Soc.*, A142, 699, 1933.
62. G. Breit and E.P. Wigner, Capture of slow neutrons, *Phys. Rev.*, 49, 519, 1936.
63. R.K. Bardin, P.J. Gollon, J.D. Ullman, and C.S. Wu, A search for the double beta decay of ^{48}Ca and lepton conservation, *Nucl. Phys.*, A158, 337, 1970.
64. P.K. Kuroda, Nuclear fission in the early history of the earth, *Nature*, 187, 36, 1960.
65. M. Neuilly, J. Bussac, C. Frejacques, G. Nief, G. Vendryes, and J. Yvon, Geochimie nucleaire, *Acad. Sci. C. R. Ser. D*, 275, 1847, 1972.
66. O. Hahn and F. Strassmann, Neutron-induced radioactivity of uranium, *Naturwissenschaften*, 27, 11, 1939.
67. P. Abelson, Cleavage of the uranium nucleus, *Phys. Rev.*, 55, 418, 1939.
68. L. Meitner and O. Frisch, Disintegration of uranium by neutrons: A new type of nuclear reaction, *Nature*, 143, 239, 1939.
69. O.R. Frisch, Physical evidence for the division of heavy nuclei under neutron bombardment, *Nature*, 143, 276, 1939.
70. W. Jentschke and F. Prankl, Untersuchung der Schweren Kernbruckstuecke Beim Zeitall von Heutron en bestrahltem Uran und Thorium, *Naturwissenschaften*, 27, 134, 1939.
71. J. Halpern, Nuclear fission, *Ann. Rev. Nucl. Sci.*, 9, 245, 1959.
72. E.K. Hyde, *The Nuclear Properties of the Heavy Elements*, Vol. 3, Prentice Hall, Englewood Cliffs, NJ, 1964.
73. N. Bohr and J.A. Wheeler, The mechanism of nuclear fission, *Phys. Rev.*, 56, 426, 1939.
74. S.G. Nilsson, Binding states of individual nucleons in strongly deformed nuclei, *Kgl. Dan. Vidensk. Selsk. Math.-Fys. Medd.*, 29(16), 1, 1955.
75. S.M. Polikanov, V.A. Druin, V.A. Karnaukhov, V.L. Mikheev, A.A. Pleve, N.K. Skobelev, V.G. Subbotin, G.M. Ter-Akop'yan, and V.A. Fomichev, Spontaneous fission with an anomalously short period, *Sov. Phys. JETP*, 15, 1016, 1962.
76. H.C. Britt, Properties of fission isomers, *At. Data Nucl. Data Tables*, 12, 407, 1973.
77. V.M. Strutinsky, Influence of nucleon shells on the energy of a nucleus, *Sov. J. Nucl. Phys.*, 3, 449, 1966.
78. V.M. Strutinsky, Shell effects in nuclear masses and deformation energies, *Nucl. Phys.*, A95, 420, 1967.
79. V. McLane, C.L. Dunford, and P.F. Rose, *Neutron Cross Sections*, Vol. 2, Neutron Cross Section Curves, Academic Press, Boston, MA, 1988.
80. J.R. Nix, Calculation of fission barriers for heavy and superheavy nuclei, *Ann. Rev. Nucl. Sci.*, 22, 65, 1972.
81. M.G. Mustafa, U. Mosel, and H.W. Schmitt, Asymmetry in nuclear fission, *Phys. Rev.*, C7, 1519, 1973.
82. R. Vandenbosch and J.R. Huizenga, *Nuclear Fission*, Academic Press, New York, 1973.

83. T.R. England and B.F. Rider, Evaluation and compilation of fission product yields, ENDF-349. Los Alamos National Laboratory, Los Alamos, NM, 1993.
84. O. Hahn and F. Strassmann, Ueber den Nachweis und das verhalten der bei der Bestrahlung des Urans mittels Neutronen entstehenden Erdalkalimetalle, *Naturwissenschaften*, 27, 89, 1939.
85. H. von Halban Jr., F. Joliot, and L. Kowarski, Liberation of neutrons in the nuclear explosion of uranium, *Nature*, 143, 470, 1939.
86. H.L. Anderson, E. Fermi, and H.B. Hanstein, Production of neutrons in uranium bombarded by neutrons, *Phys. Rev.*, 55, 797, 1939.

3 Engineering Issues

In the previous chapter, the physical properties of nuclei that govern their production and decay characteristics were discussed. They give rise to the signatures that the radiochemist interprets in the forensic analysis of nuclear materials. Although physics governs the signatures, the materials that display those signatures have been created or modified by engineers. Engineering is the technology responsible for the recovery of raw materials from the natural environment and the adjustment of their isotopic content for use in energy production. Engineering provides the methods whereby synthetic materials are produced from primordial starting materials and are isolated from undesirable byproducts. The nuclear forensic analyst often interprets physical signatures, which are defined by differences in nuclear, mechanical, and chemical engineering techniques, to identify the source of a given sample.

3.1 NATURAL *VERSUS* SYNTHETIC MATERIALS

There exist three common nuclear fuels: ^{233}U, ^{235}U, and ^{239}Pu. These nuclides have a high probability of undergoing fission when irradiated with neutrons of any energy. Of the three fuels, only ^{235}U occurs in nature, where it constitutes about 0.72% of naturally occurring U. Most of the balance of natural U consists of ^{238}U, which fissions efficiently only through interactions with high-energy (>1.2-MeV) neutrons. There is also a small admixture of ^{234}U ($t_{1/2}$ = 2.46 × 10^5 years) in natural U, in secular equilibrium with the longer-lived ^{238}U isotope. Although ^{233}U and ^{236}U are long-lived ($t_{1/2}$ = 1.59 × 10^5 years and 2.34 × 10^7 years, respectively), neither exists in nature in readily detectable concentrations. The presence of significant admixtures of ^{233}U or ^{236}U in a uranium sample can only arise through anthropogenic nuclear processes.

Reactors have been designed to consume natural U as fuel. These designs are bulky, and considerations of fuel and moderator purities are paramount. Engineering issues are simplified if the initial fuel charge is enriched in ^{235}U relative to the natural isotopic composition. Most of the 470 commercial nuclear power reactors, operating or under construction, require enriched U as fuel. However, the enrichment of ^{235}U is a difficult process, conducted on a very large scale. The production of U that is enriched in ^{235}U must necessarily be accompanied by the production of U that is depleted in ^{235}U. Enriched and depleted U are synthetic materials, derived from natural uranium through industrial processes.

The Pu isotopes are derived from ^{238}U by irradiation with neutrons. Similarly, ^{233}U results from the neutron irradiation of natural Th. Thus, Pu and ^{233}U are also synthetic materials. The mixtures of nuclides that make up a sample of synthetic fuel are defined by the engineering processes that produce the material. The nuclear forensic scientist must therefore understand not only the origins of the natural materials that give rise to the synthetic materials, but also the processes by which the synthetic materials are produced and fabricated into useful forms.

Although the focus of this chapter is on processes involving synthetic heavy-element materials, other synthetic materials have important applications, particularly in the production and operation of nuclear weapons. Most important among these are mixtures and compounds of the isotopes of H and Li. These materials will be discussed later.

3.2 RECOVERY OF ACTINIDES FROM EARTH

The starting point for the production of synthetic nuclear materials is the recovery of the natural starting materials (U and Th) from the earth. Though U was discovered by Klaproth in 1789 [1], production of it began on an industrial scale only in 1853, after it was found that U salts had practical application as colorants for glasses and ceramics [2]. Thorium was discovered by Berzelius in 1828 [1]. Interest in U and Th compounds increased considerably following the discovery of radioactivity in 1898, at which point efficient mining and milling operations for both elements became important. Details on typical ore bodies, beyond that summarized later, can be found in Refs. [2–7].

Uranium and Th are widely distributed in the earth's crust, with average concentrations of ~2 and 8 ppm, respectively. Although more than 100 different U-based minerals are known, deposits of high-grade ore (like uraninite and pitchblende, with as much as 70% U by weight) are rare. Those deposits of major economic importance are found in Zaire, Canada, and South Africa. Most of the world's supply of recovered U has been extracted from ores with a U content of only ~0.2%. Conversely, although the mineralogy of Th is somewhat less varied, deposits of its principal mineral, monazite, are fairly common. Thorium is mined in India, Brazil, Scandinavia, Russia, and the United States, and the Th content of the pure mineral is between 5% and 10%. Prospecting operations are often carried out with portable survey instruments, which can detect large deposits of ores with U and Th concentrations as low as 100 ppm.

Both U and Th have nonnuclear applications that account for some incentive to recover them from the earth. Uranium is used in counterweights for aircraft control surfaces, as projectiles in armor-piercing bullets, and as ballast in ship keels. However, most of the applications that use U are related to the nuclear industry. By volume, most of the Th produced in the world goes into the production of gas mantles for lamps, although there is increasing application as a hardener for Mg-based alloys. Thoria (ThO_2) has a very high melting point and a low dissociation pressure, making it very useful for hot filaments.

Dilute U ores are crushed and subjected to heavy media separation, followed by flotation. The resultant dense fraction is thickened and filtered, producing a cake in which U has been concentrated to ~1%. Few chemical reagents are sufficiently inexpensive to economically process an ore of this low grade. For U processing, dilute sulfuric acid is the most common reagent used. However, sodium carbonate is an alternative when the U is mixed with large quantities of limestone and the alkalinity would result in a significant increase in acid reagent (as for some deposits in the western United States).

In the acid-leach process, the ore is usually processed in plastic-coated vessels that are agitated, either with stirrers or by an air stream supplied from the bottom of the vessel. An oxidizing agent is necessary in the process to ensure that the U in solution is in the fully oxidized uranyl (hexavalent) state:

$$2\ U_3O_8 + O_2 + 6\ H_2SO_4 \rightleftarrows 6\ UO_2SO_4 + 6\ H_2O.$$

Often, pyrolusite (a MnO_2 mineral) is added to accomplish the oxidation. The U-bearing liquid is separated from the spent ore through decantation or filtration. This "pregnant liquor" is passed through diatomaceous earth to separate it from finely divided particles in suspension, after which the material is further purified by ion exchange or solvent extraction. In the ion-exchange procedure, the U-bearing liquid is passed through a cation-exchange resin, taking advantage of the fact that U is one of the few metallic elements that forms anions in a sulfate medium and is thus not bound by the exchanger. Iron also passes through the exchanger with the U. The Fe is removed by precipitation through the addition of lime. At a pH of 3.8, iron hydroxide precipitates and U remains in solution. Several organic solvent systems have been studied as alternatives to ion exchange. The extractant is usually a mono- or dialkyl ester of phosphoric acid, with kerosene as the diluent. Uranium is back-extracted from the organic solvent with a strong acid. Independent of whether ion exchange or solvent extraction is used, however, U is precipitated as a diuranate salt by addition of either magnesia or ammonia.

The carbonate-leach process offers the advantage that few impurities accompany the U into solution from the broken ore. The disadvantage is that it is not as efficient as the acid-leach process, although its effectiveness is improved at elevated temperatures. The carbonate leach is usually carried out in mild steel vessels, with a 5%–10% solution of sodium carbonate that contains a little bicarbonate to control the alkalinity and prevent the precipitation of U. The ore is sometimes roasted with NaCl before leaching to increase the reactivity of the ore with the carbonate solution. As before, an oxidant is required to convert U to the hexavalent state in solution, and air (O_2) is often used for this purpose:

$$2\ U_3O_8 + O_2 + 18\ Na_2CO_3 + 6\ H_2O \rightleftarrows 6\ Na_4UO_2(CO_3)_3 + 12\ NaOH.$$

Normally, U is precipitated from the filtered leach liquor as sodium diuranate ($Na_2U_2O_7$) upon the further addition of NaOH.

Higher-grade ores, such as pitchblende, are crushed and treated with a mixture of sulfuric and nitric acids. These more expensive reagents can be used economically when the grade of the ore is high:

$$U_3O_8 + 3\ H_2SO_4 + 2\ HNO_3 \rightleftarrows 3\ UO_2SO_4 + 2\ NO_2 + 4\ H_2O.$$

The resulting solution can be treated with either magnesia or ammonia to precipitate magnesium or ammonium diuranate.

At this point, whether arising from the acid-leach or the carbonate-leach process, U has been concentrated to the level of ~70%, as a diuranate salt, with a corresponding reduction of metal-ion impurities. Further purification is usually accomplished by dissolving the concentrate in sulfuric acid and adding Ba carrier to remove Ra through coprecipitation with $BaSO_4$. Uranium peroxide is then precipitated through the simultaneous addition of hydrogen peroxide and ammonium hydroxide, maintaining pH above 1.5:

$$UO_2SO_4 + H_2O_2 + 2\ NH_4OH \rightleftarrows UO_4 + (NH_4)_2SO_4 + 2\ H_2O.$$

The uranium peroxide precipitate is washed several times with water to remove impurities, after which it is dissolved in nitric acid (with the evolution of oxygen) to form a uranyl nitrate solution. The solution is evaporated under reduced pressure until semisolid uranyl nitrate hexahydrate is formed. This salt is melted and filtered to remove impurities, which arise mostly from the action of nitric acid on settling agents. The salt is then dissolved in water, and U is extracted into an organic solvent, back-extracted into water, and precipitated by the addition of ammonia to produce ammonium diuranate. This compound is a common storage form of purified U, often called "yellow cake." Historically, the organic solvent in this last step was diethyl ether, which is extremely flammable. More recently, ether has been replaced by safer solvents such as hexone, tributyl phosphate (TBP) in kerosene, or long-chain aliphatic esters.

Further processing of the U, either for the production of metal or for isotope separation, proceeds through fluoride species. Yellow cake is spread on trays and heated to 350°C to decompose it to uranium trioxide, with the emission of ammonia and steam. The trioxide is then reduced to the dioxide at 700°C by the addition of hydrogen:

$$UO_3 + H_2 \rightleftarrows UO_2 + H_2O.$$

The material is then reacted with HF at 450°C to convert it to UF_4. The UF_4 product can be processed to the metal or further processed to UF_6 by reaction with ClF_3 or BrF_3:

$$3\ UF_4 + 2\ ClF_3 \rightleftarrows 3\ UF_6 + Cl_2$$

or by reaction with air:

$$2\ UF_4 + O_2 \rightleftarrows UF_6 + UO_2F_2.$$

However, the latter process is not favored because it results in the loss of half of the product as nonvolatile uranyl fluoride. The chemistry of UF_6 will be discussed in more detail later.

The most important mineral of Th is monazite. It is nominally 6% ThO_2, 0.2% U_3O_8, 60% oxides of the rare earths (principally cerium), and 25% P_2O_5. The rare-earth content of monazite is sufficiently large that they are often also recovered from the mineral along with Th, which affects the chemical procedures used for the process. Monazite is paramagnetic, and dilute deposits can be concentrated electromagnetically. More often, a gravity concentration procedure (like wet tabling) is used, taking advantage of the high specific gravity of monazite (4.9–5.3). Other important Th minerals are thorite, thorianite, and uranothorite, which are silicate-based rather than phosphate-based and contain variable amounts of U in addition to Th. No chemical equations are included in the following paragraphs because the oxidation state of Th throughout the procedures is invariant (Th^{4+}).

The first step in working a Th deposit involves "opening" the ore, which can involve reacting coarsely crushed material with sulfuric acid at 200°C. In the most common procedure, Th and the lanthanides (Ln) are converted to sulfates with the liberation of phosphoric acid. After opening, the ore is agitated with water, and Th, U, and the rare earths pass into solution. It is sometimes possible to convert >99% of the metals in a pure monazite sample to water-soluble sulfates by this step.

Water is added to dissolve the sulfates. Thorium sulfate is more soluble than the Ln sulfates, so some separation may be effected if the volume of water is minimized. Any P in the ore remains as a phosphate contaminant in the Th solution. Phosphate can be removed by the addition of solid sodium chloride, which results in the precipitation of crystalline sodium double-sulfates of Th and the rare earths. These sulfate salts can then be filtered from the phosphate-bearing mother liquor. Another technique involves the precipitation of anhydrous Th and Ln sulfates from a concentrated solution by the addition of concentrated sulfuric acid. The precipitate is filtered onto a metal mesh and washed with concentrated sulfuric acid to remove excess phosphates. The solid sulfates are then dissolved in water and neutralized by the addition of sodium hydroxide. A separation of Th from the rare earths is effected by precipitation at intermediate pH: at pH 5.5, Th precipitates while the rare earths remain in solution. For this process to be efficient, however, it is necessary to avoid high local concentrations of alkali upon addition of the base, and either high-speed stirring or the addition of a low-solubility base (e.g., magnesia) is effective.

Another method of recovering Th from ores involves opening the ore with sodium hydroxide, which converts the Ln and Th phosphates into insoluble oxides and soluble sodium phosphate. This technique offers the advantage that no special phosphate separation is required before separation of Th from the rare earths. However, care must be taken to avoid the formation of refractory phases of the oxides. A 50% sodium hydroxide solution is used at a temperature of about 140°C, but the ore must be finely ground for this treatment to be effective at this low temperature. The efficiency of recovery of Th from the ore can be >98%. After this reaction, an equal volume of water is added and the mixture filtered without cooling (to prevent clogging the filter with sodium phosphate). The precipitate is then washed with water or dilute sodium hydroxide, after which it is dissolved in hydrochloric or nitric acid. Separation of Th from the Ln is again effected by a controlled-pH precipitation of thorium hydroxide.

Other methods for extracting Th from ore include fusion with a mixture of petroleum coke, lime, and feldspar, with the treatment continued until phosphoric acid is no longer evolved. The resultant product is then cooled, and the carbides are decomposed by the addition of water, which produces the evolution of acetylene. Additionally, chlorination of a mixture of monazite and charcoal with Cl_2 or HCl at 700°C may be used (volatile $ThCl_4$ is subsequently collected by distillation at 900°C).

Both of these procedures have been a focus of some industrial development, but they are not considered mainstream and will not be discussed further here. However, see Ref. [8].

Other methods for separating Th from the rare earths involve the selective dissolution of Th in either sodium carbonate or ammonium oxalate. In both cases, the lanthanides are efficiently precipitated, whereas the soluble Th complexes, $Na_6Th(CO_3)_5$ and $(NH_4)_4Th(C_2O_4)_4$, remain in solution. After filtration, the complexes can be destroyed by the addition of sulfuric or hydrochloric acid.

The goal of most of the procedures discussed is the concentration of Th into an intermediate fraction, with the removal of phosphate, sulfate, and the bulk of the lanthanides. Reactor-grade Th must be completely free of the rare earths (several of whose nuclides have large neutron-capture cross sections), so further purification is required. Oxalate or hydroxide precipitates are calcined to the oxide and then dissolved in 8 M HNO_3, after which Th is extracted into TBP in a kerosene diluent. Tetravalent Ce is also extracted under these conditions, but it can be selectively back-extracted into 8 M HNO_3 containing $NaNO_2$ or H_2O_2, which reduces the Ce(IV) impurity to Ce(III). Thorium is then back-extracted into dilute nitric acid, leaving any residual U behind in the organic phase, from which it can be recovered with a water wash.

Thorium is recovered from the solution by the addition of oxalic acid and precipitation of the oxalate. The precipitate is filtered, washed with water, dried in an oven in stainless steel trays, and calcined to ThO_2 at 700°C. ThO_2 is a common storage form for purified Th, and it may be converted to the tetrafluoride by reaction with anhydrous HF at 350°C. Thorium does not have a volatile hexafluoride.

3.3 SEPARATION AND ENRICHMENT OF U ISOTOPES

Uranium enrichment is a special case of the general problem of isotope separation. There are many applications in which small quantities of separated stable isotopes are used, most notably in biology and medicine. Nuclear research also has a need for separated isotopes. In both cases, a gram of a purified isotope is considered a large quantity. Only the isotopes of U and H have been separated in larger quantities, with process feeds on the order of millions of kilograms.

As will be discussed later, even natural U can be used as fuel in a nuclear reactor. However, as the proportion of ^{235}U increases, the ease with which a fission reactor can be used as an energy source increases. Modern light-water-moderated reactors are fueled by U enriched in ^{235}U from 0.71% (natural) to about 3%. At greater enrichments, the size of a reactor for a given power level can be decreased. Reactors for ship propulsion use initial fuel loadings enriched to at least 10% to minimize the size and weight of the power plant; to minimize refueling, submarines implement fuels enriched to 90% or more, which are considered weapons grade.

Enrichment of ^{235}U in samples of U is one of three methods for producing fission fuel for a nuclear explosive device. The other two methods involve the operation of nuclear reactors and the capture of neutrons by either ^{232}Th (to make ^{233}U) or ^{238}U (to make ^{239}Pu). Production of ^{233}U and ^{239}Pu will be discussed later. It was realized very soon after the discovery of fission that it might be possible to design a nuclear explosive if ^{235}U could be extracted and concentrated from natural U. For ^{235}U enrichments of <10%, metallic U cannot be "chemically assembled" in such a way as to produce a nuclear explosion; the fast-fission critical mass that must be assembled is effectively infinite at these enrichments. Above 10% enrichment, however, the size of the critical mass quickly decreases with increasing enrichment, especially if the U is surrounded by a good neutron reflector. The boundary between low-enriched uranium (LEU) and highly enriched uranium (HEU) is arbitrarily set at 20% ^{235}U, where the well-reflected critical mass is reduced to <100 kg.

The number of neutrons in the U nucleus exerts only a miniscule effect on the chemistry of the atom in which it is located. As a result, separation of U isotopes cannot be accomplished by the standard chemical techniques used to purify or separate chemical substances. Isotope separation must take advantage of those nuclear properties that are modified by changes in the number of neutrons in the nucleus. These include the nuclear size, shape, and mass, as well as the nuclear magnetic moment and angular momentum. Most enrichment technologies currently used, or under study, rely on differences in mass, manifested through differences either in inertia (electromagnetic isotope separation) or in velocity at thermal equilibrium (gaseous diffusion, centrifugation, and aerodynamic techniques). All of these methods use volatile compounds of U (such as UF_6) as feed material.

Uranium hexafluoride was first produced by Ruff [9], who anticipated its existence through his studies of other volatile Group VIA hexafluorides (MoF_6 and WF_6). He first produced it by reacting molecular fluorine with U metal or carbide:

$$U + 3\,F_2 \rightleftarrows UF_6$$

and

$$UC_2 + 7\,F_2 \rightleftarrows UF_6 + 2\,CF_4.$$

Ruff's work received little attention until 1939, when the discovery of fission focused attention on UF_6 as the only known, stable, gaseous compound of U. A substantial experimental effort has gone into a search for preparative methods suitable for industry. It appears that all U compounds will form UF_6 if heated with fluorine to a sufficiently high temperature; however, fluorine is not a pleasant compound to handle, and the most useful reaction is that requiring the minimum quantity of F_2:

$$UF_4 + F_2 \rightleftarrows UF_6,$$

which appears to proceed in two steps through a UF_5 intermediate.

At room temperature, UF_6 is a colorless, volatile solid with a density of 5.1 g/mL. It sublimes without melting at atmospheric pressure, and at room temperature, it has a vapor pressure of approximately 100 torr. The temperature at which UF_6 has a vapor pressure of one atmosphere (the "boiling point" of the solid) is 56.5°C. The triple point of UF_6 occurs at 64.1°C and 1133 torr (see Figure 3.1). In the gas phase, the structure of the UF_6 molecule is that of a perfect octahedron (point-group O_h), where all U–F bond distances have an identical length of 2.00 Å. In the vapor phase, molecules of UF_6 are

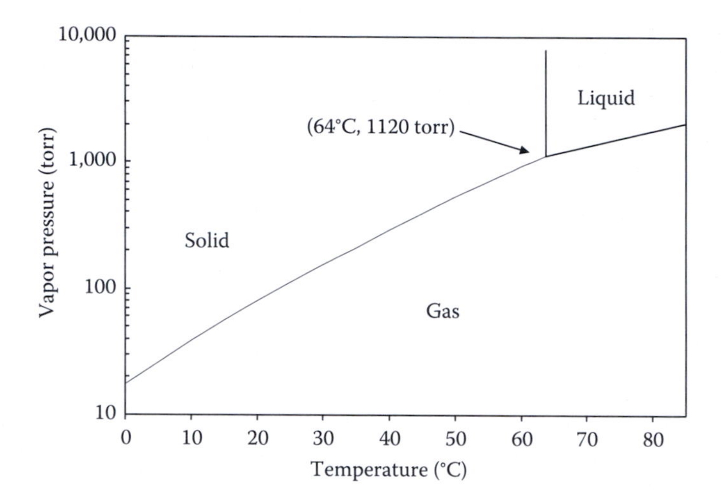

FIGURE 3.1 Phase diagram of UF_6, showing the triple point.

unassociated, giving it perfect-gas behavior, with viscosity and diffusivity properties that are favorable for gas-phase isotope separation techniques. At 69.5°C, the viscosity of UF_6 is 7 mP, and its self-diffusion coefficient is 1.9×10^{-5} cm^2/s.

The main disadvantage of working with UF_6 is its high chemical reactivity: UF_6 reacts vigorously with water, evolving considerable heat:

$$UF_6 + 2\,H_2O \rightleftarrows UO_2F_2 + 4\,HF.$$

However, it is not very reactive with dry air. UF_6 also reacts with several metals, but Ni, Cu, and Al are resistant. Chemical inertness is only valid for pure UF_6, because the presence of even small quantities of HF increases the rate of attack on even the resistant metals. The tails of the enrichment process (UF_6 depleted in ^{235}U) are stored in steel drums on-site at the enrichment plants. These drums are effectively self-sealing, as exposure of the UF_6 through penetrations caused by corrosion results in reaction with moist air, forming a plug of UO_2F_2.

3.3.1 Electromagnetic Isotope Separation

Differential deflection in electric and magnetic fields is the most general method for the separation of isotopes. It was first accomplished by Nier in 1940 in an attempt to separate minute amounts of U isotopes for fission experiments [10]. Uranium ions were produced by an electron bombardment source in which a vapor of UBr_4 from a heated furnace was ionized. The ions were accelerated by a 1000-V potential and collimated, after which the resultant ion beams traversed semicircular paths through a vacuum chamber mounted between the poles of an electromagnet. Mass-235 U ions deposited on a nichrome collection plate at a different location than did mass-238 U ions. This simple apparatus is the basis for all electromagnetic isotope separators (EMIS), which use the inertial resistance to acceleration of the atomic mass to effect isotopic separations.

Using EMIS techniques, a U-bearing volatile compound is introduced as a gas into an ion source in which the U molecules are destroyed and +1 atomic ions are produced. Ions obtained in this way move with a Maxwellian distribution of kinetic energies such that some ions will have velocities many times greater than others and will not be deflected to the same extent by a magnetic field, even if they have the same mass. This difficulty is overcome by accelerating the ions by means of a DC potential of a few kilovolts, which imparts identical kinetic energies much greater than the original thermal energies. The initial thermal energy differences become relatively less important, so that ions of the same mass are effectively deflected to the same extent. In a uniform magnetic field, the path of an ion beam is circular, and its radius ρ is given by

$$\rho = (mcv)/(qB), \tag{3.1}$$

where
- q is the charge of the ion
- v is its velocity
- m is its mass
- c is the velocity of light
- B is the magnetic field strength

If the acceleration potential is V, the velocity of the ions will be

$$v = (2Vq/m)^{1/2}. \tag{3.2}$$

Substituting Equation 3.2 into Equation 3.1 yields

$$\rho = (c/B)(2Vm/q)^{1/2},$$

and differentiating both sides results in

$$\mathrm{d}\rho/\rho = \mathrm{d}m/(2m). \tag{3.3}$$

It follows that the path radius of a particular ion is proportional to the square root of its mass. The separation of the masses is proportional to the separation of the ion orbits at the point of collection.

An EMIS apparatus with a collection position 180° from the ion source along the path of the particles offers certain advantages. The homogeneous magnetic field produces first-order focusing, such that if products with the same mass are emitted from the ion source over a small range of angles, they arrive at the same location at the 180° collection position. If the products are emitted from the source over too wide an angular spread, however, image aberration occurs that can increase the cross-contamination between masses unless it is corrected by shimming the magnetic field. The shift in the ion orbit is equal to twice dρ at 180°. Hence, if ρ = 0.5 m, Equation 3.3 gives the spacing between two masses of 235 and 238 as (3/235) × 50 cm = 0.64 cm.

If beam intensities in EMIS machines are kept low, very high-purity separations between adjacent masses are possible, whereas at higher intensities, space-charge effects cause beam spreading. However, the rate of production of enriched materials is proportional to the beam intensities. Another drawback to EMIS is that the efficiency of the ion source is usually fairly low, resulting in losses of feed material that require that the apparatus be opened and cleaned to recover the feed.

EMIS machines termed calutrons were used to produce the HEU used as the fission fuel in the weapon that destroyed Hiroshima in 1945. A single calutron would have taken hundreds of years to produce enough material for a single bomb, so hundreds were constructed, beginning in 1942, and run from 1944 to 1945 in a profligate use of resources possible only during wartime. Calutrons are extremely expensive and inefficient in both energy and resources; hence, the technique was abandoned in the United States in favor of gaseous diffusion. Though the calutron method is inferior to other techniques, it offers the advantage of producing small quantities of weapons-usable materials almost immediately. Even so, weapons inspectors in Iraq after the first Gulf War were surprised to find evidence of EMIS production of enriched U, long thought to be a nonviable technique for proliferant states.

3.3.2 Gaseous Diffusion

Gaseous diffusion is the most widely used method for enriching U, accounting for more than 95% of the world's enriched-U production in 1978 [11]; in 2003, gaseous diffusion accounted for approximately 40% of the world's enriched U production, having been supplanted by gas-centrifuge technology. The large gaseous diffusion plants are nearing the end of their design lives and currently supply only 25%–30% of the world's U enrichment capacity. Gaseous diffusion has a proven efficacy and reliability, demonstrated by large-scale production facilities in the United States, Great Britain, and the former Soviet Union. Both China and France also chose gaseous diffusion for their early weapons development programs. In 1973, however, France made the conservative decision to use gaseous diffusion in its Eurodif facility [12], at a time when the high-tech gas-centrifuge process was forecast as a more promising method.

Gaseous diffusion can be used as an enrichment method because the condition of thermal equilibrium requires that all molecules of a gas mixture in equilibrium with its surroundings have the same average kinetic energy. Therefore, lighter molecules travel faster and strike the container walls more frequently than do heavier ones. A minute hole in the wall of the cell, provided it is small enough to prevent bulk outflow of the gas as a whole, will allow passage of a larger proportion of the lighter molecules than of the heavier ones (weighted by their relative concentrations), a consequence of Maxwell's kinetic theory of gases. For uranium hexafluoride gas, comprised mainly of $^{235}UF_6$ and $^{238}UF_6$, the molecular masses are 349 and 352, respectively. Therefore, the ratio of average velocities is $(352/349)^{1/2} = 1.0043$, with $^{235}UF_6$ moving faster than $^{238}UF_6$ by 0.43% on average. It is fortuitous that elemental fluorine is composed of a single stable isotope, or gaseous diffusion would be much more complicated.

The diffusion barrier is the central component of the enrichment process. It must be resistant to corrosion by highly reactive UF_6 gas, and the holes in the barrier must have a characteristic diameter that is smaller than the mean-free-path of the molecule to be separated. If collisions between molecules could occur with significant frequency within the barrier, equipartition of energy would render the barrier ineffective. Finally, the barrier must be thin enough that it has adequate permeability at low pressure. There is limited information available on diffusion barriers [11,13–16].

At least two types of barrier design have been demonstrated to be effective: the film-type barrier, in which pores are created in an initially nonporous membrane, and the aggregate barrier, in which the pores comprise strings of voids between the particles that are pressed and sintered to form the barrier. In film-type barriers, pores can be created by dissolving one constituent of an alloy or by electrolytic etching. Film-type barriers tend to be brittle and can rupture during use. The United States uses an aggregate-type barrier, created by sintering powdered Ni in the form of tubes. The flow characteristics through the pores of a sintered bed of spheres are considerably different than those through a set of linear capillaries [11].

The barrier must provide an optimal enrichment while operating at a high pressure differential and high permeability. The requirement of a small pore size makes it necessary for a simple barrier to have very thin walls to achieve the required permeability. One method that has been proposed to accomplish this permeability involves the use of a composite barrier [11], combining the properties of both barrier types described earlier. The composite barrier consists of a sturdy support layer with large pores, covered with a thin integral coating perforated with small pores that control the diffusion process. This design increases the mechanical strength of the barrier to the pressure differential and reduces handling difficulties.

The diffusion barrier is usually fabricated in the form of tubes into which the UF_6 gas is introduced. These tubes are grouped together in sheets and bundles that are housed in Ni-plated cylindrical diffusers. Each cylindrical unit is associated with a compressor for manipulating the enriched gas that flows through the walls of the barrier tubes into the interstitial space in the diffuser and for providing the driving pressure. Also present is a heat exchanger for ensuring that adiabatically compressed or expanded UF_6 gas is at thermal equilibrium and never heats to the point that it becomes dangerously reactive or cools to a temperature at which UF_6 condenses. The seals in the compressors are manufactured from materials that are also closely held secrets. Gaseous diffusion plants operate at temperatures above, and pressures below, the UF_6 triple point (~70°C and 0.5 atm).

The elementary unit of a gaseous diffusion plant is the cell, which is effectively divided into two compartments by sections of cylindrical porous barrier. A pressure difference is imposed between the inside and the outside of the tubes. The gas that diffuses through the barrier is enriched in the light component, whereas the undiffused gas is depleted. The material enriched in the desired isotope is called "product," and the depleted material is called "tails." The input stream is referred to as "feed." In this section, the symbol N refers to the atom fraction of the desired product in a given stream. For instance, if the feedstock to a given enrichment cell is natural U, $N_F = 0.0072$ (equivalent to 0.711% by weight). The relative isotopic abundance of a given stream, R, is defined by

$$R = N/(1-N).$$

For an element that consists of a mixture of two isotopes (as is nearly the case for U, since the ^{234}U concentration is very small), R is the ratio of the number of atoms of the isotopes in the mixture. The action of a given enrichment cell results in a product stream characterized by R_P and a tails stream characterized by R_T, and the single-stage separation factor, q, is defined as

$$q = R_P/R_T.$$

Because the difference in average velocities between $^{235}UF_6$ and $^{238}UF_6$ is only 0.43%, the enrichment that can be obtained in any single separation cell is fairly small. However, the actual performance of a gaseous diffusion cell depends not just on the permeability, size, and geometrical configuration of the barrier, but also on the pressure and temperature of the gas. For example, if the operating pressure of the cell increases, the mean-free-path of the UF_6 atoms decreases and the probability of molecular collisions in the barrier pores increases. This situation reduces the efficiency of the barrier, even though the flow of gas through the barrier would be enhanced. Increasing the temperature of the diffusion cell would cause the molecules to strike the barrier more frequently, obviating the need for higher pressures, but it would also require the compressors to do more work on a given volume of gas. The result would be an increase in the wear of the fragile seals, as well as in the chemical reactivity of the gas.

Typically, UF_6 feed enters the diffuser at a pressure between 0.3 and 0.5 atm, and the flow is adjusted so that approximately half of the feed diffuses through the walls of the barrier tubes into interstitial space, becoming enriched (product), while the rest remains in the tubes, becoming depleted (tails). The lower-pressure streams are cooled with heat exchangers and then recompressed. Precooling is required because, when UF_6 is compressed, it heats substantially. Neutralizing compression heat is the biggest source of energy consumption at a gaseous diffusion plant.

For an ideal barrier, the relationship between the relative isotopic abundances of the feed material and the product is

$$R_P = R_F\left(352/349\right)^{1/2},$$

which corresponds to a stage separation factor of $q = 1.00429$. The temperature and pressure of the UF_6 gas, and the properties of the barrier, act to reduce the achievable separation factor to a number closer to 1.0. The real barrier efficiency, ε_B, is a number less than 1.0, such that

$$(q_{obs} - 1) = \varepsilon_B\,(q_{ideal} - 1).$$

If the total enrichment and throughput of a plant are known, it can be shown [17] that ε_B defines the total diffusion barrier area, thereby determining the necessary size of the plant. A value of $\varepsilon_B = 0.7$ is fairly standard for the industry.

Enrichment by gaseous diffusion through a barrier is a small effect. To obtain large enrichments, the process must be repeated many times. In a simple cascade (Figure 3.2), the product of one enrichment cell is used as feedstock for another cell,

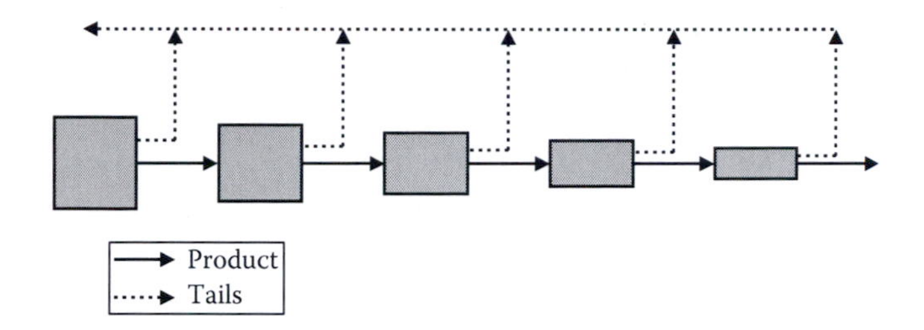

FIGURE 3.2 Diagram of a simple cascade. Product from each stage is introduced into the next stage as feed, and all tails cuts are discarded. This is a wasteful process.

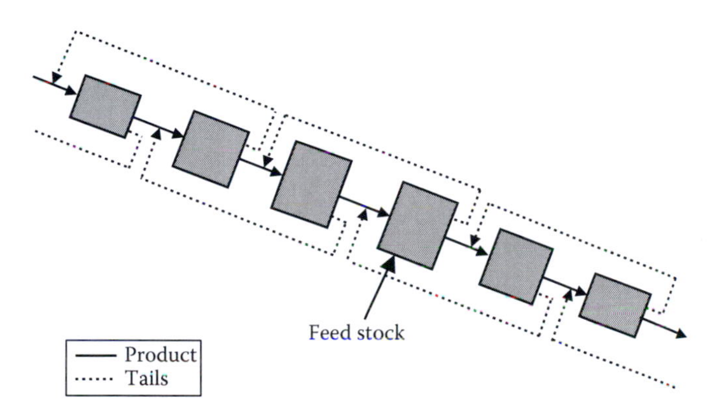

FIGURE 3.3 Diagram of a symmetric countercurrent cascade, where tails are reintroduced into the cascade as feedstock at an earlier stage of the separation.

and the tails are discarded. If the process is repeated many times, large enrichments are obtained, but the yield of final product is very low. The tails from cells near the end of the process, although significantly enriched compared to the original feedstock, are discarded. One way of improving the efficiency of the process is by using a countercurrent cascade (Figure 3.3).

If the number of atoms in the product and the number of atoms in the tails streams from a given diffusion cell are equal to each other, then the degree of enrichment in the product is equal to the degree of depletion in the tails. If the product of a diffusion cell is used as feedstock for a second diffusion cell downstream, the tails assay from the second cell will be essentially equal to the assay of the feedstock for the first cell and can be reintroduced to the system at that point. A huge linear array of such cells, called a symmetric countercurrent cascade, can be envisioned. The cascade feed is introduced somewhere in the middle of the cascade. Product is withdrawn at the end of the "enrichment" section, and tails are withdrawn at the end of the "stripping" section, in amounts sufficient to balance the cascade feed. In the middle section of the cascade, product from a given cell is used as feedstock to the next cell in line, while the cell tails are reintroduced into the cascade (Figure 3.3). The same is true in the stripping section of the cascade, which operates at lower enrichments than the feedstock. Tails from each successive cell will be of lower enrichment than the previous cell until withdrawn as cascade tails.

In practice, because the quantity of ^{238}U in the cascade feedstock far outweighs that of ^{235}U, each successive separation step must become smaller in scale as enrichments increase. Rather than change the size of the enrichment cells, a number of cells are designated to operate on UF_6 of the same degree of enrichment. This bundle is called an enrichment stage. The number of cells in a given stage decreases as the enrichment of the feedstock to that stage is more differentiated from the feedstock to the cascade as a whole, which is supplied externally. Thus, the number of atoms of product is not equal to the number of atoms of tails from a given stage, and the reintroduction of product or tails into the cascade may not occur at the next consecutive stage. This configuration is known as a nonsymmetric cascade (Figure 3.4).

The total UF_6 inventory of an "ideal" symmetric cascade as a function of stage number is shown in Figure 3.5. The stripping section of the cascade is shorter than the enrichment section because of the starting isotopic composition of U. The number of cells in each stage, and the balance between product and tails cuts, is optimized in

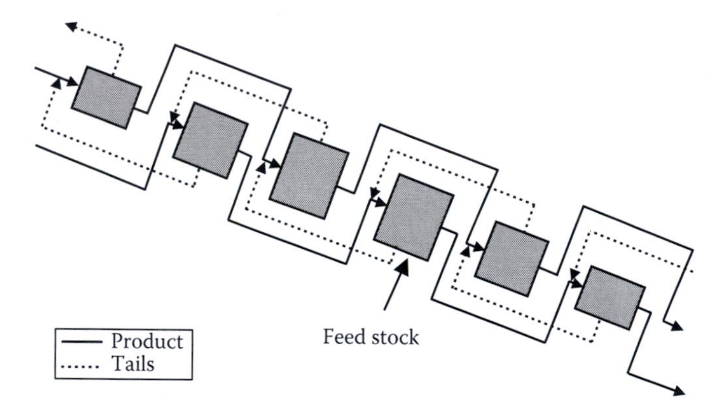

FIGURE 3.4 Diagram of an asymmetric countercurrent cascade, with product used as feed two stages upstream.

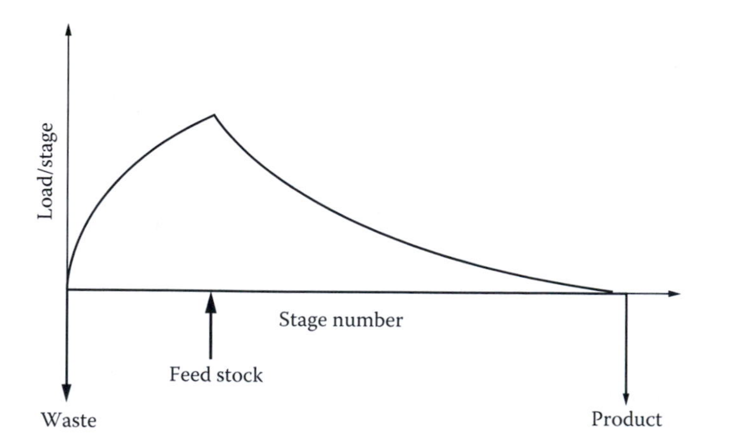

FIGURE 3.5 Load profile as a function of stage number for a symmetric countercurrent cascade.

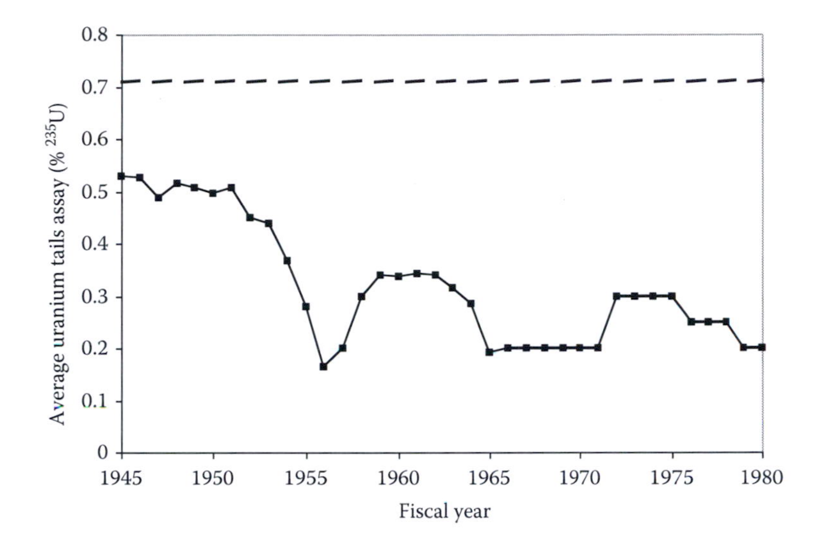

FIGURE 3.6 Average tails assays of the US gaseous-diffusion plants as a function of fiscal year, 1945–1980. Dashed line is the ^{235}U abundance of naturally occurring U.

a complicated fashion. Part of this optimization depends on the isotopic composition of the final tails material and is usually based on economics: when U is scarce, the quantity of ^{235}U in the withdrawn tails fraction is small; however, if U becomes sufficiently plentiful that electric power becomes the economic driver, the ^{235}U content of the tails increases. In practice, cells can be added or removed from the various stages by rerouting the flow of material to produce changes in the tails assay. Figure 3.6 shows the average tails assay of the US gaseous diffusion plants as a function of fiscal year [18]. The amounts of tails and product streams that can be withdrawn at the ends of the cascade are small compared to the total quantity of UF_6 circulating in the cascade as a whole, required to keep the cascade operating near optimum efficiency.

Though not an important concept for the nuclear forensic analyst, the capacity of enrichment plants is often quoted in separative work units (SWUs). SWU is a function of the amount of U processed and the degree to which it was enriched on a kilogram basis. It is a measure of the energy expended for the fractionation of the isotopes. For example, to produce 1 kg of enriched U with an assay of 3% ^{235}U requires 3.8 SWU if the process is performed with a tails assay of 0.25%, but it requires 5.0 SWU if the tails assay is reduced to 0.15%.

The discussion to this point has considered an "ideal" cascade, which equilibrates in the shortest possible time and optimizes the amount of product compared to the quantity of material circulating in the cascade. Unfortunately, the engineering of an ideal cascade is difficult because of the requirement that each stage carry a different flow than its neighbors, as well as consume a different amount of power. It is impractical from a modern engineering perspective, which necessitates the mass production of identical units. In practice, the cascade is "squared off" such that each stage has identical flow and power requirements. The result is that the relative amounts of product and tails from each cell are the same, but that streams of slightly different isotopic composition are mixed throughout the cascade. An important signature is thereby generated for forensic analysis: as the ^{235}U composition of the incompatible streams is optimized to give the best performance of the overall cascade, the result is a nonoptimum distribution of the compositions of the other isotopes of U. For example, in an ideal cascade, ^{234}U will actually be enriched (relative to its starting composition) to a much greater extent than will ^{235}U, since its relative stage separation factor is $q_{234} = (352/348)^{1/2} = 1.00573$, compared to $q_{235} = 1.00429$ for ^{235}U. In reality, the relative enrichment of the two isotopes in US material is approximately unity, only slightly favoring ^{234}U, and is characteristic of the cascade in which it was produced (see Chapter 21).

Initially, an enrichment cascade is filled with natural-composition UF_6 in all of its cells and stages, at which point it is operated without withdrawing product or tails or introducing new feed. This start-up condition is referred to as "total reflux." The enrichment of the highest stage gradually increases to the desired level, at which point it is possible to begin extracting the product. The average time that a given atom of U spends in a full-size cascade is very long—on the order of many years.

In the United States, U enrichment by gaseous diffusion was developed at three locations—Oak Ridge, Tennessee; Paducah, Kentucky; and Portsmouth, Ohio. These plants were completed between 1945 and 1956. The first plant, the 3000-stage K-25 facility at Oak Ridge, was designed to operate alone and to produce enrichments of ^{235}U in excess of 90%. Most of the enrichment capability at Oak Ridge has since been shut down. The Paducah and Portsmouth plants were designed to operate together. Paducah produces U with ^{235}U enrichments between 1% and 5%, much of it for civilian power production. Product from Paducah enriched to 1.1% ^{235}U (sometimes referred to as "slightly enriched uranium") was used as feedstock at Portsmouth, where enrichments of over 90% were obtained. Most gaseous-diffusion operations at the Portsmouth plant were shut down in 2001. At several points in the early history of the US plants, a shortage of U made it necessary to use material that had been recovered from reactor applications as feedstock. The result was cascades contaminated with the nonnatural isotopes ^{236}U, ^{233}U, and ^{232}U—a signature that persists to the present day because of the long cascade residence times for a given atom. In addition, incomplete fuel reprocessing resulted in significant quantities of Np, Pu, and Tc (all of which have volatile fluorides) being introduced into the cascade as contaminants in the feedstock. Being more reactive than UF_6, most of the NpF_6 and PuF_6 contamination reacted with the Ni diffusion barrier almost immediately, but traces of ^{237}Np are still found in the product of the US enrichment plants. It is not uniquely distinctive, however, as plants in Great Britain and the former Soviet Union had similar episodes of reactor-uranium use. The concentrations of the signatures of reactor utilization should be unique for each cascade, albeit a function of time since the contamination.

3.3.3 Thermal Diffusion

Another method that has been used to enrich U on the industrial scale is thermal diffusion, and during World War II, a large thermal-diffusion plant was constructed at Oak Ridge. The output of this plant, in which the ^{235}U concentration was increased from 0.72% (natural) to 0.86%, was used as feedstock to the calutrons that produced the HEU for the Hiroshima bomb [19,20].

Thermal diffusion is similar to gaseous diffusion in that it relies on the difference in velocities of species of different molecular weights at a given temperature. Under a temperature gradient, the heavy component of a mixture of two isotopes will concentrate at the lower temperature. The thermal-diffusion apparatus consists of two concentric, vertical tubes, where the inner tube (made of Ni) carries high-temperature steam and the outer tube (made of Cu) is jacketed and cooled by circulated water. The annular space between the two tubes, maintained by Ni spacers, is very narrow, <1 mm in thickness. UF_6 is introduced into the annulus between the two tubes under a pressure of 200 atm, which is required to maintain the UF_6 in the liquid phase near the hot inner wall. Lighter molecules move toward the inner wall and rise to the top of the tube, while heavier molecules move outward and downward.

The diffusion tubes employed at Oak Ridge were ~15 m long. Thermal expansion between the 286°C inner tube and the 64°C outer tube would have induced a difference in length of 3 cm if unconstrained. Each column was attached to a reservoir from which UF_6 was delivered, and the delivery pressure was controlled by manipulating the temperature of the reservoir. It was necessary to maintain column pressure at the hot wall greater than the vapor pressure of UF_6, or the liquid would have boiled and the resulting turbulence would have overwhelmed the much slower, weaker diffusive flow.

The spacing between the two tubes is the most crucial variable associated with the performance of a thermal-diffusion column. Both the equilibrium separation factor and the time constant associated with the diffusive flow are strongly dependent on this spacing. Both were optimized for spacing between 0.2 and 0.3 mm at the temperatures at which the Oak Ridge plant ran. Had the difference in temperature between the inner and outer tubes been greater, optimal spacing would have been smaller. In general, the most favorable performance of a thermal-diffusion column varies as a function of the operating parameters in a complicated fashion, and these parameters were determined experimentally during the war. The time constant associated with the performance of the Oak Ridge thermal-diffusion tubes to produce 0.86% $^{235}UF_6$ was on the order of 60 days, with a steady-state yield of ~60 g/day/column. As expected, because of the tolerances necessary for fabrication of the cells, and the thermal gradient under which they operated, considerable variation in performance was exhibited from column to column.

In principle, any desired degree of isotope enrichment can be attained by adjusting the length of the thermal-diffusion column. Of course, this becomes impractical with the physical dimensions of the tube, and the time constant associated with the enrichment would become very long. The thermal-diffusion tubes could be connected in cascades similar to those described for gaseous diffusion, in which the product from one stage is introduced to the reservoir associated with the next stage, with the tails fed downward. There are several technical difficulties associated with this tactic, but Abelson demonstrated a four-stage cascade of 47 columns, which he termed a "pyramid" [19].

Because of the long time constant associated with the withdrawal of product, and the high energy consumption of the process, the gaseous-diffusion process replaced thermal diffusion for the production of enriched U. The 2100-column thermal-diffusion plant at Oak Ridge was closed in 1946.

3.3.4 Gas Centrifugation

Another way to enrich U is via the gas centrifuge. UF_6 gas is subjected to an intense centrifugal force, which causes the heavier $^{238}UF_6$ species to move preferentially to the periphery of the rotor, producing a partial enrichment of $^{235}UF_6$ at the center of the apparatus. The centrifuge is considered the second most viable enrichment technology. It is under extensive development and in commercial use in Europe and Japan [11], and has replaced gaseous diffusion as the most used uranium enrichment technology.

Consider a quantity of UF_6 gas, confined in a spinning, heated cylinder of radius r, rotating about its axis with an angular velocity ω. The theory of equipartition of energy predicts that the density of particles (and consequently the pressure) at the center of the centrifuge ($r = 0$) will be less than that at a distance r:

$$N(r)/N(0) = \exp\left[m\omega^2 r^2/(2RT)\right],$$

where

m is the molecular mass
T is the absolute temperature
R is the ideal gas constant

When a mixture of two isotopes, incorporated into molecules with masses m_1 and m_2, are involved, their respective partial pressures are given by

$$p_1(r)/p_1(0) = \exp\left[\left(m_1\omega^2 r^2\right)/(2RT)\right]$$

and

$$p_2(r)/p_2(0) = \exp\left[(m_2\omega^2 r^2)/(2RT)\right].$$

The resulting distribution of light and heavy molecules in a centrifuge is shown in Figure 3.7. Because the total pressure p is the sum of the partial pressures, and the

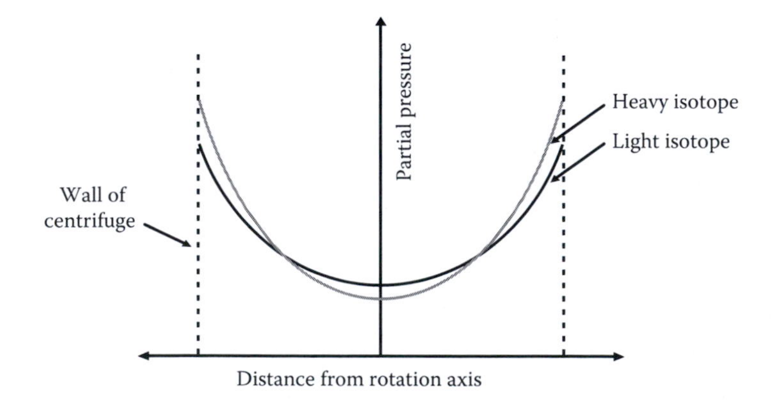

FIGURE 3.7 Distribution of light and heavy molecules in an operating gas centrifuge as a function of distance from the center.

partial pressures are proportional to their respective mass fractions in the gas, the single-stage separation factor q is given by

$$q = \exp\left[(m_2 - m_1)\omega^2 r^2/\left(2RT\right)\right].$$

The separation factor increases with the mass difference between isotopes and with the centrifugal force supplied by the motion of the centrifuge; q also decreases with an increase in temperature. The separation factor thus increases with mass difference, rather than mass ratio. Unlike diffusion, centrifugation is as effective for heavy molecules as for light ones.

Consider a hypothetical centrifuge with a radius of 10 cm, an angular frequency of 1000 rps, and an operating temperature of 35°C. Using molar masses of 349 and 352 g for $^{235}UF_6$ and $^{238}UF_6$, respectively, the value of q at the outer wall of the centrifuge is $\exp[3 \times 1000^2 \times (2\pi)^2 \times 10^2/(2 \times 8.31 \times 10^7 \times 308)] = 1.26$. This result is substantially better than the value of $q_{\text{ideal}} = 1.0043$ obtained for a gaseous-diffusion cell. In fact, although a stage in a centrifuge plant consists of vastly more centrifuges than a stage in a diffusion plant that contains converters, the centrifuge plant may require only 20 stages to reach an enrichment that would need 1000 stages of gaseous-diffusion cascade.

The separation factor for the described centrifuge is so much better than that for gaseous diffusion, it might appear surprising that the technique has not monopolized the entire industry. Of course, part of the economic decision is based on the substantial capital investment in operational diffusion plants, but technical challenges associated with operating high-speed centrifuges are considerable as well. For instance, the described centrifuge supplies an acceleration of $\omega^2 r = 4 \times 10^6$ m/s^2 (4×10^5 times the acceleration of gravity) to the wall of the rotor, so that issues of material strength are clearly very important. Similarly, even though the enrichment gain from a single centrifuge unit is more than an order of magnitude greater than that for a gaseous diffusion cell, it is still necessary to operate a centrifuge-based plant in cascade, requiring the removal of product and tails and the introduction of feedstock with the centrifuge running at speed. The advantages of centrifugation enrichment were clearly recognized by the time of World War II [21,22], but the technical difficulty of the problem required many years to address and is only now resulting in centrifuges competitive with gaseous diffusion.

The calculated ideal separation factor is actually the ratio of isotopic abundances at the center of the centrifuge to that near the wall. To accomplish this theoretical separation, the product would have to be extracted from the axis of the centrifuge and the tails from the perimeter. However, as indicated by the accelerations calculated earlier, the gas pressure at the centrifuge wall is millions of times greater than at the center. It would be futile to try to extract a product from the reasonably good vacuum that exists at the middle of the rotor.

This problem was overcome by apparatus designs similar to that of the diagram in Figure 3.8 [23]. The centrifuge spins in vacuum, powered by an electromagnetic drive. The central post does not move and makes a seal with the body of the centrifuge rotor at the top. As the centrifuge spins, the bulk of the UF_6 gas accumulates

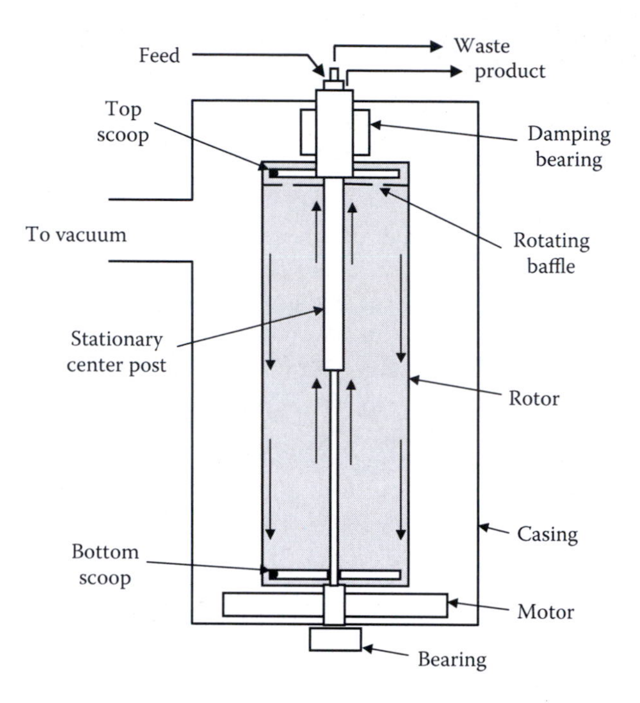

FIGURE 3.8 Diagram of a modern gas centrifuge. As the rotor spins, gas is fed into and withdrawn from the centrifuge through the stationary center post, which consists of three concentric tubes for the manipulation of the gas. The stationary scoop at the bottom provides the means by which the vertical circulation of the gas is accomplished. The top scoop is screened from the main body of the gas by the spinning baffle; otherwise, there would be no net circulation. Product is withdrawn by the top scoop, and tails are withdrawn by the bottom scoop.

along the cylindrical wall. The bottom scoop removes the tails cut, while the feedstock is supplied halfway up the body of the centrifuge near the axis. This introduction of feed and removal of tails establish a vertical countercurrent flow, which facilitates exchange between the layers of gas with differing isotopic compositions. Because most of the enrichment process takes place in the thin annulus of gas near the centrifuge wall, the gas motion through which product is formed and transported to the center would be very inefficient without countercurrent flow. Low-pressure product diffuses into the baffled region that isolates the top scoop, where it concentrates along the wall before being extracted.

The actual separative power of a centrifuge unit is sensitive to the countercurrent flow pattern. The equations governing the motion of the gas in an operating centrifuge are very complicated [11], such that much of the development of the gas-centrifuge enrichment method was accomplished by trial and error.

The vacuum vessel in which the centrifuge spins is necessary to minimize drag and eliminate environmental variables that can cause instability of the centrifuge motion. Even the low-speed centrifuges common in chemistry laboratories must be balanced to prevent a wobble that can destroy their contents. This is even more critical for high-speed centrifuges. Induced vibrations caused by resonant frequencies of the apparatus can also be a problem, and the high rotational speed of the centrifuge results in severe mechanical stresses on the outer wall of the rotor. For instance, if the 10-cm diameter centrifuge described earlier were constructed of Al, the tensile stress at 1,000 rps would be 11,000 kg/cm^2, or twice the tensile strength of the material. The centrifuge would undergo a catastrophic failure long before achieving its operating speed. Modern centrifuge rotors are constructed of high-strength materials, such as maraging steel or strong, lightweight composite materials like carbon fibers in a resin matrix.

Now that the technical barriers necessary for centrifuge operation have been overcome, the remaining limitations of the technique are the maximum load of UF_6 possible in a given unit and the low throughput caused by the slow diffusion rate of the gas. The mass load of a given centrifuge is limited by the requirement that the UF_6 remain in the gas phase under the operating conditions, or no diffusion, and consequently no enrichment, would be possible. The vapor pressure of UF_6 at room temperature is approximately 0.15 atm. Because the system must remain below this pressure at the wall of the rotor during operation, the resultant pressure at the center of the rotor makes up a good vacuum. The gas load of a stationary centrifuge with a radius of 10 cm must remain $<3.6 \times 10^{-4}$ atm, so that the pressure at the rotor wall during operation does not induce a phase change.

A cascade of gas centrifuges requires considerably less power per unit of enrichment work than does a gaseous diffusion cascade, but it operates in a very similar fashion. A typical 1000-MW_e light-water reactor (LWR) requires enriched fuel, on an annual basis, for which about 120,000 SWUs have been expended. A typical gaseous-diffusion plant consumes about 9000 MJ per SWU, whereas a modern gas-centrifuge plant consumes only 180 MJ/SWU. As a result, U enrichment by gas centrifugation is economical on a smaller scale than is gaseous diffusion. A squared-off cascade should result in the same $^{234}U/^{235}U$ isotopic signature from centrifuges as from gaseous diffusion. However, centrifuge plants have only recently competed with diffusion plants, so they are not expected to be contaminated with nonnatural uranium isotopes, which was the consequence of perceived shortages of U in the 1950s and 1960s. A sample of enriched U in which the relative enrichments of ^{234}U and ^{235}U are about the same, but in which ^{236}U is absent, is quite likely the product of a gas-centrifuge plant.

In the late 1970s, in response to a forecast that demand might exceed the production capacity of enriched U from the gaseous-diffusion cascades, construction of a gas-centrifuge facility was begun at the site of the Portsmouth plant [24]. The forecast proved to be inaccurate, and construction was halted in 1985 after an expenditure of about $3 billion. It was announced in 2002 that a consortium of nuclear utilities and related companies, including Westinghouse and Comeca, would begin construction of a gas-centrifuge plant in the United States, to be operational in 2008. This decision was in response to concerns that the US gaseous diffusion plants are reaching the end of their design lives. Centrifuge plants are being, or have been, constructed at the Portsmouth site, in Eunice, NM, and at Idaho Falls. China, France, Brazil, and Pakistan (among others) operate centrifuge plants for uranium enrichment.

3.3.5 Aerodynamic Enrichment

Aerodynamic enrichment techniques have been demonstrated at the industrial scale [25–27]. The separation effect is produced by centrifugal force applied to a stream of gas deflected by a solid barrier. Discussion here will focus on the jet nozzle (Becker) process. However, the principles employed by the vortex tube developed by South Africa (the Helikon process) are similar. Both processes use as feed a UF_6 gas diluted in a much larger quantity of a light auxiliary gas, usually He or H_2. Both techniques provide a high stage throughput with a reasonably high single-stage separation factor, and they offer the advantage that the technology is less complex than that of high-speed centrifuges.

Figure 3.9 shows a cross section through an early design separation nozzle cell. A mixture of 4% UF_6 diluted in 96% H_2 is introduced into the cell and flows along the curved wall at high velocity. The H_2 is present to increase the flow velocity of the mixture by causing a reduction in mean molecular weight, which increases the sonic velocity of the gas and minimizes turbulence in the cell. The presence of the carrier gas increases q for the cell appreciably. As in the centrifuge, the more massive molecules of UF_6 (containing ^{238}U) concentrate along the wall of the cell to a greater extent than do the lighter molecules (containing ^{235}U). At the end of the deflection, the gas jet is split into two fractions by a knife-edge, as indicated. The knife-edge is usually positioned such that a quarter of the gas is deflected into the light fraction, with the remainder going into the heavy fraction.

The best separation factors are obtained when the gas velocity is maximized and the radius of curvature is minimized, both of which actions result in an increase in the centrifugal acceleration of the molecules forced through the cell. The radius of curvature of the separation stage, designed along the lines of Figure 3.9, is approximately 100 μm. The small cell size helps in the application of a high gas pressure in the feed stream, optimally about 0.25 atm for this cell. The resulting velocity of

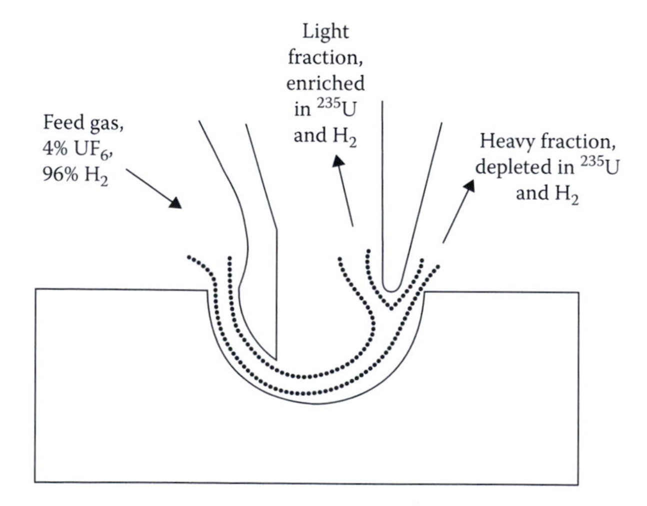

FIGURE 3.9 A separation nozzle cell. The knife-edge skimmer at the right is placed so that 25% of the UF_6 in the feed goes into the light fraction and 75% into the heavy fraction.

the gas molecules is 300 m/s. With a radius of curvature of 0.01 cm, the centrifugal acceleration at the cell wall is 9×10^8 m/s^2, which exceeds values attained in even the most powerful gas centrifuges by more than 100. q_{ideal} for the described cell is about 1.015.

The described Becker process is two dimensional, and the Helikon process extends into three dimensions. The actual separation cells are made by etching an optimized Figure 3.9 into several thin sheets of Ni, then stacking them in tandem to make a cell that extends above and below the flat plane of the page. The higher the stack, the more throughput at a given pressure. In reality, a single assembled stack of foils contains several etched separation cells.

Of course, given the q_{ideal}, enrichment plants employing the aerodynamic nozzle method must be operated in cascade. The auxiliary gas complicates the calculation of optimum cascade configuration, however. The concentration of H_2 must remain relatively constant throughout the cascade, but it is more severely depleted in the tails of each enrichment cell than is $^{235}UF_6$. Some depletion is compensated by the fundamental action of the countercurrent cascade, in which H_2-depleted tails from one stage are mixed with H_2-enriched product from another stage to form the feedstock of an intermediate stage. Nonetheless, care must be taken to balance the H_2 content of each stage feed (as well as the U isotopic content). This requires the extraction, purification, temperature adjustment, and reintroduction of H_2 at several places throughout the cascade, which increases the cost per SWU. The separation nozzle was demonstrated on an industrial scale in Brazil, but neither aerodynamic process is currently in use.

This concludes the discussion of enrichment methods that rely on differences in the masses of molecules that incorporate atoms with different neutron numbers, resulting in unequal inertias when the translational motions of the molecules are manipulated. Other methods involve subtle changes in the electronic properties of atoms and molecules that arise from the interaction of the electrons with the nucleus.

3.3.6 Laser Isotope Separation

Similar to nuclei, atoms and molecules exist in discrete states, each having a well-defined energy established by the rules of quantum mechanics. Even when the density of states is fairly high, transitions between states proceed by emission or absorption of photons of discrete energy,

$$\Delta E = h\nu,$$

where

h is Planck's constant

ν is the frequency of the photon

In an atom or ion, the energy of a given state is predominantly defined by the interactions among the electrons and between the electrons and the nucleus as a whole. However, isotopic effects appear in atomic spectra and are caused by differences in masses, shapes, and nuclear spins of different isotopes, promoted by the change in neutron number. Most of these effects are observed in the visible portion of optical atomic spectra. The situation for molecules and molecular ions is similar. Differences in isotope masses manifest themselves in the frequencies of rotations and vibrations of a molecule, observed principally in the infrared.

The differences in transition energies caused by a change in isotope are generally quite small compared to the magnitude of the transition itself, perhaps only one part in 10^5 for U atoms and compounds. Even though the concept of selectively exciting transitions in an atom or molecule of a particular isotope without exciting similar transitions in other isotopes was developed early in the history of U enrichment [28], it was not until the advent of the tunable laser that it became technically feasible [29–31].

A laser can produce large numbers of photons, all having nearly the same frequency. If these photons impinge on a collection of atoms or molecules, resonance absorption can occur in states that are matched in transition energy to a different state, ΔE, such that the energy of the photon is equal to the transition energy. The photon can then be absorbed, leaving the atom or molecule in an excited state. This process occurs with high probability only when the transition and photon energies are closely matched, relative to the energy widths associated with each. Particularly for low-energy molecular transitions, the energy associated with a given transition is broadened by collisions in a high-pressure gas and by the Doppler effect caused by motions of the molecules toward or away from the photon source. Still, with careful preparation, it is possible to selectively excite atoms or molecules of one of the isotopes of U in the presence of the others [11,17,32,33]. Given the magnitude of the isotope effect in atomic and molecular spectra, selective excitation of one isotopic species in the presence of one or more other species requires a laser that is tunable to an accuracy of one part in 10^5. Such control can be accomplished with modern dye lasers.

Laser isotope separation (LIS) techniques fall into three classes: photochemical, in which an excited atom or molecule has a higher chemical reactivity than does a similar species in its ground state and will react preferentially with a suitable partner; atomic photoionization, in which an excited atom will be more easily ionized by a second photon of a suitable energy; and photodissociation, in which an excited molecule can be more easily disassociated than when in the ground state. The two main methods for enriching U are atomic-vapor laser isotope separation (AVLIS), which is an atomic photoionization process, and molecular laser isotope separation (MLIS), based on molecular photodissociation. Both employ a multiphoton approach to selectively excite ^{235}U in the presence of ^{238}U.

In AVLIS [34], an atomic vapor of U atoms must be produced from the metal. This is a daunting technical problem because of the low volatility and high chemical reactivity of U and the high temperatures necessary to produce significant vapor pressure. Liquid U is so reactive that it is difficult to find a material to serve as a container. In practice, U vapor is produced with an electron beam, creating a molten area of limited extent in a U ingot, which then acts as its own container. The atomic vapor is then irradiated by several lasers. The energy required to remove an electron from a U atom (the ionization potential) is 6.2 eV. This energy is supplied in steps, each of which is tunable, to select ^{235}U preferentially in the presence of ^{238}U.

Ionized ^{235}U atoms are deflected electromagnetically and collected on plates, whereas the nonionic tails condense elsewhere.

In principle, MLIS [35] is similar to AVLIS. $^{235}UF_6$ molecules absorb energy from infrared lasers, and $^{238}UF_6$ molecules are unaffected. Sufficient energy is introduced into the $^{235}UF_6$ molecule (in steps) that it dissociates as follows:

$$^{235}UF_6 \rightleftarrows {}^{235}UF_5 + F.$$

UF_5 condenses into a fine powder called "laser snow," which is filtered from the UF_6 tails material. Although MLIS does not suffer the same throughput limitation as AVLIS (the difficulty of producing a concentrated atomic vapor), MLIS performance is severely degraded by the effect of molecular motion. At ordinary temperatures, collisions between UF_6 molecules result in a wide range of vibrational excitations, and the ensuing Doppler shifts of the frequencies associated with random molecular motion negate the selectivity supplied by the laser. This problem is overcome by diluting the UF_6 with a low-mass carrier gas (e.g., Ar) and letting it expand through a nozzle into a vacuum. The result is a beam of "cold" UF_6 molecules, all traveling in the same direction, which provides a suitable target for the MLIS lasers.

The isotopic signature of U enriched by LIS is more like that derived from EMIS (Section 3.3.1) than that from any of the cascade techniques. Both ^{234}U and ^{238}U should be depleted relative to ^{235}U in the product. The value of $^{234}U/^{238}U$ will depend on the frequency width of the lasers, as well as on line broadening in the target atoms or molecules determined by the experimental configuration.

Uranium isotope enrichment via laser irradiation works in principle, but to compete economically with the other enrichment methods, it must be more efficient. For example, the single-stage separation efficiency of the calutrons was 10^4 times higher than that of gaseous diffusion, but it could not compete because of the throughput limitations. The calculation of a single-stage separation factor for a LIS process is dependent on several variables, the most important of which are delivered laser power (which controls the throughput) and frequency quality (which controls the selectivity of the stage). In practice, single-stage separation factors have been so high that reactor fuel compositions have been obtained in a single-pass AVLIS separation [33]. To produce 3.2% ^{235}U reactor fuel, it is necessary to effectively remove 78% of the ^{238}U in a given sample of natural U and retain all of the ^{235}U. Economically, LIS is comparable to, or slightly better than, centrifuge techniques for the enrichment of U, but it suffers from the stigma of being too highly technical. At present, the only laser separation process still under development is separation of isotopes by laser excitation (SILEX), a MLIS technique at the prototype stage in Australia.

3.3.7 Isotope Enrichment through Chemical Exchange

Chemical-exchange methods have been explored for their utility in enriching U. These methods depend on the tendencies of different isotopes to concentrate in selected molecules when the atoms of that element can exchange between molecules. For the hypothetical uranium compounds UX and UZ, each will contain the two isotopic species ^{235}U and ^{238}U. A mixture of the molecules contains the species ^{235}UX, ^{238}UX, ^{235}UZ, and ^{238}UZ. If the U occupies a chemically active site in each molecule, then isotopic exchange can occur:

$$^{235}UX + {}^{238}UZ \rightleftarrows {}^{235}UZ + {}^{238}UX.$$

An equilibrium is established in which the relationship between the concentrations of the four species is defined by an equilibrium constant K:

$$K = ([^{235}UZ] \times [^{238}UX]) / ([^{235}UX] \times [^{238}UZ]).$$

Statistical mechanics would require that $K = 1$ for any such reaction, but quantum mechanics allows different isotopic affinities among the compounds, based on energy levels of vibrational states, tending to make lighter isotopes concentrate in less tightly bound molecules [17]. The French Chemex process, which exploited a very slight difference in the III/IV oxidation/reduction potential between the two isotopes, has been demonstrated on the pilot-plant scale.

Chemical-exchange enrichment cells involve the repeated extraction and back-extraction of U compounds in solution, transferring repetitively between chemical phases, or adsorbing and desorbing from a solid support. To date, chemical-exchange processes have not been competitive with other enrichment technologies. The isotopic signature of the process is expected to be similar to that of other processes involving cascades. However, the chemical process will not necessarily provide as efficient separation from decay products as do those processes involving feed of gas-phase UF_6.

An issue that all enrichment methods have in common is criticality safety during the latter stages of procedures generating HEU. The danger of an accidental criticality can be eliminated if the mass of HEU at a particular location in the process is maintained below the critical amount. Critical mass is also dependent on chemical form, density, and geometric configuration, as well as on the presence of neutron reflectors or moderators. For the nuclear forensic analyst, this limitation may produce a greater variability in isotopic signatures between samples of HEU than between samples of LEU. The larger batch size of LEU makes it more likely that two samples may come from the same process batch than is necessarily true for HEU.

3.3.8 Blending and Mixing

A question that arises frequently in the forensic analysis of nuclear materials is the effect on interpretation of mixing disparate samples of the same element having differing isotopic compositions. This interpretation is accomplished through principal component analysis, developed to explain mixing in geologic systems [36,37].

Consider two solutions, each of which contains different concentrations of the two analytes, A and B. If variable amounts of the two solutions are mixed with each other, a plot of the concentration of A *vs* that of B in the resulting solution will lie on a line connecting the A *vs* B data points associated with the two original solutions, also called the "end members." A series of mixtures of different proportions will map out

(A *vs* B)-space between the end members. Concentrations of the results of incomplete homogenization will also lie on the mixing line. If the only information available is the A *vs* B concentrations of the mixtures, the end-member compositions are not defined, but are constrained to lie on the mixing line outside of the concentration extremes of the mixtures, subject to the statistical limitations of the linear fit to the experimental data.

Two-component analysis, also called mixing analysis, is invalidated by nonconservative processes such as evaporation or dilution. If a third analyte C is present, three-component analysis can be made resistant to nonconservative phenomena through the correlation of concentration ratios, [A]/[C] *vs* [B]/[C]. A series of three-component concentration ratios then define a mixing line whose extrapolation encompasses the end-member concentration ratios. Three-component mixing analysis is a valuable tool when applied to isotopic ratios measured in chemically fractionated debris produced in nuclear explosions (Chapter 5) and has application to the nuclear forensic analysis of uranium.

Uranium-bearing materials are mixed on both a small scale and at industrial levels. An example of the latter is discussed in Chapter 21 in the context of isotopic enrichment. Also relevant to this discussion is the result of a recent long-term purchase agreement between the United States and Russia in which 500 t of Russian HEU has been mixed with unenriched U to produce LEU reactor fuel having a nominal ^{235}U isotopic concentration of approximately 4% [38]. The two U components are mixed in the gas phase as hexafluorides. The limitation on the batch size of HEU imposed by criticality constraints provides an avenue by which the admixture of HEU in the total product stream could be somewhat variable, and plotting the concentration ratios of [^{234}U]/[^{235}U] *vs* [^{238}U]/[^{235}U] from a set of samples of the final product would result in a mixing line that could be used to constrain the isotopic compositions of the ingoing materials.

The mixing of U samples often involves nearly simultaneous purification, either in solution or in the gas phase. We conclude that the impact of mixing on chronometry (Chapter 6) is unlikely to be important, except in the case of contamination of nuclear explosion debris with soil (Chapter 5). The daughters of the U isotopes in a soil sample are in radiochemical near-equilibrium with their parents, such that the impact of the admixture of soil containing only a few parts-per-million of aged U has an enormous effect on the chronometry of U isotopes in explosion residue.

3.4 NUCLEAR REACTORS, POWER, AND THE PRODUCTION OF Pu AND ^{233}U

A nuclear reactor is a device in which the fission process is controlled to produce either power or radionuclides (or both). On the average, each fission event releases approximately 200 MeV of energy, a consequence of the change in nuclear packing fraction. Fission is induced by absorption of a neutron in a fissile nucleus and results in the formation of two major fission products, some neutrons, and γ-rays. More than 80% of the energy released by an average fission event is manifest in the kinetic energies of the fission products. Because the range of fission fragments in matter is quite small, these nuclides stop in the immediate vicinity of their formation, with the conversion of their initial kinetic energies into thermal energy. The neutrons and γ-rays emitted promptly in fission, and the electrons and γ-rays emitted in the decay of the primary fission fragments, have longer ranges than the fission fragments, but they still undergo material interactions that result in the production of heat. Heat accompanying the fission process must be continuously removed to avoid melting the reactor medium, and this removal of heat can be used for the production of power, as in a conventional power plant.

Fission-produced neutrons may be absorbed by any of the materials present in the reactor, with the relative probabilities proportional to their concentrations and the neutron absorption cross sections of the nuclei in the reactor materials. Absorption of a neutron by a fissile nucleus results in a high probability of producing another fission event, leading to the emission of more energy and more neutrons. If, on average, one neutron produced by each fission causes one more fission, the number of neutrons in each succeeding generation will remain constant, and the neutron economy is balanced. This condition is called criticality and is the result of the chain reaction of fissions being sustained by other fissions through the agency of neutrons.

All nuclear reactors depend upon an initial load of fuel that contains fissile materials. As discussed before, the three common fissile species are ^{233}U, ^{235}U, and ^{239}Pu; absorption of a neutron of any energy produces ^{234}U*, ^{236}U*, and ^{240}Pu*, respectively, in excited states that have sufficient potential energy to overcome the fission barrier and result in fission. However, this need not necessarily happen; (n,γ) reactions compete with (n,f) reactions, even for states well above the fission barrier. Low-energy neutron irradiation of ^{235}U results in the production of ^{236}U about half as often as it induces fission. Materials containing the odd-mass fissile actinides always contain some even-mass actinide nuclides as well, which are termed "fissionable," as the neutrons present in the reactor can also cause fission in these nuclides, albeit with lower probability.

Power reactor fuel usually consists of U enriched in ^{235}U to a level of a few percent. Material that is enriched to 3% in fissile ^{235}U contains 97% fissionable ^{238}U, which does not initially contribute much to the production of fissions in the reactor. The ^{238}U in the fuel preferentially captures a neutron, resulting in the production of ^{239}U by the reaction

$$^{238}\mathrm{U}(\mathrm{n},\gamma)^{239}\mathrm{U}.$$

The nuclide ^{239}U is a short-lived species that decays to ^{239}Np, which in turn decays to ^{239}Pu, a fissile material:

$$^{239}\mathrm{U}\ (\beta^-,\ 23.45\ \mathrm{min}) \rightarrow {}^{239}\mathrm{Np}\ (\beta^-,\ 2.355\ \mathrm{days}) \rightarrow {}^{239}\mathrm{Pu}\ (\alpha,\ 24{,}110\ \mathrm{years}).$$

If the reactor runs sufficiently long that a significant fraction of the initial ^{238}U is transmuted into ^{239}Pu, the calculated neutron economy must be adjusted for fissions induced by neutron reactions with ^{239}Pu. ^{238}U is known as a "fertile" material under these circumstances, as it produces fissile fuel by neutron irradiation. The other common fertile nuclide is ^{232}Th, which produces ^{233}U in much the same way:

$$^{232}\mathrm{Th}\ (\mathrm{n},\gamma)\ ^{233}\mathrm{Th};$$

$$^{233}\mathrm{Th}\ (\beta^-,\ 22.3\ \mathrm{min}) \rightarrow {}^{233}\mathrm{Pa}\ (\beta^-,\ 26.97\ \mathrm{days}) \rightarrow {}^{233}\mathrm{U}\ (\alpha,\ 1.592 \times 10^5\ \mathrm{years}).$$

This is the origin of a second fuel cycle involving Th and ^{233}U and will be discussed later. Most of what follows immediately below is a treatment of Pu and the naturally occurring isotopes of U.

In nuclear reactor neutron economics, there are only two productive ways to lose a neutron: a fission event or a conversion event (in which a fertile nuclide becomes a fissile nuclide). An average fission produces two to three neutrons, only one of which must induce fission in another fissile nucleus (on average) to sustain a chain reaction. This balance is more difficult than it appears. One-third of the neutrons are unavoidably lost to (n,γ) reactions in the fissile component of the fuel. The accompanying fissionable materials (such as ^{238}U) compete for neutrons as well. The $^{235}U(n,f)$ cross section is two orders of magnitude higher than the $^{238}U(n,\gamma)$ cross section for thermal neutrons. However, there is 30 times more ^{238}U than ^{235}U in 3% fuel, so reactions with ^{238}U significantly deplete the number of neutrons available for subsequent fission. This competition can be reduced through increased U enrichment. In general, U is not the only component of reactor fuel, as it is usually present as a sintered oxide and encapsulated in metallic containers. Oxygen and the container materials also compete for neutrons (although the stable oxygen isotopes have very low neutron reaction cross sections). This mode of neutron competition is minimized by using collateral structural materials with small neutron reaction cross sections.

Because the fuel region in a nuclear reactor has a boundary, there are also losses caused by neutrons leaking from the reactor fuel elements. Surrounding the fuel elements with a reflector helps reintroduce escaping neutrons back into the fuel, improving the neutron economy. The reflector can be composed of almost any material with a low cross section for reactions with neutrons, but it is determined by the reactor type.

There is a wide variety of design concepts associated with nuclear reactors. One parameter by which these designs can be grouped into classes is the velocities of the neutrons causing the majority of fissions in the reactor. As discussed earlier, fission neutrons are emitted with a Maxwellian kinetic-energy distribution, with the most probable energy being near 0.7 MeV. If the reactor design requires that most of the fissions in the fuel be caused by neutrons of these kinetic energies, the reflector will consist of a dense element of high atomic number such that elastic collisions between the neutrons and the heavy nuclei do not cause the neutron to lose much energy. This design would be that of a "fast" reactor. Fission cross sections with low-energy neutrons are much higher than those involving fission-spectrum neutrons. However, the cross sections of competing reactions also tend to be higher at lower energies, and the neutron multiplicity in low-energy fission is lower than for fission at higher energies. Nevertheless, many successful nuclear reactor designs are based on the escape of neutrons from the fuel element and on their subsequent thermalization through collisions with a moderator before being reflected back into the fuel. In this case, the reflector and the moderator are likely the same material, the fuel elements are probably small in diameter and well-separated from one another, and the design is that of a "thermal" reactor. Preferred moderator materials are those with a low nuclear mass, so that the neutron-moderating collisions are most efficient. The most common thermal moderators are H_2O, D_2O, and graphite, which combine low atomic numbers with low cross sections for reactions with neutrons. This point is discussed further in the following section on LWRs.

Keeping a reactor operating at a constant power level requires that a balance be maintained between neutron production and absorption. The power level of a nuclear reactor is proportional to the number of neutrons available in the reactor and should, therefore, be kept stable. The inventory of both fissile materials (the source of neutrons) and the fission products (competition for neutrons) changes with time, and the operating temperature of the reactor, which affects the performance of the moderator and the spectrum of the neutrons interacting with the fuel, is dependent on environmental variables. Therefore, it is not possible to design a reactor in which the number of neutrons in each successive generation is exactly constant without some additional measure of engineering control. Extra fissile material over that required for initial steady-state operation is incorporated into the fuel, and adjustable control elements containing materials with high neutron absorption cross sections (neutron poisons) are included to stabilize the extra reactivity of the system and to maintain a balanced neutron economy.

The ratio of the rates with which neutrons are created and destroyed is defined as the multiplication factor, k. Another definition of k is the ratio of the number of fissions (or neutrons) in one generation to the number in the preceding generation. For a reactor operating at steady state, $k = 1$. On start-up, a reactor operates in a supercritical condition, with $k > 1$, until the desired power level is achieved, at which point the control elements are used to establish the balanced criticality condition, $k = 1$.

The reaction rate in a nuclear reactor is controlled using rods containing a material with a high neutron absorption cross section. Increasing or decreasing the length of the rod inserted into the reactor alters the number of neutrons available for fission reactions. This tactic works to control the reactivity of the core on the long term, but short-term control is another issue. In a thermal reactor, the time needed for a fission-spectrum neutron to enter the moderator, slow to thermal energy, and re-enter the fuel to induce another fission is about 0.1 ms. If the reactor neutron inventory increases beyond the steady state by 0.1% in a single generation, in 100 generations (10 ms), the power level will have increased by 10% (1.001^{100}), and in 100 ms, it would reach 2.7 times the steady-state power, a clearly unacceptable situation. For fast reactors, the situation is potentially worse. What saves the technology is the presence of delayed neutrons in the reactor core.

Although most neutrons are emitted promptly with the fission products (within 10^{-16} s of the appearance of the fission fragments), a small fraction of them is emitted in the β decays of the fission products, with time constants >1 s. This delayed emission is so much longer than the 10^{-4} s associated with the emission, moderation (if appropriate), reflection, and reabsorption of the prompt fission neutrons that control of a nuclear reactor can be achieved. A reactor operating at steady state is actually subcritical if only the prompt neutrons are considered. The total inventory, including the delayed neutrons, brings the system to criticality. The rate at which the power level of a reactor fluctuates is determined by the generation time of the delayed neutrons, provided that $k < 1$ at all times for prompt neutrons alone. In the thermal-neutron-induced fission of the various common nuclear fuels, delayed neutrons make up 0.26% of the neutron production in ^{233}U, 0.65% in ^{235}U, and 0.21% in ^{239}Pu [39]. As a result, slow-acting means of control, like adjustment of the level of insertion of a control rod into the core, are sufficient to maintain a constant or slowly changing power level, with reactors fueled by ^{235}U being the easiest to control. If a reactor becomes "prompt-critical" at any point in its operation ($k_{prompt} \geq 1$), the delayed neutrons no longer provide a means of control.

Another issue affecting the control of a reactor is the production of long-lived neutron poisons during its operation. The most important of these is the 9-h fission product ^{135}Xe, which is not only produced directly, but also arises from the β decay of the 7-h ^{135}I fission product. ^{135}Xe has an enormous cross section for capture reactions with thermal neutrons:

$$^{135}\mathrm{Xe}(\mathrm{n_{th}},\gamma)\,^{136}\mathrm{Xe}, \quad \sigma = 3\times 106\ \mathrm{b}.$$

A reactor operating at $k = 1$ will have compensated for the presence of the ^{135}Xe poison through extra reactivity (extra fuel) uncompensated by the control rods. At steady state, ^{135}Xe competes successfully for a significant fraction of the neutrons in the reactor, absorbing several percent of those that would otherwise react with ^{235}U. If the reactor power level is decreased suddenly, ^{135}Xe will continue to be supplied by β decay of ^{135}I, whose steady-state concentration was established at the old power level, and the ^{135}Xe concentration will increase even more because of reduced destruction occurring at the lower power level. If an attempt is then made to bring the reactor back to its original power level, the elevated concentration of ^{135}Xe will have to be negated through a reduced insertion of the control rods (assuming that sufficient reactivity remains in the fuel). In this case, care must be taken to assure that the reactor does not go prompt-critical. Reestablishing the former steady-state power level may require several hours.

Enough information has now been presented to attempt an initial design of a nuclear power reactor, which would probably look somewhat like the diagram shown in Figure 3.10. This particular reactor is similar to the type referred to as a

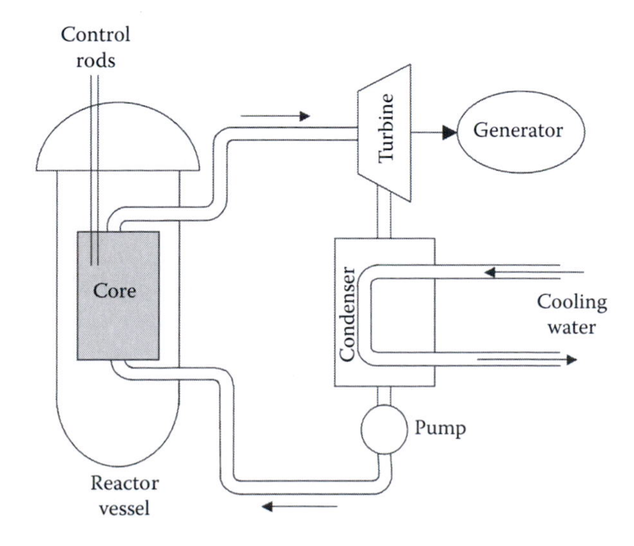

FIGURE 3.10 Conceptual block diagram of a simple nuclear reactor.

boiling-water reactor (BWR). The reactor core consists of long thin fuel rods, clad in a metal that has good thermal conductivity properties, is resistant to corrosion by high-temperature water, and does not absorb neutrons (usually a Zr alloy). The fuel itself is often pressed UO_2 pellets, filling the cladding container. Water is circulated through the stack of rods, providing both cooling and neutron moderation/reflection. Steam generated from the coolant water is used to drive a series of turbines to generate electricity prior to being condensed and reintroduced into the reactor core. Water used for neutron moderation should remain in the liquid phase, which requires high pressure within the reactor vessel. Finally, rods of neutron absorbers are strategically placed within the core and used to control the power level of the reactor.

Cooling, and the resulting temperature of the reactor core, is the fundamental limitation on the power level achievable by the reactor. Under normal operating conditions, the fuel pellets (which attain the highest temperatures in the reactor) must always be kept well below their melting points, as must the cladding material. Should the cladding (which provides structure to the core) approach its melting point, the fuel rods may distort to the extent that coolant flow is affected, causing a rapid, catastrophic failure of the system. The reactor designed in Figure 3.10 offers an intrinsic safety feature: if cooling becomes inadequate, the liquid water in the vicinity of the fuel elements will become steam; the lower density of steam makes it an inferior moderator and reflector, causing the power level of the reactor to drop. Even though nuclear reactors are designed with safety features to avoid catastrophic problems [40], most fuel assemblies and primary cooling systems are installed in heavy-walled containment buildings, providing further defense to protect the environment from the accidental release of radioactivity.

The reactor components of greatest interest to the nuclear forensic analyst are the "clean" fissionable material that constitutes fresh nuclear fuel and the irradiated fuel, which is the end-product of nuclear reactor operation. The design of the reactor determines both the ingoing fuel composition and the resulting fuel isotopic composition, so some time will now be spent reviewing history and the basic nuclear reactor types.

The first nuclear reactor was constructed in Chicago at Stagg Field in 1942, under the direction of Enrico Fermi [41,42]. It was a prototype of a reactor designed to produce Pu and was of such low power that no cooling system was necessary in the design. It consisted of a lattice of pressed natural UO_2 or U metal pellets, distributed through and supported by a large stack of high-purity graphite bricks, which acted as the moderator. To achieve $k > 1$ (corresponding to an increase in the number of neutrons in each succeeding generation), each batch of graphite had to be minimized in its neutron absorption properties (largely a function of the amount of high-cross-section boron contaminating the carbon), as did each batch of U fuel.

$$^{10}\mathrm{B}(\mathrm{n},\alpha)\,^{7}\mathrm{Li} \quad \sigma_{\mathrm{th}} = 3800\ \mathrm{b}.$$

Control rods consisted of Cd sheets nailed to flat wooden strips, which were inserted and removed manually. Moisture was excluded from the "pile" by a shroud of fabric suitable for constructing dirigibles. It was the first containment structure. The final configuration comprised approximately 380 t of graphite and 40 t of natural U in various chemical forms. The nominal operating power was expected to be 2 kW, equivalent to twenty 100-W light bulbs, so ambient air was expected to provide sufficient cooling of the pile. On the basis of his estimates of the number of delayed neutrons emitted per fission and their effective time constant, Fermi calculated that the Cd strips should provide sufficient control. Even so, he installed an automatic, gravity-powered control rod, whose release was controlled by attaining a certain reading on a neutron meter. Fermi also located students near the top of the structure with bottles of cadmium sulfate solution, just in case. On December 2, 1942, Fermi ran the reactor, achieving $k = 1.0006$ [42], operating the pile at a power level of less than 1 W before shutting it down.

The Chicago pile demonstrated the feasibility of using a natural-U-fueled reactor to produce ^{239}Pu for use in the development of nuclear weapons. The fission chain reaction was sustained by ^{235}U in the fuel, which was diluted by the vastly more ^{238}U. The ^{238}U competed for some fraction of the neutrons, capturing to form ^{239}U, which ultimately decayed to ^{239}Pu. The Manhattan Project production reactors in Hanford, Washington, were designed to produce kilogram quantities of Pu. This goal required operation of the graphite-moderated reactors at power levels that necessitated cooling, which was accomplished through the circulation of water. No practical use was made of the fission heat that was carried off by the coolant water. Another difference in design, compared to the Chicago pile, was a provision for refueling. Irradiated fuel elements could be replaced with fresh fuel when the proper quantity of Pu had been produced. The first Hanford reactor went critical in September 1944 [43]. Much of the Pu in the weapons in the US stockpile was produced at Hanford or at the production facility at Savannah River, either from natural U or from U only slightly enriched in ^{235}U. In Britain and France, the first Pu-production reactors were of a similar type, but were cooled with air or CO_2. The British variant is called a Magnox reactor. Russia used Hanford-type reactors, not only for the production of Pu, but also for power generation. The ill-fated Chernobyl unit 3 reactor was of a modified Hanford-type design [44].

The product material from a nuclear reactor is actually a mixture of nuclides in which the percentage of each nuclide is a function of the temperature of the core and moderator, the period of irradiation, the isotopic content of the starting material, the energy distribution of the neutron flux, and the radioactive half-lives of the nuclides produced. Fuel efficiency in production reactors is less important than the isotopic composition of the Pu product, for reasons that will be explained later. Particularly important is the minimization of the amount of ^{240}Pu produced relative to ^{239}Pu, as ^{240}Pu has undesirable nuclear properties for use in nuclear explosives.

With short reactor irradiation times, relatively pure ^{239}Pu is formed from natural U or LEU. The cross section for thermal neutron capture by ^{239}Pu is about 1020 b, of which 280 b results in the formation of ^{240}Pu,

$$^{239}\text{Pu}(\text{n},\gamma)\ ^{240}\text{Pu} \quad \sigma_{\text{th}} = 280\ \text{b},$$

with nuclear fission the other reaction channel. Capture-to-fission ratios in the resonance region of ^{239}Pu are also substantial. As a consequence, when a fuel element containing U is subjected to a more prolonged exposure to thermal neutrons, the quantity of ^{240}Pu formed increases relative to ^{239}Pu. The product ^{240}Pu is fissionable but not fissile [the (n,f) cross section $\sigma_{th} = 0.03$ b] and preferentially captures neutrons to form ^{241}Pu:

$$^{240}\text{Pu}(n,\gamma)\ ^{241}\text{Pu} \quad \sigma_{th} = 280 \text{ b.}$$

Higher-mass-number Pu isotopes are produced by successive captures.

A fuel element in a Pu-production reactor is usually removed when ^{240}Pu/^{239}Pu is between 0.05 and 0.07, at which point there is generally appreciable reactivity remaining in the fuel. In a natural-U-fueled reactor, this limits burn-up of the initial U content of the fuel (sum of all isotopes) to 0.1%–0.2%. In other words, at fuel discharge, approximately 99.8% of the original U remains in the form of U isotopes. Power reactors, propulsion reactors, and research reactors operate under constraints different from production reactors (because of economics of fissile isotope use, difficulty of refueling, amount of downtime, and so forth). As a result, Pu recovered from spent fuel from these applications is not very useful from the standpoint of fabricating nuclear explosives. Even so, there is much fission energy stored in the Pu isotopes present after high burn-ups, so they can be recovered for power reactor applications.

Several different reactor types were studied in the years between the start of Pu production at Hanford and the first commercial nuclear generation of electrical power (at the end of 1957). Most of these reactor types are mere curiosities for the forensic analyst, and it is unlikely that materials associated with them would ever be encountered [43]. Among these reactor types was the water-boiler reactor at Los Alamos, whose fuel was ^{235}U-enriched uranium sulfate dissolved in water, which acted as moderator. This "homogeneous" reactor was the first ever fueled by enriched U. Another reactor was the Materials Testing Reactor (MTR) at Arco, Idaho, which was designed to deliver very high neutron fluxes to targets inserted near the reactor core for isotope production and to study the effects of nuclear radiation on reactor materials. MTR used ordinary water as coolant and moderator, and fuel was an alloy of enriched U and Al, clad in Al. The first fast-neutron reactor (Clementine) was completed at Los Alamos in 1946. The fuel was ^{239}Pu, and the coolant was liquid mercury.

The demonstration of a fast-neutron reactor made possible the potential of "breeding" nuclear fuel. The concept is that a mixture of ^{239}Pu and ^{238}U, under the right conditions, might result in more production of ^{239}Pu from neutron capture by ^{238}U than there would be destruction of ^{239}Pu by fission. In other words, the breeder reactor would be used to produce electric power, yet result in a fuel charge containing more fissile material than when it started. It should be emphasized that breeder reactors do not violate any conservation laws, but instead produce usable fissile material from fertile material at a higher rate than the original fuel in the reactor is consumed. Unfortunately, the multiplicity of neutrons arising from the fission of ^{239}Pu is insufficient to accomplish breeding in a thermal reactor. In a fast reactor, however, the number of neutrons produced per fission is significantly higher than in a thermal reactor, and breeding becomes possible. The Experimental Breeder Reactor (EBR-I) was completed at Arco in 1951, using enriched U fuel and a metal Na–K eutectic as a high-temperature liquid coolant. Strictly speaking, EBR-I was not a true breeder reactor, as the initial fissile material was ^{235}U rather than ^{239}Pu. Still, EBR-I demonstrated the feasibility of breeding and was also the first nuclear reactor to generate electric power (in late 1951). Breeding can also be accomplished with a thermal reactor if ^{233}U is used as fuel, and ^{232}Th as the fertile material. The yield of neutrons from the fission of ^{233}U is relatively high at both thermal and fission-spectrum reaction energies.

It was very quickly recognized that a compact nuclear power plant would offer enormous advantages for both submarine and ship propulsion. Long journeys would be possible without a need for refueling, and submarines would not have to surface as often because oxygen would not be required for the operation of the power plant. The early reactor designs used pressurized water as both moderator and coolant, and very highly enriched U (clad in Zr) was used as the fuel to achieve a large power density in a small reactor. A prototype submarine reactor was operational at Arco in 1953, and the first nuclear-powered submarine, the USS Nautilus, underwent sea trials in early 1955. The initial fuel of a submarine propulsion reactor is nuclear-weapon-ready material. The Pu produced during its operation is much less attractive for the fabrication of explosive devices, both because of high levels of very radioactive ^{238}Pu that is produced, relative to Pu from lower enrichments of U, and because submarines are refueled only after their plants have operated for a very long time and have produced excess amounts of undesirable neutron-rich Pu isotopes.

An operating nuclear reactor contains large quantities of both short- and long-lived radionuclides that can be dangerous if not properly shielded. Power plants and production reactors are shielded with a combination of steel and concrete, and during normal operations, the only significant radionuclide emissions are those of the noble gases, which are chemically inert and thus hard to contain. Shielding a propulsion reactor must meet additional weight criteria because of the mobile nature of the vessel. This became particularly important with the development of fast, nuclear-propelled submarines [45]. Fortunately, the coolant and moderator provide an intrinsic shield that precludes neutron exposure of ships' crews. The balance of personnel protection is provided by thick steel plates, which are often mounted in the propulsion reactor's pressure vessel [46]. Secondary shielding is provided by water, whereas radioactive gases that cannot be scrubbed aboard ship must be vented to the environment.

The cladding of nuclear fuel provides the most important barrier to the release of radioactivity. The fission products tend to lodge in the matrix of the fuel pellets, but at reactor operating temperatures, some of these products are mobile and can migrate out of the pellets and into the interstitial areas in the fuel rods. Even though the metal cladding is welded closed and leak-checked before a fuel element is deployed, a small fraction of the pins or rods leak gas into the reactor coolant. Long-lived ionic radionuclides are removed from the coolant by a scrubbing system of some design, but it is ineffective for the noble-gas fission products, Kr and Xe.

Certain safety features in the nuclear industry are common among most reactor types. Some of these are intrinsic, inherent in the design of the reactor, and some are engineered systems added to the reactor design specifically for safety. Previous discussion of Figure 3.10 pointed out an intrinsic safety feature of LWRs: overheating the coolant, caused by an unplanned rise in the nuclear reaction rate, reduces the density of the surrounding water and makes it a poorer moderator of neutrons, causing a decrease in the reactivity of the core. As an example of an engineered safety feature, most reactors have both control and shutdown rods for controlling the reactivity of the core. Shutdown rods are specifically designed for rapid and fail-safe termination of the chain reaction. Although a normal operational change in power level through the agency of the control rods may take place over an hour, an emergency shutdown (or "scram") may take only seconds.

Even when the chain reaction is stopped, however, the coolant system must continue to operate because the fission products continue to generate heat through radioactive decay. Immediately after a scram, the reactor power level is still 7% of the steady-state level that was operational before shutdown. This is sufficient to require continued cooling of the core to prevent heat damage to the fuel. Because of the consequences of failure, cooling safety systems are engineered to be multiply redundant. In LWRs, there is an emergency core cooling system that is independent of the main systems.

In the United States, virtually all commercial nuclear power plants generate electricity using LWRs, and a large proportion of the reactors in operation worldwide are LWRs. Much effort has gone into the study of the safety of these reactors [47], particularly since 1979, when the Three Mile Island unit 2 pressurized-water reactor (PWR) melted down. These reactors use normal water as both the moderator and the coolant. The hydrogen atoms in water offer a significant advantage over graphite in the number of collisions required to thermalize the neutrons to reactor temperatures. However, H has a nonnegligible probability of capturing a neutron to form D,

$$^{1}\mathrm{H}(\mathrm{n},\gamma)\ ^{2}\mathrm{H} \quad \sigma_{\mathrm{th}} = 0.33\ \mathrm{b},$$

a cross section that far surpasses the capture cross sections of the mixed carbon isotopes (0.005 b for thermal neutrons). A result is that natural U cannot provide sufficient reactivity to be used as fuel in a LWR, and the initial concentration of ^{235}U in the fuel load must be enriched to >2% for such a reactor to operate.

Fuel within a LWR is provided as rods, or "pins," which are fabricated from UO_2 pressed and sintered into pellets of the proper form. The pellets are sealed under a He atmosphere in long, cylindrical cans made of a Zr alloy. The resultant fuel rods are more than 10 ft long and are incorporated into fuel assemblies whose structure is determined by the reactor design. The fuel assembly structure is open, permitting the flow of cooling water both horizontally and vertically. Fuel assemblies, along with inserted control rods, form the reactor core, which is mounted in a large pressure vessel.

There are two types of LWR: the PWR and the BWR [48]. Approximately one-third of the LWRs operating in the United States are BWRs designed and built by General Electric. In the BWR, the reactor vessel itself serves as the boiler for the steam supply system. See Figure 3.11 for a schematic representation of a BWR. Liquid water is recirculated through the reactor vessel to cool the core. Steam concentrates in the headspace of the reactor vessel, from where it proceeds to the turbine. Recondensed water is introduced into the reactor vessel from below, and steam rising directly from cooling the fuel rods drives the turbine to generate electricity.

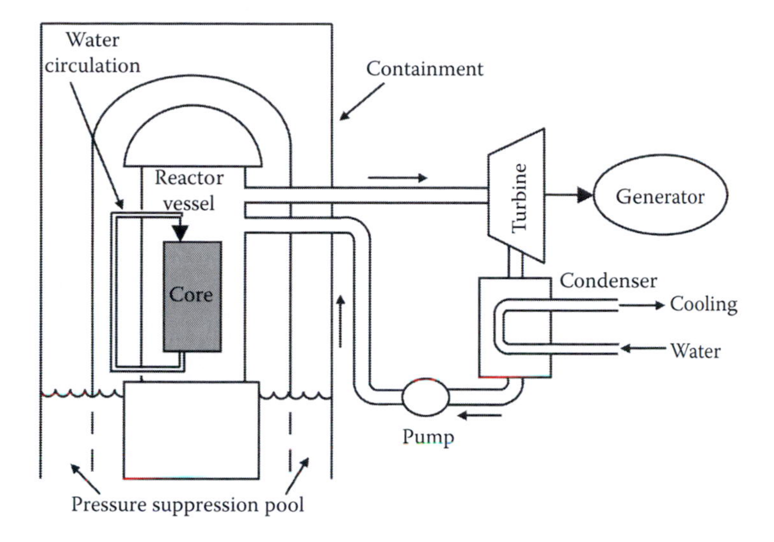

FIGURE 3.11 Schematic of a boiling-water reactor power plant. Steam from the boiling-water reactor vessel flows to the turbogenerator, after which it is condensed and returned as feedwater to the reactor vessel.

BWR fuel consists of UO_2, in which the ^{235}U content is enriched to 2.4%–3.0%. A BWR core is usually constructed of a large number of fuel assemblies, each of which consists of a square array of 49 or 64 fuel rods. Each group of four fuel assemblies is associated with a control rod fabricated of boron carbide.

At refueling, whole assemblies are removed from the central region of the reactor and replaced by partially burned assemblies from the periphery, which are in turn replaced by fresh fuel. The intent is to attain a relatively uniform distribution of power across the core. Addition of a burnable neutron poison, such as gadolinia, to the fuel elements can help accomplish this goal. The capture of neutrons by certain isotopes of Gd causes a reduction in reactivity, but these isotopes are completely depleted after <1 year of reactor operation. Typically, the spent fuel has been burned to 0.8% ^{235}U, but it also contains 0.6% Pu isotopes produced from fertile ^{238}U. Coolant water is supplied to the reactor vessel at 190°C, and water exiting the core is heated to about 290°C. The coolant is maintained at 1050 psi, so that 15% of the exiting coolant is steam. Sufficient coolant flow rate is supplied to maintain the cladding temperature <320°C.

Roughly two-thirds of the LWRs operating in the United States are PWRs, designed and built mainly by Westinghouse, but with some from Babcock & Wilcox and some from Combustion Engineering. See Figure 3.12 for a schematic representation of a PWR. In a PWR, the primary coolant is circulated in a closed cooling system. The primary system is cooled by heat conduction in a steam generator, which is used to drive a turbine to generate electricity. Water in direct contact with the fuel never leaves the primary containment building. This system is thus more isolated should problems develop with the fuel rods. The enrichment of U in fresh fuel rods is slightly higher than that of BWR fuel, about 3.2% ^{235}U. PWR fuel assemblies are much larger, being square arrays of 225 (15 × 15) or 289 (17 × 17) pins. Each assembly

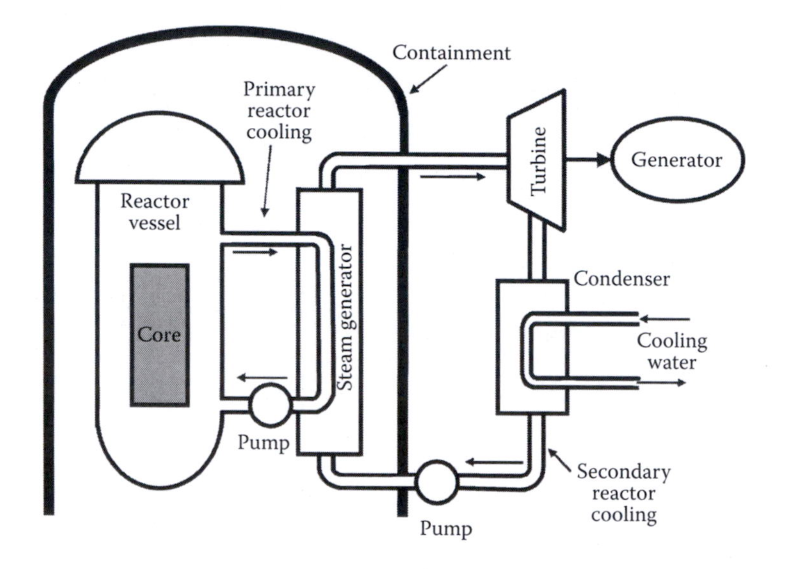

FIGURE 3.12 Schematic of a pressurized-water reactor power plant. The primary reactor system is enclosed in a steel and concrete containment building. Steam generated within the building flows to the turbine, after which it is condensed and returned to the steam generators.

includes rod guides for the insertion of one or more control rods, which are usually fabricated from an alloy of Ag, Cd, and In, but can also be boron carbide as in BWRs.

The PWR core contains three enrichment zones, rather than the two found in BWRs. As before, refueling consists of removing elements from the middle of the core, moving outer elements inward, and emplacing fresh fuel around the periphery to provide a flat power profile. During operation, coolant water is introduced into the pressure vessel at 290°C and exits the core at 325°C, at a pressure of 2250 psi, so no vapor-phase water exists. The cladding temperature is maintained at about 350°C, but the central temperature of the fuel pellets can reach 2300°C. Typically, spent fuel is burned to 0.9% ^{235}U at discharge, containing 0.6% Pu isotopes produced from fertile ^{238}U.

The quantity of fissile material produced from the fertile components of the fuel, divided by the amount of fissile material destroyed (either by fission or by capture), is the conversion ratio. For both types of LWR, the conversion ratio is about 0.6. The conversion ratio is not as important in a "once-through" fuel cycle like that in place in the United States. Elsewhere, where the Pu is extracted and recast as fuel, conversion is an important part of the economy of running a reactor. Approximately 2.0 fast neutrons are produced for each slow neutron absorbed by ^{235}U. Water moderates these neutrons (through collisions with H nuclei) to thermal energies, but before they are completely moderated, many of them react with ^{238}U at resonance energies to produce ^{239}U, which decays through ^{239}Np to ^{239}Pu. Some of the thermal neutrons are also captured by ^{238}U, but most react with ^{235}U, water, structural materials, the control rods, and fission products like ^{135}Xe. The ratio of captures by ^{238}U to total absorption by ^{235}U is 0.6. As the LWR operates, the total inventory of fissile material decreases and fission-product poisons build up, so it is necessary to decrease the insertion of control rods in the neutron environment to maintain $k = 1$.

Eventually, when there is no longer sufficient reactivity remaining in the fuel to sustain the required power level, it becomes necessary to refuel. During refueling, LWRs are not available for power production for extended periods of time—often several weeks. Typically, the region around the reactor vessel is flooded with water so that the irradiated fuel can be handled under water, providing some measure of shielding for personnel. Even so, plant workers receive most of their occupational radiation exposure dose during refueling operations.

An alternative to ordinary water as moderator in a thermal reactor is its replacement with heavy water, D_2O. Although each neutron-moderator collision is somewhat less effective at slowing the neutrons than is a similar collision with ordinary water, heavy water absorbs far fewer neutrons because the capture cross section of D is significantly lower than that of regular H (protium): 0.0005 b as opposed to 0.33 b:

$$^{2}\mathrm{H}(\mathrm{n},\gamma)\ ^{3}\mathrm{H} \quad \sigma_{\mathrm{th}} = 500\ \mu\mathrm{b}.$$

Therefore, heavy-water reactors (HWRs) can be designed to operate with natural U (0.71 atom-% ^{235}U) as fuel [49].

The design of a HWR is conceptually similar to that of a PWR: the valuable D_2O, both moderator and coolant, is circulated under pressure in a closed system, which exchanges heat to produce steam in ordinary water to drive a turbine. The D_2O remains in the liquid phase. Because the absorption cross section is low and the moderator is somewhat less effective, HWR fuel elements are mounted farther apart than those in LWRs, making it possible to cool HWR fuel elements individually. See Figure 3.13 for a schematic diagram of a HWR.

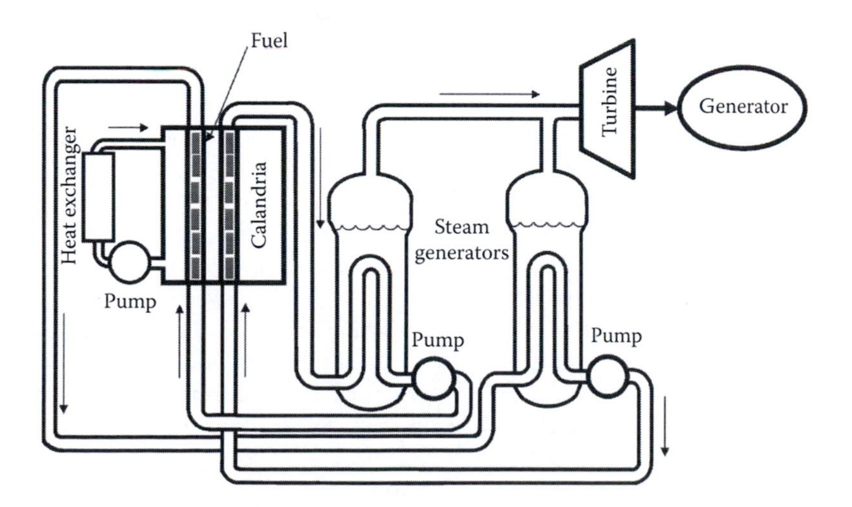

FIGURE 3.13 Schematic heavy-water reactor power plant of the CANDU type. CANDU systems are essentially PWRs, but individual fuel channels are pressurized, rather than the reactor vessel. Shown are two channels (out of many) in which steam is generated from normal water to drive the turbines. The heavy-water moderator is contained in a calandria, which has its own circulating and cooling systems.

Atomic Energy of Canada has built and operated the "Canadian Deuterium-Uranium," or CANDU reactors. CANDUs use heavy water as both moderator and coolant, but in principle, the design could be modified to use another cooling fluid such as ordinary water. Like the LWR, CANDU fuel pellets are sintered UO_2 ceramics sealed into zircaloy tubes that comprise the fuel pins. The pins are bound into short, cylindrical bundles to form the fuel elements, and the geometry is a consequence of the cooling and refueling scheme. CANDU reactors are not shut down to refuel. Rather, a few fuel elements are replaced each day of operation without reducing power. In principle, this makes it possible to operate the CANDU as a Pu-production reactor. In practice, refueling also acts as a reactivity control, increasing the fissile content of the reactor when required. However, short-term reactivity control still requires adjustable control rods.

The primary coolant system of the CANDU is similar to that of a PWR, except that the pressure vessel is replaced by hundreds of individual pressure tubes. Each tube may be opened individually for refueling. The heavy-water coolant is maintained at a pressure of about 1500 psi and reaches a temperature of 310°C before leaving the reactor core. The D_2O does not boil under these conditions. Because the coolant is pressurized, the moderator need not be. The reactor vessel, referred to as a calandria, operates at near-atmospheric pressure and employs its own cooling system to maintain the moderator at about 70°C.

The fuel cladding temperature during operation reaches 360°C, and the maximum (center) fuel temperature reaches 2100°C. When spent, the natural U fuel has burned to 0.2% ^{235}U, but contains 0.3% fissile Pu isotopes. The conversion ratio for a CANDU reactor is approximately 0.75. This value is better conversion than is achieved in LWRs, allowing more efficient fuel utilization were the Pu isotopes to be recovered and fabricated into fuel. Though refueling occurs constantly, the average time that a fuel element spends in the CANDU reactor core is >1 year.

The British have developed a variant of this reactor type, called the steam-generating HWR (SGHWR). Although the moderator remains D_2O, ordinary water is circulated through the core in vertical pressure tubes to cool the moderator. The H_2O is permitted to boil, and the steam is used to drive the turbine, similar to a BWR. The SGHWR requires fuel that is slightly enriched in ^{235}U over the natural concentration, a consequence of neutrons being captured by protium in the primary coolant.

We have discussed graphite-moderated reactors as they apply to the production of Pu in the US weapons complex. They can also be employed as power reactors. Although carbon requires many more collisions than H to slow neutrons down to thermal energies (114 collisions as opposed to 18), it is a more effective moderator because its cross section for neutron reactions is much smaller. As mentioned earlier, Russian C-moderated power reactors are cooled with water confined to pressure tubes. In Britain, a number of graphite-moderated, CO_2-cooled reactors were constructed. In principle, the high-temperature gas coolant offers potential for high thermal efficiency when driving a gas turbine. As a class, these reactors cooled with CO_2 are referred to as advanced gas reactors [50]. However, graphite-moderated reactors have fallen out of favor, largely because two of the worst reactor accidents in history, Windscale (1957) and Chernobyl (1986), involved fires that could not be extinguished, which burned in the graphite blocks of the reactor core [44,50,51].

Operation of LWRs is generating increasing amounts of Pu. In the United States, this Pu remains in the fuel elements, which are stored as waste. In Europe and Japan, breeder-reactor programs may ultimately make use of Pu recovered from spent LWR fuel. However, there have been several studies of recycling Pu through LWRs, replacing some fraction of the ^{235}U as the primary fissile fuel [52]. Many of these studies are centered on the fabrication of Pu–U mixed oxide (MOX) fuel. Significantly, an initial incorporation of Pu into reactor fuel results in a completely different Pu isotopic signature than the nuclear forensic analyst would encounter in the US materials.

The reactor types described earlier were designed to produce electric energy economically during the 1950s and 1960s, when supplies of U were relatively ensured and enrichment was inexpensive or government-subsidized; they are "slow" reactors. At the time they were created, incorporating the coolant as both the moderator and the means by which heat is transferred to steam, which in turn drives a turbine, was the most straightforward method for obtaining electric power from fission heat. For political reasons related to safeguards and nuclear nonproliferation, the United States ignores the fissile material produced in spent fuel from fertile ^{238}U, choosing to discard it with the spent fuel rather than recover it and return it to the fuel cycle; hence the profligate "once-through" fuel cycle. The only exceptions are when the product Pu isotopes are of weapons grade, or the residual U is of sufficient enrichment to make it a proliferation concern, as in propulsion or research reactors. Nevertheless, other countries have explored recycling the fissile materials. In nuclear fuel, the magnitude of the conversion factor is an important element in the economics of nuclear power in these countries. The EBR-I demonstrated that breeding can take place: it is possible to design reactors in which more fertile material is converted to fissile nuclides than fissile material is burned to fission products, for a conversion factor >1. Incorporating these concepts into the fuel cycle means that reactors can ultimately derive most of their fission energy from materials that were not fissile when they were mined (i.e., the fertile nuclides ^{232}Th and ^{238}U).

Fast reactors take advantage of the fact that, on average, absorption of a fast neutron by a fissile nucleus gives rise to a higher multiplicity of emitted neutrons in the subsequent fission than does absorption of a slow neutron. For ^{235}U, the average number of neutrons liberated per neutron absorbed (including capture) is 2.06 for thermal neutrons and 2.18 for fast neutrons. For ^{239}Pu, the values are 2.10 for thermal neutrons and 2.74 for fast neutrons. Numerically, the differences are small yet crucial, as it is the excess over two neutrons per fission that is important. A necessary requirement for breeding is that one neutron is necessary to carry on the chain reaction, while another neutron converts a fertile nucleus to one that is fissile. Because of the various modes through which neutrons are lost, there must be some significant excess over a multiplicity of 2.

The neutron multiplicity of the fission of ^{235}U with neutrons of any energy is not sufficient to make a practical breeder reactor possible. For fast neutrons on ^{239}Pu, absorption of 1 neutron releases 2.7 neutrons on average. Of these, one must be absorbed by ^{239}Pu to maintain the chain reaction (provided it is still a fast neutron), leaving 1.7 neutrons to be absorbed by other materials. If the ^{239}Pu is mixed with or used in combination with ^{238}U such that essentially all of the excess neutrons are captured by the

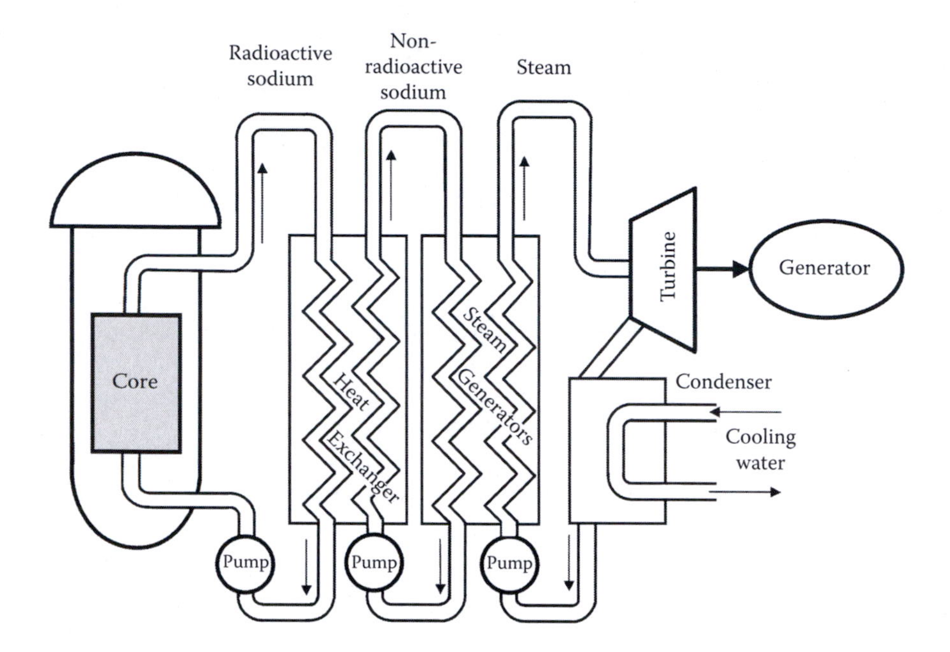

FIGURE 3.14 Schematic diagram of a liquid-metal fast breeder reactor power plant, in which mixed oxide fuel elements are cooled by circulated liquid Na metal. Radioactive Na from the reactor core exchanges heat with water indirectly through the action of an intervening loop of nonradioactive Na.

fertile material, 1.2 atoms of ^{239}Pu result, taking into consideration the unavoidable loss of 0.5 atom to fission reactions in the conversion of the fertile material.

Every program in which a fast breeder reactor has been developed has begun with the liquid-metal fast breeder reactor (LMFBR). See Figure 3.14 for a schematic diagram. The LMFBR uses a liquid metal coolant, which provides good heat-transfer characteristics and little moderation [53,54], and the coolant of choice is Na, which is liquid over a wide range of temperatures and can cool the core at atmospheric pressure. The core is a compact arrangement of fuel assemblies similar to those used in LWRs. The main differences are that the fuel pins are of smaller diameter and are made of type-316 stainless steel (which is more chemically resistant to liquid Na than is zircaloy). The coolant-core structure is designed to minimize moderation of the neutron energy, resulting in the highest possible yield of neutrons per neutron absorbed. As a result, the fuel rods are about the same diameter as a pencil, although they are several feet long.

In all LMFBR designs, the Na that cools the core is not used to directly generate the steam that drives the turbines. Rather, there is an intermediate liquid Na loop, which eliminates the possibility of releasing radioactive Na during any untoward interaction with water in the steam generator. Fortunately, the cross sections for neutron reactions with Na at fission-spectrum energies are quite small. The induced radioactivity is mainly ^{24}Na, which has a 15-h half-life and emits very penetrating γ-rays (1369 and 2754 keV) with high probability:

$$^{23}\text{Na}(n,\gamma)\ ^{24}\text{Na} \quad \sigma_{fs} = 300\ \mu\text{b}.$$

Other reactions either have high thresholds relative to the energy of the neutrons or result in short-lived species such as 11-s ^{20}F [via the (n,α) reaction]. In the primary loop, Na is pumped into the reactor vessel at a temperature of about 800°C. It leaves the reactor core at about 1050°C and passes through a heat exchanger, where it heats Na in the intermediate loop. Here, temperatures in excess of 900°C are achieved, at which point it is used to produce steam. Sodium metal is highly chemically reactive. At operating temperatures, it will burn if exposed to air and will react explosively with water vapor. Corrosion that leads to Na leakage must be scrupulously avoided.

A large-scale LMFBR contains a fuel charge of 2–3 t of fissile material and approximately 50 t of fertile material. In a true fast breeder, the fissile material would be Pu. But because Pu does not occur in nature, and the initial required charge is quite large, new LMFBRs often begin with a significant quantity of ^{235}U in the fuel. It is then replaced with Pu as it becomes available in parallel fuel-reprocessing operations. About 0.3 t of Pu can be recovered from the fuel used in the operation of a commercial LWR for a year. Of this, about 0.1 t is fissionable rather than fissile (the even-mass Pu isotopes). Approximately 20 years of LWR operation are required to provide the fissile material required by an LMFBR, hence the tactic of including ^{235}U in the initial charge of the breeder.

The heat-producing region of the LMFBR actually consists of two areas. The first is the centrally located core, in which the fuel is sintered pellets of mixed oxides of Pu and U in a proportion of about 1:5, bonded into the steel tubes with Na metal. The other is an external "blanket" region, consisting of a depleted- or natural-U ceramic oxide, again bonded into steel tubes [55]. The core sustains the chain reaction and produces a surplus of Pu from its own fertile charge. In addition, neutrons escaping the core are absorbed by U in the blanket, producing more Pu. The isotopic compositions of the Pu present in the core and in the blanket of an LMFBR at any particular time are difficult to calculate, as they are dependent on the number of previous reprocessing campaigns that were involved in fabricating new fuel elements that are returned to the reactor. Because reprocessing (and, therefore, unloading spent fuel from the reactor) is critical to a breeder fuel cycle, most designs provide simplified refueling with minimum downtime. Hence, the design discussed in detail, the loop-type LMFBR that was demonstrated in the United States at the Clinch River facility, has fallen out of favor. In a pool-type breeder, several other reactor components are contained in the reactor vessel along with the core. The first large-scale breeder reactor to operate commercially, France's Superphenix, is of this type.

At this time, there are several fast breeder reactors operating in Europe and Russia. In the United States, Congress ended the LMFBR program in 1994. Because the economic operation of these plants relies on fuel reprocessing, the possibility of diverted samples of reprocessed material is higher for LMFBRs than for a LWR. Even though LMFBRs are small in number at this point, the nuclear forensic analyst may encounter samples related to LMFBR operation with increased probability in the future. LMFBR designs are inherently less safe than those of LWRs, largely because of the chemical reactivity of the coolant. The intense radiation encountered within a fast-reactor core also puts high demands on metals and other materials, particularly when considering the necessity of cycling the reactor through large changes in the refueling process. The core of the

Fermi-1 LMFBR melted in 1968, and an LMFBR in Monju, Japan, suffered an accident in 1995 in which 3 t of Na coolant were released.

There also exists a fuel cycle that has not yet been discussed: the ^{232}Th–^{233}U cycle. It is analogous to the ^{238}U–^{239}Pu fuel cycle used in fast-flux reactors. As described earlier, ^{232}Th is a fertile nuclide that is three times as abundant in Earth's crust as U. It captures a neutron to form short-lived ^{233}Th, which β-decays twice to form 1.592×10^5-year ^{233}U, a fissile nuclide. The ^{232}Th–^{233}U fuel cycle offers many potential advantages for the generation of power when compared to reactors fueled with ^{235}U or ^{239}Pu. Even so, it is but a small factor in the nuclear power industry [56,57]. This is primarily because it does not provide new fuel as rapidly as the other fuel cycles and because Th does not exist in a fissile form in nature. A reactor cannot be fueled with Th alone; Th must be used in conjunction with one of the other fissile materials as initial fuel. To start the ^{232}Th–^{233}U fuel cycle, ^{233}U must be bred in an original charge of ^{232}Th. This is accomplished either by placing a blanket of ^{232}Th around a breeder reactor or by incorporating ^{235}U or ^{239}Pu into a ^{232}Th matrix and fabricating fuel. This fuel must then be reprocessed after use to recover ^{233}U.

Another problem with the ^{232}Th–^{233}U fuel cycle is the simultaneous production of ^{232}U. This isotope arises via ^{231}Th through three avenues. First, a neutron reaction on ^{232}Th that competes with capture for neutron energies over the neutron-binding energy is

$$^{232}\mathrm{Th}\left(\mathrm{n,2n}\right)\,^{231}\mathrm{Th} \quad E_{\min} = 6.44\ \mathrm{MeV}.$$

There are very few neutrons even in a fast reactor with sufficient energy to produce this reaction, but there are some. Second, natural Th ores almost always contain some small admixture of U. The daughter of the α decay of ^{234}U is 75,400-year ^{230}Th, which persists in the Th that is chemically purified from the ore even after U is separated. Even though it is only a small component of any natural Th sample, it undergoes the capture reaction,

$$^{230}\mathrm{Th(n,}\gamma)\ ^{231}\mathrm{Th} \quad \sigma_{\mathrm{th}} = 23\ \mathrm{b}$$

Third, any ^{235}U present in the fuel produces ^{231}Th by radioactive decay:

$$^{235}\mathrm{U}(7.04 \times 10^8\,\mathrm{years}) \rightarrow ^{231}\mathrm{Th} + \alpha$$

Produced by any of these three paths, the ^{231}Th nuclide β-decays to long-lived ^{231}Pa:

$$^{231}\mathrm{Th}\left(25.4\ \mathrm{h}\right) \rightarrow ^{231}\mathrm{Pa}\ \left(32{,}760\ \mathrm{years}\right) + \beta.$$

By whatever pathway it forms, ^{231}Pa can capture a neutron and β-decay to ^{232}U:

$$^{231}\mathrm{Pa(n,}\gamma)\ ^{232}\mathrm{Pa} \quad \sigma_{\mathrm{th}} = 260\ \mathrm{b}$$

$$^{232}\mathrm{Pa}\ (1.31\ \mathrm{d}) \rightarrow\ ^{232}\mathrm{U} + \beta.$$

In fuel in which the ^{232}Th–^{233}U cycle is initiated through the admixture of ^{235}U, the third route is very important. However, it cannot be neglected even if a blanket of pure ^{232}Th is being considered, as industrial reagents from which the blankets are fabricated contain U at levels of several ppm.

Uranium-232 is deleterious. It has a half-life of 69 years, so its intrinsic radioactivity is quite high. A sample of freshly separated ^{233}U that contained 100 ppm ^{232}U would emit about 20% more α radiation than would a pure ^{233}U sample. More important, though, is the result of ^{232}U decay daughters growing into the sample. Within a few years, six α emitters attain secular equilibrium with the ^{232}U, one of them being a noble gas with a 1-min half-life (^{220}Rn) and an inhalation hazard. About 36% of the decays of ^{232}U in an equilibrium mixture result in the emission of a 2.6-MeV γ-ray (from decay of ^{208}Tl), which is difficult to shield because of its very high energy. Even though the dose associated with freshly isolated ^{233}U (containing ^{232}U) is low, a large sample quickly produces dangerous amounts of daughter activities, some of which are gaseous and some of which emit very penetrating radiation. An aged ^{233}U fuel element is thus a personnel exposure hazard even before it is irradiated in a reactor.

Although there are drawbacks in certain situations, the advantages of the ^{232}Th–^{233}U fuel cycle may outweigh them. Thorium oxide has a much higher melting point than do the oxides of U and Pu. This stability allows higher fuel temperatures and higher burn-ups, and naturally recommends ^{232}Th-based fuel for applications requiring high reactor core temperatures. The fission of ^{233}U following neutron capture emits a high multiplicity of neutrons, even at thermal energies. The fact that the ^{233}U(n,γ) cross section is very small compared to the (n,f) cross section gives the favorable situation that the number of neutrons emitted per neutron absorbed by ^{233}U is 2.27 at thermal energies and 2.60 for fast neutrons. In addition, the ^{232}Th capture cross section is 3× greater than that of ^{238}U (7.4 *vs* 2.7 b). As a result, more ^{233}U is produced than ^{239}Pu under equivalent circumstances. This means that there is a possibility of breeding ^{233}U, even in a reactor operating with thermal neutrons, and the possibility of several new reactor designs has resulted in increased interest in a ^{233}U-fueled reactor economy.

Research on the ^{232}Th–^{233}U fuel cycle has been conducted in the United States, Germany, Japan, Russia, the United Kingdom, and India. Between 1967 and 1988, the experimental pebble-bed reactor at Julich, Germany, operated for more than 700 weeks at a power level of 15 MW_e with Th-based fuel. The fuel consisted of approximately 10^5 spherical fuel elements the size of tennis balls, containing 1.4 t of Th mixed with HEU. The 20 MW_t Dragon reactor at Winfrith, United Kingdom, demonstrated that a mixed fuel of Th and HEU could be fabricated such that ^{233}U bred from the Th replaced the ^{235}U burned as fuel at about the same rate. Dragon also demonstrated that the long half-life of ^{233}Pa (27 days), intermediate in the capture and decay processes forming ^{233}U from ^{232}Th, results in a reactivity surge in the fuel after the reactor is shut down and must be considered for reactor operation. The General Atomic Peach Bottom reactor operated between 1967 and 1974 at 110 MW_t, again using a mixture of Th and HEU as fuel. These reactors were necessary intermediates in the development of the high-temperature gas-cooled reactor (HTGR), discussed later. [The power levels MW_e and MW_t refer to the production of megawatts electrical and megawatts thermal, respectively.] The electric power output of a reactor is necessarily less than its thermal output by a factor that is the efficiency of the power plant.

The production of ^{233}U for weapons applications has many parallels with the production of Pu. Irradiations tend to be of short duration at high flux to minimize the buildup of undesirable U isotopes (including ^{232}U). A cooling period is required before ^{233}U can be recovered from the source material: it is necessary to permit 27-day ^{233}Pa decay to ^{233}U before chemical processing. And criticality safety limits the batch size of the final product in any reprocessing operation.

It has been demonstrated that Th-based fuels can be used in LWRs, giving a significantly better conversion ratio than do U-based fuels. A light-water breeder reactor operating on Th fuel has been described [58]. India is in the unique position of having much more extensive resources of Th than U [59] and has made use of Th as a reactor fuel a major goal of its energy posture. India eventually plans to use CANDU-type HWRs to accomplish this goal.

HTGRs use high-pressure (30–50 kg/cm^2) He gas as coolant (see Figure 3.15). There have been as many as 30 HTGRs in operation, and more are under construction. Helium, although expensive, is chemically inert and has good heat-conductivity properties. Its cross section for nuclear reactions with neutrons is very small, arising entirely from the 0.00014% component of ^{3}He present in natural He. HTGRs come in two varieties, with pebble-bed (Julich) and prismatic fuel elements [60]. Independent of type, fuel pellets for these reactors are about 1 mm in diameter, composed of mixed U and Th oxides or carbides, and coated with gas-tight layers of carbon or silicon carbide to contain the fission products. Uniformity of composition, size, and spatial distribution of the pellets is important to avoid the development of hot spots in the fuel elements during reactor operations.

These pellets are incorporated into graphite moderator elements with a graphite binder, in units called "compacts." Some compacts are fabricated into tennis-ball-sized spheres, through which the He gas percolates. Others are fabricated into

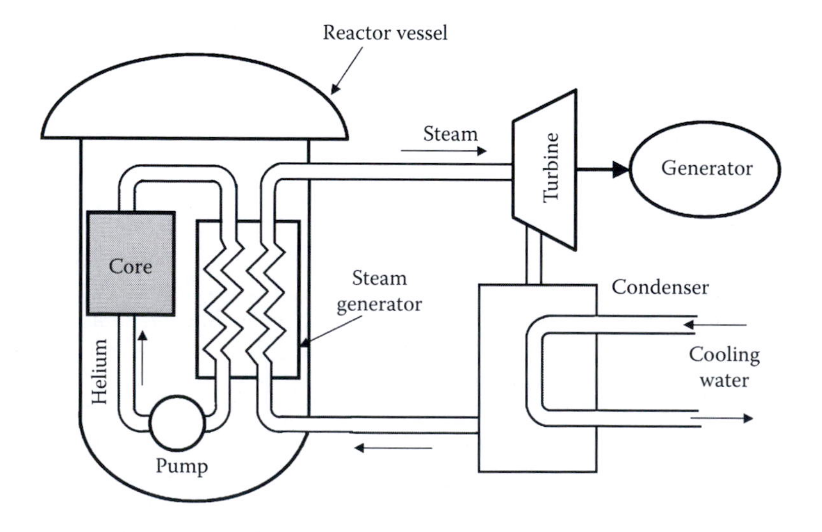

FIGURE 3.15 Schematic diagram of a high-temperature gas-cooled reactor, in which the core, containing mixed Th-based fuel, is cooled by circulating He gas. High-temperature He produces steam in a secondary cooling loop.

hexagonal cylindrical units with holes through which the cooling gas circulates. In the HTGR, the valuable He coolant is circulated in a closed system and is used to boil water, which drives the turbine. The coolant enters the reactor at about 350°C and exits at 750°C. The reactor core tends to operate at a higher temperature than do the cores of other thermal reactor types, all of which are cooled by water. The overall plant efficiency of a typical HTGR is approximately 40%, comparing quite favorably with the 30%–35% efficiencies obtained with commercial LWRs. HTGRs are refueled on a 4-year cycle.

The molten-salt breeder reactor (MSBR) is not as well developed conceptually as are most of the other reactor types discussed in this section, but test reactors have been successfully operated [58,61]. This reactor type, considered by engineers to be quite elegant, was conceived at Oak Ridge in the early 1950s. In the MSBR (see Figure 3.16), U fuel is dissolved in a high-temperature molten salt and circulated without any external coolant. The reactor is located in a loop in the circulation system where a critical mass can assemble. Some social resistance to this reactor design results from the radically different concept of intentionally melting the reactor fuel. The reactor fluid is a mixture of the tetrafluorides of fissile U and fertile Th, dissolved in a carrier salt of Li and Be fluorides. The salts, being chemically unreactive, allow reactor operation at high temperatures and low pressures, resulting in high thermal efficiencies. The design is intrinsically safe, as a leak in the reactor circulation system would not result in either a criticality or a chemical reaction. The neutron economy is also sufficient to breed ^{233}U from Th. At the heat exchangers indicated in the figure, heat is transferred from the primary reactor fluid to a molten sodium borate loop, which in turn transfers heat to steam generators. Small amounts of the reactor fluid are periodically diverted to a chemical process, in which long-lived neutron poisons that would adversely affect the conversion ratio are removed. Because this chemical process is incorporated into the function of the reactor, it does

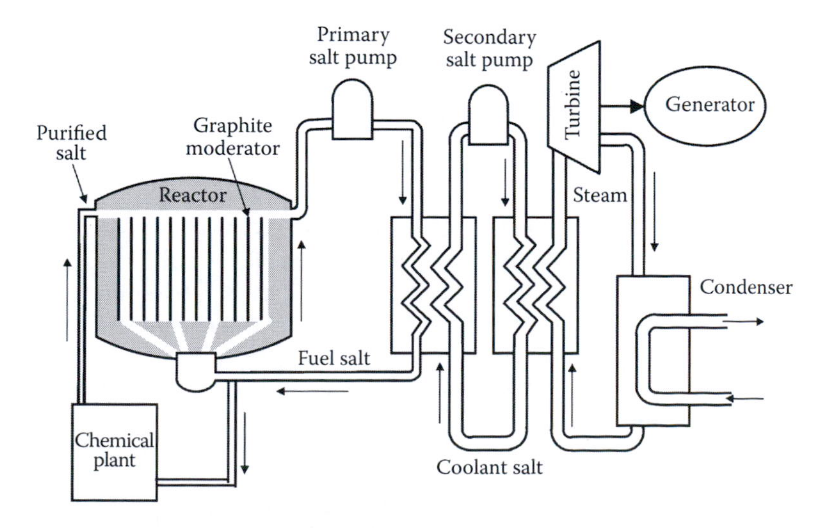

FIGURE 3.16 Schematic diagram of a hypothetical molten-salt breeder reactor, in which the fissile nuclides are dissolved in a molten salt that circulates through the reactor core.

not technically qualify as fuel reprocessing, although the techniques developed for the MSBR could be used for that purpose [62].

Although low-power MSBRs were tested in the 1960s, proving their efficacy, the United States abandoned the project in 1976 in favor of the LMFBR, which showed promise in breeding weapons-usable Pu in addition to power production. Nevertheless, the MSBR design offered sufficient promise that it is still being studied in Russia, France, Korea, and the Czech Republic. The United States is assessing its feasibility as a space reactor, and France is studying its utility as an incinerator of long-lived fission products. It is possible that, in time, ^{233}U specimens arising from the operation of a MSBR may be seized and become samples to be interrogated by a nuclear forensic analyst.

In spite of the advantages provided by its nuclear properties, it is unlikely that ^{233}U will ever compete with ^{235}U or ^{239}Pu as a principal reactor fuel. Nevertheless, Th is being used in a number of reactors, most notably in India, and the ability to breed ^{233}U will always be an economic driver. As will be discussed later, ^{233}U is also a possible fuel for nuclear explosives. The forensic analyst, when considering a sample containing ^{233}U, will have a difficult task identifying its source on the basis of the isotopic content of the material. The reason is the rapidly changing time dependence of the U isotopic composition early in the history of the reactor operation, when ^{233}U is gradually replacing ^{235}U in the fuel, but may be mixed with more pristine ^{233}U from a Th blanket.

Many of the world's nuclear reactors are not used for the production of power or weapons-usable materials. They are research reactors and are used for materials testing, engineering studies, production of radioisotopes, and training. Their primary purpose is to provide an intense source of neutrons. They are generally small relative to power reactors, reaching power levels of 100 MW_t (although pulsed research reactors achieve very high instantaneous power levels). They vary as widely in design as their applications, but their shared feature is that they operate at far lower temperatures and consume much less fuel than do power reactors. They also use fuel that is considerably more enriched than that used in power reactors in order to achieve the necessary power densities to produce external beams of neutrons or activate targets to high specific activities [63].

Of the approximately 280 research reactors operating in the world today (including critical assemblies), almost half of the older ones are fueled by U enriched to >20% ^{235}U, the designated lower limit for weapons-usable enriched U [64,65]. Use of such intermediate- and high-enriched U allows more compact cores, high neutron fluxes, and longer times between refueling. However, many research reactors are located at universities and have more relaxed security than a military or commercial site, raising concerns about safeguarding the fuel. As a result, many research reactors are undergoing modification to operate with fuel containing less ^{235}U. The United States is no longer exporting >20% reactor fuel. Samples of U and Pu obtained from research reactor fuel are likely to require assessment by the nuclear forensic analyst.

Consider in detail the production of Pu isotopes in a nuclear reactor. Plutonium is formed either in the U fuel of the reactor core or in an unenriched U blanket external to the core (as in an LMFBR). The U target may initially possess almost any mixture of the two major isotopes of U and may also include small quantities of U isotopes that do not occur in nature if the material was recovered from previous reactor applications. As discussed previously, the production of Pu is initially through capture of a neutron by ^{238}U, followed by two successive decays to form ^{239}Pu. As the reactor irradiation proceeds, the ^{239}Pu product also becomes a target, capturing neutrons to form the heavier plutonium isotopes:

$$^{239}\mathrm{Pu(n,\gamma)}\ ^{240}\mathrm{Pu} \quad \sigma_{\mathrm{th}} = 280\ \mathrm{b},$$

$$^{240}\mathrm{Pu(n,\gamma)}\ ^{241}\mathrm{Pu} \quad \sigma_{\mathrm{th}} = 280\ \mathrm{b},$$

$$^{241}\mathrm{Pu(n,\gamma)}\ ^{242}\mathrm{Pu} \quad \sigma_{\mathrm{th}} = 360\ \mathrm{b},$$

and so forth. Some of the neutrons present in the material have sufficient energy to induce a fast-neutron reaction, producing small amounts of ^{238}Pu:

$$^{239}\mathrm{Pu}\left(\mathrm{n,2n}\right)\ ^{238}\mathrm{Pu} \quad E_{\mathrm{min}} = 5.7\ \mathrm{MeV}.$$

However, with long irradiation times, there is another pathway to ^{238}Pu, via three successive neutron-capture reactions on the ^{235}U-derived component of the fuel:

$$^{235}\mathrm{U(n,\gamma)}\ ^{236}\mathrm{U} \quad \sigma_{\mathrm{th}} = 97\ \mathrm{b},$$

$$^{236}\mathrm{U(n,\gamma)}\ ^{237}\mathrm{U} \quad \sigma_{\mathrm{th}} = 24\ \mathrm{b},$$

$$^{237}\mathrm{U}\left(6.75\ \mathrm{d}\right) \rightarrow ^{237}\mathrm{Np} + \beta^-,$$

$$^{237}\mathrm{Np(n,\gamma)}\ ^{238}\mathrm{Np} \quad \sigma_{\mathrm{th}} = 180\ \mathrm{b},$$

$$^{238}\mathrm{Np}\left(2.12\ \mathrm{d}\right) \rightarrow ^{238}\mathrm{Pu} + \beta^-.$$

The Pu recovered from irradiated U fuel or blankets is always composed of a mixture of Pu isotopes, providing a basis for fingerprinting the material. At very low burn-ups (low neutron fluence), ^{239}Pu dominates the mixture. At higher burn-ups, the other Pu isotopes become more important. The most efficient use of the U starting material is through high burn-up. However, the even-mass Pu isotopes have significant probabilities of decaying by spontaneous fission, which is not conducive to using the material in nuclear weapons due to neutron emission inducing preinitiation of the device. For that application, a compromise is made between the production of unwanted Pu isotopes and inefficient use of the U starting material, and it has historically resulted in a ^{240}Pu content that is approximately 6% by mass in US weapons-grade Pu. Other countries have chosen other Pu isotopic compositions.

Neutron fluence is not the only variable determining the isotopic composition of a sample of Pu. Neutron-capture reactions can be very energy sensitive, particularly for neutrons with a kinetic-energy distribution defined by the reactor operating temperature. The identity of the moderator (e.g., light water, heavy water, or graphite), the temperatures of the moderator and fuel during reactor operations, and the dimensions of the fuel elements (distance between the point of neutron release and the moderator) all affect the production of Pu isotopes. The starting composition of the reactor fuel also influences the spectrum of neutrons available for Pu production reactions, as the capture of neutrons in the resonance region by ^{235}U alters the spectrum of residual neutrons available for reaction with ^{238}U to produce Pu.

The flux (or intensity) of the neutrons in the reactor fuel also affects the production of Pu isotopes. Consider the formation of ^{240}Pu: a neutron is captured by ^{238}U to form ^{239}U, which decays within minutes to ^{239}Np, which in turn decays with a 2.36-day half-life to ^{239}Pu. At low flux, a ^{239}Np atom decays to ^{239}Pu before it will encounter and react with another neutron, and ^{240}Pu production is solely through the capture of a neutron by ^{239}Pu. However, at higher flux, the probability of a neutron being captured by ^{239}Np before it can decay increases. In this case, ^{240}Pu production has a contribution from the subsequent decay of short-lived ^{240}Np. Even though the neutron energy spectrum can be independent of flux, the relative capture probabilities of ^{239}Np and ^{239}Pu are not, and they result in a flux dependence of the relative amounts of ^{239}Pu and ^{240}Pu in the final product. In Figure 3.17, a section of the chart of the nuclides outlines the main neutron-capture reactions and decay processes by which Pu isotopes are produced from U in reactor fuel.

The ORIGEN2 code [66] can calculate the nuclide contents of nuclear fuels at specified burn-up intervals for several types of reactor. It is available from the RSIC Computer Code Collection, CCC-371, revised in 1991, and is both easy to use and well documented. The energy dependencies of the various reactions that create and destroy each product in a reactor irradiation are integrated over the spectrum of neutrons expected for a given reactor type. This results in a set of scalar values for the

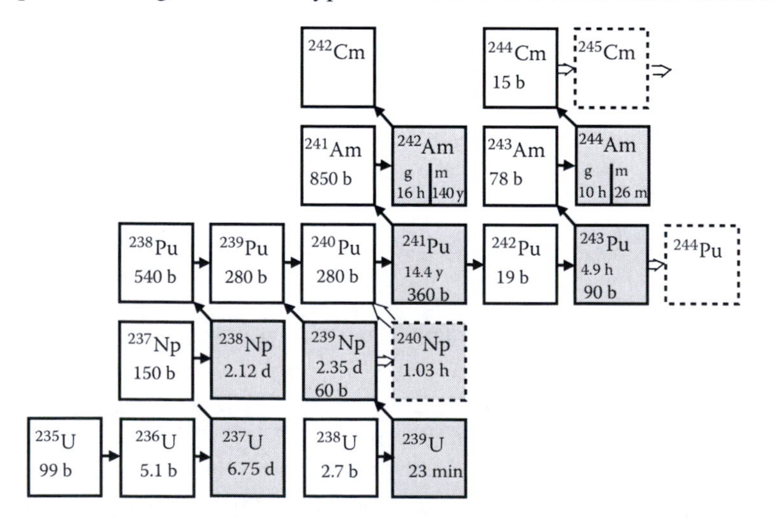

FIGURE 3.17 A section of the chart of the nuclides showing the production pathways for Pu isotopes and collateral heavy-element signatures in reactor fuel, starting with the U isotopes. Pathways indicated by dashed boxes and unfilled arrows become more important at high flux. Capture cross sections are given for thermal neutrons.

net formation of each product nuclide that is specific for a given reactor. These values (the library for that reactor) and the decay constants of the reaction products make up the parameters of a set of coupled, linear, first-order differential equations that are a function of reactor power. Starting from an initial composition of fuel, defined by the user, the code solves these differential equations at a specified exposure time and reactor power level to produce a total nuclide inventory.

What is not modeled in the ORIGEN2 calculation, however, is the inhomogeneity in the production of Pu isotopes at different locations in the reactor fuel based on local variations in neutron fluence [67]. During operation of a thermal reactor, fission-spectrum neutrons are born within the fuel. Some of these neutrons react with nearby fissile nuclei, whereas others enter the moderator and lose energy before encountering fissile material again. This results in the U of a given fuel rod being exposed to neutrons whose average energy is a function of the radial depth at which the reaction takes place within the rod. Uranium and Pu atoms near the surface of the fuel rod are, on average, exposed to a softer neutron spectrum than are the U and Pu atoms near the center of the rod. This results in a distribution of product Pu atoms as a function of depth that is different from the distribution of fission products. It also results in the relative isotopic composition of the Pu atoms being a function of depth as well. On a larger scale, the neutron spectrum in a given fuel rod is dependent on its proximity to a control element, as well as on its proximity to the outer edge of the reactor core. However, irradiated fuel is rarely reprocessed on the scale of a single fuel rod, and ORIGEN2 successfully models the average isotopic inventory of a reactor core as functions of operating time and power level.

With the set of ORIGEN2 libraries, it is possible to calculate the nuclide inventories of common reactor types at several reactor power levels and burn-ups. Under realistic operating conditions, reactor power and operating temperatures have smaller effects on the final Pu composition than do changes in moderator type, fuel enrichment, and burn-up. However, even more subtle effects are important in establishing the isotopic fingerprint of a reactor. A freely available ORIGEN2 library for the class of graphite-moderated reactor that has produced most of the world's weapons-grade Pu does not exist. For this application, an article on the isotopics of Calder–Hall-type reactors provides a starting point to construct a pertinent comparison library [67].

The calculated correlations between the mass ratio $^{240}Pu/^{239}Pu$ and the activity ratio $^{238}Pu/(^{239}Pu + ^{240}Pu)$ for three standard reactor types and an external blanket are given in Figure 3.18. In all cases, reactor power was held at 37.5 MW_t/t of fuel (or blanket). These parameters are two of the fundamental radiochemical observables: the $^{238}Pu/(^{239}Pu + ^{240}Pu)$ activity ratio is measured directly by α spectroscopy of a chemically separated sample, and the $^{240}Pu/^{239}Pu$ mass ratio is determined by mass spectrometry. The value of ^{239}Pu used in the calculation of these ratios is actually the sum of the number of atoms of ^{239}U, ^{239}Np, and ^{239}Pu given in the code output at the appropriate irradiation length. Because irradiated fuel is allowed to decay for an extended length of time before it is reprocessed, all of the short-lived ^{239}Pu precursors will decay to ^{239}Pu. If focused on Pu of nominal US weapons-grade composition (6% ^{240}Pu by mass, or $^{240}Pu/^{239}Pu = 0.064$, shown as the dashed vertical line in the figure), a variation of more than two orders of magnitude in $^{238}Pu/(^{239}Pu + ^{240}Pu)$ is observed: 2.5×10^{-4} for Pu produced in a blanket, 5×10^{-3} for production in a CANDU reactor, and 5×10^{-2} for Pu produced in LWRs operating with fuel enriched

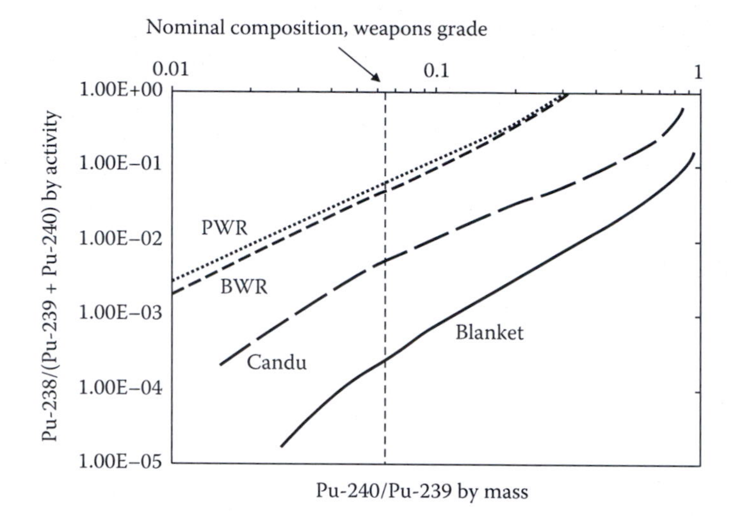

FIGURE 3.18 Results of ORIGEN2 calculations for several reactor types, correlating two Pu observables that arise in a nuclear forensic analysis: $^{238}Pu/(^{239}Pu + ^{240}Pu)$ by activity and $^{240}Pu/^{239}Pu$ by mass.

to 3.2% ^{235}U. Much of the small difference between the BWR and PWR curves is caused by the different operating temperatures and density of the coolant/moderator. Clearly, the Pu isotopic fingerprint is determined by the reactor type used for Pu production. In implementing Figure 3.18, however, a small correction must be made for the decay of ^{238}Pu (87.7-year half-life) between the time of fuel discharge and the time that the activity ratio was measured.

Figure 3.19 shows the production of heavier Pu isotopes relative to ^{239}Pu in a 37.5-MW_t/t BWR, with 3.2% ^{235}U-enriched fuel, as a function of irradiation length.

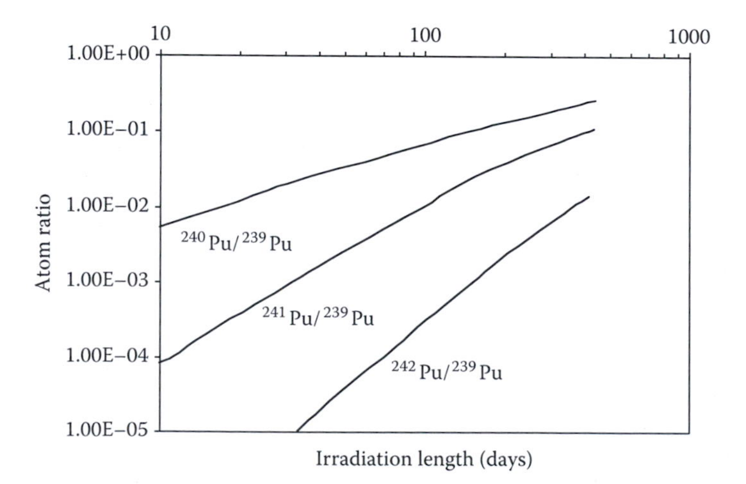

FIGURE 3.19 ORIGEN2 calculation of the indicated Pu isotope ratios as a function of irradiation time, at a power level of 37.5 MW_t/t in a BWR fueled with U enriched to 3.2% in ^{235}U.

If the isotopic signature of a Pu sample arising from the operation of a BWR is measured, as determined from Figure 3.18, the $^{240}Pu/^{239}Pu$ value and Figure 3.19 define the neutron fluence through that sample, which is 37.5 MW_t/t times the irradiation length obtained from the plot. For weapons-grade Pu of about 6% ^{240}Pu by mass, Figure 3.19 specifies an irradiation time of approximately 100 days, which results in an integrated flux of 3750 MW-days/t. Figure 3.19 can also be used to estimate the time elapsed since the fuel from which the sample was derived was discharged from the reactor: whereas $^{240}Pu/^{239}Pu$ and $^{242}Pu/^{239}Pu$ are the relative concentrations of long-lived radionuclides, the half-life of ^{241}Pu is 14.4 years, making the $^{241}Pu/^{239}Pu$ value sensitive to elapsed time. The downward deviation of a measured $^{241}Pu/^{239}Pu$ datum in comparison to the calculated value from the $^{241}Pu/^{239}Pu$ curve in Figure 3.19, at an irradiation time determined by $^{240}Pu/^{239}Pu$ or $^{242}Pu/^{239}Pu$, is a result of the radioactive decay of ^{241}Pu from the time the radionuclides in the sample were last exposed to reactor neutrons. Chronometry will be discussed in more detail later.

Figure 3.20 depicts the two plutonium parameters most sensitive to changes at high reactor flux for fixed weapons-grade ^{240}Pu content, as a function of reactor power, in a natural-U-fueled, graphite-moderated reactor. Note that the measured Pu parameters are plotted on a linear scale in the figure. At the high-flux extreme of reactor performance, there is a small power dependence of these parameters. However, it is not sufficiently strong to preclude the use of Figure 3.18 (logarithmic) to determine production reactor type, although it does show how flux can subtly affect the Pu isotopic fingerprint of a reactor.

Unlike the Pu isotopes, the relative concentrations of ^{242m}Am, ^{243}Am, ^{242}Cm, and ^{244}Cm in irradiated fuel are very sensitive to changes in reactor power. This fact is primarily due to the moderately short half-life of ^{241}Pu. The probable irradiation interval in a natural-U-fueled reactor for the production of weapons-grade Pu is

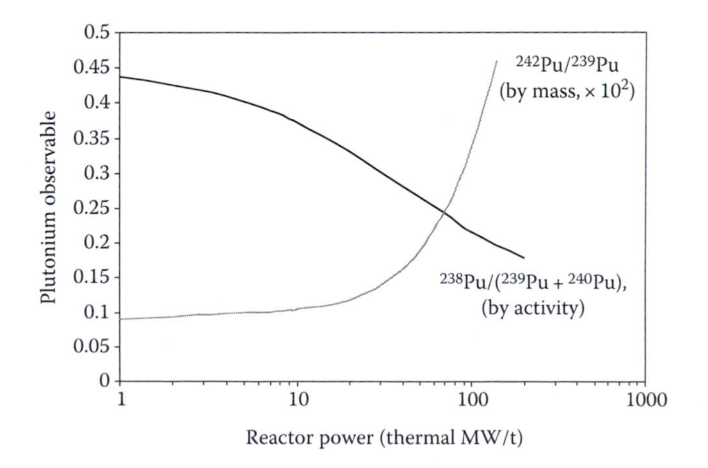

FIGURE 3.20 ORIGEN2 calculation of the reactor-power dependence of two Pu observables, $(^{242}Pu/^{239}Pu)_{mass}$ and $[^{238}Pu/(^{239}Pu + ^{240}Pu)]_{activity}$, for a fixed Pu-product ^{240}Pu concentration of 6% by mass, in a graphite-moderated reactor with natural-U fuel.

between 10 days and 2 years. The half-life of ^{241}Pu is 14.4 years, so its decay during the much-shorter irradiation interval does not significantly affect the production of species arising directly or indirectly from neutron reactions with it. These include ^{242}Pu, ^{243}Am, and ^{244}Cm:

$$^{241}\text{Pu}(\text{n},\gamma)\ ^{242}\text{Pu} \quad \sigma_{\text{th}} = 360\ \text{b},$$

$$^{242}\text{Pu}(\text{n},\gamma)\ ^{243}\text{Pu} \quad \sigma_{\text{th}} = 19\ \text{b},$$

$$^{243}\text{Pu}(4.96\ \text{h}) \rightarrow ^{243}\text{Am} + \beta^-,$$

$$^{243}\text{Am}(\text{n},\gamma)\ ^{244\text{g}}\text{Am} \quad \sigma_{\text{th}} = 74\ \text{b},$$

$$^{243}\text{Am}(\text{n},\gamma)\ ^{244\text{m}}\text{Am} \quad \sigma_{\text{th}} = 4\ \text{b},$$

$$^{244\text{g}}\text{Am}(10.1\ \text{h}) \quad \text{and} \quad ^{244\text{m}}\text{Am}(26\ \text{min}) \rightarrow ^{244}\text{Cm} + \beta^-.$$

Although irradiation length has little effect on the quantity of undecayed ^{241}Pu available as a target, the amount of ^{241}Am that is present in a given quantity of Pu increases almost linearly with time. This means that the effective ^{241}Am target available for neutron reactions during a long irradiation at low flux is larger than that present during a shorter irradiation at higher flux, even if the burn-up values are equivalent. This behavior results in lower relative concentrations of the ^{241}Am reaction products, ^{242m}Am and ^{242}Cm, at higher reactor power than at lower power.

$$^{241}\text{Am}(\text{n},\gamma)\ ^{242\text{m}}\text{Am} \quad \sigma_{\text{th}} = 73\ \text{b},$$

$$^{241}\text{Am}(\text{n},\gamma)\ ^{242\text{g}}\text{Am} \quad \sigma_{\text{th}} = 780\ \text{b},$$

$$^{242\text{g}}\text{Am}(16\ \text{h}, 83\%) \rightarrow ^{242}\text{Cm} + \beta^-.$$

Figure 3.21 plots several concentration ratios of transplutonium nuclides at discharge, as a function of reactor power, for a natural-U-fueled, graphite-moderated reactor at burn-ups that result in Pu containing 6% ^{240}Pu by mass. As expected, there is strong flux dependence in the ratios of product concentrations arising from ^{241}Am reactions relative to those arising from ^{241}Pu reactions. The value of $^{244}Cm/^{243}Am$ is roughly flux-independent because both products arise primarily through multiple neutron capture by ^{241}Pu.

The nuclides ^{242m}Am, ^{243}Am, ^{242}Cm, and ^{244}Cm do not result from decay processes in a sample of Pu isotopes. If present in such a specimen, they must be residual radionuclides remaining after incomplete fuel reprocessing, discussed in detail later.

The half-life of ^{242}Cm is only 163 days. If the concentrations of ^{242m}Am, ^{243}Am, ^{242}Cm, and ^{244}Cm in a Pu sample are measured, and if the reactor type and total

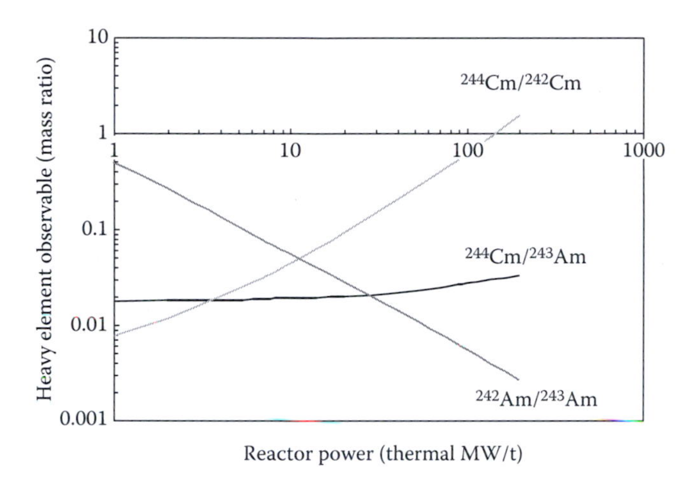

FIGURE 3.21 ORIGEN2 calculation of the effect of reactor flux on the production of heavy elements in irradiations tailored to produce 6% ^{240}Pu by mass in a graphite-moderated reactor with natural-U fuel.

burn-up are known (e.g., from the Pu isotopes), so that the equivalent of Figure 3.21 can be constructed for that sample, then ^{242m}Am/^{243}Am can be used to determine the reactor power and set the value of ^{244}Cm/^{242}Cm at the time of fuel discharge. The difference between $(^{244}\mathrm{Cm}/^{242}\mathrm{Cm})_{\mathrm{discharge}}$ and $(^{244}\mathrm{Cm}/^{242}\mathrm{Cm})_{\mathrm{observed}}$ results from the radioactive decay of the Cm isotopes and defines the time elapsed between reactor discharge and the measurement, similar to the use of ^{241}Pu/^{239}Pu and Figure 3.19. In cases in which Pu has been recovered and returned to the reactor, the transplutonium isotopes reflect a discharge signature that is characteristic of the last fuel cycle, whereas the ^{241}Pu content is a complicated function of the prior history of the Pu sample through all reactor production and subsequent purification phases.

Although not discussed in detail, the ^{232}Th–^{233}U fuel cycle results in similar isotopic signatures. The equivalent of Figure 3.17 can be constructed for the production of actinides from Th. Similar to Pu produced in reactors, ^{233}U will contain small amounts of all other U isotopes. This admixture of heavier uranium isotopes is particularly important when ^{235}U has been used to jump-start the breeding process. The heavier U isotopes ultimately provide an avenue to the production of the Pu isotopes, which should be dominated by ^{238}Pu. However, the ORIGEN2 libraries are less complete for reactors operating on Th-based fuels than they are for those burning U-based fuels.

3.5 RECOVERY AND PURIFICATION OF HEAVY ELEMENTS FROM REACTOR PRODUCTS

Reprocessing is a somewhat generic term in nuclear technology that refers to the chemical recovery of product nuclei following irradiation of fertile target material within an isotope production apparatus. A nuclear reactor is most often used for this purpose, with thermal, fission-spectrum, and epithermal neutrons inducing the desired nuclear transformations. For example, ^{237}Np targets are reprocessed for ^{238}Pu, and ^{232}Th blankets around breeder reactors are reprocessed for ^{233}U. However, by far, the most important reprocessing efforts, contemporary and historic, have focused on the isolation of ^{239}Pu from depleted U blankets or spent reactor fuel. The development of radiochemical reprocessing protocols for ^{239}Pu in irradiated nuclear targets began in 1944 with the Manhattan Project, and the technology has evolved ever since [14,68]. The first determinations of the chemical properties of the new element Pu were performed in 1941. The fact that industrial procedures were in place for separating kilogram quantities of Pu from tons of intensely radioactive U and fission products in just 3 years was an impressive accomplishment of the twentieth-century technology.

Fission products build up in reactor fuel more quickly than do Pu isotopes, and they are the neutron poisons that, combined with the loss of reactivity associated with the consumption of the fissile fuel, ultimately make it impossible to maintain reactor power. Most of the radioactivity in freshly discharged reactor fuel of any significant burn-up is a result of the decay of the fission products (except for Th-based fuels, for which a significant fraction of the external dose is caused by emissions from the daughters of ^{232}U decay). The fission yield is distributed over hundreds of different nuclides, covering a wide range of atomic numbers and masses.

Most fission-product nuclides are radioactive, with half-lives ranging from subsecond to millions of years. Other fission products are stable, contributing nothing to the radiation environment. During reactor irradiation, the quantity of a given short-lived species quickly reaches an equilibrium value, where its rate of loss as a result of decay (or neutron capture) balances its rate of production. Products with half-lives comparable to, or longer than, the reactor irradiation interval continue to build up in the fuel, as per the Activation Equation. When reactor operations stop (e.g., with the discharge of the fuel), production of significant amounts of fresh fission products ceases, and radioactive decay results in a substantial reduction in total radioactivity over the next several months, leaving only longer-lived and stable species to be removed by fuel reprocessing.

Table 3.1 shows the results of an ORIGEN2 calculation of selected long-lived and stable fission-product yields in a graphite-moderated, natural-U-fueled reactor, operated at a power level of 37.5 MW_t/t. The calculation was run for an irradiation interval to produce Pu containing 6% ^{240}Pu by mass, and the fission-product yields given in Table 3.1 are for the fraction of the fuel containing 1 g of the mixed Pu isotopes. The values in Table 3.1 are quantities present soon after the end of irradiation, and after short-lived fission products have decayed to the observed nuclides. The shorter-lived species (e.g., ^{106}Ru, ^{125}Sb) would need correction for decay to some later time in a typical investigation.

The fission products listed in Table 3.1 span a wide range of chemical properties, from alkali metals (^{135}Cs, ^{137}Cs) to alkaline earths (^{90}Sr), lanthanides (^{143}Nd, ^{155}Eu), transition metals (^{93}Zr, ^{99}Tc), noble metals (^{106}Ru), halogens (^{129}I), and amphoteric elements (^{125}Sb). Chemistry designed for the recovery of Pu from irradiated fuel must provide a separation from all of these elements, other fission and activation products, and the actinides, including a large quantity of unburned U. The U is still present at the level of 510 g/g of Pu product; 99.77% of the U in the initial reactor charge

TABLE 3.1
Long-Lived Fission Products Present at Discharge in Reactor Fuel Used to Produce 1 g of Pu Containing 6% ^{240}Pu by Mass

Nuclide	Half-Life (Years)	Mass (mg)	Activity (dpm)
^{90}Sr	28.9	11	3.4×10^{12}
^{93}Zr	1.64×10^{6}	14[a]	8.0×10^{7}
^{99}Tc	2.1×10^{5}	16	6.1×10^{8}
^{106}Ru	1.02	4.1	3.0×10^{13}
^{125}Sb	2.76	0.45	1.0×10^{12}
^{129}I	1.61×10^{7}	3.4	1.3×10^{6}
^{135}Cs	2.0×10^{6}	2.4	7.1×10^{6}
^{137}Cs	30.1	22	4.2×10^{12}
^{143}Nd	Stable	20	0
^{155}Eu	4.76	0.23	2.4×10^{11}

Note: ORIGEN2 calculation, natural-U-fueled, graphite-moderated reactor, 37.5 MW_t/t.

[a] Lower limit; ^{93}Zr is also produced in claddings.

remains in the fuel; and the ^{235}U content has been reduced from natural (0.72%) to 0.62% by mass. Reprocessing must remove all of these materials and still provide a quantitative recovery of Pu. These same issues also arise for the recovery of U from spent Th fuel. A further complication is that most of the separation processes must be performed remotely because of the intense radiation field of both the mixture and (to a lesser extent) the final product itself. In the later steps of a reprocessing procedure, when the fissile product has become more concentrated, batch size is limited by the constraints of criticality safety. Nuclear reprocessing is a daunting prospect: there must be a balance between maximizing the yield of precious fissile product while minimizing the content of impurity species that can result in a contaminated final product. These residual contaminants, which can be detected at levels of <1 disintegration per minute (dpm) with standard radiochemical techniques, provide a chemical "fingerprint" profile of the industrial process used to recover the material.

A number of diverse reprocessing strategies are possible for the recovery of ^{239}Pu, and some are still undergoing active research. These techniques have included coprecipitation (e.g., with LaF_3, $BiPO_4$), molten-salt extraction (in KCl, NaCl, $MgCl_2$), ion exchange, and fluoride volatility. However, the most developed analytic methods have been based on solvent extraction and are designated by names such as Purex, Redox, Halex, and Butex. Detailed discussion of these varied approaches is beyond the scope of this treatment, but in-depth reviews with primary reference citations are available [69–71].

Consider the contaminant fingerprints remaining after the two reprocessing procedures implemented during the early days of nuclear technology. The Butex process, at one time employed at the Windscale works in Great Britain, uses solvent extraction with dibutyl carbitol. This extraction protocol provides decontamination factors of more than 10^6 for all of the fission products listed in Table 3.1 except Ru, for which the decontamination factor is only about 10^3. Thus, a 1-g sample of Pu at the first purification steps in early Windscale operations also contained approximately 10^7 dpm of ^{106}Ru, an amount that is directly observable by γ spectroscopy. Although later purification steps remove much of the residual Ru, Pu recovered from irradiated fuel by the Butex process is characterized by a significant ^{106}Ru impurity. The hexone process, used at Hanford in the early 1950s, employs solvent extraction with methyl isobutyl ketone (MIBK). The hexone process is more successful at removing Ru than is Butex (decontamination factor >10^4), but it provides decontamination factors of only 10^5 for Zr, Nb, and Ce. A sample of hexone postprocessing solution with 1 g of Pu would therefore also contain about 1000 dpm of ^{93}Zr, easily observable via modern radiochemistry.

Variants of the Purex (Pu–U reduction extraction) process are the most widely used reprocessing schemes in the world. Purex on the industrial scale began at the US Savannah River Plant in 1954, and every country that has produced significant quantities of Pu has exploited the method. The heart of Purex is the extractant, TBP. In addition to optimum complexation properties for nuclear analytes of interest, TBP is chemically and radiolytically stable and has low aqueous solubility. It is also completely miscible with common organic solvents (e.g., kerosene, n-dodecane) at ordinary temperatures.

One variation of the Purex process for Pu and U isolation from neutron-irradiated U target material is described later. The Purex process is incorporated into the full separation scheme depicted in Figure 3.22 for reprocessing spent U fuel. The discussion that follows here serves strictly as a relevant example and lacks comprehensive experimental detail. The method relies on the strong transport of Pu(IV) and U(VI)

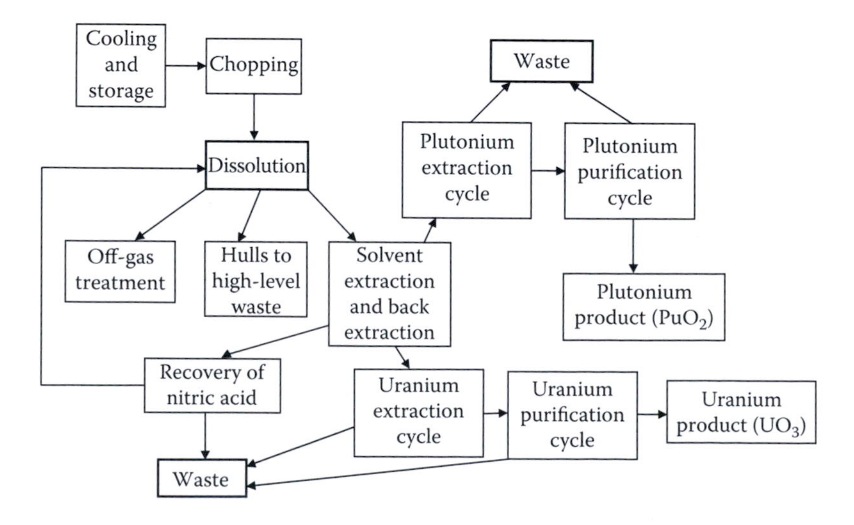

FIGURE 3.22 Flow diagram for a U–Pu reprocessing facility. In a typical reprocessing operation, the fuel is chopped and dissolved, and U and Pu are extracted from the resulting nitric acid solution, usually by the Purex process.

into TBP by countercurrent solvent extraction, efficiently eliminating fission products and actinide by-products such as Th, Am, and Cm. Purex also effects clean separations of the desired radioelements (Pu, U, and Np) from each other.

Before reprocessing begins, a suitable delay time allows the decay of intermediate product species (e.g., 6.75-day ^{237}U, which would result in the ingrowth of Np in the recovered U) and short-lived, volatile fission products (especially 8-day ^{131}I and 5.2-day ^{133}Xe, which might migrate into the effluent of the reprocessing plant). The outer cladding of a fuel rod or blanket element may then be removed by chemical dissolution (e.g., if Al-clad) or may be mechanically chopped to expose the U (e.g., if clad with corrosion-resistant Zr alloy). The bare U-ceramic pellets are then dissolved in hot, high-concentration HNO_3.

The Pu in solution, initially in the +6 oxidation state, is reduced to Pu(IV) by the addition of a reagent such as $NaNO_2$, N_2O_4, or hydroxylamine. Pu(IV) and U(VI) are then extracted as $Pu(NO_3)_4 \cdot 2(TBP)$ and $UO_2(NO_3)_2 \cdot 2(TBP)$, respectively, into an organic phase consisting of ~30% TBP in a nonpolar, inert diluent (often kerosene). This first step provides initial decontamination from most of the fission products. Ferrous sulfamate then reduces Pu(IV) to Pu(III), which is back-extracted into another aqueous solution. This step separates U from Pu. Another TBP extraction of the aqueous Pu product stream provides additional fission-product decontamination. The Pu(III) is then oxidized to Pu(IV) to prepare it for concentration by ion exchange, after which it is reduced and precipitated as PuF_3. Subsequent roasting in an oxygen atmosphere converts the Pu to a mixture of PuF_4 and PuO_2. Finally, reduction with elemental Ca produces Pu metal.

Uranium is stripped from the initial TBP phase into a low-acid medium, concentrated, and subjected to a second extraction cycle for further separation from Pu and fission species. Multiple evaporation steps concentrate the purified U solutions to $UO_2(NO_3)_2 \cdot 6H_2O$. Thermal denitration then produces UO_3 as a final product for storage.

The recovery of U from spent Th-based fuel is accomplished using a very similar process, called Thorex, also based on a TBP extraction [72,73]. The principal difference compared to Purex is that in Thorex, the valuable fissile material is U, which is the primary product stream, rather than an ancillary side stream. In addition, the complementary actinide (Th) can be neither oxidized nor reduced from the IV state. As with Purex, there is a necessary decay time before Thorex chemistry, required by the long half-life of ^{233}Pa, which decays to the fissile product. An outline of a reprocessing scheme incorporating the Thorex process is shown in Figure 3.23.

Performance parameters for the Purex process are impressive. For a two-cycle procedure, the recoveries of both Pu and U are on the order of 99.9%, whereas the Pu/U separation factor is 10^6. Plutonium decontamination factors from the fission products are approximately 10^8, and the only impurities from Table 3.1 that are sufficiently extracted and detected in the final product are ^{93}Zr, ^{99}Tc, and ^{106}Ru. However, our experience is that the light rare earths, Th, Np, and the trivalent actinides Am and Cm, which undergo nonnegligible complexation in nitrate media [74], are present at concentrations detectable in a gram-sized Pu sample by radiochemical means. This observation is important because, as discussed earlier in conjunction with Figure 3.21, the relative concentrations of Am and Cm isotopes are important indicators of reactor flux.

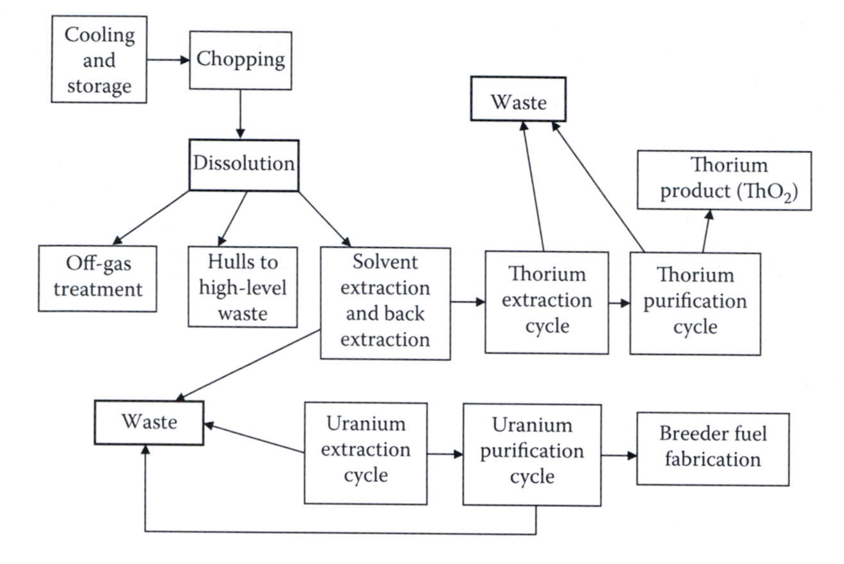

FIGURE 3.23 Flow diagram for a Th–U reprocessing facility. The basic operations are similar to those outlined in Figure 3.22 for the U–Pu process. The main difference is that U is the fissile component rather than the fertile component, so more care must be taken for efficient recovery. The usual process is Thorex.

At low fuel burn-ups, such as those employed for the production of weapons-grade Pu, the concentrations of most of the stable fission products are reasonably constant relative to each other and to ^{239}Pu, independent of reactor power. In Table 3.2, results are given of ORIGEN2 calculations of the production of stable Nd isotopes in a graphite-moderated, natural-U-fueled reactor, as a function of reactor power. In each case, the calculation was run for an irradiation interval to produce Pu containing 6% ^{240}Pu by mass. The required irradiation time is 775 days at 1.25 MW_t/t, but only 12.15 days at 125 MW_t/t. These values are reasonable extremes for the reactor power, spanning minimum production requirements and the limitation on reactor performance. Values are given relative to ^{143}Nd production, which, in turn, is given relative to the production of total Pu. The isotopic composition of naturally occurring Nd is also listed for comparison.

From Table 3.2, the isotopic composition of fission-product Nd (except for the minor isotope ^{142}Nd, whose production is largely through neutron capture by fission-product ^{141}Pr) is observed to be relatively insensitive to neutron flux over a broad span of reactor power. Similar behavior is seen for the production of Nd relative to total Pu of fixed composition. However, the distribution of isotopes in fission-product Nd is significantly different from the composition of natural Nd. If the distribution of isotopes in a Nd fraction separated from a Pu sample is measured, the flux independence of the fission yields can be used to deconvolute the relative amounts of natural and radiogenic Nd. The quantity of natural Nd provides more information about the reprocessing chemistry.

Reagent-grade solvents contain impurity levels on the order of 100 parts per billion (ppb) of Nd (by mass). Neodymium levels in industrial solvents are likely higher.

TABLE 3.2
Neodymium Isotope Ratios (by Mass) and the Production of ^{143}Nd vs Total Pu Production (6% ^{240}Pu) as a Function of the Power Level of a Natural-U-Fueled, Graphite-Moderated Reactor

	Reactor Power Level (MW_t/t)			Natural
Mass Ratio	**1.25**	**12.5**	**125**	**Nd**
^{142}Nd/^{143}Nd	0.00172	0.00081	0.00019	2.212
^{144}Nd/^{143}Nd[a]	0.964	0.951	0.949	1.973
^{145}Nd/^{143}Nd	0.704	0.697	0.699	0.692
^{146}Nd/^{143}Nd	0.571	0.567	0.572	1.478
^{148}Nd/^{143}Nd	0.343	0.342	0.351	0.487
^{150}Nd/^{143}Nd	0.160	0.159	0.160	0.484
^{143}Nd/^{tot}Pu	0.0202	0.0202	0.0194	—

Note: Natural Nd isotopic composition is included for comparison.
[a] Assumes complete decay of 285-day ^{144}Ce.

The solubility of reactor fuel (mostly uranium oxide) in nitric acid is less than 500 g/L, and the weapons-grade Pu content of the fuel is no more than a few tenths of a percent of the U. This means that the total chemical-reagent volume used in the early steps of an aqueous fuel-reprocessing procedure is on the order of 10 L/g of Pu. Therefore, impurities in the reagents introduce roughly a milligram of natural Nd into the process for each gram of Pu, which is associated with 75 mg of fission-product Nd (Table 3.2). Even if decontamination factors for Nd in the reprocessing are 10^8, approximately 1 ng of the initial Nd content present in the Pu sample will remain and is adequate for mass-spectrometric analysis. Natural Nd introduced as contaminants in the smaller quantities of reagents used in later chemical steps, or in metallurgical processes, will also be present in the final sample, and at higher chemical yield. The relative contributions of natural and fission-product Nd are defined by the purity and volumes of the reagents used in the chemistry and are likely unique to the processing plant, a valuable forensic signature.

Table 3.3 gives the concentrations of selected heavy-element nuclides, produced in natural-U reactor fuel, from the same ORIGEN2 calculation that produced the fission-product data of Table 3.1. Once again, the results are scaled to the production of 1 g of Pu that is 6% ^{240}Pu by mass. As before, short-lived species are assumed to have completely decayed to longer-lived species during the cooling period, but no corrections for radioactive decay since fuel discharge were made. The activities of the transplutonium impurities in Table 3.3 are similar in magnitude to the activities of the long-lived fission products in Table 3.1.

Detection techniques based on α spectroscopy are more sensitive than those involving the counting of photons or electrons, as α-detector backgrounds can be very low. One might therefore conclude that the signatures of incomplete fuel reprocessing would be easier to detect through heavy-element contaminants than through

TABLE 3.3
Production of Selected Heavy-Element Nuclides in Reactor Fuel (at Discharge) Used to Produce 1 g of Pu Containing 6% ^{240}Pu by Mass

Nuclide	Half-Life (Years)	Mass (g)	Activity (dpm)
^{237}Np	2.14×10^{6}	5.6×10^{-3}	8.8×10^{6}
^{241}Am[a]	432	9.7×10^{-6}	7.4×10^{7}
^{242m}Am	141	5.3×10^{-8}	1.2×10^{6}
^{243}Am	7370	3.9×10^{-6}	1.7×10^{6}
^{242}Cm	0.446	4.4×10^{-7}	3.2×10^{9}
^{244}Cm	18.1	7.6×10^{-8}	1.4×10^{7}

Note: From the same ORIGEN2 calculation used to generate Table 3.1 data.
[a] Assumes no ingrowth from postdischarge decay of ^{241}Pu.

fission products. However, many of the nuclides listed in Table 3.3 are also progeny species that grow into a purified sample of Pu from radioactive decay. This reduces the effective sensitivity of the technique for these nuclides by several orders of magnitude. However, the ingrowth of progeny species is the basis of chronometry and will be discussed further in this work. The nuclides ^{242m}Am, ^{243}Am, ^{242}Cm, and ^{244}Cm are not progenies of decay processes in reprocessed Pu samples, and their use as reactor-flux indicators was discussed earlier.

Nuclear reprocessing operations are prime targets for nonproliferation and intelligence surveillance measures. Fugitive emissions of signature species may be detected in situ or collected to provide potential information on classified or covert weapons-related activities [75–77]. Such specimens can be measured using nuclear forensic techniques, and the analytes of interest here are predominantly inorganic and radioisotopic in nature. They include airborne signatures like isotopes of Kr and Xe; other volatile species such as I, Te, and Sb; various fission-product activities within condensed-phase samples; and actinide isotope ratios.

However, organic analyses can also provide forensic data pertinent to investigations of reprocessing facilities. The intense radiation fields accompanying reprocessing have marked effects on the more sensitive chemical components of extraction processes, including Purex. Radiolysis gives rise to energy- and charge-transfer reactions as well as to electron scavenging, and the primary products of such radiation chemistry are excited-state and ionic species of the chemical reactants.

Products of the autoradiolysis of solvent systems can seriously affect the performance of the extractions. At large doses, the concentrations of radiation-induced primary species build up and participate in secondary reactions that can result in higher-molecular-weight and polymeric products. Although previously used TBP reprocessing solutions are washed before recycle, at an integrated dose of about 10^{8} R, the radiation chemistry of TBP induces polymerization to produce compounds that extract Pu(IV) irreversibly.

Radiation decomposition of the TBP–HNO_3 system results in the formation of phosphoric acid, monobutyl and dibutyl phosphoric acids, n-butanol, and 1-nitrobutane, among other products. In addition, the presence of HNO_3 increases the radiolytic degradation of TBP. TBP breakdown also varies as a function of the protecting properties of whatever solvent is used as the diluent. A number of different compounds have been used as solvent for TBP in Purex, including n-dodecane, refined kerosene, branched alkanes, and various n-paraffin oils. The choice of diluent can affect the performance of the extraction separation. For example, using linear hydrocarbons rather than kerosene diminishes the retention of fission-product Ru by radiolysis interactions.

The radiation chemistry of the diluent can also affect reprocessing operations in significant ways. Although such degradation products are not as well understood, the formation of carboxylic acids, esters, ketones, nitro-organic compounds, and other species results in effective emulsifiers, complexants, and surfactants that are liable to obstruct efficient process separations. Such reactions can cause the loss of actinide products to waste streams, the enhanced retention of fission products, the formation of emulsions, and the development of interfacial precipitates.

However, the decomposition products of TBP and its organic solvents can also potentially provide intelligence data for nuclear nonproliferation surveillance. Many of these species can only be formed as products of high-energy chemical reactions, and some can be formed only by high-energy radiation chemistry, thereby providing unambiguous signatures of nuclear reprocessing. Because of their volatility, a number of these products can be collected in air; others are soluble in aqueous and other condensed-phase samples to greater or lesser extents. Many of these compounds can be detected at trace and ultratrace concentrations using modern analytic instrumentation.

Once collected and analyzed, the quantities or multiplicity of reprocessing signature species may possibly be referred back to material quantities, irradiation times, process dynamics, or specific chemical protocols. For organic and organophosphate analytes, the optimum forensic technique is gas chromatography–mass spectrometry. This methodology permits extremely sensitive and specific determinations of chemical compounds of interest. A more detailed discussion of the technique is presented elsewhere in this book.

3.6 HEAVY-ELEMENT METALS AND ALLOYS

The metallurgy of U, Th, and Pu is a rich field of study, in part because of the unusual chemical and physical properties of the metals. All three metals are nominally members of the actinide series of elements, lying just below the lanthanide series in the periodic table. In the lanthanides, the 4f-electron shell is filled in with increasing atomic number. One would expect that the 5f-electrons would be similarly populated across the actinides, but this is only true for the heavier members of the series. The larger atomic radii of the actinides relative to the lanthanides cause the 5f electrons to bind less strongly to the nucleus, resulting in 6d electrons becoming more tightly bound in the lighter actinides. The gas-phase valence electron configuration [78] of Th is $5f^0 6s^2 6p^6 6d^2 7s^2$, very similar to that of Hf. The gas-phase valence electron configuration of U is $5f^3 6s^2 6p^6 6d^1 7s^2$; note the shift in electronic character from the 6d shell, filling the 5f shell instead. In solution, Th and U have properties very similar to Hf and W, respectively, and before 1945, they led to their classification as group IVA and VIA transition metals [79].

With the discoveries of Np [80] and Pu [81] in 1940, it became evident that lanthanide-like behavior was being introduced with increasing atomic number, so Th and U were moved from their positions as eka-Hf and eka-W to the positions they now occupy in the periodic table. The valence electron configuration of Pu in its ground state is $5f^6 6s^2 6p^6 7s^2$, which is exactly analogous to that of Sm. However, because of the electronic screening of the nuclear potential, the ionization potentials of the valence electrons are similar enough that Pu has a much richer and more varied chemistry than its lanthanide homolog.

The increase in 5f-character of the electronic structures of the actinides with increasing atomic number, and the associated decrease in 6d-character, leads to wildly variable alloying behavior from one element to the next, which is not observed in the lanthanides. It is particularly noteworthy that the large size of the Th atom relative to those of U and Pu (caused by the 6d valence electrons) results not only in it having a very different set of alloys and intermetallic compounds, but also in its very limited solubility in liquid U [3].

Even though there are substantial differences among the metals, there are also a large number of similarities, both in chemical properties and in methods of production. All three metals build up surface oxide coatings upon exposure to air, resulting in their chemical passivation. Strong nitric acid will not attack any of the three metals without the presence of a complexing agent; in the case of Th, fluoride is required. Uranium and Pu are pyrophoric in bulk, burning with an intense white light similar to Mg metal. Fine powders of all three metals can spontaneously ignite unless protected from air, and all can be reduced from bulk metal to a powder form via a hydriding/dehydriding process. The metals exist in more than one solid allotropic form. (Plutonium displays six distinct solid phases between room temperature and its melting point of 640°C.) The metals form a very limited number of intermetallic solutions, so powder metallurgy techniques have been developed to produce useful alloys from materials that are insoluble in the molten actinides. For details beyond the summary discussion presented later, reference can be made to [3,8,82–86].

The formation of the halides of Ca, Mg, and Na is favored over the formation of the halides of U at all temperatures up to 1500°C. Therefore, the reaction of a uranium tetrahalide with Ca^0, Mg^0, or Na^0 will proceed exothermically and produce U metal. However, there are some limitations: UCl_4, UBr_4, and UI_4 share the properties of being deliquescent and air-reactive, and Na metal boils at only 880°C. Therefore, in pragmatic terms, the production of U metal generally results from the reaction of UF_4 with either Mg^0 or Ca^0:

$$2\,Ca^0 + UF_4 \rightleftarrows U^0 + 2\,CaF_2.$$

The reaction of Ca^0 with UF_4 generates sufficient heat to melt the reaction products. The equivalent Mg^0 reaction is less exothermic so that reaction must be initiated at a higher starting temperature to result in molten product. CaF_2 and MgF_2 are not soluble in molten U; thus, the reaction produces two phases: a dense metallic U phase and a lighter salt phase (in which the unreacted starting materials are more soluble and therefore concentrate). After cooling, the salt cake is physically separated from the metal product and discarded.

The container in which this reaction is conducted (usually mild steel) must be lined with chemically resistant material. CaF_2 and MgF_2 are commonly utilized, and BeO and thoria have also been used for this purpose. For reaction with Mg^0, a graphite lining can be used as well. However, Ca^0 reacts with graphite to form CaC_2, which is soluble in liquid U metal.

A small impurity of UO_2 or U_3O_8 in the UF_4 starting material has no significant effect on the reaction with Ca^0, but it can poison the reaction with Mg^0. As a consequence, Ca reduction can take place in an open container, whereas the Mg reduction must be performed in a closed vessel under an Ar atmosphere (a "bomb") to prevent the competing reaction of UF_4 with oxygen. The Ca^0 charge in an open-vessel reduction must contain no Mg^0, as the formation of Mg vapor (boiling point 1090°C) can blow the reactants out of the crucible. This is another reason to use a closed bomb for the reduction with Mg^0. Even so, the reaction of oxides of U with Ca^0 can be quite violent, such that the reacting system is often diluted by the addition of $CaCl_2$.

The reaction of Ca^0 or Mg^0 with UO_2 produces U metal powder. However, the CaO or MgO reaction products are too refractory to melt from the generated heat, so a metal billet does not separate. Ca^0 is usually preferred, as Mg^0 results in the formation of a U^0 powder of finer mesh size after a lower chemical yield, and the smaller powder size makes pyrophoricity a bigger problem. Metal powder is also formed in the electrolysis of UCl_3 or UF_4 dissolved in a molten KCl/NaCl eutectic, or KUF_5 or UF_4 dissolved in $CaCl_2$/NaCl. Grain size is governed by temperature and current density. In both cases, the electrolysis must take place under an inert atmosphere. Uranium powder can also be produced through a hydriding/dehydriding process, but the pyrophoricity of the final product makes it virtually unusable.

Further purification of bomb-produced U metal is accomplished by remelting in vacuo or under an Ar atmosphere. This step is usually carried out in graphite crucibles lined with CaO, MgO, or ZrO_2. Casting of U is performed in graphite or mild steel faced with Al_2O_3, as below 1600°C, the reaction of molten U with graphite is slow. Uranium can be hot-worked in air, but it is usually handled in a molten salt bath to avoid corrosion.

The preferred method for producing Th metal involves the reaction of ThF_4 with metallic Ca in a bomb. The reactions of $ThCl_4$ and ThO_2 with Ca^0 also work, but the Th metal product is created in powder form and dispersed in the other products of the reaction (see the following text).

$$ThF_4 + 2\,Ca^0 \rightleftarrows Th^0 + 2\,CaF_2.$$

ThF_4 is nonhygroscopic and readily produced. Because the reactions are less exothermic than those with UF_4, and because the Th melting point (1750°C) is higher than that of U (1135°C), Mg^0 is a poor reductant, and even the Ca^0-driven reaction requires a "boost" to yield Th metal in bulk form. This can be accomplished by either preheating the bomb before ignition or adding a thermal booster like I_2 or S to the reaction system. Excess Ca metal reacts with the booster, increasing the temperature of the bomb vessel and creating products that are insoluble in the dense, molten

Th reaction product. If an application requires particularly pure Th metal, $ZnCl_2$ can be used as the booster. Zn^0 and Th^0 form a low-melting alloy of biscuit metal, from which Zn^0 can be distilled, leaving only approximately 100 ppm in the Th. Also, the $CaCl_2$ reaction product acts as a flux with the CaF_2 slag, giving a better phase separation following the reaction. Thorium metal can as well be produced in a bomb reduction of $ThCl_4$ with Na^0, under an inert atmosphere.

As with U, powdered Th metal can be produced in the reaction of Ca^0 with ThO_2. The reaction is usually performed in a Ni boat under a H_2 atmosphere to prevent premature ignition. This reaction is also exothermic, but is not sufficient to form a molten billet. Following reaction, the product is immersed in water, which dissolves product CaO and residual Ca^0 to form $Ca(OH)_2$. Nitric acid is added to the solid residue to clean the powder and dissolve any residual Ca salts up to a concentration of 0.2 M. However, the time for this process must be limited to avoid redissolution of passive Th^0 powder. The metallic product is filtered, rinsed with water, and vacuum-dried, resulting in a 300-mesh powder suitable for sintering or pressing. Thorium metal powder can also be obtained by electrolyzing $ThCl_4$ dissolved in molten KCl/NaCl eutectic. A further method of obtaining powdered Th is via a hydriding/dehydriding process (see the following text).

Thorium metal is produced and cast in BeO or BeO-coated graphite containers. Other refractory materials often used as liners in molds (e.g., Al_2O_3, MgO, and CaO) dissolve rapidly in molten Th. Pure graphite molds have been used, but they must be cooled quickly before attacked by the melt. Ingots contain sizable quantities of oxide and slag inclusions, as well as metal discontinuities such as cold shunts, shrinkage pipe, and porosity. Machining impure Th is very difficult because of abrasion caused by the oxide inclusions, and very-high-purity Th can be produced via formation of the iodide: iodine reacts with Th^0 at high temperatures to produce ThI_4 vapor, which is then condensed on a cooler surface. Subsequent heating decomposes the salt, leaving Th^0. High-quality Th metal ingots can also be produced via consumable-electrode arc-melting in a low-pressure inert gas.

The production of Pu metal is similar to the production of both U and Th metals. The best reactants are PuF_4 and either Ca^0 or Mg^0. PuF_3 can also be reduced by the alkaline earth metals:

$$3\,Ca^0 + 2\,PuF_3 \rightleftarrows 3\,CaF_2 + 2\,Pu^0$$

Optimal results are obtained with a 25% excess of Ca^0 over the stoichiometric amount. To ignite the reacting materials, about 10 g of I_2 per kg of Pu are added as a booster. Reaction bombs are lined with magnesia and CaF_2 and filled with Ar gas. Small quantities of O_2 are not a problem, but they can lead to inclusions of Pu_2O_3 in the metal billet. The reaction is initiated in the booster by heating it to around 325°C. During reaction, the pressure in the bomb increases to 35 atm, and the temperature reaches 1600°C.

Plutonium metal can also be produced by the reaction of Ca^0 or Mg^0 in 50% excess over the stoichiometric amount needed to reduce PuO_2:

$$2\,Ca^0 + PuO_2 \rightleftarrows 2\,CaO + Pu^0.$$

This chemical system offers the advantage of being fluoride free, which reduces the neutron background in the laboratory caused by the $^{19}F(\alpha,n)$ reaction. The reaction is generally carried out in CaO crucibles. The final product is small beads of Pu metal, but they are coated with CaO reaction product that prevents formation of a billet. The beads are recovered by dissolving the slag in dilute acetic acid and then washing the product with nitric acid.

Another method of producing Pu metal is reduction of $PuCl_3$ with Ca^0. However, even though the melting points of the products are lower, the reaction is considerably less exothermic and proceeds only through the action of a booster. The reaction of graphite with PuO_2 has been used to produce a mixture of Pu metal and Pu carbide, but losses caused by the formation of the intractable carbide make this technique infrequent.

The initial Pu metal product contains impurities of Ca^0 and colloidal inclusions of CaF_2, PuO_2, and so forth. It may also contain fission products from incomplete fuel reprocessing. There are several techniques for improving the purity of the product: the first is vacuum refining, in which the billet is melted in a MgO or CaO crucible in good vacuum. The temperature of the melt is increased until the partial pressure of Pu metal vapor is between 10^{-5} and 10^{-3} torr, at which point residual alkalis, alkaline earths, and their halide salts are purged from the melt. Plutonium metal only sparingly wets CaO or MgO, so the product can be efficiently recovered from the crucible after cooling. Second is oxidation refining, in which a small quantity of O_2 is introduced into the melt, moving reactive refractory contaminants into the slag. Third is electrolytic refining, in which the Pu sample is immersed in a molten chloride salt under an inert atmosphere, where it acts as the anode of an electrolytic cell. At a high current density, liquid Pu metal is collected at the surface of a W cathode and drips into a collector. Transition-metal impurities tend to remain in the anodic sludge, rare earths and other actinides concentrate in the electrolyte, and the yield of purified Pu metal can be as high as 97%. Finally, there is zone melting, in which the Pu billet is fabricated into a bar along which a high-temperature zone is passed. Many impurities stabilize δ-phase Pu. As a melt zone is moved along the bar, impurities that form a eutectic with Pu concentrate in the melt, while impurities that raise the melting point effectively move in the opposite direction. After repeated passes, the ends of the bar contain most of the impurities, and they are removed.

Plutonium metal powder can be produced through the hydrogenation and dehydrogenation of Pu metal. Plutonium reacts with H_2 at room temperature, disintegrating the metal (via an enormous volume change) to a $PuH_{2.7-3.0}$ powder. Heating this product in a vacuum furnace at 200°C results in the formation of plutonium hydride, PuH_2. Increasing the furnace temperature to 420°C reforms the metal in the α phase upon cooling. The apparatus in which this reaction takes place must be stringently free from oxygen. Similar techniques exist for powdering Th and U through hydriding/dehydriding.

Uranium is a dense, silvery metal, malleable, ductile, and slightly paramagnetic. It forms a wide range of intermetallic compounds but, because of its unusual crystal structures, does not form many solid solutions. Uranium metal is a poor electrical conductor, with a conductivity only about half that of Fe. It is also semiplastic, tending to yield under load. The metallic radius of the U atom is 1.542 Å.

Uranium metal exists in three allotropic forms between room temperature and its melting point of 1135°C. The α phase exists at temperatures up to 668°C, and it has an orthorhombic atomic arrangement and is quite easy to work. Uranium metal is rolled only in the α phase; otherwise, cracking occurs. Uranium foils are cold-rolled, with periodic reannealing at 600°C. At temperatures above 500°C, the metal starts to soften, but heat generated in the rolling process can cause the formation of the complex, tetragonal, β-phase U, which is hard and brittle. Beta phase exists at temperatures between 668° and 776°C, although the phase change at 668°C is not sharp and occurs over a span of about 10°C. Above 776°C, U metal converts to its body-centered-cubic γ phase. Gamma-phase U is quite soft, deforming even under its own weight at these temperatures. As a consequence, it is the phase that is easiest to extrude. However, hot U metal chemically attacks steel dies and rollers, so other materials must be used to handle it.

Impurities in a U sample can strongly affect the transition temperatures between phases. Both α and β phases of uranium are complex, with neither one favoring the formation of solid solutions. However, the γ phase does form extensive solid solutions. Thus, for instance, addition of more than 2.3% by weight of Mo to U metal causes the β phase to completely disappear and stabilizes the γ phase to the extent that it can be preserved in a sample at room temperature through rapid quenching. Important alloys of U with transition metals include U–Mo, which is fairly chemically inert and has applications as ballast in the aircraft industry; U–Nb and U–Nb–Zr, with various applications in the nuclear industry; and U–Ti, a very strong alloy with application to nuclear weapons.

Colloidal inclusions are often found in U metal. These consist mainly of small grains of the carbide, but they sometimes include significant amounts of the oxide or nitride. Intermetallic compounds are also represented in inclusions—mainly U_6Fe, UAl_2, and U_4Si.

Thorium is a silvery metal that darkens on contact with air, and it is a good reducing agent, falling between Al and Mg in the electrochemical series. The metallic radius of the Th atom is 1.798 Å. Thorium metal is dimorphic, changing at 1400°C from the face-centered-cubic α phase to the body-centered-cubic β phase. Thorium melts at 1750°C, and the physical properties of Th are strongly influenced by the oxygen content of the sample. Pure α-phase Th can be rolled to reductions of 99% without intermediate annealing. However, hot-working of Th cannot be performed in a reducing atmosphere, as Th is an efficient getter of both hydrogen and carbon. In working with macroscopic Th, it is much better to heat the material in air and machine off the surface (containing both oxides and nitrides) later. Otherwise, the metal must be worked in an inert atmosphere or in a molten salt bath consisting of mixed Ba, K, and Na chlorides. The molten salt does not attack Th at temperatures less than 900°C.

Thorium alloys containing up to 8% by weight of Ti, Zr, or Nb are ductile. However, Ti and Nb increase the chemical reactivity of the metal with water. The addition of small amounts of Be, C, Al, or Si reduces the ductility of the metal, but increases its corrosion resistance.

Plutonium metal is a poor conductor of both heat and electricity. Its electrical resistivity is nearly that of a semiconductor, and α-phase plutonium has the highest resistivity of all metals. The metallic radius of the Pu atom is 1.523 Å, and Pu metal has six solid allotropic forms below its melting point of 640°C. This property is the main reason that solid, unalloyed Pu metal cannot be used in applications in which heat is generated (e.g., nuclear fuel elements). The volume expansion that occurs in proceeding from α phase to δ phase is greater than that occurring in any other metal except Sn. Alpha-phase Pu is stable at temperatures up to 115°C, and it has a 16-atom unit cell with a simple monoclinic arrangement of atoms and density of 19.8 g/cm^3. Alpha-Pu is hard, brittle, and difficult to machine. The β phase is stable from 115° to 185°C. Beta-Pu has a 34-atom unit cell and body-centered monoclinic atomic arrangement. From 185° to 310°C, the γ phase is stable and is face-centered orthorhombic. Phase changes between α- and β-Pu and between β- and γ-Pu are sluggish, taking place over several degrees. From 310° to 452°C, δ phase is stable. Delta-Pu has a face-centered-cubic structure, making it easy to machine, roll, and extrude. It also has a negative linear expansion coefficient, making it one of the few materials that contracts upon heating. From 452° to 480°C, a δ′ phase is stable, and it has a tetragonally deformed, cubic close-packed structure, as well as a negative expansion coefficient. Finally, between 480°C and the melting point (640°C), the ε phase is stable and has a body-centered-cubic structure.

Plutonium forms high-melting-point compounds with all A-subgroup metals and metalloids, but with the exception of δ-phase and ε-phase mixtures, solid solubilities are limited. Most δ-phase solid solutions are stable at room temperature or can be retained by rapid cooling. However, the ε phase is not retained at room temperature in any Pu mixture. Significant solid solutions of the other Pu phases are rare. Neptunium and Pu are mutually soluble in the α phase, and Th and U mix with Pu in both the β and γ phases. Alpha-phase Pu is highly reactive with O_2, whereas delta-phase Pu is less so. Alloys that stabilize the δ phase are more corrosion-resistant than the others and are also more workable. This situation is similar to that with the body-centered-cubic γ phase of U.

Plutonium alloys are usually prepared by adding the constituents as metals to the melt. However, it is also possible to introduce oxides or halides of Pu into a melt of the alloying element if it is sufficiently reducing. This offers the advantage of avoiding the very reactive Pu metal and may obviate the need for an inert atmosphere. In practice, important δ-phase weapons alloys are made by adding PuF_3 to molten Ga or Al.

3.7 SUMMARY

The engineering processes that result in the production of Th, U, or Pu can be summarized in Figure 3.24. Material can be diverted at any point in the progression. Forensic signatures in the early steps derive mainly from the deposits of the ores in nature and are characteristic of the ore body. As the process outlined in Figure 3.24 continues, the natural signatures are destroyed and the signatures imposed by engineering dominate. Using the information outlined in this chapter, an experienced nuclear forensic analyst can interpret the isotopic signatures of the sample matrix and the trace radionuclidic impurities to infer conclusions about the source material and processes that resulted in any sample under examination.

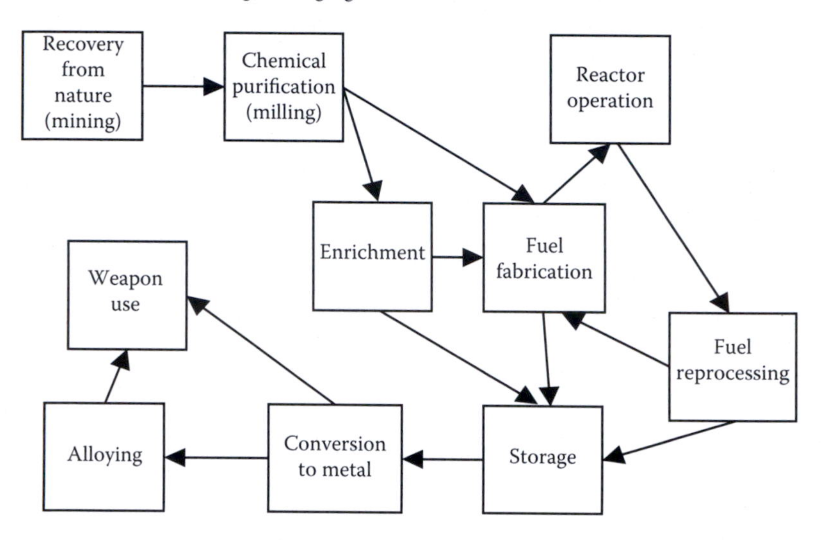

FIGURE 3.24 Block diagram of the engineering processes resulting in samples of Th, U, and Pu. Samples examined by the nuclear forensic analyst can arise from diversion at any point in the sequence.

REFERENCES

1. D.R. Lide, ed., *CRC Handbook of Chemistry and Physics*, 81st edn., CRC Press, Boca Raton, FL, 2000.
2. N.P. Galkin, B.N. Sudarikov, U.D. Veryatin, Yu.D. Shishkov, and A.A. Maiorov, *Technology of Uranium*, Atomizdat, Moscow, Russia, 1964.
3. L. Grainger, *Uranium and Thorium*, George Newnes, London, U.K., 1958.
4. R.W. Boyle, *Geochemical Prospecting for Thorium and Uranium Deposits*, Elsevier, Amsterdam, the Netherlands, 1982.
5. J.W. Frondel, M. Fleischer, and R.S. James, Glossary of uranium and thorium bearing minerals, 4th edn., Geological Survey Bulletin 1250, Government Printing Office, Washington, DC, 1967.
6. B.A. Wills, *Mineral Processing Technology*, 6th edn., Butterworth/Heinemann, Oxford, U.K., 1997.
7. International Atomic Energy Agency, Uranium extraction technology, IAEA Technical Reports Series 359, International Atomic Energy Agency, Vienna, Austria, 1993.
8. H.A. Wilhelm, ed., *The Metal Thorium*, American Society for Metals, Cleveland, OH, 1956.
9. O. Ruff and A. Heinzelmann, Ueber das Uranhexafluorid, *Z. Anorg. Chem.*, 72, 63, 1911.
10. A.O. Nier, E.T. Booth, J.R. Dunning, and A.V. Grosse, Nuclear fission of separated uranium isotopes, *Phys. Rev.*, 57, 546, 1940.
11. S. Villani, ed., *Uranium Enrichment, Topics in Applied Physics*, Vol. 35, Springer, Berlin, Germany, 1979.
12. T. Wilkie, Tricastin points the road to energy independence, *Nucl. Eng. Intl.*, 25(305), 41, 1980.
13. H.deW. Smyth, *Atomic Energy for Military Purposes*, Princeton University Press, Princeton, NJ, 1945.

14. L.R. Groves, *Now It Can Be Told*, Harper, New York, 1962.
15. M. Gowing, *Britain and Atomic Energy (1939–1945)*, McMillan, London, U.K., 1964.
16. D. Massignon, *Proceedings of the 2nd UNO Geneva Conference on Peaceful Uses of Atomic Energy*, Vol. 4, UNO, New York, 1958.
17. A.S. Krass, P. Boskma, B. Elzen, and W.A. Smit, *Uranium Enrichment and Nuclear Weapon Proliferation*, International Publications Service Taylor & Francis, New York, 1983.
18. T.B. Cochran, *Nuclear Weapons Databook*, Vol. II, U.S. Nuclear Warhead Production, Harper Business, Cambridge, MA, 1987.
19. J. Kistemaker, J. Bigeleisen, and A.O.C. Nier, eds., *Proceedings of the International Symposium on Isotope Separation, Amsterdam, 1957*, North-Holland, Amsterdam, the Netherlands, 1958.
20. G. Vasaru, *Separation of Isotopes by Thermal Diffusion*, Rumanian Academy, Bucharest, Romania, 1972 [distributed by USERDA, report ERDA-tr-32, 1975].
21. H. London, *Separation of Isotopes*, George Newnes, London, U.K., 1961.
22. D. Irving, *The German Atomic Bomb*, Simon and Schuster, New York, 1967.
23. D.R. Olander, The gas centrifuge, *Sci. Am.*, 239(2), 37, 1978.
24. J.B. Merriman and M. Benedict, eds., *Recent Developments in Uranium Enrichment*, AIChE Symposium Series 78, American Institute of Chemical Engineers, New York, 1982.
25. R.R. Eaton, R.L. Fox, and K.J. Touran, Isotopic enrichment by aerodynamic means: A review and some theoretical considerations, *J. Energy*, 1(4), 229, July/August 1977.
26. E.W. Becker, P. Noguiera-Batista, and H. Voelcker, Uranium enrichment by the separation nozzle method within the framework of German/Brazilian Cooperation, *Nucl. Technol.*, 52, 105, 1981.
27. W.L. Grant, J.J. Wannenburg, and P.C. Haarhoff, *The Cascade Technique for the South African Enrichment Process*, AIChE Symposium Series 73, American Institute of Chemical Engineers, New York, 1977.
28. R.L. Farrar Jr. and D.F. Smith, Photochemical isotope separation as applied to uranium, Union Carbide Report KL-3054, Oak Ridge, TN, 1972.
29. D.K. Evans, ed., *Laser Applications in Physical Chemistry*, Marcel Dekker, New York, 1989.
30. C.B. Moore, ed., *Chemical and Biochemical Applications of Lasers*, Vol. III, Academic Press, London, U.K., 1977.
31. A. Ben-Shaul, Y. Haas, K.L. Kompa, and R.D. Levine, eds., *Lasers and Chemical Change (Springer Series in Chemical Physics)*, Vol. 10, Springer, Berlin, Germany, 1981.
32. H.D.V. Boehm, W. Michaels, and C. Weitkamp, Hyperfine structure and isotope shift measurements on ^{235}U and laser separation of uranium isotopes by two-step photoionization, *Opt. Commun.*, 26(2), 178, 1978.
33. V.S. Letokhov, Laser isotope separation, *Nature*, 277, 605, 1979.
34. J.I. Davis, Atomic vapor laser isotope separation at Lawrence Livermore National Laboratory, Report UCRL-83516, Livermore, CA 1980.
35. P. Rabinowitz, A. Stein, and A. Kaldor, Infrared multiphoton dissociation of UF_6, *Opt. Commun.*, 27, 381, 1978.
36. F. Albarede, *Introduction to Geochemical Modeling*, Cambridge University Press, Boston, MA, 1996.
37. G. Faure, *Principles and Applications of Inorganic Geochemistry*, McMillan, New York, 1991, Chap. 18.
38. R.L. Garwin and G. Charpak, *Megawatts and Megatons*, Knopf, New York, 2001, p. 6.
39. G.R. Keepin, *Physics of Nuclear Kinetics*, Addison-Wesley, Reading, MA, 1965.
40. A.E. Green, *Safety Systems Reliability*, Wiley, New York, 1983.

41. A.H. Compton, *Atomic Quest*, Oxford University Press, Oxford, U.K., 1956.
42. E. Fermi, *Collected Papers*, University of Chicago Press, Chicago, IL, 1965.
43. S. Glasstone and A. Sesonske, *Nuclear Reactor Engineering*, Van Nostrand, Princeton, NJ, 1967.
44. B.G. Levi, Cause and impact of Chernobyl accident still hazy, *Phys. Today*, 39, 21, July 1986.
45. Joint Committee on Atomic Energy, Congress of the United States, *Nuclear Submarines of Advanced Design*, Government Printing Office, Washington, DC, 1968.
46. International Atomic Energy Agency, *Nuclear Ship Propulsion*, IAEA Proceedings Series, Vienna, Austria, 1961.
47. J.N. Lillington, *Light Water Reactor Safety*, Elsevier, Amsterdam, the Netherlands, 1998.
48. International Atomic Energy Agency, *Review of Fuel Element Development for Water Cooled Nuclear Power Reactors*, Technical Report Series 299, IAEA, Vienna, Austria, 1989.
49. H.C. McIntyre, Natural-uranium heavy-water reactors, *Sci. Am.*, 233(4), 17, 1975.
50. A. Cottrell, *How Safe Is Nuclear Energy?* Heinemann, London, U.K., 1981.
51. V.M. Chernousenko, *Chernobyl, Insight from the Inside*, Springer, Berlin, Germany, 1991.
52. *A Plutonium Recycling Scenario in Light Water Reactors*, Commission of the European Communities Symposium Series, Harwood Academic, New York, 1982.
53. G.T. Seaborg and J.L. Bloom, Fast breeder reactors, *Sci. Am.*, 223(5), 13, 1970.
54. G.A. Vendryes, Superphenix: A full-scale breeder reactor, *Sci. Am.*, 236(3), 26, 1977.
55. W. Mitchell and S.E. Turner, Breeder reactors, U.S. Atomic Energy Commission Office of Information Services, Washington, DC, 1972.
56. R.G. Wymer, ed., Thorium fuel cycle, *Proceedings of the Second International Thorium Fuel Cycle Symposium*, Gatlinburg, TN, 1966, U.S. Atomic Energy Commission/Division of Technical Information, Oak Ridge, TN, 1968.
57. J.M. Dukert, *Thorium and the Third Fuel*, US Atomic Energy Commission/Division of Technical Information, Oak Ridge, TN, 1970.
58. A.V. Nero, *A Guidebook to Nuclear Reactors*, University of California Press, Berkeley, CA, 1979.
59. N. Patri, Thorium-burning Kakrapar-1 declared commercial by NPC, *Nucleonics Week*, May 13, 1993, p. 4.
60. Electric Power Research Institute, Development status and operational features of the high temperature gas-cooled reactor, EPRI Report NP-142, Palo Alto, CA, April 1976.
61. U.V. Ignatiev, R.Y. Zakirov, and K.F. Grebenkine, *Proceedings of the Workshop on Advanced Reactors with Innovative Fuels*, Villingen, Switzerland, October 1998, Organization for Economic Co-operation and Development, Nuclear Energy Agency, Paris, France, 1999, p. 381.
62. F.T. Miles, R.H. Wiswall Jr., R.J. Heus, and L.P. Hatch, A continuously separating breeder blanket using thorium fluoride, in *Nuclear Engineering*, Vol. 2, American Institute of Chemical Engineers, New York, 1954.
63. International Atomic Energy Agency, *Proceedings of the Symposium on Irradiation Facilities for Research Reactors*, Teheran, November 1972, IAEA, Vienna, Austria, 1973.
64. C.D. West, Research reactors: An overview, *ANS Nuclear News*, 40, 50, 1997.
65. O. Bukharin, Making fuel less tempting, *Bull. Atomic Sci.*, 58(4), 44, July–August 2002.
66. A.G. Croff, Origen2: A versatile computer code for calculating the nuclide compositions and characteristics of nuclear materials, *Nucl. Technol.*, 62, 335, 1983.
67. L.J. Barbieri, J.W. Webster, and K.T. Chow, Plutonium recycle in the Calder Hall type reactor, *Nucl. Sci. Eng.*, 5, 105, 1959.

68. S. Groueff, *Manhattan Project*, Little, Brown, Boston, MA, 1967.
69. F. Weigel, J.J. Katz, and G.T. Seaborg, Plutonium, in *The Chemistry of the Actinide Elements*, Vol. 1, 2nd edn., Chapman & Hall, London, U.K., 1986, Chap. 7.
70. D.O. Campbell and W.D. Burch, The chemistry of fuel reprocessing: Present practices, future trends. *J. Radioanal. Nucl. Chem.*, Articles, 142(1), 303, 1990.
71. D.D. Sood and S.K. Patil, Chemistry of nuclear fuel reprocessing: Current status, *J. Radioanal. Nucl. Chem.*, Articles, 203(2), 547, 1996.
72. A. Palamalai, S.V. Mohan, M. Sampath, R. Srinivasan, P. Govindan, A. Chinnusamy, V.R. Raman, and G.R. Balasubramanian, Final purification of uranium-233 oxide product from reprocessing treatment of irradiated thorium rods, *J. Radioanal. Nucl. Chem.*, 177, 291, 1994.
73. J.T. Long, *Engineering for Nuclear Fuel Reprocessing*, American Nuclear Society, La Grange Park, IL, 1978, Chap. 3.
74. L.I. Guseva and G.S. Tikhomirova, Separation of transplutonium elements on macroporous anionites via the use of aqueous alcoholic solutions of a mixture of monocarbonic and nitric acids, *J. Radioanal. Nucl. Chem.*, 52, 369, 1979.
75. D.E. Jenne and J.W. Healy, Dissolving of twenty-day metal at Hanford, Report HW-17381-DEL, May 1, 1950.
76. K. Ebert and R.V. Ammon, eds., *Safety of the Nuclear Fuel Cycle*, VCH Publishers, New York, 1989.
77. P. Chiotti, ed., *Symposium on Reprocessing of Nuclear Fuels, Nuclear Metallurgy*, Vol. 15, Report CONF-690801, U.S. Atomic Energy Commission/Division of Technical Information, Oak Ridge, TN, 1969.
78. F.A. Cotton and G. Wilkinson, *Advanced Inorganic Chemistry*, 5th edn., Wiley, New York, 1988.
79. G.T. Seaborg, The chemical and radioactive properties of the heavy elements, *Chem. Eng. News*, 23, 2190, 1945.
80. E. McMillan and P.H. Abelson, Radioactive element 93, *Phys. Rev.*, 57, 1185, 1940.
81. G.T. Seaborg, E.M. McMillan, J.W. Kennedy, and A.C. Wahl, Radioactive element 94 from deuterons on uranium, *Phys. Rev.*, 69, 366, 1946.
82. J.C. Warner, ed., *Metallurgy of Uranium and Its Alloys*, U.S. Atomic Energy Commission Technical Information Service, Oak Ridge, TN, 1953.
83. A.N. Vol'skii and Ya.M. Sterlin, *Metallurgy of Plutonium*, Israel Program for Scientific Translation, Jerusalem, Israel, 1970.
84. F.A. Rough and A.A. Bauer, *Constitutional Diagrams of Uranium and Thorium Alloys*, Addison-Wesley, Reading, MA, 1958.
85. A.S. Coffinberry and W.N. Miner, eds., *The Metal Plutonium*, University of Chicago Press, Chicago, IL, 1961.
86. O.J. Wick, ed., *Plutonium Handbook*, Vol. 1, Gordon and Breach, New York, 1967.

4 Chemistry and Nuclear Forensic Science

Although engineering-scale chemical operations create the forensic signatures that the analyst interprets, chemistry on the laboratory scale establishes the methods used by the analyst to isolate those signatures and defines the accuracy with which they can be measured. Chemical effects on a small scale control the fine structure of the matrix of a sample down to micrometer dimensions. This is particularly important in the study and characterization of particles. Chemical effects also control the interactions of the sample surface with the environment and may provide clues to the path over which the sample moved from the point of fabrication to the point of interdiction. In this chapter, the analytical chemistry of the heavy elements in the laboratory, and the processing of samples on the microscale, are considered.

4.1 TRACERS IN INORGANIC ANALYSIS

The basic method of radiochemical analysis involves the use of tracers to follow the behavior of elements in a chemical reaction. A tracer may consist of a short-lived radionuclide, measured by radiation counting, or of a long-lived or separated stable nuclide that can be measured by mass spectrometry (MS). A tracer is usually a species added to a reacting system by the analyst, but there are cases in which it may arise in situ through decay processes (discussed later).

In traditional radiochemistry, where the tracer technique is based on radiation counting, most radionuclides used as tracers have half-lives between several hours and thousands of years. The identity and quantity of tracer that should be used for a particular application depends on several factors (e.g., if the chemical procedure used to isolate the traced element and produce a counting source requires several days to perform, a very short-lived activity would be an inappropriate tracer). The analyst must be careful to add enough tracer activity that it can be accurately measured in the final sample. This includes considerations of decay during the chemical procedures before counting, the efficiency of the radiation counter for detecting the characteristic emissions of the tracer nuclide, the reduction of tracer by chemical-yield losses, and the level of activity of any specimen radioanalytes that might interfere with the observation of the tracer. Similarly, the analyst must not add excessive tracer that will overwhelm the signature of the analyte. In historic radiochemistry, quantities of radionuclides that are quite undetectable by ordinary instrumental methods emit easily detectable levels of radiation. Indeed, a collateral meaning of the word "tracer" in radioanalytical chemistry is "weightless." For instance, 10 cps of ^{14}C (quite easily detected with a gas-filled, proportional β counter) corresponds to a sample of but 2.5×10^{12} atoms (i.e., 6×10^{-11} g)—quite undetectable by gravimetry or many other analytic techniques.

The nuclear forensic analyst's main use of the tracer technique applies to the determination of the efficiency of a series of general chemical operations, such as precipitation, filtration, solvent extraction, ion-exchange chromatography, and volatilization. It results in establishing the chemical yield of the procedure—both overall and stepwise. A chemical yield is defined in terms of the number of atoms of the tracer activity in a purified chemical fraction relative to the number of atoms initially added to the analytic sample (assuming that the specimen contained none of the tracer activity initially).

Use of a radioactive tracer to determine a chemical yield is part of a broad collection of techniques known as isotope dilution. A goal is often to analyze a sample for a stable element X, from which a pure chemical fraction of X may be but incompletely separated. A tracer aliquot containing a known mass of X (M_X), labeled with a radioactive isotope of X characterized by the emission of radioactivity of a known intensity (D), is added to the questioned sample, and the resultant material is thoroughly mixed and equilibrated. A chemical separation is performed to obtain a pure sample of X, of mass M_S, and a measured radioactive intensity D', which is $<D$ because of losses in the separation procedure. The variable M_S is determined by any suitable, standard, quantitative-analysis method (e.g., gravimetry of a stoichiometric compound of X). The specific activity of the chemical fraction, D'/M_S, is equal to the specific activity of the element X in the mixture just before separation, $D/(M_X + M_0)$, where M_0 is the mass of element X in the original sample. This yields the isotope dilution formula:

$$M_0 = (D/D')M_S - M_X.$$

In the case in which the tracer aliquot contains only a "massless" quantity of X (a "carrier-free" tracer), the isotope dilution equation reduces to the classical tracer equation:

$$M_S/M_0 = D'/D;$$

the fraction of the initial activity (D) present in the final sample is equal to the fraction of the initial mass (M_0) in that sample.

With the introduction of sensitive MS techniques, the tracer concept has been expanded to include the use of long-lived radionuclides and stable materials. Often an element may not have an isotope with properties suitable for radiation counting. For instance, the longest-lived radioactive isotope of oxygen is ^{15}O, with a half-life of 122 s, and the radioactive-tracer technique is consequently precluded for most analytical applications involving the determination of oxygen. Only its three stable isotopes provide the means by which oxygen can be determined through isotope dilution.

In forensic radiochemistry, it is frequently required to measure the isotopic content of a mixture of naturally occurring isotopes in a U fraction by MS. This is a consequence of the half-lives of the U isotopes being so long as to make the radiation counting of small samples difficult. Because MS measurements will be made anyway, one can often avoid adding tracer aliquots of 69-year ^{232}U or 6.7-day ^{237}U, even though both are easily measured through their radioactivity, but to add instead 1.6 × 10^5-year ^{233}U or 2.34 × 10^7-year ^{236}U. The yield is then measured by MS. If a known quantity of ^{233}U (M_{233}) is added to the original sample, which contained no ^{233}U initially, then the amount of ^{238}U (M_{238}) in the original sample can be determined from the isotopic ratio of $^{238}U/^{233}U$ in the chemical fraction:

$$M_{238} = M_{233} \times (^{238}\mathrm{U}/^{233}\mathrm{U})_{\mathrm{atom}}(238/233),$$

where the factor (238/233) = 1.0215 converts the $^{238}U/^{233}U$ atom ratio, determined by MS atom counting, to a mass ratio.

Even when the sample contains ^{233}U, the use of ^{233}U as a tracer is still viable. In this case, the techniques of isotope dilution are termed a "spiked/unspiked" procedure. The sample is subdivided quantitatively into two or more aliquots. One of these aliquots is traced with a known quantity, M_T, of ^{233}U (the spiked fraction, $\mathscr{S}$), whereas a second aliquot is untraced (the unspiked fraction, $\mathscr{U}$). Each sample is chemically processed to yield a purified U fraction, which is analyzed by MS. The unspiked isotope ratio, $(^{238}\mathrm{U}/^{233}\mathrm{U})_{\mathrm{atom},\mathscr{U}}$, and the spiked isotope ratio $(^{238}\mathrm{U}/^{233}\mathrm{U})_{\mathrm{atom},\mathscr{S}}$, combined with the mass of ^{233}U in the tracer aliquot, can be used to determine the mass of ^{238}U in the spiked aliquot, M_{238}:

$$M_{238} = M_T(238/233)\frac{\left[\left(^{238}\mathrm{U}/^{233}\mathrm{U}\right)_{\mathrm{atom},\mathscr{U}} \times \left(^{238}\mathrm{U}/^{233}\mathrm{U}\right)_{\mathrm{atom},\mathscr{S}}\right]}{\left[\left(^{238}\mathrm{U}/^{233}\mathrm{U}\right)_{\mathrm{atom},\mathscr{U}} - \left(^{238}\mathrm{U}/^{233}\mathrm{U}\right)_{\mathrm{atom},\mathscr{S}}\right]}.$$

The other isotope ratios measured in the unspiked sample can then be used with M_{238} to determine the masses of each of the other U isotopes in the sample.

An example of tracer use in nuclear forensic analysis (NFA) is as follows. The nuclear forensic analyst is often interested in measuring the concentration of ^{230}Th in a solution containing natural U. The relative concentrations of ^{230}Th and ^{234}U constitute a primary chronometer for determining the age of a sample of U (discussed fully later in this book). In general, the Th concentration is on the order of 10^{-9} of the U concentration in such a sample. As a consequence, the Th must be isolated from U before its concentration can be measured. Unfortunately, the chemical separation processes are not 100% efficient, and the analyst must correct for chemical yield. This can be accomplished by adding a known quantity of ^{228}Th at the beginning of the chemical procedure, followed by measuring the amount of ^{228}Th remaining in the final sample. If the recovery of the ^{228}Th tracer is 50%, for example, then the recovery of ^{230}Th would also be 50%, and its concentration is corrected accordingly.

Another tracer application: Gallium is commonly alloyed with Pu for weapons applications. The admixture of a small quantity of Ga stabilizes the δ-Pu metallic phase, which is more workable than α-phase Pu (α is the phase of pure Pu that is stable at room temperature). Even though appropriate radioactive isotopes for classical tracer applications exist (e.g., 78-h ^{67}Ga and 14-h ^{72}Ga), the availability of these nuclides is limited. Natural Ga consists of two stable isotopes, ^{69}Ga (60.1%) and ^{71}Ga (39.9%), making isotope dilution with stable isotopes possible. Assume the availability of a supply of separated ^{71}Ga. The Pu analytical sample is quantitatively dissolved. One aliquot is traced with a known amount of ^{71}Ga, while a second aliquot is untraced. Gallium is then chemically isolated from the aliquots, and the ^{71}Ga/^{69}Ga isotope ratios for both the spiked and unspiked Ga fractions are measured by MS. To first approximation, the analyst treats the ideal case in which the tracer had no ^{69}Ga present:

$$\left({}^{71}\mathrm{Ga}/{}^{69}\mathrm{Ga}\right)_{\mathcal{S}} - \left({}^{71}\mathrm{Ga}/{}^{69}\mathrm{Ga}\right)_{\mathcal{U}} = \left[\text{atoms}\,{}^{71}\mathrm{Ga}\text{ in spike}\right] / \left[\text{atoms}\,{}^{69}\mathrm{Ga}\text{ in sample aliquot}\right]$$

The equation is solved for ^{69}Ga, which is then multiplied by $[1 + ({}^{71}\mathrm{Ga}/{}^{69}\mathrm{Ga})_{\mathcal{U}}]$ to obtain the total atoms of Ga in the sample aliquot.

Unfortunately, it is impossible to purchase ^{71}Ga that does not contain a small residuum of ^{69}Ga. For better accuracy, therefore, the ^{69}Ga content of the tracer must be taken into account. In this case,

$$({}^{71}\mathrm{Ga}/{}^{69}\mathrm{Ga})_{\mathcal{S}} = [{}^{71}\mathrm{Ga}_{\mathcal{U}} + {}^{71}\mathrm{Ga}_{\text{in spike}}]/[{}^{69}\mathrm{Ga}_{\mathcal{U}} + {}^{69}\mathrm{Ga}_{\text{in spike}}].$$

If a Taylor-series expansion is performed around $^{69}\mathrm{Ga}_{\mathcal{U}}$ and truncated after two terms, the following relationship is obtained:

$$\text{Atoms }{}^{69}\mathrm{Ga}\text{ in aliquot} = \Big\{{}^{71}\mathrm{Ga}_{\text{in spike}} - \left({}^{71}\mathrm{Ga}/{}^{69}\mathrm{Ga}\right)_{\mathcal{U}} \times {}^{69}\mathrm{Ga}_{\text{in spike}} - \left[{}^{71}\mathrm{Ga}_{\text{in spike}} \times {}^{69}\mathrm{Ga}_{\text{in spike}}\right] / \left[\text{atoms }{}^{69}\mathrm{Ga}\text{ in aliquot}\right]\Big\} / \left[\left({}^{71}\mathrm{Ga}/{}^{69}\mathrm{Ga}\right)_{\mathcal{S}} - \left({}^{71}\mathrm{Ga}/{}^{69}\mathrm{Ga}\right)_{\mathcal{U}}\right]$$

where the "atoms ^{69}Ga in aliquot" embedded in the new equation is derived from the first formula given earlier (no ^{69}Ga in tracer). Applying the unspiked ratio gives the total quantity of Ga in the spiked aliquot to good accuracy.

The tracer technique is valid only if the tracer aliquot and analytical aliquot are well mixed. The tracer isotope must also be in the same chemical form as the analyte element or isotope, accomplished by isotopic exchange usually involving a redox cycle of the mixed solutions. If significant time elapsed between the characterization of the concentration of the tracer nuclide and the measurement of its concentration in the final sample, an appropriate decay correction must be applied.

When an isotope is used to trace the fate of its element in a chemical reaction, the fundamental assumption is that the chemical properties of all isotopes of the same element are identical. Equivalent chemical speciation is ensured through mixing and isotopic-exchange reactions, and in tracer chemistry, it is important that isotopic fractionation be absent. In general, in a simple chemical apparatus, fractionation of this type can occur via two mechanisms. The first is isotopic disequilibrium, in which the mass difference between isotopes causes either a significant difference in the equilibrium constants of reactions of two isotopic species or a difference in the rates with which the two isotopic species react. The magnitude of isotopic fractionation is greater for traced elements in which there is a large relative difference between the isotope mass number of the tracer and that of the traced element. For example, fractionation effects between the isotopes of oxygen (primarily mass numbers 16 and 18) may well be of greater magnitude than those between the isotopes of nitrogen (masses 14 and 15). In addition, the magnitude of the effect decreases with increasing mass, being considered negligible for all elements with atomic numbers greater than 10 [1]. The nuclear forensic analyst considering heavy-element samples can ignore isotopic disequilibrium, but it could be very important for the measurement of the weapons-relevant isotopes of H, He, and Li.

The second mechanism by which isotopic fractionation occurs is through recoil or "hot-atom" chemistry. In a decay process, the daughter nuclide is often formed in an excited electronic state and will attain a valence state different from both its precursor atom and other isotopes of the same element already in solution or in the sample matrix [2]. This phenomenon is perhaps best demonstrated by the following example: the principal isotope of Np in a heavy-element sample is 2.14×10^6-year ^{237}Np. It is usually present at a very low concentration that precludes it from being measured without a procedure for chemical separation and concentration, which requires use of a yield tracer to relate the concentration in the final fraction to that of the original sample. Often, 2.36-day ^{239}Np is used for this purpose. A common way of tracing with ^{239}Np is to add an aliquot of α-decaying 7370-year ^{243}Am, in which the ^{239}Np decay daughter is in secular equilibrium. In this way, the analyst can cheat time in the sense that the concentration of ^{239}Np will not begin to decay with its characteristic half-life until chemical separation from Am disrupts the equilibrium. During this time, the analyst can ensure that exchange has taken place between the atoms of the sample and the atoms of the tracer. However, the separation must be effected immediately after the isotopic exchange because of hot-atom effects. The ^{239}Np concentration in equilibrium is constant, but is not stagnant. The ^{239}Np atoms that decay are constantly replaced by ^{239}Np atoms supplied by the ^{243}Am decay process and may not be in the same oxidation state or chemical form as the other Np atoms. This can result in unequal chemical yields of ^{239}Np and ^{237}Np in the final sample. For truly precise tracing of Np, the ^{239}Np decay daughter should be chemically purified from ^{243}Am and its concentration characterized, and it should then be added directly to the experimental sample before isotopic exchange.

At low concentrations, the chemical behavior of trace elements can be erratic and unpredictable [3]. These phenomena are often ascribed to adsorption on container walls or on microparticles present in solution. To avoid these effects, carriers are added to the sample. For instance, an old radiochemical trick is to add a "pinch" (micrograms to milligrams) of CsCl to any solution in which tracer-level ^{137}Cs is present in order to maintain the radionuclide concentration in solution constant. Another application of a carrier is similar to that of a tracer: if several milligrams of an element are added to a solution containing a radionuclide that is one of its isotopes, and a stoichiometric compound exists such that the element concentration can be determined by weight, the gravimetric yield of the carrier can be applied to the radionuclide [4]. This tactic was a popular technique for determination of the yields of rare-earth fission products. Aliquots containing known quantities of stable rare-earth carriers were added to dissolved reactor targets, after which the individual rare earths were isolated, precipitated as oxalates, and fired to stoichiometric oxides. The lanthanide chemical yields, determined from the weights of the samples, were applied to the radionuclide concentrations determined by radiation counting to give final fission yields.

Tracer and carrier solutions for nuclear forensic applications must be prepared from high-purity reagents to minimize the addition of interfering analytes to samples with complex matrices. The acid concentration of a tracer solution should be maintained fairly high (≥2 M) to prevent losses to the walls of the vessel in which the tracer is stored. This is particularly important for solutions of heavy elements such as Pa and Pu. The most stable tracer and carrier solutions are based on an HCl medium. For the transuranic elements, solutions should be stored in glass; for the lighter actinides, Teflon should be used. The U and Th contents of glass can be sufficiently large as to interfere in a MS analysis, and Pa irreversibly sorbs on glass in the absence of fluoride.

The concentration of carrier in a final sample can also be determined in other ways. For instance, the metal-ion concentration of a liquid fraction can often be characterized spectrophotometrically. Neutron activation analysis has also been employed for this purpose in the past [5]. However, the use of separated stable isotopes and MS has become much more common.

4.2 RELEVANT CHEMICAL PROPERTIES

In general, chemical methods used to separate radioactive substances from one another are identical to those used for stable substances. Just as in classical separations, the merit of a radiochemical separation is judged in terms of both the yield and the purity of the separated material. As remarked earlier, the goals of high purity and high recovery are often inconsistent. Radiochemical procedures involving the presence of an isotopic carrier are often simpler to design than carrier-free procedures. Losses from adsorption on vessel walls and suspended particles, which are usually negligible on the macroscale, may be the fate of a significant fraction of a tracer-level sample.

In any chemical separation scheme, the original sample becomes divided into at least two fractions through application of a driving force across a separation boundary, in such a way that the final fractions differ in chemical composition from the initial sample. Typical driving forces are solvent flow, electromotive potential, or concentration gradients, while the usual separation boundaries include liquid–liquid or liquid–solid interfaces. The effectiveness of the separation of contaminant A from desired material B can be expressed in terms of a decontamination factor, $\mathcal{D}$:

$$\mathcal{D} = \{[A]_i \times [B]_i\} / \{[A]_f \times [B]_f\},$$

where

$[A]_i$ and $[B]_i$ are the concentrations of A and B in the initial sample
$[A]_f$ and $[B]_f$ are their concentrations in the isolated chemical fraction of B

A decontamination factor can be calculated for a single step in a chemical process or for a sequence of different processes.

Separation by precipitation is a common procedure that is familiar to most chemists and is one of the most commonly used classical methods of analytical chemistry. All of the same limitations that apply in the classical use of the technique (temperature, excess of precipitating agent, rate of formation, etc.) also apply when radioactive materials are involved. At the tracer level, where the quantity of radioactive material is small, the concentration in solution before precipitation is too low to permit precipitation by exceeding the solubility of even the most insoluble of compounds. It is necessary that a carrier be present that can be removed from solution by precipitation. This carrier is not merely a substance that can be precipitated—it must also transport the desired radioactive material with it.

There are two types of carrier. The first is called an isotopic carrier and is a compound of the element of which the radionuclide is an isotope. For instance, 12.8-day ^{140}Ba can be carried from solution by adding a soluble barium salt, mixing to give a uniform solution, and precipitating $BaSO_4$ by the addition of a soluble sulfate. For the heavy elements and specific practical applications, nonisotopic carrying—or coprecipitation—is more important. Nonisotopic carriers are often used to separate "weightless" radioactive materials when it is desirable to retain high specific activity. For some elements, such as Po or At, no isotopic carrier is available because there are no stable isotopes of those elements.

An example of nonisotopic carrying is the following: La-140 is a 40-h activity that grows into transient equilibrium in samples containing ^{140}Ba. It is also carried on $BaSO_4$, which is a nonisotopic carrier for tracer La. The amount of ^{140}La carried on $BaSO_4$ can be decreased by the addition of stable La carrier, which functions as a holdback carrier. The addition of the isotopic carrier to the system does not prevent carrying the contaminating element; rather, it may even increase the total number of La atoms sequestered by the sulfate. However, because the fraction of radioactive atoms in the total number of La atoms present is decreased, the fraction of radioactive contaminant carried is also decreased. A holdback carrier is used to diminish contamination by a radioactive substance carried nonisotopically.

This leads to the concepts of chemical purity and radiochemical purity, which are not equivalent. Consider a sample of ^{140}Ba (with ^{140}La in equilibrium) to which an isotopic barium carrier has been added. Addition of a soluble sulfate results in the precipitation of $BaSO_4$, with which the ^{140}La is effectively coprecipitated. The $BaSO_4$ precipitate contains very few atoms of La because only the radiotracer nuclides were present. Thus, it is very chemically pure but very radiochemically impure, as no significant separation of ^{140}Ba from ^{140}La was effected. Now consider the same system with the addition of a macroquantity of La holdback carrier before addition of the sulfate. The resulting $BaSO_4$ precipitate may still be chemically pure, though not quite as pure as before, but it is now also radiochemically pure because the small number of La atoms carried by the precipitate is now overwhelmingly dominated by the stable carrier.

In some instances [6], ions of the nonisotopically carried element replace some of the carrier ions in isomorphous compounds and form mixed crystals. The trace element is distributed through the carrier, depending on the precipitation conditions, and can result in considerable heterogeneity in each crystal. In other instances, carrying can be effected by adsorption of the tracer-level ions on the surface of the precipitate. There is also a mixed mechanism by which surface adsorption takes place and the precipitate crystals grow to cover and trap the tracer ions. An elevated temperature during precipitation results in larger particles and a precipitate easier to manipulate. In addition, at higher temperatures, coprecipitation tends to be more selective because distribution of the trace element in isomorphous compounds is favored, while surface adsorption and trapping are reduced because of recrystallization of the outer layers of the particles. However, overall yield is increased at lower temperatures, a condition under which solubilities tend to be smaller. The most effective coprecipitation conditions involve an initial mixing and digestion of the precipitate and solution at high temperatures, followed by cooling before separation of the precipitate from the supernatant liquid. An increase in the ionic strength of the mother solution favors the addition of otherwise colloidal particles to the precipitate.

For nuclear forensic analytical samples, isotopic carrying is disadvantageous because it interferes with subsequent MS and radiation-counting measurements. Nonisotopic carrying, however, has many applications in the radiochemical analysis of the actinides. Iron hydroxide, $Fe(OH)_3$, is very effective in carrying the actinides, and lanthanum fluoride is often used to concentrate Pu, Am, and Cm from large solution volumes. Typically, these precipitates are separated from their associated supernatant liquids by centrifugation and decantation, rather than by filtration, unless a counting source is being prepared.

Another useful class of separation methods for the forensic radiochemist is solvent extraction. Unlike precipitation, solvent extraction does not require visible amounts of analyte materials for success. In general, liquid–liquid extraction systems involve intimate contact between immiscible liquid phases, which leads to a partitioning of inorganic or organic solutes between the two phases. When inorganic substances are being separated, one of the phases is usually aqueous.

For an inorganic substance to distribute appropriately between two liquid phases, it must exist in a form that has a characterized affinity for both phases. Typically, this consists of a complex ion or compound. Extraction systems for cations can be based on the nature of the extractable species—either coordination complexes or ion-association complexes [7]. Coordination complexes are formed through the interaction of an electron-acceptor cation and electron-donor ligand. The tendency for complex formation correlates well with the strength of the ligand as a Lewis base and is dependent on the electronic configuration of the cation. Chelating agents are also potential coordination ligands. Ion-association complexes are aqueous ionic species that are forced to seek the organic phase through the formation of electrically neutral clusters stabilized by the incorporation of organic solvent molecules into the cluster.

The extent to which a given solute (A) distributes between organic and aqueous phases is usually represented by a partition coefficient, p:

$$p = [\mathrm{A}]_{\mathrm{org}}/[\mathrm{A}]_{\mathrm{aq}},$$

where $[\mathrm{A}]_{\mathrm{org}}$ and $[\mathrm{A}]_{\mathrm{aq}}$ are the concentrations of the solute in the organic and aqueous phases, respectively. For an ideal system, in which the solute does not react with or alter the solvents, and in which the immiscibility of the solvents is not affected by the results of the extraction, the partition coefficient is independent of the quantity of solute present. With macro amounts of solute, there are very few extracting systems that approach this ideal [8,9]. The partition coefficient is a function of environmental variables. For example, p tends to decrease with an increase in temperature. The use of volumes of different size also affects p, which is particularly important in partition chromatography.

The value of p also changes if an inert diluent is used. For example, the extraction of U into tri-n-butylphosphate is the basis of the PUREX process. Carbon tetrachloride can also be added to the organic phase to decrease its viscosity and make the physical separation of the liquid phases easier. Although CCl_4 does not participate in the separation, it does affect p through the change in solubility of the solute species. Similarly, the presence of an unextractable bulk impurity in the aqueous phase may cause an increase in p by changing the ionic strength of the aqueous phase, thereby altering the activity of the solute. As an example, the addition of a salting-out agent such as $Al(NO_3)_3$ to the aqueous phase can increase the partition of U into diethyl ether. Conversely, the presence of interfering substances that form unextractable compounds with the solute acts to decrease p. The addition of small amounts of fluoride or phosphate in the previous example will interfere with the extraction of U into most organic solvents.

The classical method of performing a solvent extraction, in which both phases are placed in a separatory funnel and agitated by hand to effect the partition, is only seldom used in NFA, as typical sample sizes are small and the resulting phase volumes need be no larger than a few mL. Usually, phases are mixed in a capped centrifuge cone using a vortex mixer. This technique is particularly convenient in a glove box, where the loss of manual dexterity makes the manipulation of a separatory funnel and stopcock cumbersome. Any subsequent difficulties in phase separation are easily remedied through use of a centrifuge.

The nuclear forensic analyst may often employ one or several applications of solvent extraction as mass-reduction steps before performing chromatographic procedures. For example, if an iron hydroxide precipitation has been used to concentrate actinides from a large volume, it is useful to dissolve the precipitate in a minimum volume of HCl and extract Fe into methyl isobutyl ketone. Most of the Fe can be removed in a single step, and the actinide analytes have low affinity for the organic phase in the chloride system. The final aqueous phase must then be boiled to expel organic residues before any ensuing ion-exchange steps.

Although extractions involving organophosphorus compounds are widely used in industry, they have little application in the radioforensic laboratory. Residual phosphates interfere with most subsequent purification steps for higher-valence actinides, and similar considerations hold for organic amines. However, there are frequent applications of extraction involving the chelating agent thenoyltrifluoroacetone (TTA) in an organic solvent, particularly for the isolation of U, Np, and Pu from HNO_3 solutions. Some of the radiochemical milking experiments that are performed to determine the concentrations of small quantities of ^{236}Pu, ^{232}U, ^{233}U, or ^{243}Am in the presence of much larger quantities of other isotopes of the same elements begin with a TTA extraction. Unfortunately, most of the literature on the TTA/HNO_3 system is based on benzene as the TTA diluent, and there is a surprisingly large effect when the inert diluent is changed from carcinogenic benzene to the more environmentally friendly toluene [10]. Because the volumes used in these procedures are small, benzene continues to be used.

A class of separation methods of particular value to the nuclear forensic analyst is column chromatography. This term is usually applied to the selective partition of analytes between a flowing fluid and an insoluble solid, referred to as the support. Because of the limited chemical capacity of most supports, an increase in the size of the sample being processed often requires an increase in the size of the chromatographic apparatus. In the forensic laboratory, this effectively limits sample sizes for chromatographic steps to <1 g. However, the intrinsically high separation factors and high chemical yields attainable with column chromatography make this technique one of the most powerful available to the radioanalyst [11].

There are three basic types of chromatographic separation: reversed-phase partition chromatography, ion-exchange chromatography, and adsorption chromatography. Partition chromatography is a special case of solvent extraction, discussed earlier. An extractant is attached to the immobile support, and an immiscible mobile phase flows past, leading solute molecules to partition between the two phases. It is a well-known technique in organic chemistry, where a filter paper supports the aqueous phase and an organic solvent moves across the paper by capillary action, creating a lateral separation of organic compounds. In the radioanalytical laboratory, it is usually the organic phase that is attached to the support and the aqueous phase that flows [12], and the standard solvent extraction processes are carried out on the column [13]. The organic phase is distributed on the surface of small, equally sized particles that are placed in a glass tube—the chromatographic column. This provides the maximum surface area for chemical exchange and the means by which the mobile phase is prevented from taking selected paths through the column bed (channeling). The driving force that supplies the flow is either externally applied pressure (positive or negative) or gravity.

There are several products available commercially that provide ion-specific organic materials on a solid support. Although the advertised selectivities of some of these materials are most impressive, these products have found little utility in the nuclear forensic laboratory. Part of their limitation lies in their narrow capacities, while part lies in the variety of interfering ions that may be present in unknown samples. In addition, trace amounts of the organic phase that elute from these columns can interfere in subsequent purification steps.

The techniques of ion exchange, however, are of great utility for the radiochemical isolation of most species of interest to the nuclear forensic analyst [14,15]. Good basic reviews of ion-exchange methods can be found in references [16,17].

Ion-exchange resins consist of an insoluble polymeric hydrocarbon to which ionizable functional groups are attached. The structural backbone is usually a cross-linked styrene divinylbenzene polymer. These polymers are fabricated into small porous beads that may be narrowly sized and are almost perfectly spherical, over which an aqueous solvent is passed. Optimal separations are obtained when the beads are all approximately the same size, which helps prevent channeling of the solvent through a column packed with the resin. Resin beads are available in most mesh sizes from about 18 mesh (1-mm diameter spheres) to colloidal size.

Most resins used in the radioanalytical laboratory encompass 100 to 200 mesh, and dry resin beads swell when immersed in an aqueous medium. The extent of this swelling is controlled by the degree of cross-linking of the polymer strands, which is a function of the percentage of divinylbenzene used in fabricating the polymer. Typical resins used in the radioanalytical laboratory are cross-linked between 4% and 12%. The higher cross-linked resins are generally more selective, but the reaction kinetics are slower. Lower cross-linked resins exchange quickly and are particularly useful in sorbing large ions. However, a drawback is that the volume of the low-cross-linked beads can vary greatly when the solvent concentration is changed, resulting in variable column length.

These products are classified as either cation-exchange or anion-exchange resins, depending on the identity of the incorporated functional groups, which can consist of strong or weak acids or bases. A typical, strong cation-exchange resin contains sulfonic acid groups; the SO_3H group is completely ionized in water and can exchange the coordinated hydrogen ion for another cation under the right conditions. For example,

$$-SO_3^-H^+ + K^+ \leftrightarrows SO_3^-K^+ + H^+.$$

Similarly, a typical, strong anion-exchange resin contains quaternary amine bases; the =N–OH group can rapidly exchange hydroxyl ions with anions from the solution; for example,

$$=N^+OH^- + Cl^- + H^+ \leftrightarrows =N^+Cl^- + H_2O.$$

The capacity of a resin for adsorbing analyte ions depends on the structure of the polymer and the nature of the functional group. The strong anion- and cation-exchangers that are commonly used in radioanalytical chemistry have capacities that are effectively independent of environmental variables (such as pH) and that are usually expressed in terms of milliequivalents (meq) per mL of wet resin. For a standard, strong anion-exchange resin like DOWEX-1, a typical capacity is 1.2 meq/mL; for the DOWEX-50 strong cation-exchanger, a typical capacity is 2.0 meq/mL.

An ion-exchange column is comprised of a glass tube that supports a plug of glass wool (or a glass frit) to prevent the resin beads from flowing out of the column. For applications in which fluoride is a component of the mobile phase, the glass-wool plug should be replaced with Teflon or saran wool. The resin should never be poured into the column dry but, rather, always wet with water to ensure that swelling is complete. After the column is packed, the resin bed should be thoroughly washed with water and then the initial eluent to reduce the level of manufacturing impurities.

There are two techniques for separating ions by means of an ion-exchange column, and they are termed "elution" and "breakthrough." In the elution technique (or classical ion-exchange chromatography), ions are adsorbed from a dilute solution that is passed through the column. Some ions have no affinity for the resin and pass through the column unhindered, whereas others interact strongly with the resin and are adsorbed in a narrow band at the top of the column. A free-column volume is defined as the volume of solution retained between the resin particles of the column.

A truly unbound analyte will be predominantly removed from the resin bed after passage of a single free-column volume through the column. However, some straggling exists, caused by the random motions of solute molecules, which makes it necessary to collect a somewhat greater volume to ensure good recovery. After unwanted species are washed from the resin, the elution solution is changed such that the desired product is no longer bound to the resin, and it then elutes from the column. Figure 4.1 is a photograph of an anion-exchange column (DOWEX-1 × 8) on which a gram of Pu is adsorbed. The red-brown hexachloro-Pu(IV) complex, favored in the 9 M HCl column solution, is tightly retained at the top of the column and sequestered on a volume of resin sufficient to supply the meq of capacity necessary to match the number of meq of analyte. Later, a solution of 0.5 M HCl was passed through the column. The chloride concentration was then no longer adequate to maintain an anionic complex with Pu(IV), which consequently eluted from the column.

In breakthrough (or differential) ion exchange, as the solution of ionic species is passed through the column, the ions migrate down the column at rates dependent on their affinities for the functional groups of the resin. The most weakly adsorbed ions are the ones to appear first in the effluent. Many cation-exchange procedures operate on the breakthrough principle: the analytical sample is loaded in a minimum volume of solution, and the product fractions are collected from the column in the order of increasing affinity. As an example, in a 13 M HCl solution, Am has less affinity for cation-exchange resin than do the lanthanide fission products. Essentially 100% of the Am can be collected from a DOWEX-50 column before the rare earths begin to break through.

FIGURE 4.1 An anion-exchange column containing approximately 1 g of adsorbed Pu (left). The solution medium is 9 M HCl.

Adsorption chromatography involves the repeated partitioning of an analyte between a solvent and a solid adsorbent. Adsorbents consist of substances such as silica gel, alumina, or activated carbon. The technique is most effective for organic analytes and is of limited use in the nuclear forensic analytical laboratory.

Isolating a chemical fraction from a sample of questioned nuclear material is just the first step of a radioforensic analysis. The relative concentrations of analytes must be determined, a chemical-yield correction applied, and the absolute concentrations of the analytes in the original sample derived. This is accomplished either through MS or by radiation counting. Samples submitted for MS typically require no special handling beyond requirements of analytic cleanliness. They are submitted in small volumes of liquid or as dry deposits, preferably in Teflon containers. However, for radiation counting, the geometric extent of the radionuclide sample, whether liquid or solid, affects the efficiency of the detector and the accuracy with which the concentrations of radionuclides can be determined.

Sources for radiation counting can be prepared as gases, liquids, or solids, but the need for gaseous sources has largely disappeared with the advent of more sensitive MS techniques [18]. Tritium and ^{14}C are no longer measured in this way, although it is still necessary to measure the radioactive isotopes of the rare-gases as noncondensed sources. In general, these samples are purified and collected using cryogenic or gas-transfer techniques and are delivered to a thin-walled pressure vessel of well-defined geometry for γ-ray measurements [19].

The analyst may have two different types of liquid samples to prepare. Liquid scintillation counting is particularly useful for measuring very-low-level β activities, or β activities with low-energy end-points. Liquid scintillation offers very high sensitivity and efficiency, even for large samples, but these fractions must be meticulously purified from all traces of a radioactive sample matrix or the signal from the analyte of interest will be lost in background events. The analyst must prepare either a solid or liquid sample that is soluble in the scintillating liquid, or a finely divided solid that can be suspended in the scintillator. The analyst should thus be familiar with acceptable solubilizing agents. For example, aqueous samples can be introduced into toluene-based scintillator liquids through solubilization with ethanol.

Liquid sources may also be counted for γ-rays. As long as the radionuclide analytes are in true solution, Ge photon detectors can be calibrated to accurately determine their concentrations in well-defined geometries. In our laboratory, a standard sample counting configuration consists of 10 mL source solution confined in a cylindrical polymeric vial that has a cross-sectional area of exactly 10 cm^2. In those cases in which the chemical yield is determined from the mass of an added carrier, this method is consistent with the requirements of spectrophotometry or other optical techniques.

Solid sources can be either "thick" or "weightless." Gamma-ray and some β-particle counting can be fairly forgiving of the mass of the sample, and corrections for self-attenuation can be calculated or measured with standards. When the chemical yield of an element is determined gravimetrically by the weight of a stoichiometric compound, sources are often prepared by filtration. The precipitating agent is added to a liquid sample, and after digestion, the supernate is drawn by vacuum through a flat filter paper mounted on a frit. With care, this method can result in a uniform deposit of well-defined extent, which is necessary for accurate counting.

Weightless sources are required for α spectrometry, with the method of preparation and the substrate dependent on the counting requirements. If an absolute α-particle disintegration rate is required, it is normal to prepare thin sources on Pt. The atomic number of the substrate affects the counter efficiency through α-particle backscatter, and most calibration sources are prepared on Pt. A common method of performing a radioassay is by stippling a liquid from a volumetric micropipette: the active solution is placed on the counting plate in a series of distributed small drops, which are then evaporated to dryness under a heat lamp. The micropipette (which is likely "to contain," rather than "to deliver") must then be rinsed to the counting plate with fresh solvent. The plate is then heated in a Bunsen-burner flame to an orange glow to fix the activity and volatilize any unwanted material. However, samples prepared in this fashion are generally not of good spectroscopic quality because of the inhomogeneity of the "nearly weightless" radionuclide deposit.

Alpha sources used for the determination of isotope ratios must be thin and uniform, and they are produced either by volatilization or electrodeposition. The former is a wasteful technique in that about half of the final purified sample does not transfer to the counting plate. Thus, an HCl solution of the purified material is evaporated to dryness on a W filament, which is placed in an electrical fixture. The counting plate is suspended facedown over the filament, and the entire assembly is mounted in vacuum. When the pressure is <20 mtorr, current is discharged through the filament, heating it white-hot and transferring the adhering radionuclides to the counting plate. Although this method is wasteful of sample, it is fast and produces spectroscopic-quality α sources.

If the sample size is limited, α sources can be prepared by electrodeposition, the methods of which are varied and strongly dependent on the chemical nature of the analyte. For the actinide elements of most interest to NFA, the analyte is dissolved in a minimum volume of dilute HNO_3 and transferred into an electroplating cell with isopropanol. The counting plate comprises the bottom surface of the cell and is electrically connected as the cathode, while a Pt wire anode is suspended in the liquid. The actinides are deposited on the cathode as hydrous oxides or hydroxides, induced by the reduction of the H^+ concentration of the solution near the cathode [20,21]. After plating, the source disk is lightly flamed to fix the activity. Sources prepared in this way are thin and uniform, and the efficiency of the plating procedure can be near unity.

These are the underlying chemical properties upon which radiochemical analysis is constructed. Methods involved in specific nuclear forensic separations are outlined in other sections.

4.3 RADIONUCLIDES IN MEDICINE AND INDUSTRY

The focus of the development of NFA has been on U, Pu, and the other actinides, either as the principal component of purified materials or as a residual nuclear fuel dispersed in a more complicated matrix. However, the widespread use of radioisotopes in industry and medicine provides opportunities for their diversion into unsanctioned uses. While these materials are not of interest from the standpoint of nuclear proliferation, their potential impact on public health and safety is significant. Identifying their origin is a problem that may be addressed by applying the techniques outlined in this book, primarily through the investigation of trace chemical signatures.

The methods of analysis of these "non-fuel-cycle" materials are as varied as the materials themselves, and several of the real-world case studies included in the last chapters of this book are examples of these types of samples. Rather than produce a comprehensive record of procedures that we have utilized for such analyses, however, we focus mainly on signatures that can be measured with well-designed protocols. The reader may then utilize this background discussion to develop his/her own forensic methods, both radiochemical and mass spectrometric.

The specimens to be most likely encountered by the nuclear forensic analyst are sealed (e.g., industrial irradiation sources), dispersible (e.g., medical preparations or tracers for analytical chemistry), or dispersed (either accidently or through deliberate inclusion of radioactive materials in an explosive device). High-level sources of long-lived radionuclides are greater concerns than short-lived sources, but samples of any activity level are potentially encountered, perhaps requiring the use of shielding materials or remote operations.

The uses of radionuclides in diagnostic medicine, in which biodistribution of small quantities of short-lived radionuclides are used to assess transport processes in the body to identify and localize a variety of ailments, are possible sources of nuclear forensic samples. Signatures indicative of a source or production method are difficult to develop in the diagnostic aliquots themselves due to the conservative and consistent methods used to produce materials for medical applications. However, the radiochemical generators (or "cows") that are often sources of short-lived diagnostic nuclides can incorporate a variety of signatures. For example, the 3.9-h positron-emission tomography nuclide, ^{44}Sc, is "milked" from 59-year ^{44}Ti; the ^{44}Ti may be produced by a variety of charged-particle reactions (including spallation), which also produce other long-lived or stable nuclides indicative of the target, projectile, and energy of the production reaction, as well as of the chemical techniques employed to produce the generator parent activity. The other applications of radionuclides in medicine, those supplying therapy, tend to use larger quantities of a wide variety [22] of longer-lived radionuclides (e.g., 8.02-day ^{131}I and 73.8-day ^{192}Ir), and these are other potential sources of nuclear forensic samples. More than half of all cancer patients receive radiation therapy at some point during their treatment, and dozens of different radioisotopes have been so utilized. Again, characteristic signatures of isotope production and the subsequent chemical isolation can remain within the materials.

One of the most important industrial uses of radionuclides is as tracers in analytical chemistry (see Section 4.1). Other industrial uses include (1) measurement of density and hydrogen content through neutron scattering, via use of ^{252}Cf or mixed ^{241}Am/Be sources; (2) observation of flaws in welded joints and cast-metal parts by γ-ray radiography, using a variety of sources; (3) sterilization of bulk products that cannot be heat-treated, often using ^{60}Co or ^{137}Cs sources; and (4) tracing transport of materials through complex pathways, including environmental and industrial systems.

The main sources of radionuclides for medical or industrial applications are charged-particle accelerators and nuclear reactors [23]. The short path length of charged particles in matter, and the relatively low particle fluxes compared to those available in a reactor, limit the use of accelerators to applications in which the product cannot be produced by low-energy neutron reactions (e.g., the ^{44}Ti mentioned earlier) or where product mass purity is required. Since capture of the neutron produces an isotope of the element being irradiated, the target and the reaction product are inseparable by chemical means, unless the desired product is the result of β decay or fission. For example, the isotope most frequently used in diagnostic nuclear medicine is 6.0-h ^{99m}Tc, which is eluted from generators containing sources of 66.0-h ^{99}Mo (produced by fission of ^{235}U). The no-carrier-added molybdenum parent is chemically isolated from fission products, in which the concentrations of stable Mo isotopes are about the same as that of the ^{99}Mo. There have been recent concerns about the use of fissile HEU as the reactor target material for isotope production, as well as about the reliable availability of ^{99}Mo, and research into alternate methods for its production is ongoing. Reactor irradiation of natural Mo produces ^{99}Mo via the $^{98}Mo(n,\gamma)$ reaction, but also makes the neutron-deficient isotope, 4000-year ^{93}Mo, which is not present in fission products. Presence of ^{93}Mo in a ^{99}Mo generator is thus indicative of production through irradiation of stable Mo targets. Absence of ^{93}Mo is an indicator of production by fission, even if the stable isotopes of Mo are also present in excess (signature, in this instance, of the use of Mo carrier in the postirradiation chemical separation procedure to purify the ^{99}Mo).

Enriched stable isotopes are often used as starting materials in radioisotope production, and a stable isotopic signature significantly different from natural abundances for that element can be a key forensic discovery. In certain cases (e.g., U and Pu), the distribution of isotopes is characteristic of the enrichment process. The electromagnetic isotope separation capabilities at Oak Ridge are currently more limited than in the past; therefore, the commercial and scientific needs for small quantities of enriched materials drive a variety of enrichment technologies, each with its own characteristic signature.

Sources for industrial applications tend to be more intensely radioactive than medical sources, whose external dose rates are generally constrained to be sublethal. As a result, chemical procedures that can be performed prior to irradiation are generally preferred to those performed after irradiation, where remote handling and shielded enclosures may be required. Interpretation of minor isotopes detected in high-level γ-ray sources (e.g., ^{60}Co) can be treated like a neutron-activation analysis measurement; the distribution of isotopes in the products of fission can be indicative of the nuclear fuel giving rise to them (see Chapter 26); and sources of the heavy elements ^{241}Am and ^{252}Cf contain minor isotopic signatures characteristic of production (^{243}Am and ^{250}Cf, respectively) and age (^{237}Np and ^{248}Cm, respectively).

The wide variety of radioisotopes used in medicine and industry can lead to a complicated suite of forensic problems. However, while possibly helpful, the nuclear forensic analyst does not require expertise in the applications from which a radionuclide sample was diverted. With some focused thought, general radiochemical principles provide sufficient background to perform a forensic analysis.

4.4 AUTOMATION OF RADIOCHEMICAL PROCEDURES

Many of the radiochemical separation procedures that have been performed by nuclear forensic analysts can be very labor intensive and time consuming, particularly those applied to samples of interdicted materials. In those situations, the number of analytical aliquots is generally limited, and the time requirements for data return are often on the order of weeks or months. A small number of radiochemists and mass spectrometrists can then produce high-quality forensic information, including appropriate quality assessment, on a schedule that meets the needs of law-enforcement investigators.

However, if faced with an incident in which the questioned-sample load increases dramatically, this situation may no longer be true. Imagine a scenario requiring the forensic analysis of debris from a nuclear explosion: requests for analytical information relating to questions of national security cannot be deferred for weeks, and the number of samples and number of analytes in each sample are expected to be considerably higher than for a predet case of interdicted special nuclear material. An appropriate political response should be formulated only after the results of a reliable forensic analysis and assessment are available. Compounding this problem of available manpower, the attribution experts that interpret the analytical results are often the same analysts that must perform the laboratory measurements.

Automation refers to the replacement of human labor by machines, operated automatically or by remote control. The automation of chemical procedures is an avenue by which the rate of return of nuclear forensic information to the analysts can be maximized, with increased reproducibility and reduced radiation dose to the radiochemical staff. The idea of automating radiochemical procedures is not novel, but dates back to the 1940s and the remote handling of irradiated reactor materials for the isolation of Pu. The modern radiochemical analyst often employs automated aids such as drop counters, fraction collectors, and sample changers to introduce "hands-off" periods during procedures, thereby permitting parallel operations and increased productivity. Accelerator-based production of short-lived radionuclides for medical applications is often highly automated [24,25], and heavy-element chemists have performed chemical separations on short-lived transactinide isotopes that are essentially unattended [26,27]; both applications take advantage of a defined and reproducible starting material.

Rapid radiochemical procedures are often categorized as two types: continuous and discontinuous [25]. Nuclear forensic samples are intrinsically of the second type, as they arise from a discrete event like an interdiction by law enforcement or a nuclear detonation. A series of sequential chemical operations is then performed on the sample, starting at some reference time.

It is the view of some expectant program managers that NFA can be automated to the same degree as nuclear medicine or heavy-element separations, with a debris sample inserted into one end of a shoebox-sized automaton and radionuclide concentration data expelled out of the other. We do not believe this will ever be the case, partly because of the fundamental progression of a nuclear forensic inquiry: information obtained from the sample often defines the direction of future analytical measurements. Another issue arises from the complicated nature of the sample matrices that we have encountered, both with interdicted materials and in explosion debris (see Chapter 5). The early phases of NFA investigations will always involve sample handling, including the creation of homogeneous radiochemical solutions that can be aliquoted for multiple analyses. The sample matrix can be anything (powders, metals, glasses, soils, liquids, gases, organic matter, and so forth), which can require a variety of processing techniques (wet or dry ashing, pulverization, fusions with caustic compounds, dissolution by different combinations of strong acids, etc.). It seems unlikely that the initial strategies for NFA efforts will ever be absent expert oversight by an experienced radiochemist.

Even with these reservations, though, it is clear that increased application of automation to forensic procedures is desirable. Many of the radiochemical methods applied to the analysis of forensic samples have changed little since the days of Hahn and Strassmann, when they provided the only means for disentangling complex mixtures of radiochemical species for element identification and the determination of decay properties. The half-lives of the analyte nuclides of interest, the speed at which samples are processed, and imposed reporting deadlines constrain sample throughput. Processing rate is the only variable controlled by the laboratory analyst, and automation could have a significant impact on its optimization.

Dissolution of samples of nuclear explosion debris at Livermore during the 1980s provides an example of an application of chemical automation to forensic–like samples. Each postdetonation sample was crushed and placed in a platinum crucible, which was then passed into a glove box–like enclosure. The crucible was placed on a hot plate, which also incorporated a weighing balance. Delivery tubes for the introduction of acids, as well as Pt sensor wires, were suspended a few cm above the debris sample. A set of microprocessor-controlled pumps delivered volumes of preselected acids from reservoirs located outside the enclosure. A given acid was added until stopped via the liquid level causing electric current to flow between the chemically resistant platinum wires, after which the solvent was evaporated through the action of the hot plate. When minimum weight was detected by the balance, the next acid solution of the procedure was added. The process continued for 5–6 h, culminating in an HCl solution containing a variable quantity of insoluble material, a consequence of the fact that even such products of a nuclear explosion can be chemically distinct. At this point in the procedure, laboratory radiochemists replaced the automated sequence; their first action was to filter the solution (containing most of the fission products) into a glass bottle placed in a lead-lined recess in the back of the gloved enclosure. The result of this application of automation was radiochemical completion of the dissolution with only the small, intractable fraction of the debris sample, in a much lower radiation environment than would otherwise have been possible. This "semi-attended" automated procedure saved more than one man-week of effort in the historic analyses of LLNL Test-Program debris (6 h × 9 samples).

Traditional NFA tends to be a very sequential endeavor. The sample set is acquired in the field and is sufficiently characterized to allow it to be shipped to the analysts. Any sample down-selection can take place before and/or after shipment. Sample subdivision takes place after receipt by the laboratory, typically resulting in the isolation of small samples for material characterization, isotope-ratio determination, and specialty analyses (e.g., organic extraction). Parallel analyses generally do not take place until after the sample is homogenized, usually by dissolution.

One goal of automation is to shorten each individual sequential step in order to accelerate the analysis timeline, and another goal should be to facilitate parallel processes. As an example, in the sequential process outlined earlier, the shipping of nuclear forensic samples to the analytic facility constitutes a gap during which no information is generated and no processing is occurring. At a minimum, automated radiation detection and other nondestructive interrogation could be conducted during this time (e.g., qualitative XRF surface screening and/or organic headspace sampling). Better yet, there is no scientific reason that the samples could not be subjected to automated dissolution during shipment, provided that any necessary sample-splitting occurred in the field.

Following dissolution, parallel processing of multiple samples is a common scenario for NFA. For instance, if several interdicted U samples have been dissolved, it is common to perform ion-exchange chromatography for the chronometric nuclides (Chapter 6) on aliquots of each solution in parallel, and often in replicate. Four analytical solutions can result in 18 ion-exchange separations (two traced and two untraced aliquots from each solution, and two process blanks). Assuming the conscientious prevention of cross-contamination, the analyst may typically run these 18 columns simultaneously. There are thus obvious advantages in both timeliness and reproducibility to automate repetitive processes such as column packing, column conditioning, sample loading, and fraction collection.

In automated chemistry, solutions are often moved through the process by an application of pressure (or vacuum). One might naïvely assume that a series of sequential chemical steps performed on a given sample can be accelerated by increasing pressures or reducing volumes, but this is not necessarily the case. Each chemical step must take into account the rate at which the chemical reactions occur (kinetics), the solubility of the analytes and the matrix in the solvent, the chemical affinity of the analytes for the various phases, and the means of detection of the final product. Chemical methods include coprecipitation, solvent extraction, ion exchange, and volatilization, all of which can be impacted by the concentrations of matrix ions [28]. Any automated processes must be constrained so that the limitations of the chemical properties of the analytes are taken into account.

Truly precision analyses will always require the ultimate return of samples to the controlled environment of fixed-location analytic laboratories, but generation of key "indicator" data in the field or in transport can guide scientific analyses and drive preliminary forensic conclusions. Revealing signatures based on isotope ratios of a single element (e.g., U and Pu) do not require complete sample dissolution, thereby enabling unattended operations. The time-consuming, total-dissolution procedures can be deferred until the fixed-location radioanalytical laboratory following shipment. Another potential benefit of automation is a reduced footprint for the analytical instrumentation, an important consideration in the development of field-deployable radiochemical apparatus. In certain cases, a decrease in size can result in the reduction of reagent volumes and other consumables, minimizing the generation of hazardous waste. However, the processing of samples in the field is of little value without on-site MS or radiation detection to generate real-time reportable results.

Radiation detection is particularly problematic in the nuclear-detonation scenario: high radiation backgrounds would require enormous amounts of shielding, making photon counting impractical. Counting techniques such as conversion-electron spectrometry with position-sensitive detectors may provide an alternative in some cases [29]. Additionally, although preparation of thin sources for α spectrometry would be a significant challenge for automated apparatus, research into appropriate γ- and β-insensitive scintillators may eventually permit rough measurements of α-decay-derived isotope ratios in the field [30].

More often than not, the matrix of a nuclear forensic sample is incompletely defined at the start of an analysis, meaning that feedback information would have to be generated during an automated process, much of it determined "by eye" by a nonrobotic radiochemist. Information on turbidity, color, and ionic strength must be obtained, as well as radiation assay information at the location of the analyte. These considerations seriously complicate the design and construction of a nuclear-forensic automaton and ensure the continued employment of radiochemical analysts for the foreseeable future.

REFERENCES

1. J.F. Duncan and G.B. Cook, *Isotopes in Chemistry*, Clarendon, Oxford, U.K., 1968.
2. D. Brune, B. Forkman, and B. Persson, *Nuclear Analytical Chemistry*, Studentlitteratur, Lund, Sweden, 1984.
3. A.K. Lavrukhina, T.V. Malysheva, and F.I. Pavlotskaya, *Chemical Analysis of Radioactive Materials*, Chemical Rubber Co., Cleveland, OH, 1967.
4. J. Ruzicki, *Stoichiometry in Radiochemical Analysis*, Pergamon, New York, 1968.
5. F. Girardi, *Modern Trends in Activation Analysis*, U.S. National Bureau of Standards Special Publication 312, Washington, DC, Vol. 1, p. 577, 1969.
6. O. Hahn, *Applied Radiochemistry*, Cornell University Press, Ithaca, NY, 1936.
7. G.H. Morrison and H. Freiser, *Solvent Extraction in Analytical Chemistry*, Wiley, New York, 1957.
8. D.C. Grahame and G.T. Seaborg, The distribution of minute amounts of material between liquid phases, *J. Am. Chem. Soc.*, 60, 2524, 1938.
9. R.J. Myers, D.E. Metzler, and E.H. Swift, The distribution of ferric iron between hydrochloric acid and isopropyl ether solutions, *J. Am. Chem. Soc.*, 72, 3767, 1950.
10. D.L. Heisig and T.E. Hicks, The extraction mechanism of plutonium (IV) TTA chelate in sec-butylbenzene-nitric acid-uranyl nitrate mixtures, Lawrence Radiation Laboratory Report UCRL-1664, Berkeley, CA, 1952.
11. F. Girardi and R. Pietra, Multielement and automated radiochemical separation procedures for activation analysis, *At. Energy Rev.*, 14, 521, 1976.
12. L.S. Bark, G. Duncan, and R.J.T. Graham, The reversed-phase thin-layer chromatography of metal ions with tributyl phosphate, *Analyst*, 92, 347, 1967.
13. E.K. Hulet, An investigation of the extraction-chromatography of Am(VI) and Bk(IV), *J. Inorg. Nucl. Chem.*, 26, 1721, 1964.
14. J.J. Katz and G.T. Seaborg, *The Chemistry of the Actinide Elements*, Wiley, New York, 1957.
15. E.K. Hyde, Radiochemical separation methods for the actinide elements, in *Proceedings of the International Conference on the Peaceful Use of Atomic Energy*, Geneva, Switzerland, 1955, A/CONF. 8/7, United Nations, New York, 1956, p. 281.
16. W.C. Bauman, R.E. Anderson, and R.M. Wheaton, Ion exchange, in G.K. Rollefson and R.E. Powell, eds., *Ann. Rev. Phys. Chem.*, 3, 109, 1952.
17. W. Rieman and H.F. Walton, *Ion Exchange in Analytical Chemistry*, Pergamon, Oxford, U.K., 1970.
18. R.T. Overman and H.M. Clark, *Radioisotope Techniques*, McGraw-Hill, New York, 1960.

19. M.B. Reynolds, Techniques for counting radiokrypton, *Nucleonics*, 13(5), 54, 1955.
20. D.C. Aumann and G. Muellen, Preparation of targets of Ca, Ba, Fe, La, Pb, Tl, Bi, Th and U by electrodeposition from organic solutions, *Nucl. Instrum. Methods*, 115, 75, 1974.
21. G. Muellen and D.C. Aumann, Preparation of targets of Np, Pu, Am, Cm and Cf by electrodeposition from organic solutions, *Nucl. Instrum. Methods*, 128, 425, 1975.
22. S.J. Adelstein and F.J. Manning, eds., *Isotopes for Medicine and the Life Sciences*, National Academy Press, Washington, DC, 1995.
23. M.D. Kamen, *Isotopic Tracers in Biology*, Academic Press, New York, 1957.
24. R.A. Ferrieri and A.P. Wolf, The chemistry of positron emitting nucleogenic (hot) atoms with regard to preparation of labelled compounds of practical utility, *Radiochim. Acta*, 34, 69, 1983.
25. G. Herrmann and N. Trautmann, Rapid methods for identification and study of short-lived nuclides, *Ann. Rev. Nucl. Part. Sci.*, 32, 117, 1982.
26. N. Trautmann, Fast radiochemical separations for heavy elements, *Radiochim. Acta*, 70/71, 237, 1995.
27. M. Schaedel, W. Bruechle, and B. Haefner, Fast radiochemical separations with an automated rapid chemistry apparatus, *Nucl. Instrum. Methods Phys. Res.*, A264, 308, 1988.
28. J. Inczedy, *Analytical Applications of Ion Exchangers*, Pergamon Press, Oxford, U.K., 1966.
29. J. van Klinken and K. Wisshak, Conversion electrons separated from high background, *Nucl. Instrum. Methods*, 98, 1, 1972.
30. J.W. McKlveen and W.J. McDowell, Liquid scintillation alpha spectrometry techniques, *Nucl. Instrum. Methods Phys. Res.*, 223, 372, 1984.

5 Principles of Nuclear Explosive Devices and Debris Analysis

We believe that the discussion that follows presents sufficient information about nuclear explosive devices that the nuclear forensic analyst can benefit from it. For obvious reasons, this is a sensitive subject. The reader is referred to Glasstone [1] for a good, comprehensive discussion of the performance of nuclear explosives. Further, if a suitable US security clearance is held, references [2,3] are quite good.

5.1 ONE-STAGE FISSION EXPLOSIVE (ATOMIC BOMB)

A nuclear explosive device can produce a large amount of energy from a small quantity of active material. The yield-to-mass ratio of a nuclear device is orders of magnitude greater than that obtainable from chemical explosives and is the principal advantage of a nuclear weapon. A disadvantage is that the residual fission products, activation products, and any unconsumed fissile fuel are dispersed over a large area in an atmospheric explosion. They constitute a significant and long-lasting population health hazard over a significant downwind distance.

Some of the principles developed previously for criticality and nuclear reactors are equally applicable to nuclear explosives. In a reactor, the neutron-multiplication factor, k, is maintained at slightly >1 until the desired operating level is attained. At that point, control rods or neutron poisons are used to limit k to a value of 1.0, so that the reactor operates at steady state. Fission-generated heat is conducted away from the reactor core to perform work and produce electricity, as well as to prevent the core from being damaged by reaching high temperatures. A nuclear explosive attempts to achieve supercriticality and generate the largest possible value of k—that is, the greatest increase in the number of neutrons (and consequently fissions) from one generation to the next. This reaction takes place on such a short timescale that the only mechanism through which the weapon's "core" can cool itself is explosive disassembly, which drives k to drop to zero. Most of the engineering tactics involved in the design of an efficient nuclear explosive focus on maintaining the assembled fissile material sufficiently long for the neutrons to do their work to produce increasingly more fissions.

Neutrons released in fission arise predominantly from deexcitation of the primary fission products and possess a distribution of energies. The average laboratory-frame neutron kinetic energy is ~2 MeV, a combination of the fission-fragment excitation energy and the energy given to the neutron through the kinetic energy of the emitting fragments. The neutron spectrum is approximately Maxwellian and characterized by an apparent temperature near 1 MeV, with a distribution peaking at 0.7 MeV and extending to energies >10 MeV [4,5]. These are the fast neutrons that sustain the runaway chain reaction and produce the nuclear yield of the device.

The generation time is roughly equal to the average time elapsed between the release of a neutron and its capture by a fissile nucleus. If a neutron energy of ~1 MeV and a reasonable capture cross section are assumed, a neutron generation in a supercritical system lasts ~10^{-8} s (a unit termed a "shake" in weapons jargon). It requires between 50 and 60 generations to produce 1 kt of fission yield [1], necessitating that the device stay assembled on the order of 0.5 μs. Actually, as the production of energy is exponential with time at constant k, most of the explosive yield arises in the last few generations. A difference in assembly time of 30 ns results in approximately an order-of-magnitude change in explosive yield.

The speed at which a nuclear explosive device performs entails another important difference compared to the function of nuclear reactors. In a reactor, some of the neutrons that escape the core are returned to the fuel after being moderated by surrounding materials. This moderation process is time-consuming. In a graphite-moderated reactor, >100 collisions with C atoms are required (on average) to reduce the energy of a 2-MeV incident neutron to thermal energy (0.025 eV) [6]. The moderation requires more time than the critical mass of a nuclear explosive device can remain assembled, so moderated neutrons play only a trivial role during the initial, high-k burn of the fissile fuel. However, reflection of fast neutrons is important to the operation of a nuclear device, as this process allows a decrease in critical mass (see later). Moreover, elements of high density are good reflectors of high-energy neutrons and also provide inertia, which helps slow down the expansion of the exploding fuel and maintains device assembly. In this case, the reflector also acts much like a tamper in conventional ordnance, providing more efficient use of the explosive fuel.

For a nuclear explosion to occur, the explosive device must contain a sufficient quantity of fissile material to exceed the critical mass during the high-k phase of its operation. The critical mass depends on several factors, including the shape of the part, its composition (including both impurities and nonfissile isotopes that compete for neutrons), the density, and reflection.

The most important requirement of the fissile fuel of an explosive device is that it be of weapons-grade isotopic composition, meaning that the nonfissile isotopes of the same element (mainly ^{238}U in ^{235}U parts and ^{240}Pu in ^{239}Pu parts) be kept to a practical minimum. The rate at which neutrons must be produced, to result in generations propagating sufficiently fast to keep the device assembled and deliver an explosive yield, requires that (n,f) reactions with fission-spectrum neutrons be the drivers of device performance. The (n,f) reaction cross sections with these fast neutrons are orders of magnitude lower than the equivalent cross sections with thermal or resonance-energy neutrons. The (n,γ) reaction cross sections of the nonfissile isotopes are also lower with fast neutrons, but only by an order of magnitude. Thus, materials that function well as moderated reactor fuel are completely inappropriate for fission explosive fuel.

There are other reasons why the nonfissile content of weapons-usable fission fuels must be reduced. One is dilution: the more nonfissile components in the fuel, the greater the mass that must be assembled to achieve criticality, and the more difficult it is to achieve a large k. For Pu, there is the additional, special problem associated with the spontaneous-fission (SF) decay probability of the even-mass isotopes, which is discussed in detail here.

The minimization of nonfissile, heavy-element isotopes leads to the definition of weapons-grade materials. The isotopic signature of this material is the nuclear forensic analyst's first indicator of a sample from a weapons-development program. Weapons-grade Pu is characterized by a ^{240}Pu content of <7%, whereas weapons-grade U is characterized by enrichment in ^{235}U of >20%. There is no formal definition of the isotopic content of weapons-grade ^{233}U, but from a standpoint of radiation exposure during handling, it is desirable that the ^{232}U content be <10 ppm. Reference [7] provides a discussion of the classes of Pu and U isotopic mixtures, and how the boundary isotopic concentrations between weapons-grade and fuel-grade materials are somewhat arbitrary. US weapons-grade Pu is ~5%–6% ^{240}Pu, and US weapons-grade U (Oralloy) is 93% ^{235}U [8]. However, the analyst must also consider that some research reactors and naval propulsion reactors operate on charges of weapons-grade U.

The presence of low-Z materials in fissile fuel is undesirable for some of the same reasons as nonfissile isotopes: they can consume neutrons, and they lower the density of the fissile nuclides. In addition, they provide moderation. This leads to another signature of weaponization that can be measured by the nuclear forensic analyst: the most desirable material form of fuel in a nuclear explosive device is the metal. A sample of weapons-grade material in the form of an oxide, fluoride, or carbonate/oxalate is a storage form and was not recovered directly from a weapon. The presence of small admixtures of medium-mass elements (with low neutron reaction cross sections) as alloys of U and Pu may also be indicators of weaponization. The addition of Ga to Pu, or Nb to U, results in alloys that are easier to fabricate into precise shapes than are the pure metals, in addition to providing extra corrosion resistance.

Even with a large quantity of fissile fuel, if the physical dimensions of the assembly were such that there was a large surface area relative to the mass, the proportion of neutrons escaping would be so great as to preclude achieving $k \gg 1$. In this case, the production of explosive yield is impossible. However, if the mass were redistributed to a more spherical configuration (minimizing surface area/mass), k would increase, along with the possibility of achieving supercriticality. Addition of a reflector also results in an increase in k, as does compression of the nuclear fuel.

The design of a fission explosive requires assembly of a critical mass from a subcritical configuration, followed by an introduction of neutrons to initiate the chain reaction. Because of the presence of stray neutrons in the environment, and from the SF of the even–even actinide isotopes, it is impossible to construct a supercritical mass and leave it "on the shelf" until implemented. The accidental introduction of a single neutron could initiate the fission chain. The design of the explosive must therefore maintain the fissile fuel in a distributed configuration, such that the assembly is subcritical. When the device has been delivered to its destination, at the appropriate time, the material must be assembled into a supercritical configuration to induce an explosion.

There are two general methods for the rapid assembly of a supercritical mass in a nuclear explosive device. The first method involves bringing two or more pieces of

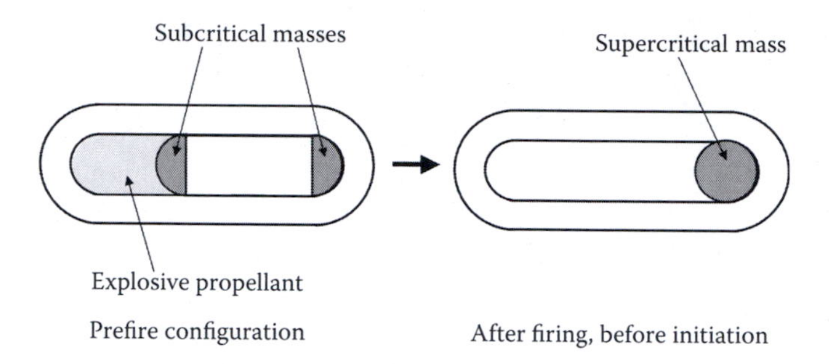

FIGURE 5.1 Schematic diagram of a gun-assembly nuclear explosive device. Two subcritical masses of U are forced together by a HE propellant; upon assembly, neutrons are introduced to start the chain reaction.

fissile material together, each of them subcritical, to form a single unit that exceeds the critical mass. This is known as a gun-assembly explosive, and the weapon that destroyed Hiroshima was of this type (see Figure 5.1). In the gun-assembly device, an explosive propellant is used to move one subcritical piece of fissile material, down what is effectively a gun barrel, into another subcritical piece held firmly in the gun's muzzle. As the supercritical mass is assembled, neutrons are introduced from an external source to initiate the chain reaction. In early stockpile designs, neutrons were generated by crushing a pellet containing segregated zones of powders of Be and a high-energy α-emitter (usually ^{210}Po). When the powders became intimately mixed, ^{9}Be(α,n) reactions introduced neutrons into the nuclear fuel. The nuclear forensic analyst can consider the presence of such initiator nuclides as an indicator of proliferant weaponization activities.

The design of a gun-assembled nuclear weapon was considered so straightforward that it was not tested as a system prior to use in war. Other than issues of actinide metallurgy, it was considered to be a standard ordnance problem [9]. However, despite its simplicity, there are major drawbacks to gun assembly: it can only be utilized with the U isotopes as fuel, and relatively large quantities of fissile material are necessary. Just before the supercritical mass is assembled, the two subcritical masses approach one another at a rate of a cm every few μs. The SF half-life of ^{238}U is 8.2×10^{15} years. Therefore, a 10-kg quantity of U that is ~50% ^{238}U experiences SF at the rate of 10^{-4}–10^{-3} during that period of time. It is therefore unlikely that a SF neutron would be introduced into this fuel at an inappropriate time during assembly of the supercritical mass. However, use of weapons-grade Pu as fuel in the same device is unfeasible: the material is ~6% ^{240}Pu, which has a SF half-life of 1.34×10^{11} years. This means that a 10-kg quantity of weapons-grade Pu experiences 2.5 SFs in 10 μs, making it very likely that a neutron would be introduced into the explosive fuel before assembly was complete. Early initiation of the chain reaction can result in fission energy that would produce backpressure and cause the supercritical mass to incompletely assemble. It would therefore drastically reduce the number of generations of neutrons, along with the explosive yield of the device. A nuclear explosive that fails because of premature neutron initiation is termed a "fizzle."

The gun-assembly design would also be appropriate for an explosive fueled with ^{233}U. In principle, ^{233}U fuel bred from ^{232}Th can be fabricated with only small quantities of even-mass U isotopes as diluents. The major contaminant, ^{234}U, has a SF half-life of 2×10^{16} years, making a ^{233}U-fueled, gun-assembled weapon even less likely to preinitiate than a ^{235}U-fueled device. The United States did not produce sufficient ^{233}U to construct a fission weapon until the mid-to-late 1950s, and by that time, the gun-assembly design was considered obsolete [10]. However, the nuclear forensic analyst must be aware of this possibility as a potential proliferant weapon design.

The second method of assembling a supercritical mass in an explosive device is, incongruously, implosion. A subcritical quantity of nuclear fuel is surrounded by a high explosive (HE) that is detonated. The fissile material is explosively compressed, which radically increases its density and decreases the surface area. This tactic results in a decrease in the mass required for a supercritical assembly. A multiplying chain reaction may then be subsequently initiated in a mass that had been subcritical in an uncompressed state.

The compression may be achieved by means of a spherical arrangement of specially fabricated shapes of ordinary HE. A subcritical sphere of fissile material is positioned at the center of the explosive, which is initiated by means of a number of detonators fired simultaneously on the outside of the explosive to create an inwardly directed compression wave. When the implosion front reaches the fissile material, it is driven inward, resulting in an increase in density, a corresponding decrease in the mean-free-path of the fission neutrons, and the formation of a supercritical assembly (see Figure 5.2). The introduction of neutrons from a suitable source is then used to initiate the nuclear explosion. If sufficient neutrons are introduced in a short pulse during initiation, the time associated with several generations of neutron production can be saved at the front-end of energy production.

The Nagasaki bomb was an implosion device. Its design was sufficiently novel that a similar device was first tested near New Mexico's Alamogordo Air Base: the Trinity test. From the standpoint of modern weapons safety, the Nagasaki device was

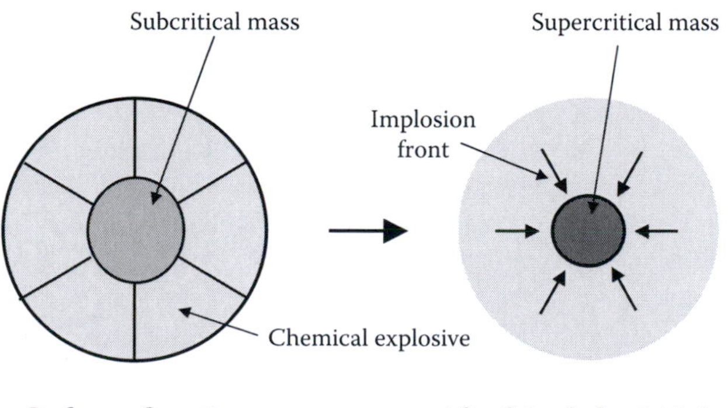

FIGURE 5.2 Schematic diagram of an implosion-type nuclear explosive device. A subcritical mass of Pu is surrounded by appropriately configured HE charges that are detonated simultaneously. An ensuing implosion front drives the Pu to higher density, converting the mass in its compressed state to a supercritical assembly.

frightening. The best HE available at that time could not induce the same compression of the fissile fuel that can be accomplished today. As a result, the Pu fuel of the device, placed at the center of the moderating and reflecting HE, was perilously close to a $k = 1$ arrangement, even in its uncompressed configuration. There were concerns that, had the B-29 carrying the weapon crashed on takeoff from Tinian Island, there could have been sufficient nuclear yield to cause heavy damage to the military installations [11]. This led eventually to the concept of "one-point safe": any US stockpile weapon will not produce appreciable nuclear yield unless the HE is detonated in the prescribed fashion [12].

Using the implosion technique, a supercritical mass may be assembled in a time sufficiently short that Pu can be used as the nuclear fuel. However, as the ^{240}Pu isotopic content of the Pu fuel increases, so does the chance of device preinitiation induced by SF neutrons. This property is the reason for the effective limitation on the concentrations of even-mass Pu isotopes in weapons-grade materials.

Another indication of weaponization activities for consideration by the nuclear forensic analyst is the presence of HE residues or by-products in any sample of U or Pu. An actinide sample in the chemical form of a metal, or of weapons-grade isotopic composition, should be subjected to an organic analytic protocol.

Importantly, in addition to allowing Pu as an explosive fuel, implosion designs also accommodate nuclear boosting techniques.

5.2 BOOSTING

As discussed previously, nuclear energy can be derived from each extreme of the periodic table, a consequence of the nuclear packing fraction (Figure 2.4). For the heaviest elements, division of nuclei into two roughly equal parts in the fission process results in the release of energy. Fission is initiated by the absorption of a neutron. The nuclear chain reaction that drives the device performance arises from the fact that the fission process is accompanied by a release of neutrons, and the multiplicity of these neutrons is sufficient that >1 additional fission is produced by the neutrons emitted in a single fission.

At the other extreme of the packing diagram, the nuclear reactions in which two very light nuclei fuse may also result in a release of energy. An important limitation of the fusion process as a source of energy is that the multiplicity of neutrons arising from the fusion of low-Z nuclides is small, and no combination of light materials will sustain a fusion chain reaction. The fission chain reaction is the means by which an application of a relatively small quantity of chemical energy can trigger the release of a large amount of fission energy in the explosion of a nuclear device. However, direct application of HE cannot produce the release of a significant quantity of fusion energy.

The release of fusion energy arises through direct action on each nucleus that undergoes a nuclear reaction. In particle accelerators or the fusion reactors under development, this action is accomplished by acceleration: reactants are propelled at one another with velocities sufficiently large to overcome the electrostatic Coulomb barrier to their reaction. However, there is no self-propagating contribution from a chain reaction, and if the accelerator is turned off, the fusion reaction stops. Fusion reactions also occur through high pressures and temperatures, as encountered in stars. In that case, gravity provides the force that confines the nuclei in close proximity. The high temperature of a stellar environment creates a Maxwellian distribution of velocities of the ions in the plasma that constitute the stellar fuel. The upper extreme of this distribution extends to velocities that correspond to kinetic energies greater than the Coulomb barrier to nuclear reactions. These reactions occur at a rate such that the introduction of energy into the stellar interior balances the energy lost through radiation, thereby maintaining the equilibrium configuration of the star.

Generation of high temperatures and pressures is the key to attempts to use thermonuclear fusion as a controlled source of energy. For devices of Earth-bound dimensions, energetic light particles in gases and plasmas at low densities have long path lengths and will encounter the walls of the apparatus before reacting with another light particle. The confinement of the fast particles in a magnetic field is one method proposed as a basis for a fusion power reactor, the tokamak [13]. The tokamak takes advantage of the fact that a hydrogen isotope plasma is electrically conducting and can be manipulated with magnetic fields, "pinching" the plasmas to the temperatures and pressures required for fusion. Another path to fusion energy is inertial confinement [14,15]. Thermonuclear fuel is greatly compressed (up to 1000× normal liquid density) through the deposition of energy on the outer surface of a fuel pellet. This energy must be applied with sufficient symmetry that the pellet is compressed spherically. During compression, the fuel must not be prematurely heated, since the resulting backpressure works against achieving maximum density; as with the initial neutrons in a fission explosive, heating must be introduced when the fuel is already compressed. Confinement times are so short that extremely high densities and temperatures are required to achieve significant burn-up of the fuel.

In fission explosive devices, the fusion process is important, not principally because of the energy produced, but as a source of high-energy neutrons. Fusion neutrons are introduced into the fissile fuel of a device, at a late stage in the functioning of those fuels, in order to maintain and enhance the progress of the fission reactions [2]. This process is referred to as boosting [1–3].

Fusion reactions involving the lightest elements can be induced by means of the application of very high temperatures; hence, they are referred to as thermonuclear processes [1]. These temperatures are so high that, in a terrestrial environment, they can only be produced through the generation of fission energy. Four fusion reactions occur sufficiently rapidly, in a compressed volume of mixed deuterium (D) and tritium (T) gas at the temperatures in the center of a fission weapon, to contribute to energy production:

$$^{2}\mathrm{H} + {}^{2}\mathrm{H} \rightarrow {}^{3}\mathrm{He} + \mathrm{n} + 3.2\ \mathrm{MeV},$$

$$^{2}\mathrm{H} + {}^{2}\mathrm{H} \rightarrow {}^{3}\mathrm{H} + {}^{1}\mathrm{H} + 4.0\ \mathrm{MeV},$$

$$^{2}\mathrm{H} + {}^{3}\mathrm{H} \rightarrow \mathrm{n} + {}^{4}\mathrm{He} + 17.6\ \mathrm{MeV},$$

and

$$^{3}\mathrm{H} + {}^{3}\mathrm{H} \rightarrow {}^{4}\mathrm{He} + 2\mathrm{n} + 11.3\ \mathrm{MeV}.$$

Of these four reactions, the third occurs with the greatest probability at the operational temperatures of tens of millions of degrees Kelvin. In this reaction, a single neutron, with a laboratory reference-frame energy of 14.1 MeV, is produced.

The number of neutrons released when a nucleus undergoes fission varies with the particular mode of fission, but the average number is well defined [4]. Fusion neutrons, having a higher energy than neutrons from the fission process, cause fission events to occur with higher neutron multiplicity. For example, the reaction of ^{235}U with thermal neutrons yields an average of 2.4 neutrons per fission, and the reaction with fission-spectrum neutrons yields 2.5–2.6 neutrons. However, the reaction of ^{235}U with 14-MeV fusion neutrons produces 4.2 neutrons per fission [5]. In addition, under irradiation by 14-MeV neutrons, even-mass Pu and U isotopes fission very efficiently and give similarly high neutron multiplicities. The number of neutrons per generation of fissioning species thus increases. The introduction of fusion neutrons into a fissioning part boosts the neutron multiplicity in fission, which accelerates the multiplication rate in subsequent generations [1–3].

Consider a modification to the Nagasaki implosion device (depicted in Figure 5.2), and replace the solid ball of Pu with a Pu shell filled with a mixture of D and T, as shown in Figure 5.3. When the HE charge is detonated, it creates an inwardly directed compression wave, driving the Pu shell inward and compressing and heating the volume of gas. At some point, the gas molecules are at such high pressure that they successfully push back against the imploding Pu, slowing its progress. During this compression, the Pu fuel is initiated by the introduction of neutrons. When the gas filling the Pu is near maximum compression, the runaway fission process deposits enough energy into the gas that it "lights," and the D–T mixture is rapidly consumed. The fusion reaction produces He and a pulse of 14-MeV neutrons that is introduced into the Pu, causing a rapid acceleration in the production of neutrons and, consequently, the number of fissions.

The increase in yield of the primary resulting from the energy released by the fusion process is relatively small. The main effect of boost is to increase the number of fissions obtained in the surrounding supercritical mass of fissile (and fissionable) material. Boosting is an important component for increasing the fission efficiency of a fission fuel. However, hydrogen isotopes react rapidly with both Pu and U to form

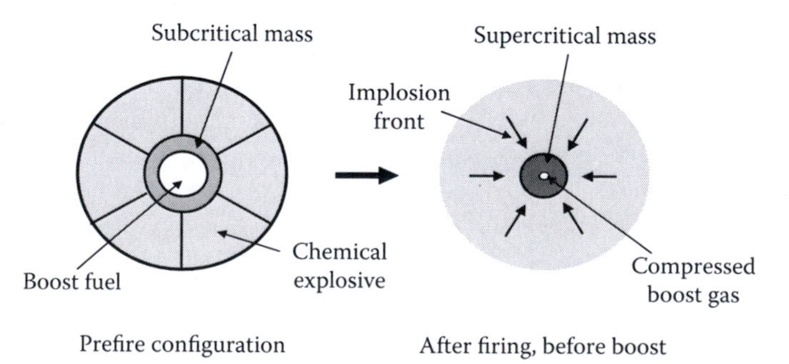

FIGURE 5.3 Schematic diagram of a boosted, implosion-type nuclear explosive device. Similar in design to the unboosted implosion explosive, it incorporates a volume of D and T as boost fuel. The compressed boost fuel is triggered by heat generated by fission in the surrounding supercritical mass.

hydrides, and these reactions require that the fissile pit not be filled with the boost gas until just prior to detonation. Forensic signatures of a (unexploded) boosted, single-stage fission device include gas-manipulation hardware contaminated with T and the presence of low-level T and D in samples of U or Pu. The latter can result from valve leakage from the hydrogen reservoir.

5.3 TWO-STAGE NUCLEAR EXPLOSIVE (HYDROGEN BOMB)

During the time between the firing of the Trinity device in 1945 and the detonation of the first Soviet nuclear explosive ("Joe-1") in 1949, much thought went into the idea of using the output of an atomic bomb to drive a "superbomb," but there was little progress. This was partially because it was virtually impossible to perform relevant laboratory-scale experiments. In the case of single-stage fission explosives, the bulk behavior of fissile materials can be studied in the operation of fast reactors and critical assemblies, and the critical material properties of U can be studied by imploding nonfissile U isotopes with chemical explosives. As a result, these laboratory measurements, combined with a growing body of theory, made it possible to reasonably estimate the performance of a new fission explosive before it was constructed. In the case of a staged nuclear device driven by the output of a fission explosion, however, the reaction processes could not be effectively studied in the laboratory. In the absence of high-speed computers, the only means to analyze the development was to field experiments on single-stage nuclear explosive devices.

Such experiments were difficult and expensive. Nevertheless, by mid-1949, two concepts were proposed for a hydrogen bomb, a nuclear explosive that would release energy produced by the fusion of pairs of D nuclei (the first two reactions in the previous section): the "alarm clock" and the classical "super" [16].

It was not until after the Joe-1 test that work began in earnest on the hydrogen bomb. Although the concepts had been formulated, no designs existed. The design of a single-stage fission explosive is fairly basic; the difficulty arises in producing the requisite nuclear fuel. For a hydrogen bomb, the D was relatively easy to obtain, but a design to use the output of a fission explosive to heat and compress thermonuclear fuel in a separate stage to produce explosive energy is complex, with performance difficult to predict [17]. The alarm-clock design, in which the fission trigger is centrally located, was soon abandoned because of the size constraints imposed on potential delivery systems [18]. The classical-super design (see Figure 5.4), in which the output of a fission primary is applied to a secondary stage from outside, was first tested on November 1, 1952, with the explosion of the 10-Mt Ivy Mike device. Mike used liquid D_2 as thermonuclear fuel and required unwieldy application of refrigeration equipment. Although not a weaponizable design, it did spectacularly demonstrate the principle of the super concept.

Later hydrogen bombs were based on the use of ^{6}LiD as the thermonuclear fuel. Lithium deuteride is a grayish-white, hygroscopic salt that can be pressed into shapes, making it functional for use in deployable weapons [17,19]. It performs as a fuel because of the reaction of ^{6}Li with neutrons to produce energetic, recoil T:

$$^{6}\text{Li} + \text{n} \rightarrow {}^{4}\text{He} + {}^{3}\text{H} + 4.8\ \text{MeV}.$$

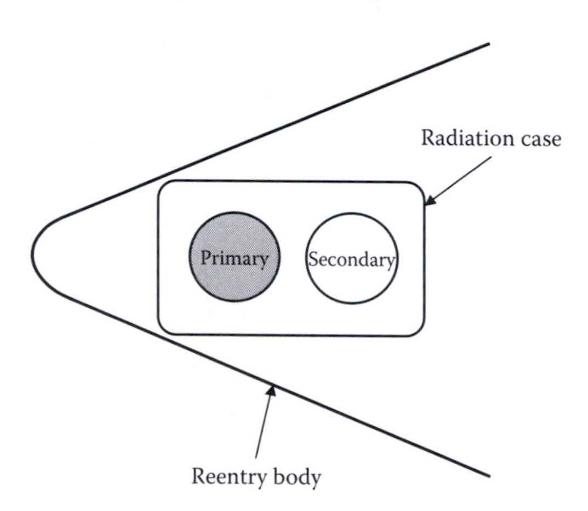

FIGURE 5.4 Schematic diagram of a two-stage thermonuclear explosive device, also known as a hydrogen bomb. The primary (a fission explosive) is detonated, and its output is used to drive the thermonuclear secondary.

In fact, T for use in boost fuel is produced in nuclear reactors by irradiation of targets of enriched ^{6}Li (which is nominally 7.5% of natural Li). In the secondary stage of a thermonuclear explosive, T is produced in situ with sufficient energy to overcome the Coulomb barrier, and it reacts with the D component of the salt to produce 14-MeV fusion neutrons.

Figure 5.4 depicts a two-stage thermonuclear device. High-energy neutrons from the thermonuclear fuel can induce fissions in surrounding heavy elements, even those that are fissionable but not fissile. Signatures of weaponization activities that link a sample of U or Pu to use in a thermonuclear secondary include the presence of heavy isotopes of H or enriched ^{6}Li in materials that may not be of ordinary weapons-grade compositions. Lithium reagents for normal applications have often been obtained from commercial vendors with a depleted ^{6}Li signature, the "tails" of Li enrichment activities supporting the production of T in reactors.

5.4 FORENSIC ANALYSIS OF NUCLEAR EXPLOSIVE DEBRIS

A requirement of the Nonproliferation Initiative of 1992 is the technical ability to determine the source of an exploded nuclear device from the residual debris. This capability has since been extended to include the debris from radiologic dispersal devices (RDDs), in which radionuclides (not necessarily associated with a weapons program) are dispersed by a conventional HE detonation. (RDDs are discussed in Chapter 13.) In other parts of this book, we discuss the differences between source and route evidentiary aspects of interdicted actinide materials. Signatures of source are radiochemical in nature, point to the origin of the material itself, and to methods used in its production. Signatures of route are derived from traditional forensic measurements and lead to information on persons of interest and the real-world pathways by which a sample traveled from a point of origin to the interdiction location.

Similarly, the nuclear forensic analysis (NFA) of explosion debris may also be divided into the two combined aspects of device design and device delivery. Signatures of design point to the construction of the weapon and to the entity that fabricated it. Signatures of delivery relate to the deployment of the device and are indicative of who detonated it, rather than who built it. If the explosion debris arose from firing a stolen stockpile device, the difference between design and delivery would be a critical forensic focus.

Forensic analysis of nuclear explosion debris for design signatures derives from, and is benchmarked against, radiochemical diagnostic measurements that were developed during the Nuclear Test Program, that is, between the Trinity test of 1945 and the end of US underground testing in 1992. Debris collected from foreign atmospheric tests provided the opportunity to adapt these methods to limited forensic analysis. We emphasize that, while related, diagnostics and forensics are not equivalent. Nuclear diagnostic radiochemistry was designed to determine the performance of a device whose design was known. However, one of the objectives of nuclear forensic radiochemistry is to use performance information (e.g., fission efficiency) to extract design information, which may then lead to conclusions about the source of the weapon.

During the era of nuclear testing, designers of a nuclear explosive device were also the agents of its deployment and detonation, so signatures of delivery were not explicitly developed. Everything within the vicinity of a nuclear explosion of any magnitude is vaporized, making conventional forensic measurements such as fingermarks or tooling impressions no longer applicable. Only radiochemical and mass-spectrometric indicators persist. The radiochemical signatures are particularly challenging; like many physical effects, the activation of surrounding materials decreases proportionally as the inverse square of distance from the source of the neutrons. The radionuclides induced within the delivery and deployment apparatus are consequently diluted by the overwhelming signatures of the device performance, making them very difficult to measure. In addition, incorporation of soil or urban structural materials in the explosion debris can interfere with subtle mass-spectrometric indicators. NFA of delivery signatures is expected to be complicated and difficult, and would involve extensive radiochemical recovery and purification procedures.

5.4.1 Diagnosis of Nuclear Performance

When considering debris from an exploded nuclear device, there are markers of performance of the device within the debris matrix. As the most basic example, by measuring the concentrations of the fission products produced by the explosion and comparing them to the concentration of unconsumed nuclear fuel in a homogeneous debris specimen, it is possible to determine the efficiency of the fissile fuel in the explosive device. The fission fuel is never completely consumed, and 1 kt of fission yield is $\cong 0.25$ mol of fissions (arising from the complete fission of ~55 g of ^{239}Pu or ^{235}U). Such measurement of fission efficiency was first proposed by Manhattan Project radiochemists prior to the Trinity test. The fission products in a given homogeneous sample that is representative of the debris field (Section 5.4.2) define a number of fissions, which is related to the number of atoms of fuel that was consumed in the explosion.

A high efficiency is indicative of a more sophisticated design and perhaps acquired by theft from the stockpile of an established nuclear weapons state.

A common misconception is that radiochemical analysis of explosion debris gives an independent measurement of the fission yield of the device. Such is not the case, as radiochemistry provides only an accurate measurement of efficiency and serves to highlight a significant difference between diagnostics and forensics. In traditional diagnostics, the initial quantity of fission fuel in a device was known, and application of the fission efficiency resulted in a determination of the fission yield. In a forensic analysis, however, the inventory of fuel in a weapon is most likely unknown, but if an independent measurement of yield by another means exists, the reverse calculation can result in a determination of the quantity of nuclear fuel in the predetonation configuration of the device.

Many performance metrics for nuclear explosive devices are strongly correlated with fission efficiency, and the reader is encouraged to review the physics of fission in Section 2.8 before proceeding. The radiochemical analyst does not measure the number of fissions in a debris sample directly. Peak-yield fission products are observable in γ-ray spectra obtained from unseparated debris samples, while γ-rays emitted by off-peak products with cumulative fission yields as small as 0.0001% have been observed in isolated radiochemical fractions. Thus, the concentrations of the fission products are measured, and the number of fissions associated with the sample is deduced from those data. However, while this process is simple in concept, it is more complicated in execution.

The fission of actinide nuclei by exposure to neutrons leads to hundreds of fission products, most of which are radioactive and have half-lives ranging from less than one second to millions of years. Each fission product is produced with an individual probability, called the independent fission yield. The actinide fuels are neutron-rich relative to stable isotopes of the elements that comprise the fission products, making nearly all fission products unstable to the processes of β^- decay. The emission of β particles provides a decay connection between nuclei in the same isobar (having the same atomic mass number), and Section 2.3 provides a more thorough discussion. Beta-delayed neutron emission provides some connection between neighboring isobars, but it is ignored in the following discussion here for the sake of simplicity.

Immediately after the explosion, the prompt fission products begin to decay to nuclei closer to the valley of β stability, which tend to have longer half-lives. Long after the detonation, all short-lived fission products will have decayed to their longer-lived (or stable) daughters; these shorter-lived parent nuclides are also termed fission-product precursors. If the analyst measures the concentration of a long-lived product and corrects for its decay from explosion time without considering the unobserved precursors, he/she will obtain a fission yield that is approximately equal to the sum of the independent yields of the nuclide itself and all of its precursors. This sum is known as the cumulative fission yield for that nuclide. The approximation becomes quite accurate if the half-life of the observed nuclide is much longer than those of its precursors. The rate laws in Section 2.3 allow a determination of the magnitude of the error introduced by longer precursor half-lives, but for most fission products, it is less than the intrinsic uncertainties fundamental to radiochemical measurements.

As an example, Figure 5.5 shows a partial β-decay chain for the A = 140 nuclides produced by the fission of ^{235}U with neutrons. The half-life of each member of the isobar is shown [20], as is its independent yield when induced by the fission-spectrum

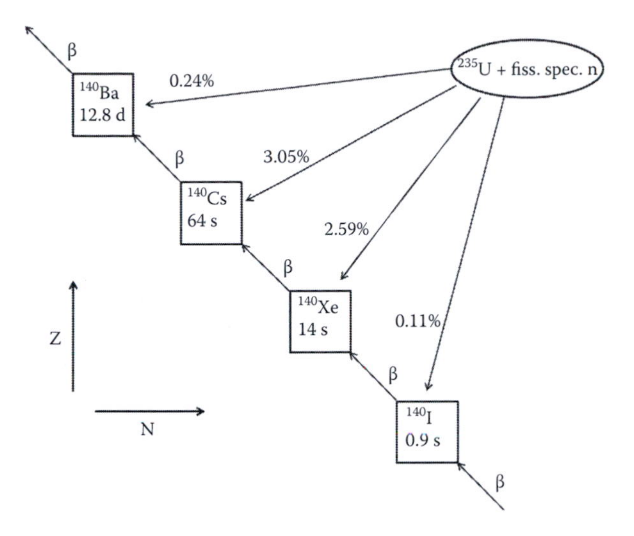

FIGURE 5.5 The A = 140 β-decay chain arising in the fission of ^{235}U with fission-spectrum neutrons. The cumulative fission yield of ^{140}Ba is 5.99%, even though very little of the fission process results in its direct production.

neutrons [21] from an unboosted nuclear device. All of the A = 140 isotopes that are more neutron-rich than ^{140}Ba have half-lives that are short compared to the time required for debris recovery and radiochemical characterization. These three precursor nuclides will have completely decayed to 12.8-day ^{140}Ba prior to measurement. While very little of the A = 140 fission yield is born as ^{140}Ba (i.e., only 0.24%), a radiochemical determination of its cumulative fission yield is approximately 0.11% + 2.59% + 3.05% + 0.24% = 5.99%. The information in Figure 5.5 leads directly to Figure 5.6,

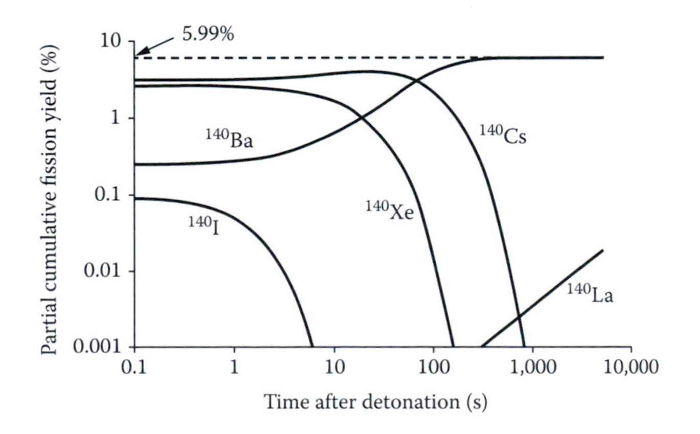

FIGURE 5.6 Evolution of the concentrations of the A = 140 fission chain nuclides, arising from the reaction of ^{235}U with fission-spectrum neutrons, as a function of time after detonation. After 1000 s, the precursors of ^{140}Ba no longer contribute. The dashed line is the extrapolation of the ^{140}Ba concentration at later times to the detonation time, ignoring the precursors, and resulting in an approximation to its cumulative yield.

in which the amount of each A = 140 nuclide is shown as a function of the time after detonation. Clearly, by 1000 s after explosion, no appreciable concentrations of any of the precursor nuclides remain. A decay correction of the measured ^{140}Ba activity to detonation time (i.e., with the assumption that all of the A = 140 fission yield was born as ^{140}Ba) leads to a discrepancy of <0.01% in the cumulative fission yield, relative to a more complete decay analysis that includes the precursors.

The decay pattern in Figure 5.5 is typical of most of the fission-product isobars. Though there are many exceptions, the nuclides that are most neutron-rich tend to be shorter-lived than are those that lie closer to the stable isotopes in Z and N. The cumulative yields of the fission products likely to remain in a sample after the time required for its collection and analysis will be close to the total fission yields for those masses and related to the double-humped fission distribution (such as shown in Figure 2.23). The shielded nuclides are exceptions to this rule; the decay of precursor nuclides in the isobar piles up in a very long-lived or stable nuclide that is more neutron-rich than the observed shielded species, leaving only the independent-yield contribution to the concentration of the latter in the debris. The nuclides ^{126}Sb, ^{136}Cs, and ^{148}Pm are shielded fission products that are commonly observed in nuclear explosion debris (see later).

The distribution of products arising from the fission of each nuclear fuel is known (e.g., [21]). This means that it is possible, in principle, to obtain the fission efficiency of a device from the measurement of the concentration of a single fission product, relative to that of the residual fuel, in a homogeneous sample that is representative of the debris field. The probability for the formation of the observed product is extracted from a set of tabulated cumulative fission yields (approximately the value of the double-humped distribution at that mass), which leads to the number of fissions associated with that debris sample. In the earlier example, the concentration of ^{140}Ba in a sample of debris derived from the explosion of an unboosted ^{235}U device is converted into the number of fissions associated with that sample by dividing by 0.0599. In practice, however, the problem is complicated by several factors, two of which are the spectrum of neutron energies in the device and the possible presence of more than one nuclear fuel. Additionally, the effects of sample inhomogeneity will be discussed in the next section.

If more energetic neutrons are generated by the device, they produce more highly excited fissioning nuclei, overcoming some of the energetic advantage favoring asymmetric fission over symmetric fission [4,5]. The probability of symmetric fission therefore increases, as does the probability of unusually asymmetric fission. That is, the shape of the double-humped fission distribution as a whole changes, with the "valley" between the humps filling in and the "wings" broadening (see Figure 2.23).

Consequently, the distribution of fission products from the explosion of a boosted, single-stage ^{235}U weapon can be significantly different from that arising from the explosion of an unboosted ^{235}U device. The cumulative yield of ^{140}Ba in the 14-MeV neutron fission of ^{235}U is 4.50% [21], significantly lower than from the unboosted spectrum (5.99%) and a consequence of the spreading of the fission mass yield at higher neutron energies. When considering (e.g.) a homogeneous debris sample produced by a hypothetical, ^{235}U-fueled device in which 75% of the fissions were from fission-spectrum neutrons and 25% were from thermonuclear neutrons, the cumulative fission yield of ^{140}Ba would be $(0.75 \times 5.99\%) + (0.25 \times 4.50\%) = 5.62\%$.

Moreover, neutrons of a given energy interacting with two different fission fuels give rise to different distributions of fission products. The cumulative yield of ^{140}Ba from the reaction of fission-spectrum neutrons with ^{239}Pu is 5.32%, while from fission induced by thermonuclear neutrons, it is 3.70% [21]. Both are significantly different from the equivalent cumulative yields generated by the fission of ^{235}U. The ^{140}Ba concentration in debris generated by a mixed-fuel device can be interpreted as a number of fissions only if the relative contributions from each fuel can be determined. This consideration leads to the concept of the "fission split" and the construction of a device-specific, fission mass-yield distribution.

There are three fission fuels commonly used in US nuclear explosive devices [2,22]: fissile ^{235}U and ^{239}Pu, and fissionable ^{238}U. In the following development, our assumption is that these fuels are the only sources of fission yield in a hypothetical device; however, the nuclear forensic analyst must always consider that other fission fuels are also possible (e.g., ^{233}U). During the progress of a detonation, the energy distribution of the neutrons present in the device is complicated, consisting initially of fission-spectrum neutrons with a possible admixture of thermonuclear boost neutrons. Eventually, the neutrons scattered to lower energies through elastic interactions with device materials [1,23] also contribute as the device disassembles. Even so, in US debris analysis, the cumulative yields arising from device fission may be convoluted with reasonable fidelity from six sources: the interaction of two neutron energies (fission-spectrum and thermonuclear) with the three fission fuels. The relative contribution of each of these six components to the fission-product inventory of a debris sample is termed the fission split.

If the fission split is known, each fission-product concentration in a homogeneous debris sample may be converted to a number of fissions. More relevant to forensic analysis, if the concentrations of a large number of fission products can be measured, the convoluted data can define the relative contributions of each component of the fission split.

Figure 5.7 shows the cumulative fission yields for products on the high-mass side of the fission mass distribution for all six components of the fission split for US

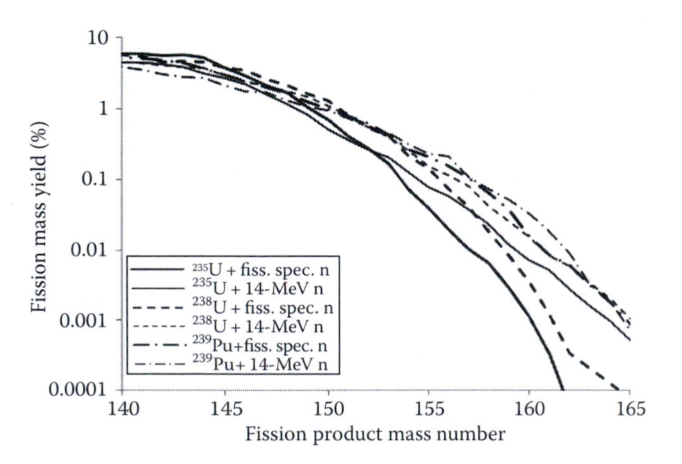

FIGURE 5.7 The fission product mass-yield distributions for the six components of the radiochemical fission split, for nuclides on the high-mass side of the double-humped mass distribution. The variation between the wing-product concentrations increases with mass.

TABLE 5.1
Cumulative Yields of Two Fission Products [21] Arising in the Fission of Six Potential Fission-Split Components

Fission-Split Component	^{147}Nd (%)	^{161}Tb (%)
^{235}U + fission-spectrum n	2.14	0.00033
^{235}U + 14-MeV n	1.62	0.0053
^{238}U + fission-spectrum n	2.59	0.00122
^{238}U + 14-MeV n	2.09	0.0085
^{239}Pu + fission-spectrum n	1.99	0.0086
^{239}Pu + 14-MeV n	1.71	0.0184

devices. In Table 5.1 are listed explicit cumulative-yield values for ^{147}Nd and ^{161}Tb for each fission-split component. For products near the top of the heavy hump (A = 140 – 147), the spread in cumulative yields among the six components of the fission split is < ×2 (e.g., see the ^{147}Nd values in Table 5.1).

During nuclear testing, heavy reliance was placed on these peak-yield products for radiochemical diagnosis of fission performance, because a small error in the determination of the relative contributions of the six split components had only minor impact on the convolution of a device-specific set of cumulative fission yields for these nuclides. Such is not the case for the wing products, however. An examination of the cumulative-yield values of ^{161}Tb in Table 5.1 reveals a spread of almost two orders of magnitude among the split components. A small error in the convolution of these data into a device-specific fission yield for ^{161}Tb can result in a large error in the interpretation of ^{161}Tb concentration for a number of fissions. However, this limitation of the diagnostic use of ^{161}Tb in a debris sample is key to its value as a forensic indicator of fission split. Measurement of the concentrations of wing fission products, along with constraints provided by the actinide isotope content of the samples (see later), may be used to infer the relative contributions of the six components of the fission split to the fission-product inventory of that sample. If the fission split has been constructed correctly, the number of fissions derived from the concentrations of each fission product should be the same.

In the absence of collateral information, the tabulated cumulative fission yields, along with the relative concentrations of six or more fission products in a homogeneous sample representative of the debris field, may be used to extract the fission split. The fidelity of the analysis is improved if the concentrations of split-sensitive wing and valley products are available, rather than those of the split-insensitive peak-yield products. The analysis is usually performed using a table-driven computer code that minimizes a weighted χ^2.

Although the cumulative yields of the peak-yield fission products are insensitive to the fission split, the independent yields of their fission precursors are not. In Figure 5.8, the independent yields of the near-peak, A = 148 prompt fission products for each of the split components are plotted. The independent yields presented in this fashion show the charge dispersion at A = 148 arising from each of the six fissioning systems. The peak yield of a smooth functional fit to a given data set is located

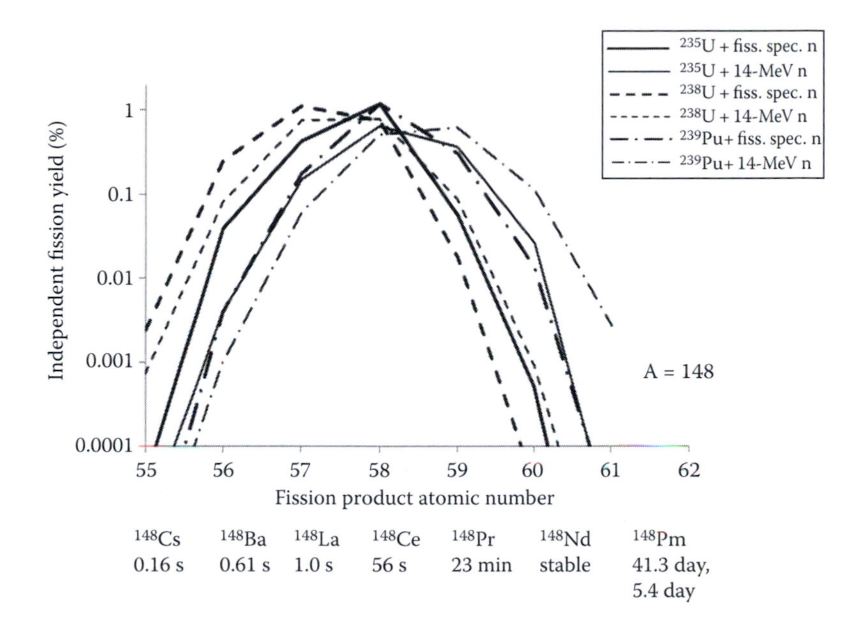

FIGURE 5.8 The independent yields of A = 148 nuclides arising in the six components of the fission split. ^{148}Nd is stable; therefore, ^{148}Pm is shielded, receiving no decay intensity from short-lived precursors.

at the "Z-most-probable" for that mass, with the specific position controlled by the neutron richness (N/Z) of the fuel and the multiplicity of the neutrons emitted in fission (Section 2.8). While there are substantial differences in the neutron richness of the six charge dispersions, the cumulative fission yields (sums of the independent yields) vary over less than ×2, from 1.17% for ^{235}U + 14-MeV neutrons to 2.11% for ^{238}U + fission-spectrum neutrons.

The importance of the A = 148 charge dispersion lies in the yield of ^{148}Pm, which is shielded from the β-decay process. The ^{148}Nd isotope is stable, and decay of the short-lived precursors results in most of the A = 148 fission yield accumulating in the Z = 60 isobar within a few minutes of detonation, and not further carried into ^{148}Pm. Any ^{148}Pm in a debris sample was born directly as ^{148}Pm; its independent yield is equal to its cumulative yield, and it can be used to infer a number of fissions from the postdetonation ^{148}Pm concentration. Thus, the ^{148}Pm concentration in a debris sample is sensitive to the fission split, even though the total A = 148 fission yield is not.

The ^{148}Pm produced in explosion debris exists in two isomeric states, a 41.3-day metastable state and a 5.4-day ground state, both of which may be measured through the γ-rays emitted by a chemically purified Pm sample, along with the γ-rays emitted by the fully cumulative ^{149}Pm (53-h) and ^{151}Pm (28-h) nuclides. In Table 5.2, cumulative fission yields are listed for ^{148m}Pm and ^{149}Pm for each of the six fission-split components, along with their ratios.

An example of a limited fission-split analysis is the following. In an unspecified nuclear test, a ^{148m}Pm/^{149}Pm concentration ratio of $>1.0 \times 10^{-4}$ was measured.

TABLE 5.2
Cumulative Fission Yields for Two Pm Fission Products

Split Component	$^{149}Pm_c$ (%)	$^{148m}Pm_i$ (%)	$^{148m}Pm/^{149}Pm$
^{235}U + fiss.-spec. n	1.037	3.41e–8	3.29e–8
^{235}U + 14-MeV n	0.813	9.42e–6	1.16e–5
^{238}U + fiss.-spec. n	1.625	1.66e–9	1.02e–9
^{238}U + 14-MeV n	1.458	5.73e–7	3.93e–7
^{239}Pu + fiss.-spec. n	1.241	1.12e–5	9.02e–6
^{239}Pu + 14-MeV n	1.055	2.25e–3	2.13e–3

Note: The ^{148m}Pm nuclide is shielded, making the cumulative yield equal to the independent yield.

Examination of Table 5.2 leads to the conclusion that this concentration ratio cannot result from any combination of the first five fission-split components alone. An admixture of the sixth component, ^{239}Pu + 14-MeV neutrons, is required to produce a concentration ratio as high as was observed. The device must have therefore incorporated a boosted Pu stage.

During the explosion of a nuclear device, the fission products are created within a bath of free neutrons. The fission products may undergo reactions with these neutrons, changing their relative concentrations and potentially affecting the determination of the fission split. Over most of the mass range that spans the fission products, the fission yield changes slowly as a function of mass number. The assumption is therefore made that the loss of fission products of mass A due to neutron reactions during the explosion is balanced by the production of mass-A fission products from reactions with mass A-1 nuclides; that is, "burn-in = burn-out." Nevertheless, the use of published cumulative fission yields for the fission-split components can result in systematic errors in the relative magnitudes of the extracted fission-split fractions.

While most of the prompt fission yield at a given mass is distributed among nuclides that are considerably distant from the (N,Z) of the stable nuclei, the measured cumulative products that result from multiple β decays lie considerably closer to the valley of β stability (Chapter 2). As a result, many of these products can also be produced by neutron reactions on medium-mass elements in the device or in the soil. For peak-yield cumulative products, it may be demonstrated that the contribution from these reactions is almost always negligible compared to their production by fission. For the wing and valley products that are important for a fission-split analysis, however, such may not be the case. Correction factors for the contributions from neutron reactions on soil are usually calculated from an inorganic analysis of the elemental composition of the debris, the measurement of an activation product that is not produced by fission (nor likely to arise from structural materials in the device), and a set of neutron-reaction cross sections.

Once the number of fissions associated with a sample is known, the concentration of residual fuel is required to determine a fission efficiency. In measuring the concentration of residual fuel(s), the analyst must factor that the interactions of neutrons with the different actinide isotopes produce fission with different probabilities. For instance, a ^{235}U atom that encounters a low-energy neutron is more likely to undergo fission than is a ^{238}U atom. In fission fuel containing both ^{235}U and ^{238}U, the action of an unboosted device causes the depletion of ^{235}U relative to ^{238}U in the residuum, with the value of $^{235}U/^{238}U$ decreasing with increasing fuel efficiency. Furthermore, the interaction of neutrons with actinide isotopes does not always result in fission. Neutron capture competes with fission, even in reactions with the fissile isotopes, and for a device with a thermonuclear contribution to the yield, significant actinide isotope production via (n,2n) reactions is also observed.

An assumption in the analysis of fission efficiency is that neutron reactions resulting in other isotopes of the element constituting the fuel are the only important transmutation alternatives to fission (i.e., [perhaps multiple] (n,γ) and (n,xn) reactions). Any actinide isotope that does not fission continues to be an actinide isotope, even if it changes in mass number. In debris from a U-fueled device, the fission efficiency is calculated from the number of fissions and the total inventory of device-derived U isotopes present in the sample. It includes long-lived isotopes like ^{232}U, ^{233}U, ^{234}U, ^{235}U, ^{236}U, and ^{238}U, and shorter-lived isotopes such as 6.7-day ^{237}U, 23.5-min ^{239}U, and 14-h ^{240}U. Quantitating the concentration of ^{239}U is usually accomplished by measuring the concentration of its daughter, 2.36-day ^{239}Np, or its granddaughter, long-lived ^{239}Pu. In debris from a Pu-fueled device, the total inventory of ^{238}Pu, ^{239}Pu, ^{240}Pu, ^{241}Pu, and ^{242}Pu is used as the residual fuel concentration in the debris sample. Since aged Pu always contains 433-year ^{241}Am ingrown from the β decay of 14-year ^{241}Pu, the concentrations of ^{241}Am and its capture product, 16-h ^{242}Am (calculated from the concentration of its 162-day ^{242}Cm decay daughter), are included as part of the Pu inventory.

The actinide inventory of a debris sample can be used to constrain the calculation of the fission split. Absence of one of the three fissile fuels in a debris sample allows the analyst to exclude two of the six fission-split components from the χ^2 analysis of the fission-product wing concentrations.

If the fission efficiency of the device is low, the isotopic composition of the residual fuel will be nearly unchanged relative to that of the original fuel in the device. But at higher efficiencies, changes caused by the different fission probabilities of individual isotopes and their transmutation into one another can be significant. If the efficiency and the fission split are determined experimentally, nuclear cross sections may be used to “reverse-engineer” the starting isotopic composition of the fuel from the isotopic signatures in the debris. This calculation requires application of an explosion-code computer simulation (unavailable outside the weapons laboratories) to model the physics of device performance, including isotope production. The fidelity of this computation decreases as the fission efficiency increases.

Radiochemical analysis of the residual fuels can be complicated if more than one fission fuel was incorporated in the device. For instance, capture of a neutron by ^{238}U produces short-lived ^{239}U, which decays in two steps to ^{239}Pu. This “extra” ^{239}Pu should not be included in the inventory of Pu isotopes in the debris sample for the efficiency calculation; since it began as ^{239}U, it should be included in the U inventory instead. In any debris sample, it is important to measure the concentration of 2.36-day ^{239}Np so that the residual ^{239}Pu concentration can be corrected for the action of neutrons on U. The ^{239}Np concentration of a good debris sample may be measured for at least 2 weeks after the explosion. The time at which the Np and Pu fractions are isolated from one another must be known, and the completeness of the separation well-defined, so that the decay equations can be applied (Section 2.3).

The presence of more than one fission stage also complicates the calculation of efficiency. The total number of fissions represented by the sample must be separated into the fractional number of fissions associated with each fuel. The analyst will find a fission-split determination quite helpful in this calculation. Even so, there are instances where a more complicated analysis based on computer simulations is required. For instance, if the device contained two HEU stages of differing efficiency, a straightforward treatment of the post-explosion debris concentrations could mislead the analyst into postulating a single HEU component of intermediate efficiency.

Any debris sample will contain a significant quantity of natural-isotopic Th and U due to the incorporation of soil, which is typically 2–5 ppm U and 10–20 ppm Th. Due to the vast energy released in the explosion and the resultant dilution of the device residuum (see Section 5.4.3), the soil contribution to the U inventory of a debris sample can be >0.5. The soil nuclides must be subtracted from the U inventory of the debris sample before the fission efficiency can be calculated, since these isotopes were not associated with the functioning of the device and did not contribute to its fission yield. The subtraction is usually performed under the assumption that the U in the soil is of natural isotopic composition and is unperturbed by neutrons from the device. There are several ways to accomplish this subtraction, one of which takes advantage of the fact that, given a collection of several debris samples, it is unlikely that the U from the soil and U from the device are present in the same proportion in each sample. (Fractionation is discussed in the next section.) If natural U is a significant component of the device itself, the soil correction becomes more difficult and is realized in other ways.

Incorporation of soil U and Th in explosion debris makes it very difficult to accurately calculate the age of the device components from the analysis of a post-explosion sample. As discussed in Chapter 6, chronometry of heavy elements is achieved by measuring the ingrowth of long-lived daughter nuclides in the processes of radioactive decay. Except for ^{241}Pu (which decays to ^{241}Am), the Pu isotopes decay to the U isotopes that are commonly found in soil or are produced by neutron irradiation of the natural U isotopes (e.g., ^{236}U). In most cases, the soil signature overwhelms the radiogenic signature, making calculation of an age problematic. However, if the Pu fission efficiency is not too high, the $^{241}Pu/^{241}Am$ concentration ratio in the debris can be used to calculate the age of the Pu in the original device. Further, if the device contained no U and there was limited interaction between the soil and neutrons output by the device, interpretation of the $^{240}Pu/^{236}U$ chronometric pair as an age might be possible.

Chronometry of the U isotopes is also complicated by the introduction of soil, even when the device U signature far outweighs that of the background. This is best demonstrated by the following example. From Table 6.2, we know that the chronometric ratio $^{231}Pa/^{235}U$ in 20-year-old U is ~ 2×10^{-8}. Since the half-life of ^{231}Pa (33,000 years) is very short compared to the age of the earth, the concentration of

^{231}Pa in undisturbed soil is defined by its secular equilibrium with ^{235}U (Chapter 2) and is roughly proportional to the ratio of the two half-lives ($33{,}000/7.0 \times 10^8 = 4.7 \times 10^{-5}$). If 1% of the ^{235}U inventory in a debris sample arising from a device with 20-year-old HEU fuel is due to the contribution of soil, then only 4% of the ^{231}Pa content of the debris originated from the weapon.

During nuclear testing, the problem of determining thermonuclear (or fusion) efficiency received at least as much attention as did the issue of deducing fission efficiency. However, our successes in fusion diagnostics do not translate to a forensic method as readily as those involving fission diagnostics. Even though the yield of the fusion fuel is the result of the application of the most extreme temperatures, pressures, and particle energies in the exploding device, the products of fusion are of limited variety and difficult to detect. From the reactions given in Sections 5.2 and 5.3, fusion fuel consists of nuclides of lower Z and A than ^{9}Be, and the main fusion product is the noble-gas stable isotope, ^{4}He. While it might be possible to measure the concentration of residual fuel, which usually consists of nonnatural isotopic mixtures of Li and/or hydrogen, the numerator of the fusion-efficiency fraction cannot be measured. The concentration of ^{4}He in air is ~5 ppm by volume, so that 1 km^3 of air at STP contains 2.2×10^5 mol of ^{4}He. The complete reaction of 1 kg of ^{6}LiD in thermonuclear fusion results in the production of only 250 mol of ^{4}He (two ^{4}He atoms per ^{6}Li atom consumed). Even were an air sample obtained before the fusion products completely diffused into the atmosphere, the device adds a quantity of ^{4}He to the air that is insignificant compared to the background level. The thermonuclear reactions in Sections 5.2 and 5.3 indicate that hydrogen enriched in the heavy isotopes may also be observable in a debris sample, but they are present both as fuel residuals and as fusion products, making their interpretation as a performance indicator problematic.

Fusion efficiency may be obtained only indirectly. One common analyte is 53-day ^{7}Be, produced by the reactions of fast hydrogen ions with the Li atoms in solid thermonuclear fuel. These ions arise predominantly through elastic collisions between fuel atoms and 14-MeV neutrons. If the residual concentration of device-originated Li in a representative debris sample can be measured, the relative concentration of ^{7}Be can be interpreted as a fusion efficiency via explosion-code calculations. The most important issue is the subtraction of background: Li has no radioactive isotopes with half-lives amenable to forensic analysis [20], and the average terrestrial soil abundance of Li is relatively high at 20 ppm. Device-residual ^{6}Li is observable over the soil blank only as a stable-isotope shift arising from the use of the enriched isotope in the device. Cosmic-ray interactions with air molecules in the upper atmosphere continuously produce ^{7}Be [24], which is readily observable in filter samples collected by environmental air sampling. This ^{7}Be environmental blank must be subtracted from the radionuclide inventory of a debris sample before it can be used for fusion diagnostics.

During the nuclear test era, diagnosis of fusion efficiency was accomplished through the activation of well-characterized contaminant species that were intentionally added in various locations in a device. These detector elements were exposed to very high fluxes of energetic particles, principally 14-MeV neutrons, which activated them to generate neutron-deficient radionuclides quite distinct from the neutron-rich fission products. Measurement of the products of multiple-order reactions (i.e., (n,2n) reactions on the products of preceding (n,2n) reactions) defined the fast-neutron flux in a device part, which was interpreted as a thermonuclear yield. Since dilution of the thermonuclear fuel by addition of detector elements cools the fusion plasma and negatively impacts performance of the fuel, it is unlikely that an organization that fabricated such an explosive device would add detectors unless they were planning to perform fusion diagnostics. This technique is therefore of no relevance in NFA.

Forensic analysis of the fusion performance of a nuclear explosive is best accomplished through measurement of the nuclides arising from the interactions of thermonuclear neutrons with the fission fuels. The relative contributions of 14-MeV and fission-spectrum neutrons to the fission split can be interpreted (with the computer simulations) as the ratio of fusion yield to fission yield. Isotope production via (n,2n) reactions is dominated by fusion neutrons when present, the most important example of which entails the measurement of the $^{238}Pu/^{239}Pu$ atom ratio in bomb debris. Significantly more ^{238}Pu is produced by the $^{239}Pu(n,2n)$ reaction in a boosted Pu primary than in an unboosted device.

The earlier description is intended to convey an impression of the radiochemical signatures that are pertinent to the NFA of explosion debris. It emphasizes interpretations that may be performed without access to classified information or computer codes. Other aspects of such analyses cannot be given here, partly due to their connections to the historic diagnoses of nuclear tests. For NFA analysts who work "outside the fence," however, improved methods for returning radionuclide signatures with greater sensitivity and timeliness are always needed, as is the development of new signature species (see Section 5.4.4).

5.4.2 Fractionation of the Debris Field

In the previous section, discussion focused on analyses that may be performed to support a forensic evaluation of the design of a nuclear weapon based on the debris produced in its explosion. The interpretation of the resulting data is straightforward, providing the device residuum in the debris field is homogeneous and the analytical samples are representative of the debris field as a whole. The issue of representative samples is vitally important to the analysis, since it is unlikely that any individual specimen will contain $>10^{-8}$ of the explosion residuum unless the nuclear yield of the device is low or a prohibitively large quantity of debris is collected. Formulating global conclusions based on billionths of a sample set would make most analysts uneasy.

Obtaining representative debris specimens was recognized as a crucial issue from the beginning, even prior to the Trinity test. In detailing the radiochemical determination of the nuclear efficiency of the Trinity device, H.L. Anderson and N. Sugarman stated in 1945:

> There was some concern whether the fission products would deposit in constant proportion with the 49. Even though the fission products have different physical and chemical properties from the 49, we hoped that in the violence and suddenness of its explosion the nuclear bomb would not contrive to produce a separation of all the fission products from the 49.

In this quote, "49" is the wartime shorthand for Pu. Anderson and Sugarman recognized before the test that the validity of the radiochemical determination of fission efficiency would be compromised if the fission products were introduced into a debris sample with a chemical yield different from that of the residual fuel.

The debris field includes the melt deposit near ground-zero, ejecta deposited further downrange, fallout and airborne particles, gases, and liquid solutions arising from a single nuclear explosion. To this point, we have never encountered an individual sample of nuclear explosion debris that was representative of the entire debris field. Further, it is very unlikely that the chemical and physical processes that result in debris formation would allow the existence of such a specimen. A substantial fraction of the effort applied to the radiochemical diagnosis of nuclear tests involved processing a collection of nonrepresentative samples and, through extrapolation, calculating the relative concentrations of a limited set of analytes for a hypothetical representative sample.

The fractionation of radionuclides in the debris from a nuclear explosion is divided into two classes of mechanism. Physical effects result in "geometrical fractionation," manifest immediately after the detonation of the device, in which the device residuum never becomes well mixed [25]. "Chemical fractionation" occurs seconds to minutes later and involves the segregation of chemical species during condensation and solidification of the hot debris [26–29]. While a debris sample may not exhibit any effects of geometric fractionation, chemical fractionation is ubiquitous.

Geometrical fractionation arises from the momentum imparted to the individual parts of the device through energy production in the live components. It is common to assume that the energy released by a nuclear explosive would provide a path to homogenization, but there is no mechanism that guarantees complete mixing; diffusion and turbulence act to homogenize the debris only until it begins to condense. During nuclear testing, it was not unusual to find that debris arising from the individual stages of a weapon was represented to different extents in each debris specimen. Geometric fractionation may eventually provide the means for isolating signatures of delivery from samples that might otherwise be rejected because of a poor fission-product inventory (see Section 5.4.4).

At zero-time, everything in the immediate vicinity of the nuclear fuel of the weapon is vaporized, attaining temperatures of tens of millions of degrees and pressures of millions of atmospheres in <1 μs. The primary thermal radiation is absorbed by the surrounding medium, producing a plasma of completely and partially stripped ions to which chemistry does not effectively apply [30]. The device residuum, consisting of fission products, unconsumed fuel, and structural materials, moves outward from the center(s) of energy production at high velocity. Transfer of thermal energy within the plasma mass is rapid, and as the fireball expands, the energy is distributed over an ever-increasing volume, cooling the nascent debris. As the temperature falls, the mean-free-path of the energy emitted by the fireball decreases. The expanding fireball transfers energy to the surrounding medium by impulse, so that explosion energy becomes partitioned between radiation and shock. The shock front eventually catches the radiation front when the temperature of the fireball is still $>10^5$°K [1,31,32]. At this point, geometrical fractionation processes are considered to be frozen, while chemical fractionation has not yet begun, even though the initial plasma has begun to quench. Collision processes in the fireball damp the initial relative motion that results in geometric fractionation long before the debris has cooled sufficiently for chemical processes to occur. Geometric and chemical fractionations are therefore treated as separable: all product atoms in a geometrically fractionated sample are assumed to have undergone the same chemical processes, regardless of the device component from which they arose.

Even those debris samples that do not exhibit an observable degree of geometric fractionation will be chemically fractionated. Chemical fractionation was first observed in the debris of the Trinity test, confounding the radiochemists' desire for a homogeneous debris field. The chemists noted that elements forming compounds with high melting points tended to be distributed closer to ground-zero than those elements that persisted in the gas phase at lower temperatures. The classification of these elements as "refractory" or "volatile" is terminology that remains in use to the present day [29,33,34].

The fireball cools to the melting point of iron oxide (3500°C) in 20–30 s [25,30], where the vaporized device components, delivery structures, and entrained soil begin to form liquid droplets. The nucleation and condensation of these particles is an interesting problem because of the high temperatures and low concentrations involved. Chemical fractionation is driven by the different properties of the elements in all of their chemical states, which evolve as the explosion debris cools. These properties include vapor pressure, chemical affinity for surfaces, diffusivity, and solubility, all of which are considered together and termed "volatility" for simplicity. Although subsequent large-scale, shock-driven mixing and turbulence caused by the buoyancy of the rising fireball is present, the low-temperature debris never becomes uniform on any scale. For purposes of device diagnostics, as remarked by J. Shearer in 1966, *everything that goes on between the explosion of the device and the collection of a chemical sample is a nuisance*. Chemical fractionation is the principal source of that nuisance.

The early diagnostic radiochemists developed a pragmatic model for debris formation based on two chemical groups, the volatile elements and the refractory elements [25,33,35]. As the debris cools, the refractory elements condense first, followed by the volatile elements. Since the debris cools quickly initially and more slowly later, it was thought that the refractory elements might not be as badly fractionated as the volatile elements, which would have sufficient time to display a wide variety of chemical properties. It was assumed for the sake of convenience that the refractory elements would be unfractionated from each other during debris formation.

The two-chemical-group model was tested during the 1940s and 1950s. It was found that consistent results for the fission yield of a device could be obtained from several different refractory fission products, even when their chemical and thermodynamic properties were dissimilar, provided the analytical samples were selected for their refractory character. Mixtures of actinide elements were placed proximate to nuclear weapons prior to detonation; the refractory components of these "tracer packages" were not fractionated from one another [36] and were distributed with the refractory fission products and residual Pu fuel (see later). It was established that refractory diagnostic indicators (including residual Pu fuel, fission products, and actinide tracers) tended to concentrate in glassy melt debris, which was recognized as being the most representative of the mixture of radionuclides of device origin [25,26]. While acknowledged that differences in the high-temperature and high-pressure chemical properties of refractory species make some degree of chemical fractionation inevitable, scatter of the chemical concentration ratios among the refractory species is thought to be $< \pm 5\%$ in most cases.

Based on years of experience in nuclear testing, the refractory analytes in bomb debris include Zr, Ta, Th, Np, Pu, Am, Cm, and lanthanide isotopes with $A > 143$ [33]. Historically, several debris samples from each test were processed in order to increase the probability of obtaining a representative and refractory radionuclide mixture. Due to the refractory nature of Pu distributed within debris, the fission efficiency of a Pu-fueled device determined from the refractory fission products is thought to be little influenced by the effects of chemical fractionation.

The two-chemical-group model applies only to the processes that produce nuclear explosion debris, however, and should not be confused with chemical effects that are observed in the laboratory. For example, the radionuclide ^{140}Ba is a volatile, peak-yield fission product that has decay properties making it straightforward to observe and quantitate in the field. The degree of depletion of its expected concentration, relative to those of the refractory fission products, is used to preselect refractory debris samples for radiochemical diagnostic analyses. Although the chemical and thermodynamic properties of elemental Ba lead to an expectation that it should be a refractory element (the melting point of BaO is 1920°C), its volatility is well-established. Reference to Figure 5.5 explains this dichotomy: most of the $A = 140$ fission yield is born as the fission precursors, rather than as ^{140}Ba directly. As previously discussed, chemical fractionation and immobilization of the radiochemical analytes occur in tens of seconds after detonation. From data in the figure and the Bateman equations, it requires ~70 s for half of the $A = 140$ mass chain to decay to ^{140}Ba, dictating that the incorporation of $A = 140$ nuclides into the debris is dominated by the chemical properties of an inert gas (Xe) and compounds of the alkali metal, Cs. Xenon is a noble gas at ambient temperature, and the cesium oxides melt in the range of 400°C–500°C. Very little of the ^{140}Ba measured in a debris specimen was actually Ba during debris formation. The distribution of ^{140}Ba in the debris field at sampling time is most strongly influenced by the chemical properties of volatile Cs.

Another example of the subtle effects of fission precursors on volatility involves the cumulative fission products, ^{141}Ce and ^{144}Ce, both of which are observed in post-explosion debris. The $A = 141$ and $A = 144$ fission-product precursors, their half-lives, and independent yields are shown in Figure 5.9. The chemical properties of Ba, La, and Ce are such that they would be predicted to behave as refractory species during debris formation. As with the ^{140}Ba chain, though, the isotopes of I, Xe, and Cs are expected to be volatile. In the figure, it is seen that a substantial fraction of the $A = 141$ fission chain is born as 25-s ^{141}Cs and its precursors. The half-life of this Cs isotope is comparable to the time corresponding to the onset of chemical fractionation of the debris field, so the volatility of Cs will influence the ultimate distribution of ^{141}Ce. Because the half-life of ^{141}Cs is shorter than that of 64-s ^{140}Cs, ^{141}Ce should be less volatile than ^{140}Ba, which is usually observed (see later). Very little of the $A = 144$ mass chain is born as volatile isotopes, and those nuclides have very short half-lives compared to the timescales of chemical fractionation; ^{144}Xe

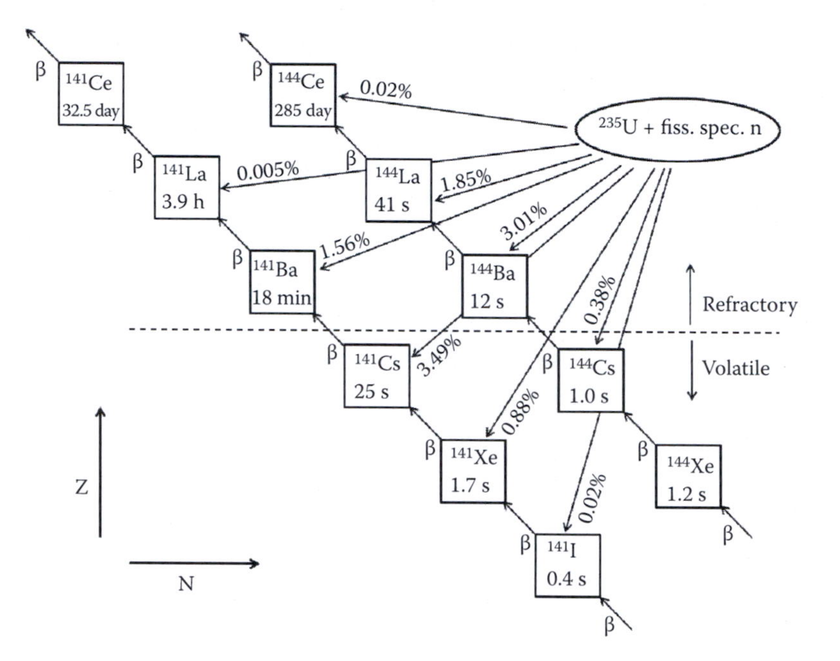

FIGURE 5.9 The A = 141 and A = 144 decay chains from the fission of ^{235}U with fission-spectrum neutrons. Chemical and nuclear-decay properties of the precursor nuclides combine to result in ^{141}Ce and ^{144}Ce being chemically fractionated from each other during debris formation, even though they are isotopes of the same element.

and ^{144}Cs will have thus completely decayed to ^{144}Ba before fractionation processes become evident. Consequently, ^{144}Ce should be refractory, with the distribution of its concentration in debris proportional to that of the other refractory products. These phenomena explain the interesting observation that, although ^{141}Ce and ^{144}Ce are chemically identical in the laboratory, they are distributed in bomb debris as if they had different chemical properties. This result is strictly an artifact of the differences in the elemental distributions of their fission precursors.

Another interesting exercise is the study of the distribution of the ^{99}Mo fission-product precursors. The 66-h ^{99}Mo isotope is itself somewhat volatile, while its precursors are refractory. Much of the precursor yield piles up in 2.6-min ^{99m}Nb, whose half-life is long compared with the time of onset of chemical fractionation. As a result, in an atmospheric explosion, ^{99}Mo can be treated as a well-behaved, refractory fission product. In underground tests, however, ^{99}Mo is found to be chemically fractionated from the refractory fission products, a consequence of the fact that underground debris remains hotter longer than does atmospheric debris. This feature gives the ^{99m}Nb precursor greater opportunity to decay to the volatile end-member fission product before the fission products become completely immobilized.

Discussion has focused on varying degrees of radioanalyte volatility as a consequence of the decrease in slope of the temperature profile during latter stages of debris formation (see Section 5.4.3). But consider also the neighboring fission products, 13-day ^{136}Cs and 30-year ^{137}Cs. The volatile fission product ^{136}Cs is shielded (has no precursors), and its distribution in debris is defined at all times by the chemical properties of Cs. The volatile fission product ^{137}Cs is not shielded and has a noble-gas fission precursor, 3.8-min ^{137}Xe, that receives most of the A = 137 fission yield and dominates its distribution in debris. While both ^{136}Cs and ^{137}Cs are members of the chemical group of volatile elements, and are fractionated from the refractory elements, they are also significantly fractionated from each other. In underground testing, partial ^{137}Cs inventory was often found to be widely distributed in the soil medium above the device, significantly removed from the rest of the fission products (including ^{136}Cs).

The chemical behaviors of the fission precursors, and the differences in their half-lives, provide a means of estimating the time dependence of the processes involved in the formation of debris. A thorough knowledge of the high-temperature chemical behavior of the precursor elements could potentially lead to a method for defining the temperature/pressure history of a specific sample isolated from the debris field. Such information could allow the analyst to make fundamental corrections to the radionuclide inventory for the effects of chemical fractionation. Other than studies of the volatilities of shielded fission products, very little relevant chemical information currently exists. As a result, conversion of the concentrations of volatile elements to a refractory basis is generally performed through semiempirical correlations between isotope ratios [25,28].

Fractionation in bomb debris is studied by means of a mixing analysis, discussed in detail in Section 3.3.8. In nuclear debris analysis, it is sometimes referred to as the "two-pocket" model. Incomplete homogenization of two distinct debris components, each containing radionuclides A, B, and C, results in concentration ratios [A]/[C] and [B]/[C], which, if measured across a collection of samples, maps out a mixing line. Unless sampling is biased in some way, the composition of a hypothetical sample representative of homogeneous debris should lie somewhere on this mixing line [36,37]. For the case of geometrical fractionation, mixing is assumed to involve two debris components that encompass the structure of the device, assuming that the radionuclides arising from each component are independently homogeneous and that only the part-to-part mixing ratio varies. A two-pocket analysis of geometric fractionation is most successful when applied to refractory analytes where chemical effects are expected to be minimal. For the case of chemical fractionation, mixing is assumed to be between the two chemical groups comprising volatile and refractory species.

Due to the complicated interplay between the half-lives and the chemical forms of their precursors, some question arises about the validity of performing a mixing analysis on radionuclide mixtures that include the end-member fission products. Analysis of chemical fractionation with a mixing model is also questioned by the observation that the volatile group itself displays internal fractionation among its members. Nevertheless, a three-isotope plot, such as described in Section 3.3.8, can be used to treat fission-product data involving radionuclides having different degrees of volatility. The justification is pragmatic: more often than not, conclusions may be drawn from this treatment that can be validated against other device-performance measurements, or with the results from large-scale computer calculations.

Figure 5.10 is a three-isotope plot involving one refractory (^{144}Ce) and two volatile (^{140}Ba and ^{141}Ce) fission products. Specimens were collected downrange from the test of a nuclear weapon that was detonated in the atmosphere in 1962, and the concentrations

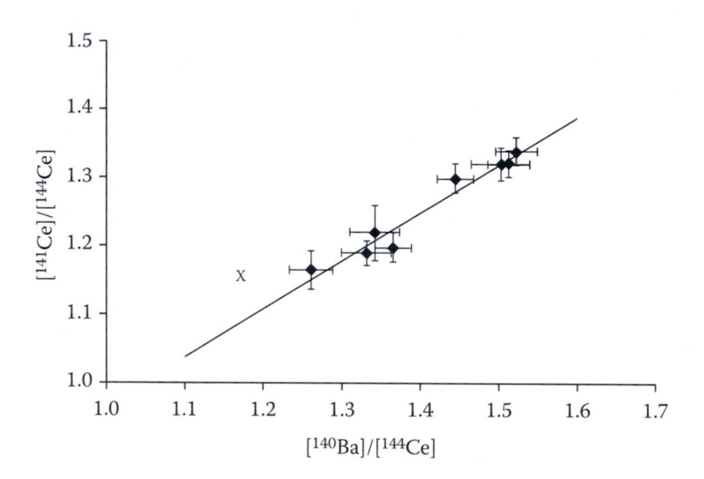

FIGURE 5.10 A three-isotope plot used in a mixing analysis of eight samples of fractionated debris collected downrange from a 1962 nuclear test. Diamonds are the experimental data, the diagonal line is a linear fit, and "x" is the result of a calculation of homogeneous debris based on device performance.

of the three fission products were determined independently in each sample. Eight concentration ratios of [^{141}Ce]/[^{144}Ce] *vs* [^{140}Ba]/[^{144}Ce] are depicted as diamonds in the figure, with each point associated with its statistical counting errors in x and y. The solid diagonal line is a regression analysis of the data. The plotted "x" is the result of a computer model of the performance of the device, in which the fission split was calculated. It is noted that each of the experimental samples was more enriched in volatile debris than refractory debris, since each [volatile]/[refractory] ratio is greater than expected from the fission split. From the spread in the isotope ratios, we would claim that ^{140}Ba is more volatile than is ^{141}Ce, as expected. If the application of a mixing analysis was justified, it would be anticipated that the results of the calculation, which mimics unfractionated debris, should lie on the mixing line. However, there are several reasons why the results of the split calculation may not be supported by the empirical data: (1) the computer simulation (performed in 1962) may have resulted in an imperfect reproduction of device performance; (2) the samples may have been geometrically fractionated, as well as chemically fractionated; (3) the validity of the method is most reliably tested against samples enriched in refractory species, rather than in volatiles; and/or (4) the nuclear data used to convert radiation counting data into the measured concentrations of fission products may not be referenced to the same basis as were those used to construct the member sets of the six-component fission split. Inaccuracies associated with points 1 and 4 are more likely for older experiments than for more modern work.

Figure 5.10 utilizes data that imperfectly display the power of a mixing analysis on chemically fractionated debris; however, there exist many examples of a three-isotope plot that accurately reconstruct a representative homogeneous debris specimen. For debris that is not geometrically fractionated, the expectation is that the three-isotope correlation of three refractory species would produce a scatter of data points closely clustered around a single point in (x, y), since the refractory nuclides are taken as unfractionated from each other. Deviation from such behavior is generally attributed to geometrical fractionation, but the method has also been used to examine collections of refractory nuclides to determine their relative volatilities [33,38]. Historical results from three-isotope examinations of refractory species support the premise that the refractory nuclides are not significantly fractionated from one another in appropriate samples.

It is interesting and somewhat surprising that the list of refractory elements given earlier does not include U. From Section 5.4.1, the calculation of fission efficiency relies on measurements of both the fission-product inventory and the concentration of residual fuel in a representative sample. Early in the history of nuclear testing, it became apparent that fission efficiencies determined from analyses of refractory-rich samples were apt to be biased high when based on the measurement of residual U, implying that U was depleted in the refractory samples. The issue was exacerbated when US nuclear testing moved underground in 1962. This effect was better quantified in experiments in which mixtures of actinide elements, including U, were emplaced with nuclear test devices: uranium is found to be a somewhat volatile element that chemically fractionates from the other actinides and the refractory fission products, complicating the determination of its relative concentration in a representative sample. It is not known which chemical species control(s) the volatile behavior of U in the nascent debris field.

Eventually, the volatility of U relative to the refractory species was measured routinely through the use of actinide tracer packages. The U isotope, ^{233}U, which was not present to any great extent in the post-shot device residuum, was paired with a refractory isotope such as ^{244}Cm (also not present or produced in the device), and the assembly was emplaced near the device. Intimate placement of the two actinide tracers minimized the possibility of geometric fractionation between them. Knowledge of the quantity of both actinide materials in the tracer package, and the measurement of [^{233}U]/[^{244}Cm] in each debris specimen, provided a direct measurement of the chemical depletion of U in that sample. An added benefit of the use of "volatility packages" was the comparison of the ^{244}Cm concentration with that of a chemically similar refractory product in device residuum to test for the effects of geometric fractionation.

It is unlikely that an organization that would deploy and detonate a clandestine nuclear explosive device would include a calibrated volatility package for our convenience, yet another limitation of applying nuclear diagnostic techniques to NFA. Fortunately, it was determined that the depletion of U in a debris sample may be reliably correlated with the concentrations of the volatile fission products via a three-isotope plot. This discovery was rather unexpected, since the chemical volatility of fission products is a complicated dance between the chemical properties of precursors and their half-lives, while U is always U during debris condensation. The onset of heterogeneity during debris formation, and the nonreversible condensation of refractory products, lead to a speculation that the unknown chemical species responsible for U volatility may exist for only a short time prior to the immobilization of radionuclides in the nascent debris [39,40].

Reconstructing Figure 5.10 using [^{239}U] in place of [^{141}Ce] results in a similar linear mixing diagram. If the fission split associated with the weapon is known (either measured or calculated), it is possible to compute the value of [^{140}Ba]/[^{144}Ce] for unfractionated debris for its position along the *x*-axis. Using the linear fit to the experimental mixing data, the value of [^{239}U]/[^{144}Ce] associated with unfractionated debris can then be read off the *y*-axis of the plot, leading to the representative concentration of ^{239}U and part of the residual fuel inventory.

This same process has also been used to extract the unfractionated concentrations of the other U isotopes, as the chemical correction for the volatility of U in device residuum should be the same for each U nuclide. Unfortunately, the long-lived isotopes, ^{234}U, ^{235}U, and ^{238}U, in a debris specimen may have substantial contributions from the incorporation of soil (see previous discussion). For a near-surface explosion, prompt incorporation of soil into the fireball will result in a near-natural mixture of U isotopes undergoing the same chemical processes as residual device U. However, soil that is subsumed into the mushroom cloud at later times is not vaporized before being incorporated into the nascent debris. The complex result is a U inventory for a given sample that includes a large natural soil component, both chemically and geometrically fractionated from the device-residual U, along with a smaller natural soil component that is well-mixed with the weapon residue, but is still chemically fractionated from refractory species. While a mixing analysis may be helpful in deciphering the relative contributions from soil and device to the ^{234}U, ^{235}U, and ^{238}U inventories of debris samples, correcting the U concentrations in debris for the incorporation of soil before performing the volatility analysis is always preferred. For incorporation of U, each specimen has its own unique and peculiar condensation history, which is nonsystematic and could therefore invalidate a mixing analysis.

Any protocol for recovering samples from a debris field after a nuclear explosion should always include sufficient specimens to support a fractionation analysis. Such action could entail discarding samples whose field assays indicated rich collections of fission products, in favor of weaker samples possessing more diverse radiochemical compositions. In underground testing, Livermore generally processed six principal samples (along with a number of collateral specimens) that were collected from widely distributed locations in the debris field. That experimental purpose was diagnostic, so that samples rich in refractory species and representative of a single chemical phase (i.e., puddle glass) were preferred. This approach usually resulted in the determination of the composition of a homogeneous sample as an extrapolation toward more volatile debris, rather than as a more accurate interpolation between separated datasets of relatively refractory and relatively volatile specimens. Economics drove the number of acquired samples, and often a single outlying datum affected the definition of the mixing line as much as did all of the other data points, were they closely clustered in "fractionation space."

5.4.3 Debris Morphology and Processes of Debris Formation

In the earlier section, focus was on the processes of debris formation and distribution following a nuclear explosion, that is, those leading to fractionation effects that distort the radiochemical inventories of individual samples such that they were not representative of the performance of the device as a whole. Inhomogeneity of explosion debris on a larger scale also impacts NFA, even though it may not affect the radiochemical signatures. The turbulent incorporation of soil at late times, and its potential impact on the fractionation analysis of U isotopes, has already been mentioned. In addition, inhomogeneity at the macroscale drives the selection of sampling sites in a debris field and affects the radiation dosimetry at those locations. Subsequently, the chemical composition of nuclear explosion debris strongly influences the selection of methods for sample preparation and dissolution prior to ensuing NFA.

Section 5.4.2 discussed the early stages in the evolution of a fireball—those leading to geometric and chemical fractionation. During the second of time immediately following nuclear explosion, the environment in which the weapon was detonated does not strongly affect the evolution of debris formation [30]. We have practical experience in all aspects of debris analysis associated with nuclear testing, ranging from far below ground to exo-atmospheric, as well as explosions within many diverse media, including air, soil, water, and salt [26,29,34]. For the following discussion, however, we exclude here nuclear detonations outside the atmosphere, in the more exotic condensed media, or far below the surface of the earth. They are outside the scope of this book. The remaining explosions may be subdivided into two groups, depending on whether the fireball touches the ground.

Consider first the situation in which the fireball does not touch the ground. The two weapons used against Japan in 1945 were of this type, as were the US Dominic tests in 1962 [41]. After the first 20 s, the fireball cools from 3500°C, where the refractory elements begin to condense, to much lower temperatures. The first particles to condense consist of the refractory materials used in the construction of the device and its delivery system, and the initial submicrometer-sized droplets aggregate into larger liquid drops. The oxygen and other light elements that comprise air react with the explosion residuum to form thermodynamically stable salts, which are soluble in the liquid media of the droplets. As cooling continues, droplets begin to solidify, and other agglomeration processes occur [42–44]. At temperatures too low to facilitate dissolution or diffusion, volatile elements tend to condense at the surfaces of particles. Small particles are likely to carry an electric charge, which promotes the formation of threads and ropes of particles under low-density conditions. Eventually, larger spherical aggregates form, and after approximately 10 min, the debris cloud cools to near-ambient temperatures. This complicated nature of the condensation and agglomeration processes is predisposed to result in debris particles that incorporate both volatile and refractory elements. However, the refractory species are likely to be distributed throughout the bulk of the material, while the volatiles are concentrated on the surfaces of individual grains.

This behavior is the fundamental source of chemical fractionation in the distribution of debris downrange from the explosion. Fallout particles from a high-altitude detonation vary in size from <1 μm to ~1 mm in diameter, with a distribution that peaks at 50 μm [45,46]. For individual debris particles, the refractory radionuclides are distributed throughout the volume of the material (proportional to the cube of the radius of the particle), while the volatiles are distributed on the surface (proportional to the square of the radius). Therefore, smaller particles tend to be richer in volatile elements than larger particles. The settling rates of descending debris particles will depend on their dimensions, their shapes and mass, the viscosity and drag of the air through which they move, and meteorological conditions such as wind and precipitation. The debris cloud is buoyant initially and will tend to carry small particles upward, while the largest particles will have a net downward velocity. This development implies that refractory species are likely to be distributed closer to ground-zero than are the volatile species, which are strongly affected by wind and weather conditions. Most of the downrange radiation exposure from weapons testing at the Nevada

Test Site during 1951–1957 was caused by devices detonated on tall steel towers, ~150 m above ground level; the radiation inventory of off-site fallout from those tests was dominated by volatile radionuclides [25].

However, the more likely nuclear forensic scenario entails a nuclear explosion near, on, or slightly below the surface. In this case, a considerably larger fraction of the debris from the weapon is deposited locally, rather than downrange, when contrasted to the distribution of debris from a high-altitude explosion. The thermal pulse and shock from the explosion melt the ground surface, and some of the device residuum is incorporated into the medium almost ballistically [34]. This phenomenon was the source of the "sea of green glass" reported after the Trinity test [30], which consisted of a synthetic mineral called trinitite. Trinitite is chemically related to the puddle glass analyzed by nuclear diagnostics during underground testing, and it was first collected for radiochemical analysis shortly after the Trinity explosion. The forensic examination of trinitite has recently become a popular research topic (e.g., [47–49]), work that has illuminated effects of the complicated processes that result in nuclear melt debris.

In near-surface nuclear explosions, great quantities of the adjacent environment are swept up and incorporated into the incandescent fireball. As much as a kiloton of debris per kt of device yield [30] is added ballistically or suspended in the cold air that eventually quenches the fireball. Smaller environmental particles can be vaporized or melted, but most of the environmental adduct persists as solid items of variable size and shape that can act as condensation sites for both the refractory and volatile components of device residuum. The processes of near-surface debris formation are much more complicated than those involved in high-altitude debris, and they were extensively studied during the 1950s and 1960s (e.g., [50–58]). Early melting of particles and particle surfaces favors the incorporation of smaller, vapor-condensed droplets containing refractory elements, while volatile elements tend to condense after particle surfaces have solidified [34,35]. This action leads to the enrichment of volatile elements in small particles and to chemical fractionation of the debris field, similar to that observed following high-altitude explosions. The concentration of device residuum in debris from a near-surface explosion is significantly lower than that in high-altitude debris, a consequence of dilution by the environmental matrix.

In a near-surface detonation, the peak of the particle-size distribution is significantly larger than that from a high-altitude explosion. The result is a shift favoring the incorporation of the radionuclide content into local fallout (near ground-zero), relative to that deposited downrange or distributed worldwide. Residual radiation at ground-zero following a high-altitude explosion may contain little fallout and can be strongly affected by activation of the surrounding environment (including air) induced by neutrons escaping the device. The Trinity explosion was a near-surface test (approximately 30 m in altitude), and the radiation field at ground-zero was dominated by local fallout and the trinitite melt zone. The Nagasaki explosion derived from a similar weapon, but was detonated at higher altitude; although local fallout was reduced, significant activation of the local medium was observed [45], complicating radiation dosimetry calculations based on the time-dependent power-law decay of mixed fission products [25].

The mass of a deployable nuclear explosive device would likely be measured in tons or less. Kilotons of the local medium are incorporated within the debris from a near-surface explosion. Therefore, even those samples that are richest in the radionuclides produced by the device will likely have a stable-element matrix dominated by the environment in the vicinity of the detonation. From the mechanisms discussed, the individual debris particles will be no more homogeneous than the original environmental matrix. While it is possible to locate isolated regions of a debris field that strongly resemble the residue from the weapon, the physical scale of these increments is small in comparison to the peak of the particle-size distribution. The specific composition of the environment in which the device was emplaced will thus strongly affect the laboratory chemistry implemented for debris dissolution.

Every few years, it is proposed that the stable isotopes that are the ultimate decay products of the fission products might be measured in aged debris specimens. For bulk samples of near-surface debris, it is relatively easy to demonstrate that this is not possible. The fission products are severely diluted within the environmental blank and can be observed only through very minor deviations from the normal isotopic ratios of the natural elements that constitute soil. However, on the microscale, it may be possible to isolate individual grains of debris-rich material, identified by the presence of high concentrations of Pu or HEU. If debris-rich grains that are sufficiently large to produce a mass-spectrometric signal can be identified, it may prove possible to measure the concentrations of aged fission products.

The validity of NFA of explosion debris strongly depends on the selection of analytical samples from the debris field. Debris deposited near ground-zero includes both puddle glass and large fallout particles. The refractory radionuclide content of specimens of this material is more representative of the performance of the device than is the radionuclide content of the more volatile samples collected further downrange. Nevertheless, even refractory-rich debris specimens must be examined for the effects of chemical fractionation, so more volatile samples should also be included in the suite of analytical collections. Acquisition of specimens from the vicinity of ground-zero is complicated by the radiation field: at 1 h after the surface explosion of a 1-kt fission device, the area around ground-zero will contain $>10^8$ Ci of fission products [45]. The Trinity debris field was sampled with army tanks that incorporated extra shielding; it is unlikely that such bulky vehicles would be available for a forensic scenario because the geographic distribution of potential targets would require the deployment of an inordinate number of collection platforms. It will therefore be necessary to sample the most representative part of the debris field remotely.

However, even nonoptimum samples will provide forensic information if an adequate number is collected and analyzed. Sufficient specimens should be obtained to support a mixing analysis to correct for the effects of both geometric and chemical fractionation. Because of the possibility of a biased collection, a diversity of specimens should be included in the sample suite (e.g., puddle glass, ejecta, and fallout particles). The laboratory analyst must also consider that an urban environment provides a rich variety of potential sample matrices, including flora, fauna, structural steels, glass, concrete, and asphalt.

5.4.4 Delivery Signatures

During the US Test Program, the activation products of materials surrounding the device were measured in the radiochemical analyses of explosion debris, but they were only considered if they impacted the diagnosis of device performance. Since structural activation was not studied in earnest, what follows is a somewhat speculative narration on the application of forensic radiochemistry to the development of signatures of device delivery.

While 90% of the neutrons generated by an exploding nuclear device are absorbed by its fission fuel [30], the fraction that escapes can provide large exposures of nearby materials. Some of these neutrons escape the fuel mass during the multiplication phase of the explosion, while others are residual after the device disassembles. The escaping neutrons make it possible, in principle, to deduce information about the container in which the device was detonated [45]. Activation of the materials surrounding the device is a signature of weapon delivery, providing clues as to the source for the deployment, rather than to the origin of the device.

The debris field thus contains not just fission products and residual fuel, but also activation products of the nonnuclear components of the weapon. The neutron fluence outside the exploding device falls off rapidly with distance from the nuclear fuel. This property means that external materials experience a much lower neutron dose than do the live device parts, leading to a lower production of radionuclides from the surrounding materials. Therefore, the activation products in debris specimens will tend to be overwhelmed by the radionuclide signatures of the fission process, necessitating radiochemical techniques to isolate the structural and other environmental activation nuclides from fission products and concentrate them to detectable levels. For example, a debris sample that contains 10^{12} atoms of a peak-yield fission product (such as ^{95}Zr or ^{144}Ce) will likely contain $<10^8$ atoms of a lanthanide activation product generated by neutron reactions with soil.

Although the radionuclide signatures of delivery are difficult to measure over the radioactivity of the fission products, their interpretation is simplified. Since they are not involved in energy production, neutron exposure of these parts is delivered from a distance, and isotope production follows a $1/r^2$ relationship, where r is the distance from the center of neutron generation. Consequently, a large-scale computer deconvolution may not be necessary to adequately assess structural-activation signatures.

To first order, energy production in an exploding nuclear device is sufficiently fast that the structural "target" can be considered to be effectively stationary during irradiation. From Section 2.6, radionuclide production is proportional to the product of flux and target mass. Given a centrally located source of neutrons, the radionuclide production from a small quantity of target material placed close to the source can be the same as that from a much larger target placed at a longer distance. To reverse-engineer the device structure, an analyst must be able to differentiate between (e.g.) 100 g of steel placed 10 cm from the device and 1 metric ton of steel placed 10 m from the device, both of which would give rise to similar activation signatures. To make such distinction, there exist several associated considerations that may be examined:

1. Limitations of density: The device is constructed from normal terrestrial materials. The spatial volume associated with a given distance, and the density of the material in question, may prevent the placement of the required number of target atoms at that distance.

2. Normal isotopics: Materials constituting the structure of the weapon outside the live nuclear components are most likely constructed of elements of normal isotopic composition. Enriched isotopes are expensive and would probably not be used, unless required by the physics that controls energy production. Of course, a significant exception to this rule involves the use of depleted U, a by-product of U enrichment that is inexpensive and available in quantities of thousands of kg. For structural elements consisting of more than one stable isotope, the assumption of a normal isotopic distribution allows the analyst to measure the neutron activation products from the entire isotopic mixture and establish the time-integrated energy distribution of the neutron flux in that material.
3. Neutron energy loss: As neutrons transit the materials surrounding the device, they interact with those materials. High-energy neutrons lose energy through elastic-scattering interactions, and low-energy neutrons are absorbed. The neutron spectrum becomes softer with increasing distance from the centers of energy production, and the production of radionuclides from reactions with an energy threshold (e.g., (n,2n) and (n,p)) decreases relative to isotope production by reactions without a threshold (e.g., (n,γ) and some (n,f) reactions). The relative concentrations of different activation products in a given material are indicators of the nature and quantity of the intervening materials. Two external elements that are exposed to the same time-integrated neutron flux were possibly colocated in the pre-explosion weapon configuration, even if their products are chemically fractionated in the explosion debris. This last issue is key, because the relative volatilities of many of the light elements are poorly known. Using radionuclide concentration ratios, the analyst may use the activation products to speculate on the compositions of alloys and other mixtures used in the construction of the device.
4. Geometrical fractionation: Fractionation of debris deriving from different live device parts has been observed, even among refractory species that should be chemically similar [25]. This result arises from energy production during detonation of the weapon, which provides enough impulse to individual device components that diffusion and turbulence are insufficient to produce uniformity before the nascent debris begins to condense. It seems reasonable to presume that geometric fractionation of the products of fission and fusion, from the products of the activation of the more distant delivery structures, would be more extreme. Elements that are intimately mixed with each other before the explosion should not be geometrically fractionated from one another. A comprehensive mixing-diagram analysis of chemically similar structural materials could therefore aid in identifying elements that were colocated before explosion.
5. Environmental blank: Neutrons leaving the vicinity of the device will eventually encounter soil. Soil is a complex composition that contains some contribution from every naturally occurring element, so the products of soil activation must be removed from the concentration data before the device structure may be assessed. The concentrations of most elements in soil are

> minor, but the medium is effectively infinite in extent; combined with the long range of neutrons in matter, there could be substantial isotope production induced in minor species in soil. Soil is exposed to a softer neutron spectrum than are the intervening structural materials, resulting principally in (n,γ) products, and it is reasonable to assume that the elemental composition of soil is pseudo-homogeneous on the large scale. It should then be possible to calculate the soil contribution to every radionuclide in a debris specimen from the analysis of the elemental composition of the matrix and the concentrations of indicator nuclides that are not likely produced in the device from neutron reactions with the fuel or structural materials.

The low concentrations of the delivery signatures, compared to the performance signatures, would require the analysis of larger samples, with concomitant difficulties associated with personnel exposure and the increased scale of chemical operations. Experience from nuclear testing dictated the requirement for the number of analytical specimens necessary for a fractionation analysis of device performance, but more samples may be required for a similar analysis of device-delivery signatures. A protocol for field selection of samples, specifically for the analysis of delivery signatures, has not been explicitly defined.

5.5 POST-EXPLOSION FORENSIC SUMMARY

Most of this book is focused on the NFA of nuclear evidentiary materials that have not been involved in a nuclear detonation. Heavy-element samples are amenable to forensic methods because the signatures of the processes that were perhaps used to create the sample are well known, and the physical extent of the sample is defined, making complete characterization a tractable problem.

NFA of post-explosion debris samples is less straightforward. The nuclear material signatures that are most familiar are severely modified by the action of the nuclear device; issues related to dilution and the inhomogeneous distribution of the product radionuclide inventory are incompletely resolved; and the analyst is unlikely to ever process a significant fraction of the total analyte inventory. The post-explosion situation is intrinsically more complicated and less defined than the predetonation problem. Nevertheless, significant progress has been made in the forensic interpretation of signatures arising in the analysis of explosion debris specimens. Such assessment is an active area of investigation.

We attempt here to give the reader a sense of what can be accomplished using first-principles information and data published in the open literature. Considerably more is known, but access is restricted and limited to the small community of forensic analysts who possess government security clearances.

In our opinion, unclassified areas of research remain where contributions can be made in the area of post-explosion nuclear forensics. Examples are experimental measurements of relevant nuclear data (including nuclear decay properties and reaction cross sections); development of methods to reduce the time for the generation of analyte concentration and isotope-ratio data; study of the condensation chemistry of materials in a cooling plasma; and advanced field applications of radiochemical techniques.

REFERENCES

1. S. Glasstone and P.J. Dolan, *The Effects of Nuclear Weapons*, Department of Defense and Department of Energy, Washington, DC, 1977.
2. S. Glasstone and L.M. Redman, An introduction to nuclear weapons, report WASH-1038 Revised, 1972 (classified report).
3. R.W. Seldon, An introduction to fission explosives, report UCID-15554, 1969 (classified report).
4. E.K. Hyde, *The Nuclear Properties of the Heavy Elements*, Vol. III, Fission Phenomena, Prentice Hall, Englewood Cliffs, NJ, 1964.
5. R. Vandenbosch and J.R. Huizenga, *Nuclear Fission*, Academic Press, New York, 1973.
6. A.V. Nero, *A Guidebook to Nuclear Reactors*, University of California Press, Berkeley, CA, 1977.
7. B. Pellaud, Proliferation aspects of plutonium recycling, *J. Nucl. Mater. Manage.*, 31(1), 30, 2002.
8. D. Albright, F. Berkhout, and W. Walker, *World Inventory of Plutonium and Highly Enriched Uranium, 1992*, Oxford University Press, Oxford, U.K., 1993.
9. S. Groueff, *Manhattan Project*, Little, Brown, Boston, MA, 1967.
10. M. Lung, Perspectives of the thorium fuel cycle, *Seminar at JRC-Ispra*, Ispra Varese, Italy, July 2, 1996. Available at: http://www.nrg-nl.com/nrg/extranet/thorium/links/.
11. L.R. Groves, *Now It Can Be Told*, Harper and Rowe, New York, 1962.
12. J. Larus, *Nuclear Weapons Safety and the Common Defense*, Ohio State University Press, Columbus, OH, 1967.
13. E. Teller, ed., *Fusion*, Vol. 1A, Magnetic Confinement, Academic Press, New York, 1981.
14. G. Velarde, Y. Ronen, and J.M. Martinez-Val, *Nuclear Fusion by Inertial Confinement*, CRC Press, Boca Raton, FL, 1992.
15. J.H. Nuckolls, The feasibility of inertial-confinement fusion, *Phys. Today*, 35(9), 25, 1982.
16. H. York, *The Advisors: Oppenheimer, Teller, and the Superbomb*, Stanford University Press, Stanford, CA, 1989.
17. J.S. Foster, Jr., Nuclear weapons, in *Encyclopedia Americana*, Vol. 20, Grolier, Danbury, CT, 1981.
18. E. Beard, *Developing the ICBM*, Columbia University Press, New York, 1976.
19. R. Rhodes, *Dark Sun: The Making of the Hydrogen Bomb*, Simon & Schuster, New York, 1995.
20. R.B. Firestone and V.S. Shirley, *Table of Isotopes*, 8th edn., Wiley & Sons, New York, 1996.
21. T.R. England and B.F. Rider, *Evaluation and Compilation of Fission Product Yields*, ENDF-239, 1993.
22. T.B. Cochran, W.M. Arkin, R.S. Norris, and M.M. Hoenig, *Nuclear Weapon Databook*, Vol. II, U.S. Nuclear Warhead Production, Ballinger, Cambridge, MA, 1987.
23. B. Davison and J.B. Sykes, *Neutron Transport Theory*, Clarendon Press, Oxford, U.K., 1957.
24. P.S. Goel, S. Jha, D. Lal, P. Radhakrishna, and Rama, Cosmic-ray produced beryllium isotopes in rain water, *Nucl. Phys.*, 1, 196, 1956.
25. H.G. Hicks, Calculation of the concentration of any radionuclide deposited on the ground by offsite fallout from a nuclear detonation, *Health Phys.*, 42, 585, 1982.
26. I.Y. Borg, Radioactivity trapped in melt produced by a nuclear explosion, *Nucl. Technol.*, 26, 88, 1975.
27. K. Edvarson, K. Low, and J. Sisefsky, Fractionation phenomena in nuclear weapons debris, *Nature*, 184, 1771, 1959.

28. E.C. Freiling, Radionuclide fractionation in bomb debris, *Science*, 133, 1991, 1961.
29. H.A. Tewes, Results of the Cabriolet excavation experiment, *Nucl. Appl. Technol.*, 7, 232, 1969.
30. H.L. Brode, Review of nuclear weapons effects, *Ann. Rev. Nucl. Sci.*, 18, 153, 1968.
31. G.W. Jackson, G.H. Higgins, and C.E. Violet, Underground nuclear explosions, *J. Geophys. Res.*, 64, 1457, 1959.
32. C.E. Adams, N.H. Farlow, and W.R. Schell, The compositions, structure, and origins of radioactive fallout particles, *Geochim. Cosmochim. Acta*, 18, 42, 1960.
33. R.W. Lougheed, J.W. Meadows, and W. Goishi, The chemical fractionation of diagnostic species, in Nuclear chemistry division annual report FY81, G. Struble, ed., Lawrence Livermore National Laboratory report UCAR 10062-81/1, Livermore, CA, 1982, p. 36.
34. E. Teller, W.K. Talley, G.H. Higgins, and G.W. Johnson, *The Constructive Uses of Nuclear Explosives*, McGraw-Hill, New York, 1968.
35. R.E. Heft, The characterization of radioactive particles from nuclear weapons tests, *Adv. Chem.*, 93, 254, 1970.
36. F. Albarede, *Introduction to Geochemical Modeling*, Cambridge University Press, Boston, MA, 1996.
37. G. Faure, *Principles and Applications of Inorganic Geochemistry*, McMillan, New York, 1991.
38. R.W. Lougheed, R.J. Dougan, and A.A. Delucchi, Reliability of actinide tracers, in Nuclear chemistry division FY1989 annual report, D.C. Camp, ed., Lawrence Livermore National Laboratory report UCAR 10062-89, Livermore, CA, 1990, p. 24.
39. H.J. Kreuzer, *Nonequilibrium Thermodynamics and Its Statistical Foundations*, Clarendon Press, London, U.K., 1981.
40. B.J. Wood and D.G. Fraser, *Elementary Thermodynamics for Geologists*, Oxford University Press, New York, 1977.
41. U.S. Department of Energy Nevada Operations Office, United States nuclear tests July 1945 through September 1992, Department of Energy Report DOE/NV-209-REV 15, December 2000.
42. J. Arnold, Condensation and agglomeration of grains, in *Comets, Asteroids, Meteorites: Interrelations, Evolution, and Origins*, A.H. Desemme, ed., University of Toledo, Toledo, OH, 1977.
43. N.A. Fuchs, *The Mechanics of Aerosols*, MacMillan, New York, 1964.
44. S.K. Friedlander, *Smoke, Dust, and Haze*, John Wiley & Sons, New York, 1977.
45. Stockholm International Peace Research Institute, *Nuclear Radiation in Warfare*, Taylor & Francis, London, U.K., 1981.
46. M.W. Nathans, R. Thews, W.D. Holland, and P.A. Benson, Particle size distribution in clouds from nuclear airbursts, *J. Geophys. Res.*, 75, 7559, 1970.
47. F. Belloni, J. Himbert, O. Marzocchi, and V. Romanello, Investigating incorporation and distribution of radionuclides in trinitite, *J. Environ. Radioact.*, 102, 852, 2011.
48. J.J. Bellucci, A. Simonetti, C. Wallace, E.C. Koeman, and P.C. Burns, Isotopic fingerprinting of the world's first nuclear device using post-detonation materials, *Anal. Chem.*, 85, 4195, 2013.
49. R.E. Hermes, A new look at trinitite, *Nuclear Weapons Journal*, Vol. 2, Los Alamos National Laboratory Report LALP-05-067, Los Alamos, NM, 2005, p. 2.
50. J. Mackin, P. Zigman, D. Love, D. Macdonald, and D. Sam, Radiochemical analysis of individual fallout particles, *J. Inorg. Nucl. Chem.*, 15, 20, 1960.
51. C.S. Cook, R.L. Mather, R.F. Johnson, and F.M. Tomnovec, Fractionation of nuclear weapon debris, *Nature*, 187, 1100, 1960.
52. J. Sisefsky, Debris from tests of nuclear weapons, *Science*, 133, 735, 1961.
53. G.R. Crocker, J.D. O'Connor, and E.C. Freiling, Physical and radiochemical properties of fallout particles, *Health Phys.*, 12, 1099, 1966.

54. K.P. Makhon'ko and S.G. Malakhov, eds., *Nuclear Meteorology*, Israel Program of Scientific Translation, Jerusalem, Israel, 1977.
55. C.F. Miller, *Biological and Radiological Effects of Fallout from Nuclear Explosions*, Stanford Research Institute, Menlo Park, CA, March 1964.
56. W. Bleeker, *Meteorological Factors Influencing the Transport and Removal of Radioactive Debris*, Technical Note #43, WMO-No.111.TP.49, World Meteorological Organization, Geneva, Switzerland, 1961.
57. J.B. Knox, Prediction of fallout from subsurface nuclear detonations, in *Radioactive Fallout from Nuclear Weapons Tests, Proceedings of the Second Conference*, A.W. Klement, Jr., ed., Germantown, MD, U.S. Atomic Energy Commission, November 1965, p. 331.
58. E.C. Freiling, G.R. Crocker, and C.E. Adams, Nuclear debris formation, in *Radioactive Fallout from Nuclear Weapons Tests, Proceedings of the Second Conference*, A.W. Klement, Jr., ed., Germantown, MD, U.S. Atomic Energy Commission, November 1965, p. 1.

6 Chronometry

Radionuclides that are linked to one another by the processes of radioactive decay have relative concentrations that can be calculated by the simple laws of radioactive ingrowth or, in more complicated cases, by the Bateman equations. If a time exists at which all descendant radionuclides have been removed from a parent material, that time can be determined through radiochemical separations and the subsequent measurement of the relative concentrations of mothers and daughters at a later time. The interval between the time that the sample was purified and the time that it was subsequently analyzed is defined as the "age" of the material at the analysis time. However, this technique does not apply when the half-lives of the descendant nuclides involved in the determination become significantly shorter than the elapsed time.

Chronometry has been successfully applied to geochemical systems [1]. For instance, under the assumption that actinides are incorporated into zircons while the lead isotopes are not, the Pb concentration can be used to determine the age of a zircon crystal, sensitive to time scales on the order of the age of the earth. In mixed-lanthanide minerals, an enhanced concentration of ^{143}Nd relative to the other Nd Isotopes can be attributed to the α decay of primordial 1×10^{11}-year ^{147}Sm over geological time scales.

Interestingly, the radiometric age-dating technique with which we are most familiar, involving the measurement of ^{14}C in biological or paleontological specimens, is not chronometry by our definition [2–4]. Neutrons that are the by-product of cosmic-ray interactions in the atmosphere react with nitrogen to produce the 5730-year ^{14}C activity, which is incorporated into atmospheric CO_2 and hence metabolized into the biosphere. Assuming that cosmogenic production occurs at a steady state and the composition of the atmosphere does not change, the fractional amount of atmospheric $^{14}CO_2$ is approximately 10^{-12}, with the isotope decaying at the same rate at which it is produced. When life ends, atmospheric exchange ends, and for a sequestered sample, the ^{14}C concentration decays with its characteristic half-life. Unlike chronometry, which depends on the ingrowth of a daughter activity, ^{14}C-dating relies on measuring the concentration of the isotope relative to its inferred reference concentration, and it is useful for measuring ages up to approximately 70,000 years. Age determination in ^{14}C-dating relies on assumptions that are different from those that must be made for chronometry. One of the most important is the assumption of steady-state production of ^{14}C relative to the other isotopes of carbon. But the increased consumption of fossil fuels caused a dilution of a few percent of ^{14}C in the first half of the twentieth century, and during the decade between the early 1950s and early 1960s, atmospheric explosions of thermonuclear bombs doubled the ^{14}C content of atmospheric CO_2.

6.1 HEAVY ELEMENTS AND FISSION-PRODUCT CHRONOMETERS

Figure 6.1 gives the "4n + 2" chronometric relationships among the heavy-element nuclides, so named because in each case the mass number divided by 4 leaves a remainder of 2. There is some overlap between the information shown in this figure and that found in Figure 2.17 on the physical basis of nuclear forensic science. The processes of α, β, and IT decays cannot produce a nuclide from a member of the grid that is not a 4n + 2 nuclide. The decay network begins with 87.7-year ^{238}Pu and proceeds through the ingrowth of long-lived ^{234}U, ^{230}Th, and ^{226}Ra. Subsequent decays of short-lived ^{222}Rn, ^{218}Po, ^{214}Pb, ^{214}Bi, and ^{214}Po result in 22.3-year ^{210}Pb. For samples older than a few weeks, the short-lived species are not useful chronometers and can be considered to be in approximate secular equilibrium with ^{226}Ra. If any member of the 4n + 2 decay chain is purified, decay processes will immediately begin to produce descendant species; in a purified U sample, the ^{238}Pu concentration is zero and remains zero because ^{238}Pu is a decay precursor of ^{234}U and not vice versa (being a higher member of the decay chain). The ratio of any two concentrations among the decaying nuclides can be calculated from the elapsed time between purification and any subsequent time (given the known half-lives and the Bateman equations). It therefore follows that the elapsed time can be calculated from the measured isotope ratio.

Table 6.1 presents the results of decay calculations among the 4n + 2 nuclides. Each section of the table starts with a different long-lived member of the decay chain presented in Figure 6.1 and gives the atom ratio of the descendant chronometric nuclides

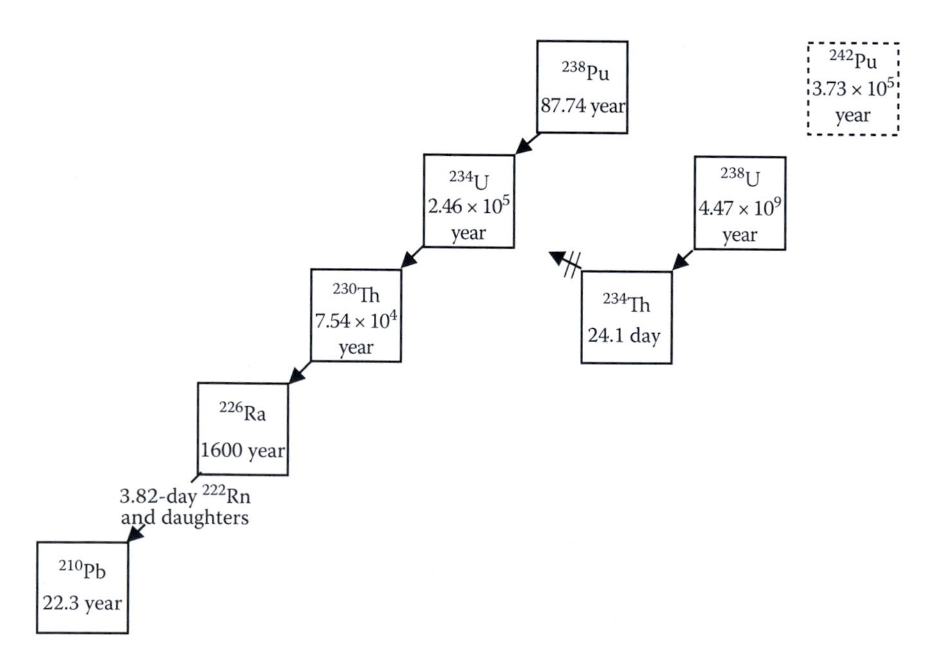

FIGURE 6.1 A radionuclide box diagram showing the chronometric relationships among the 4n + 2 nuclides. These data were used in the construction of Table 6.1. ^{242}Pu and ^{238}U are 4n + 2 nuclides, but their decay properties limit their use in chronometry for anthropogenic processes.

TABLE 6.1
4n + 2 Chronometers

^{238}Pu Chronometers, to be used with ingrown analytes in Pu samples

Elapsed Days	$^{234}U/^{238}Pu$	$^{230}Th/^{238}Pu$	$^{226}Ra/^{238}Pu$	$^{210}Pb/^{238}Pu$
0	0.0000E+00	0.0000E+00	0.0000E+00	0.0000E+00
1	2.1629E–05	8.3893E–14	0.0000E+00	0.0000E+00
3	6.4888E–05	7.5517E–13	0.0000E+00	0.0000E+00
10	2.1631E–04	8.3918E–12	0.0000E+00	0.0000E+00
20	4.3267E–04	3.3571E–11	4.9391E–18	0.0000E+00
40	8.6552E–04	1.3433E–10	4.5667E–17	0.0000E+00
70	1.5152E–03	4.1155E–10	2.4157E–16	0.0000E+00
100	2.1652E–03	8.4026E–10	7.0546E–16	2.5525E–20
150	3.2496E–03	1.8919E–09	2.3821E–15	1.1130E–19
200	4.3351E–03	3.3659E–09	5.6500E–15	3.3425E–19
300	6.5097E–03	7.5843E–09	1.9100E–14	1.6909E–18
400	8.6890E–03	1.3502E–08	4.5344E–14	5.3473E–18
550	1.1967E–02	2.5584E–08	1.1816E–13	1.9119E–17
700	1.5255E–02	4.1531E–08	2.4418E–13	5.0163E–17
850	1.8554E–02	6.1370E–08	4.3824E–13	1.0905E–16
1,000	2.1864E–02	8.5125E–08	7.1531E–13	2.0891E–16
1,500	3.2975E–02	1.9292E–07	2.4335E–12	1.0578E–15
2,000	4.4207E–02	3.4546E–07	5.8144E–12	3.3437E–15
3,000	6.7037E–02	7.8861E–07	1.9940E–11	1.6935E–14
4,000	9.0367E–02	1.4224E–06	4.8024E–11	5.3559E–14
5,500	1.2632E–01	2.7486E–06	1.2787E–10	1.9169E–13
7,000	1.6346E–01	4.5506E–06	2.7004E–10	5.0382E–13
8,500	2.0182E–01	6.8584E–06	4.9529E–10	1.0976E–12
10,000	2.4145E–01	9.7034E–06	8.2616E–10	2.1078E–12
15,000	3.8323E–01	2.3501E–05	3.0224E–09	1.0784E–11
20,000	5.4118E–01	4.4999E–05	7.7686E–09	3.4572E–11
30,000	9.1325E–01	1.1767E–04	3.0868E–08	1.8196E–10
40,000	1.3751E+00	2.4371E–04	8.6270E–08	6.0529E–10
55,000	2.2851E+00	5.8180E–04	2.8792E–07	2.3834E–09
70,000	3.5438E+00	1.1958E–03	7.6439E–07	7.0312E–09

^{234}U Chronometers, to be used with ingrown analytes in U samples

Elapsed Days	$^{230}Th/^{234}U$	$^{226}Ra/^{234}U$	$^{210}Pb/^{234}U$
0	0.0000E+00	0.0000E+00	0.0000E+00
1	7.7587E–09	9.7339E–17	0.0000E+00
3	2.3276E–08	8.7842E–16	0.0000E+00
10	7.7587E–08	9.7642E–15	0.0000E+00
20	1.5517E–07	3.9058E–14	1.4677E–19
40	3.1035E–07	1.5624E–13	1.6996E–18
70	5.4311E–07	4.7846E–13	1.0556E–17
100	7.7587E–07	9.7644E–13	3.2842E–17

(Continued)

TABLE 6.1 (*Continued*)
4n + 2 Chronometers

^{234}U Chronometers, to be used with ingrown analytes in U samples

Elapsed Days	$^{230}Th/^{234}U$	$^{226}Ra/^{234}U$	$^{210}Pb/^{234}U$
130	1.0086E–06	1.6502E–12	7.4744E–17
170	1.3190E–06	2.8218E–12	1.7188E–16
220	1.7069E–06	4.7257E–12	3.8015E–16
300	2.3276E–06	8.7873E–12	9.8133E–16
400	3.1035E–06	1.5621E–11	2.3528E–15
550	4.2673E–06	2.9532E–11	6.1651E–15
700	5.4311E–06	4.7834E–11	1.2750E–14
850	6.5949E–06	7.0527E–11	2.2850E–14
1,000	7.7587E–06	9.7609E–11	3.7197E–14
1,300	1.0086E–05	1.6494E–10	8.1512E–14
1,700	1.3189E–05	2.8201E–10	1.8128E–13
2,200	1.7069E–05	4.7220E–10	3.8964E–13
3,000	2.3275E–05	8.7778E–10	9.7357E–13
4,000	3.1034E–05	1.5598E–09	2.2637E–12
5,500	4.2671E–05	2.9474E–09	5.7142E–12
7,000	5.4308E–05	4.7714E–09	1.1441E–11
8,500	6.5945E–05	7.0313E–09	1.9898E–11
10,000	7.7580E–05	9.7260E–09	3.1488E–11
13,000	1.0085E–04	1.6418E–08	6.5425E–11
17,000	1.3188E–04	2.8030E–08	1.3621E–10
22,000	1.7066E–04	4.6850E–08	2.7116E–10
30,000	2.3270E–04	8.6841E–08	6.0602E–10
40,000	3.1024E–04	1.5378E–07	1.2451E–09
55,000	4.2652E–04	2.8901E–07	2.6802E–09
70,000	5.4277E–04	4.6539E–07	4.6905E–09

^{230}Th Chronometers, to be used with ingrown analytes in Th samples

Elapsed Days	$^{226}Ra/^{230}Th$	$^{210}Pb/^{230}Th$
0	0.0000E+00	0.0000E+00
1	2.5171E–08	8.9981E–16
3	7.5513E–08	2.2215E–14
10	2.5171E–07	6.2366E–13
20	5.0342E–07	3.6321E–12
40	1.0068E–06	1.8403E–11
70	1.7619E–06	6.2863E–11
100	2.5170E–06	1.3406E–10
130	3.2720E–06	2.3193E–10
170	4.2787E–06	4.0378E–10
220	5.5370E–06	6.8485E–10
300	7.5501E–06	1.2869E–09
400	1.0066E–05	2.3014E–09

(*Continued*)

TABLE 6.1 (*Continued*)
4n + 2 Chronometers

^{230}Th Chronometers, to be used with ingrown analytes in Th samples

Elapsed Days	$^{226}Ra/^{230}Th$	$^{210}Pb/^{230}Th$
550	1.3840E–05	4.3639E–09
700	1.7612E–05	7.0677E–09
850	2.1385E–05	1.0404E–08
1,000	2.5157E–05	1.4366E–08
1,300	3.2698E–05	2.4133E–08
1,700	4.2749E–05	4.0884E–08
2,200	5.5306E–05	6.7615E–08
3,000	7.5382E–05	1.2313E–07
4,000	1.0045E–04	2.1317E–07
5,500	1.3800E–04	3.8735E–07
7,000	1.7548E–04	6.0337E–07
8,500	2.1290E–04	8.5610E–07
10,000	2.5025E–04	1.1411E–06
13,000	3.2477E–04	1.7921E–06
17,000	4.2371E–04	2.7920E–06
22,000	5.4675E–04	4.1906E–06
30,000	7.4213E–04	6.6359E–06
40,000	9.8381E–04	9.8674E–06
55,000	1.3412E–03	1.4820E–05
70,000	1.6923E–03	1.9759E–05

^{226}Ra Chronometer, to be used with ingrown analytes in Ra samples

Elapsed Days	$^{210}Pb/^{226}Ra$
0	0.0000E+00
1	1.0560E–07
3	8.4413E–07
10	6.5422E–06
20	1.7599E–05
40	4.1148E–05
70	7.6602E–05
100	1.1197E–04
130	1.4725E–04
170	1.9415E–04
220	2.5256E–04
300	3.4549E–04
400	4.6080E–04
550	6.3195E–04
700	8.0095E–04
850	9.6785E–04
1,000	1.1326E–03
1,300	1.4561E–03

(*Continued*)

TABLE 6.1 (*Continued*)
4n + 2 Chronometers

^{226}Ra Chronometer, to be used with ingrown analytes in Ra samples

Elapsed Days	$^{210}Pb/^{226}Ra$
1,700	1.8749E–03
2,200	2.3790E–03
3,000	3.1429E–03
4,000	4.0283E–03
5,500	5.2247E–03
7,000	6.2794E–03
8,500	7.2096E–03
10,000	8.0296E–03
13,000	9.3903E–03
17,000	1.0746E–02
22,000	1.1910E–02
30,000	1.3002E–02
40,000	1.3650E–02
55,000	1.4003E–02
70,000	1.4103E–02

^{238}U Chronometer (short), to be used with ingrown analytes in U samples; useful only for short elapsed times

Elapsed Days	$^{234}Th/^{238}U$
0	0.0000E+00
0.08333	3.5353E–14
0.166667	7.0621E–14
0.291667	1.2337E–13
0.5	2.1085E–13
0.75	3.1515E–13
1	4.1869E–13
1.333333	5.5660E–13
1.666667	6.9121E–13
2	8.2551E–13
2.5	1.0246E–12
3	1.2208E–12
4	1.6049E–12
5	1.9781E–12
6.5	2.5181E–12
8	3.0354E–12
10	3.6912E–12
12	4.3104E–12
15	5.1749E–12
19	6.2174E–12
24	7.3627E–12
30	8.5364E–12

(*Continued*)

TABLE 6.1 (*Continued*)
4n + 2 Chronometers

^{238}U Chronometer (short), to be used with ingrown analytes in U samples; useful only for short elapsed times	
Elapsed Days	**$^{234}Th/^{238}U$**
36	9.5241E–12
43	1.0480E–11
51	1.1362E–11
59	1.2062E–11
67	1.2618E–11

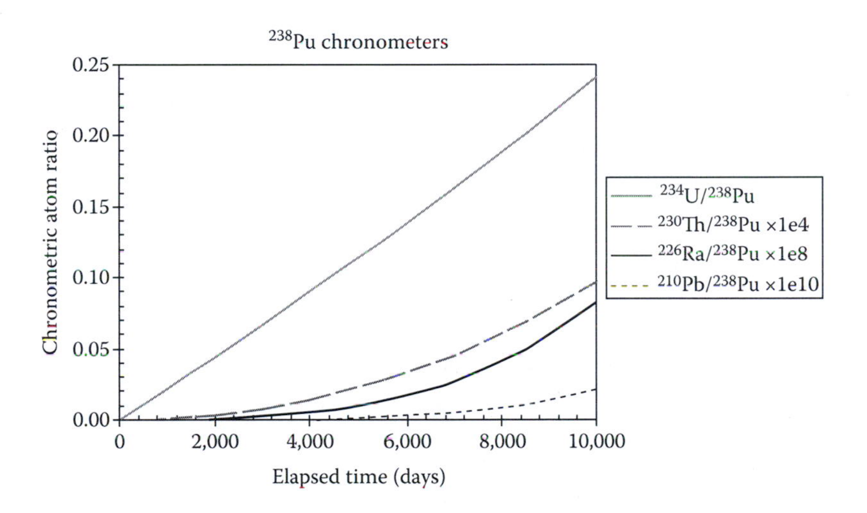

FIGURE 6.2 The concentration of each long-lived descendant nuclide arising from the decay of ^{238}Pu, expressed as an atom ratio relative to ^{238}Pu, as a function of time after separation/purification.

to the parent activity as a function of time. The time associated with any given ratio of activities may be obtained through interpolation. Figure 6.2 depicts the chronometric relationships between ^{238}Pu and its descendants as a function of age (i.e., time since last purification). The first-order chronometer ($^{234}U/^{238}Pu$) is reasonably linear with time: when the half-lives of both parent and daughter are long compared to the ingrowth interval, the number of parent atoms remains fairly constant over the interval, and the number of daughter atoms is roughly equal to the parent's constant decay rate multiplied by the elapsed time. The chronometers of the second order and higher ($^{230}Th/^{238}Pu$, $^{226}Ra/^{238}Pu$, and $^{210}Pb/^{238}Pu$) are decidedly nonlinear: the number of atoms of the radionuclide decaying to produce the daughter member of the pair is not constant, and the behavior of the chronometric pair is therefore nonlinear over time. For the purpose of interpolation, the values given in Table 6.1 can be used at face value; for improved accuracy, however, it is better to interpolate between table values for the first-order chronometric pairs in each subtable and to interpolate between the logarithms of the table values for second-order and higher chronometric pairs.

The values in the table can be used to construct a table for other isotope ratios. For example, if an analyst working with a sample of ^{238}Pu (which requires the use of the first section of Table 6.1) measures the concentrations of ^{234}U and ^{230}Th before measuring the ^{238}Pu concentration, he or she can construct the relevant table by dividing the values in the ^{230}Th/^{238}Pu column by the corresponding values in the ^{234}U/^{238}Pu column to construct a ^{230}Th/^{234}U chronometric set that is valid for Pu. This set can then be used to derive the age of the sample.

There are several minor decay branches among the A = 218, 214, and 210 members of the decay chain that are not depicted in Figure 6.1. All of these minor branches are small (~0.1% or less), all populate short-lived species, and every decay path eventually results in ^{210}Pb. When Table 6.1 was constructed, it did not include components for any of these nuclides in the Bateman equations, assuming that they would be in equilibrium with 3.8-day ^{222}Rn (which was included in the calculation).

Although the very important 4n + 2 nuclide, ^{238}U, is included in Figure 6.1, it is shown as detached from the rest of the decay chain. It decays through ^{234}Th to ^{234}U via the short-lived ^{234}Pa isomer. The nuclide ^{238}U is almost valueless as a chronometer. The ratio of 24-day ^{234}Th to ^{238}U is useful for age-dating materials that are less than 6 months old, but it is likely to have limited value for problems investigated by the nuclear forensic analyst. Even so, chronometric data for ^{234}Th/^{238}U are given at the end of Table 6.1. The ratio ^{234}U/^{238}U does not provide useful chronometric information because ^{234}U is not removed from ^{238}U by any chemical process. The quantity of ^{234}U that is residual in even the most highly depleted of U samples is much more than can grow into the material over thousands of years. We have also neglected the 4n + 2 nuclide ^{242}Pu, which is a minor component in most samples of mixed Pu isotopes. Its daughter is ^{238}U, but the long half-lives of the two nuclides, combined with the fact that ^{238}U occurs in nature in background quantities, make it reasonably likely that some fraction was introduced into the Pu material as a contaminant during purification. The ^{238}U/^{242}Pu chronometric pair is therefore of dubious value.

In Figure 6.3, the 4n + 3 chronometric relationships among the heavy-element nuclides are shown. These data share some overlap with the information in Figure 2.18. The decay network starts with 24,110-year ^{239}Pu and proceeds through the ingrowth of ^{235}U, ^{231}Th (too short-lived to provide chronometric information), ^{231}Pa, and ^{227}Ac. The chronometric sequence terminates at 22-year ^{227}Ac because subsequent decay daughters are too short-lived to be of interest for age dating, ending in stable ^{207}Pb. However, the decay-chain members 18.7-day ^{227}Th and 11.4-day ^{223}Ra are useful nuclides for the determination of ^{227}Ac through radiochemical milking (discussed later).

Table 6.2 presents the results of decay calculations among the 4n + 3 nuclides. Each section of the table begins with a different long-lived member of the decay chain presented in Figure 6.3 and gives the atom ratio of the descendant chronometric nuclides to the parent activity as a function of time. In Figure 6.4 are plotted the ^{239}Pu chronometric relationships as a function of the age of the sample. As before, the values in the table can be used to interpolate an age, either with the numbers "as-is" or logarithmically for second-order and higher chronometric pairs. Again, if the parent was not measured, the chronometric information can be assembled by constructing

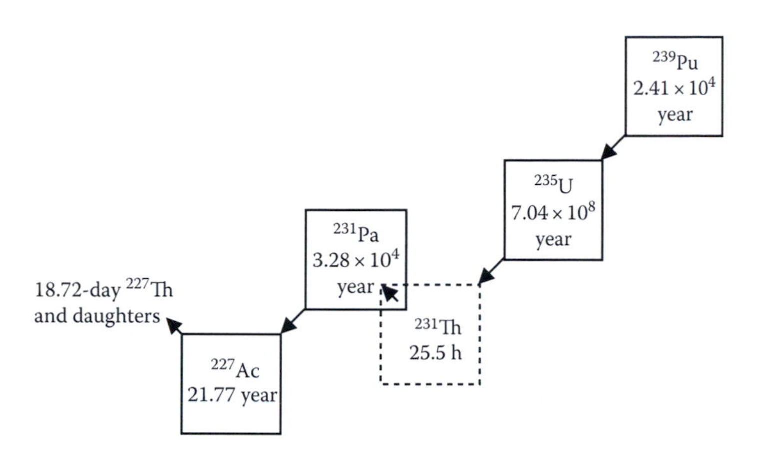

FIGURE 6.3 A radionuclide box diagram showing the chronometric relationships among the 4n + 3 nuclides. These data were used in the construction of Table 6.2.

ratios from the values in Table 6.2 for two nuclides relative to the parent [thus, e.g., for Pu material, $^{227}Ac/^{235}U = (^{227}Ac/^{239}Pu)/(^{235}U/^{239}Pu)$ from the first section of the table].

At the end of Table 6.2, chronometric values for the ingrowth of 25-h ^{231}Th in a sample of ^{235}U are included. Although not of chronometric interest, the ingrowth of ^{231}Th toward radioactive equilibrium is an important factor influencing its use as a chemical-yield indicator, linking the yield of Th to that of U. It is included here for completeness.

In Figure 6.5, the 4n chronometric relationships among the heavy-element nuclides are depicted. Again, there is some overlap with the information shown in Figure 2.19. This decay network is more complicated than those discussed previously. The grid actually comprises three semi-independent chronometric sequences. The first is the decay of ^{240}Pu to ^{236}U. The decay of ^{236}U to ^{232}Th does not provide chronometric information for the same reasons that the decay of ^{242}Pu to ^{238}U does not: the half-life of ^{236}U is very long, ^{232}Th is a naturally occurring radionuclide that was almost certainly present in the facility in which the sample was purified, and it was most likely residual in the purified mother material. The second chronometric sequence is the decay of ^{232}Th to ^{228}Ra, which decays in turn through short-lived ^{228}Ac to ^{228}Th. The ^{232}Th sequence terminates at ^{228}Ra for the same reason that the ^{238}U sequence terminates at ^{234}Th after one step. Any sample containing ^{232}Th will also contain trace amounts of ^{228}Th, and a radiochemical purification will not separate the two. However, introduction of disequilibrium between ^{232}Th and ^{228}Th by removal of ^{228}Ra can provide chronometric information, which will be discussed later.

The third independent 4n chronometric decay chain begins with 2.86-year ^{236}Pu and proceeds through the ingrowth of long-lived ^{232}U and ^{228}Th. As before, the descendants of 3.66-day ^{224}Ra are all short-lived until the decay sequence arrives at stable ^{208}Pb, so the effective end of the chronometric chain is ^{228}Th. The results of decay calculations among the ^{236}Pu descendants are included in Table 6.3; for completeness, we also include the ingrowth of ^{224}Ra in a sample of ^{228}Th, which is of value in designing radiochemical milking experiments. The values in Table 6.3 are used analogously to those in Tables 6.1 and 6.2.

TABLE 6.2
4n + 3 Chronometers

^{239}Pu Chronometers, to be used with ingrown analytes in Pu samples

Elapsed Days	$^{235}U/^{239}Pu$	$^{231}Pa/^{239}Pu$	$^{227}Ac/^{239}Pu$
0	0.0000E+00	0.0000E+00	0.0000E+00
1	7.8682E–08	2.6128E–20	0.0000E+00
3	2.3604E–07	4.1145E–19	0.0000E+00
10	7.8682E–07	7.8503E–18	0.0000E+00
20	1.5736E–06	3.6434E–17	0.0000E+00
40	3.1473E–06	1.5722E–16	0.0000E+00
70	5.5078E–06	4.9750E–16	0.0000E+00
100	7.8682E–06	1.0288E–15	0.0000E+00
130	1.0229E–05	1.7509E–15	4.3230E–21
170	1.3376E–05	3.0108E–15	9.7431E–21
220	1.7310E–05	5.0632E–15	2.1239E–20
300	2.3604E–05	9.4500E–15	5.4052E–20
400	3.1473E–05	1.6844E–14	1.2832E–19
550	4.3276E–05	3.1910E–14	3.3354E–19
700	5.5079E–05	5.1753E–14	6.8667E–19
850	6.6882E–05	7.6368E–14	1.2269E–18
1,000	7.8685E–05	1.0576E–13	1.9930E–18
1,300	1.0229E–04	1.7886E–13	4.3551E–18
1,700	1.3377E–04	3.0603E–13	9.6644E–18
2,200	1.7311E–04	5.1274E–13	2.0738E–17
3,000	2.3608E–04	9.5383E–13	5.1736E–17
4,000	3.1478E–04	1.6961E–12	1.2016E–16
5,500	4.3285E–04	3.2077E–12	3.0302E–16
7,000	5.5092E–04	5.1968E–12	6.0631E–16
8,500	6.6902E–04	7.6635E–12	1.0541E–15
10,000	7.8713E–04	1.0608E–11	1.6677E–15
13,000	1.0233E–03	1.7930E–11	3.4639E–15
17,000	1.3385E–03	3.0668E–11	7.2105E–15
22,000	1.7325E–03	5.1372E–11	1.4358E–14
30,000	2.3633E–03	9.5557E–11	3.2116E–14
40,000	3.1522E–03	1.6993E–10	6.6100E–14
55,000	4.3368E–03	3.2146E–10	1.4281E–13
70,000	5.5229E–03	5.2097E–10	2.5107E–13

^{235}U Chronometers, to be used with ingrown analytes in U samples

Elapsed Days	$^{231}Pa/^{235}U$	$^{227}Ac/^{235}U$
0	0.0000E+00	0.0000E+00
1	7.1549E–13	1.4516E–20

(*Continued*)

TABLE 6.2 (*Continued*)
4n + 3 Chronometers

^{235}U Chronometers, to be used with ingrown analytes in U samples

Elapsed Days	$^{231}Pa/^{235}U$	$^{227}Ac/^{235}U$
3	4.5386E–12	2.9928E–19
10	2.2835E–11	5.7726E–18
20	4.9793E–11	2.6769E–17
40	1.0372E–10	1.1547E–16
70	1.8461E–10	3.6511E–16
100	2.6550E–10	7.5432E–16
130	3.4640E–10	1.2827E–15
170	4.5425E–10	2.2032E–15
220	5.8907E–10	3.6996E–15
300	8.0478E–10	6.8890E–15
400	1.0744E–09	1.2243E–14
550	1.4789E–09	2.3095E–14
700	1.8833E–09	3.7294E–14
850	2.2878E–09	5.4795E–14
1,000	2.6922E–09	7.5556E–14
1,300	3.5010E–09	1.2669E–13
1,700	4.5795E–09	2.1431E–13
2,200	5.9276E–09	3.5402E–13
3,000	8.0844E–09	6.4397E–13
4,000	1.0780E–08	1.1139E–12
5,500	1.4824E–08	2.0224E–12
7,000	1.8867E–08	3.1485E–12
8,500	2.2910E–08	4.4655E–12
10,000	2.6952E–08	5.9499E–12
13,000	3.5036E–08	9.3414E–12
17,000	4.5812E–08	1.4551E–11
22,000	5.9279E–08	2.1846E–11
30,000	8.0818E–08	3.4635E–11
40,000	1.0773E–07	5.1633E–11
55,000	1.4806E–07	7.7963E–11
70,000	1.8836E–07	1.0461E–10

^{231}Pa Chronometer, to be used with ingrown analytes in Pa samples

Elapsed Days	$^{227}Ac/^{231}Pa$
0	0.0000E+00
1	5.7855E–08
3	1.7355E–07
10	5.7833E–07
20	1.1561E–06
40	2.3103E–06
70	4.0377E–06

(*Continued*)

TABLE 6.2 (*Continued*)
4n + 3 Chronometers

^{231}Pa Chronometer, to be used with ingrown analytes in Pa samples

Elapsed Days	$^{227}Ac/^{231}Pa$
100	5.7607E–06
130	7.4791E–06
170	9.7634E–06
220	1.2607E–05
300	1.7132E–05
400	2.2744E–05
550	3.1071E–05
700	3.9291E–05
850	4.7402E–05
1,000	5.5409E–05
1,300	7.1113E–05
1,700	9.1422E–05
2,200	1.1584E–04
3,000	1.5275E–04
4,000	1.9541E–04
5,500	2.5283E–04
7,000	3.0321E–04
8,500	3.4743E–04
10,000	3.8622E–04
13,000	4.5015E–04
17,000	5.1311E–04
22,000	5.6645E–04
30,000	6.1548E–04
40,000	6.4380E–04
55,000	6.5865E–04
70,000	6.6267E–04

^{235}U Chronometer (short), to be used with ingrown analytes in U samples; useful only for short elapsed times

Elapsed Days	$^{231}Th/^{235}U$
0	0.0000E+00
0.08333	2.1870E–13
0.166667	4.2584E–13
0.291667	7.1619E–13
0.5	1.1505E–12
0.75	1.5995E–12
1	1.9809E–12
1.333333	2.4018E–12
1.666667	2.7404E–12
2	3.0129E–12

(*Continued*)

TABLE 6.2 (*Continued*)
4n + 3 Chronometers

^{235}U Chronometer (short), to be used with ingrown analytes in U samples; useful only for short elapsed times	
Elapsed Days	**^{231}Th/^{235}U**
2.5	3.3252E–12
3	3.5506E–12
4	3.8307E–12
5	3.9767E–12
6.5	4.0757E–12
8	4.1129E–12
10	4.1293E–12
12	4.1337E–12
15	4.1351E–12
19	4.1354E–12
24	4.1354E–12
30	4.1354E–12

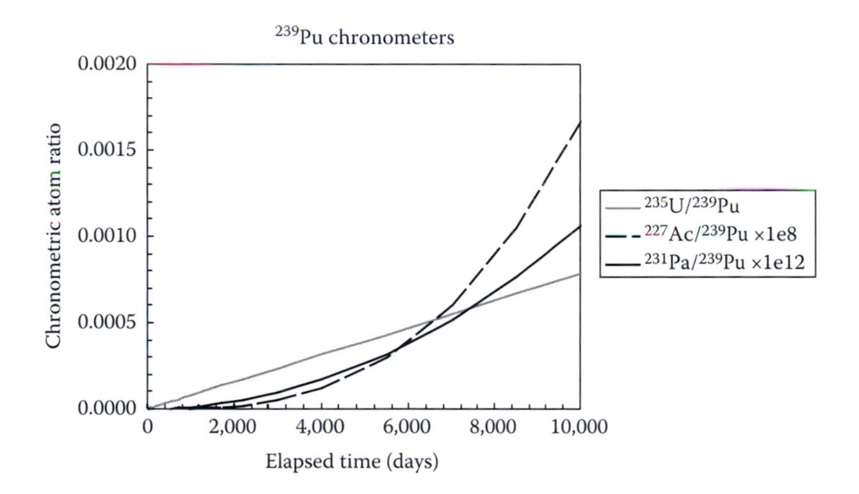

FIGURE 6.4 The concentration of each long-lived descendant nuclide arising from the decay of ^{239}Pu, expressed as an atom ratio relative to ^{239}Pu, as a function of time after separation/purification.

Figure 6.6 shows the 4n + 1 chronometric relationships among the heavy-element nuclides. There is some redundancy between the data in this figure and those of Figure 2.20. The decay network begins with 14.4-year ^{241}Pu and proceeds through long-lived ^{241}Am, ^{237}Np, ^{233}U, and ^{229}Th. The chronometric sequence ends with ^{229}Th because the subsequent decay daughters are too short-lived to be of interest to age

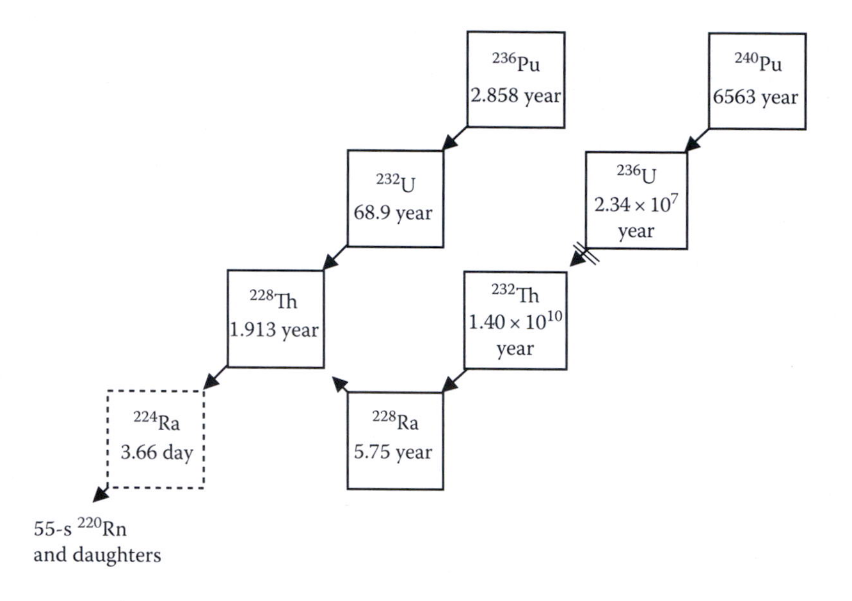

FIGURE 6.5 A radionuclide box diagram showing the chronometric relationships among the 4n nuclides. These values were used in the construction of Table 6.3. The decay of ^{236}U cannot be used as a chronometer because of parent and daughter half-lives and blank levels.

dating, and the sequence is terminated at stable ^{209}Bi. However, decay-chain members 14.9-day ^{225}Ra and 10-day ^{225}Ac are useful nuclides for the determination of ^{229}Th via radiochemical milking.

The 4n + 1 series is of special interest for several reasons. First, it contains chronometric nuclides that can be measured nondestructively; it is common to calculate the age of a Pu sample through the relative concentrations of ^{241}Am and ^{241}Pu determined by γ-ray spectrometry. Second, the decay of ^{241}Pu proceeds 99.998% of the time by β emission to 432-year ^{241}Am and (0.00245 ± 0.00002)% of the time through emission of an α particle to 6.75-day ^{237}U. In most cases, temptation would be to ignore the latter decay branch as unimportant, but the short half-life of the ^{237}U intermediate results in a significant effect on the production of ^{237}Np and on subsequent decay products at short times after the chemical purification of Pu [5]. In other words, even though atoms of ^{241}Am are being produced at a rate 41,000 times faster than the production of ^{237}U atoms, the fact that ^{237}U decays to ^{237}Np 23,000 times faster than does longer-lived ^{241}Am makes ^{237}Np production from ^{237}U comparable with that from ^{241}Am, until the approach to secular equilibrium limits the number of ^{237}U atoms in the sample. In Figure 6.7, the ratios of the number of atoms of the chronometric decay products ^{237}Np and ^{233}U, produced through the ^{237}U decay intermediate, to the total number in the sample are plotted as a function of time.

In Table 6.4, the results of decay calculations among the 4n + 1 nuclides are presented, including the effect of the ^{241}Pu α-decay branch in the first section of the table (which does not apply to the later sections of the table for purified samples of ^{241}Am, ^{237}Np, and so forth). The time associated with a given ratio of

TABLE 6.3
4n Chronometers

^{240}Pu Chronometer, to be used with ingrown analytes in Pu samples	
Elapsed Days	**$^{236}U/^{240}Pu$**
0	0.0000E+00
1	2.8915E–07
3	8.6746E–07
10	2.8915E–06
20	5.7831E–06
40	1.1566E–05
70	2.0240E–05
100	2.8916E–05
130	3.7591E–05
170	4.9157E–05
220	6.3615E–05
300	8.6750E–05
400	1.1566E–04
550	1.5905E–04
700	2.0243E–04
850	2.4581E–04
1,000	2.8919E–04
1,300	3.7597E–04
1,700	4.9168E–04
2,200	6.3634E–04
3,000	8.6784E–04
4,000	1.1572E–03
5,500	1.5916E–03
7,000	2.0261E–03
8,500	2.4608E–03
10,000	2.8957E–03
13,000	3.7660E–03
17,000	4.9276E–03
22,000	6.3816E–03
30,000	8.7123E–03
40,000	1.1633E–02
55,000	1.6031E–02
70,000	2.0447E–02

^{232}Th Chronometer, to be used with ingrown analytes in Th samples	
Elapsed Days	**$^{228}Ra/^{232}Th$**
0	0.0000E+00
1	1.3505E–13
3	4.0501E–13
10	1.3485E–12
20	2.6925E–12

(Continued)

TABLE 6.3 (*Continued*)
4n Chronometers

^{232}Th Chronometer, to be used with ingrown analytes in Th samples

Elapsed Days	^{228}Ra/^{232}Th
40	5.3673E–12
70	9.3466E–12
100	1.3287E–11
130	1.7188E–11
170	2.2330E–11
220	2.8662E–11
300	3.8580E–11
400	5.0614E–11
550	6.7937E–11
700	8.4422E–11
850	1.0011E–10
1,000	1.1504E–10
1,300	1.4278E–10
1,700	1.7574E–10
2,200	2.1126E–10
3,000	2.5721E–10
4,000	2.9995E–10
5,500	3.4263E–10
7,000	3.6865E–10
8,500	3.8451E–10
10,000	3.9417E–10
13,000	4.0366E–10
17,000	4.0777E–10
22,000	4.0898E–10
30,000	4.0924E–10
40,000	4.0926E–10
55,000	4.0926E–10
70,000	4.0926E–10

^{228}Ra Chronometer, to be used with ingrown analytes in Ra samples

Elapsed Days	^{228}Th/^{228}Ra
0	0.0000E+00
1	3.2994E–04
3	9.8916E–04
10	3.2896E–03
20	6.5574E–03
40	1.3029E–02
70	2.2577E–02
100	3.1936E–02
130	4.1112E–02
170	5.3066E–02

(*Continued*)

TABLE 6.3 (*Continued*)
4n Chronometers

^{228}Ra Chronometer, to be used with ingrown analytes in Ra samples

Elapsed Days	^{228}Th/^{228}Ra
220	6.7571E–02
300	8.9803E–02
400	1.1599E–01
550	1.5215E–01
700	1.8490E–01
850	2.1455E–01
1,000	2.4140E–01
1,300	2.8772E–01
1,700	3.3677E–01
2,200	3.8238E–01
3,000	4.3015E–01
4,000	4.6331E–01
5,500	4.8552E–01
7,000	4.9374E–01
8,500	4.9679E–01
10,000	4.9792E–01
13,000	4.9851E–01
17,000	4.9856E–01
22,000	4.9858E–01
30,000	4.9858E–01
40,000	4.9857E–01

^{236}Pu Chronometers, to be used with ingrown analytes in Pu samples

Elapsed Days	^{232}U/^{236}Pu	^{228}Th/^{236}Pu
0	0.0000E+00	0.0000E+00
1	6.6422E–04	9.1453E–09
3	1.9940E–03	8.2325E–08
10	6.6613E–03	9.1539E–07
20	1.3365E–02	3.6653E–06
40	2.6901E–02	1.4692E–05
70	4.7532E–02	4.5136E–05
100	6.8559E–02	9.2407E–05
130	8.9993E–02	1.5668E–04
170	1.1922E–01	2.6909E–04
220	1.5680E–01	4.5316E–04
300	2.1950E–01	8.5033E–04
400	3.0248E–01	1.5295E–03
550	4.3729E–01	2.9452E–03
700	5.8560E–01	4.8633E–03
850	7.4878E–01	7.3170E–03
1,000	9.2830E–01	1.0344E–02

(Continued)

TABLE 6.3 (*Continued*)
4n Chronometers

^{236}Pu Chronometers, to be used with ingrown analytes in Pu samples

Elapsed Days	$^{232}U/^{236}Pu$	$^{228}Th/^{236}Pu$
1,300	1.3431E+00	1.8287E–02
1,700	2.0350E+00	3.3403E–02
2,200	3.1885E+00	6.1339E–02
3,000	5.9978E+00	1.3508E–01
4,000	1.2263E+01	3.0797E–01
5,500	3.3526E+01	9.0916E–01
7,000	8.8763E+01	2.4829E+00
8,500	2.3226E+02	6.5789E+00
10,000	6.0505E+02	1.7224E+01
13,000	4.0896E+03	1.1674E+02
17,000	5.2172E+04	1.4899E+03
22,000	1.2576E+06	3.5916E+04

^{232}U Chronometer, to be used with ingrown analytes in U samples

Elapsed Days	$^{228}Th/^{232}U$
0	0.0000E+00
1	2.7530E–05
3	8.2509E–05
10	2.7411E–04
20	5.4558E–04
40	1.0807E–03
70	1.8644E–03
100	2.6256E–03
130	3.3652E–03
170	4.3186E–03
220	5.4598E–03
300	7.1749E–03
400	9.1410E–03
550	1.1757E–02
700	1.4020E–02
850	1.5978E–02
1,000	1.7672E–02
1,300	2.0407E–02
1,700	2.3016E–02
2,200	2.5137E–02
3,000	2.6977E–02
4,000	2.7955E–02
5,500	2.8417E–02
7,000	2.8525E–02
8,500	2.8550E–02
10,000	2.8557E–02

(*Continued*)

TABLE 6.3 (*Continued*)
4n Chronometers

^{232}U Chronometer, to be used with ingrown analytes in U samples

Elapsed Days	$^{228}Th/^{232}U$
13,000	2.8558E–02
17,000	2.8559E–02
22,000	2.8558E–02
30,000	2.8558E–02

^{228}Th Chronometer (short), to be used with in-grown analytes in Th samples; useful only for short elapsed times

Elapsed Days	$^{224}Ra/^{228}Th$
0	0.0000E+00
0.08333	8.2019E–05
0.166667	1.6276E–04
0.291667	2.8152E–04
0.5	4.7335E–04
0.75	6.9380E–04
1	9.0414E–04
1.333333	1.1695E–03
1.666667	1.4188E–03
2	1.6530E–03
2.5	1.9778E–03
3	2.2733E–03
4	2.7870E–03
5	3.2126E–03
6.5	3.7180E–03
8	4.0989E–03
10	4.4651E–03
12	4.7163E–03
15	4.9534E–03
19	5.1185E–03
24	5.2081E–03
30	5.2468E–03
36	5.2594E–03
43	5.2637E–03
51	5.2649E–03

radionuclide concentrations, as a function of time, can be interpolated as before to determine the age of a sample. We include values for the ingrowth of ^{237}U in a ^{241}Pu sample at the end of the table for completeness. In Figure 6.8, the values of the ^{241}Pu chronometers are plotted as a function of time.

A sample consisting of mixed U or Pu isotopes provides the opportunity to measure the age of the sample through as many as a dozen different chronometers. Table 6.5 lists the quantities of various heavy-element nuclides that will be present in a 1-g sample of weapons-grade Pu after an ingrowth period of 1 year. Were

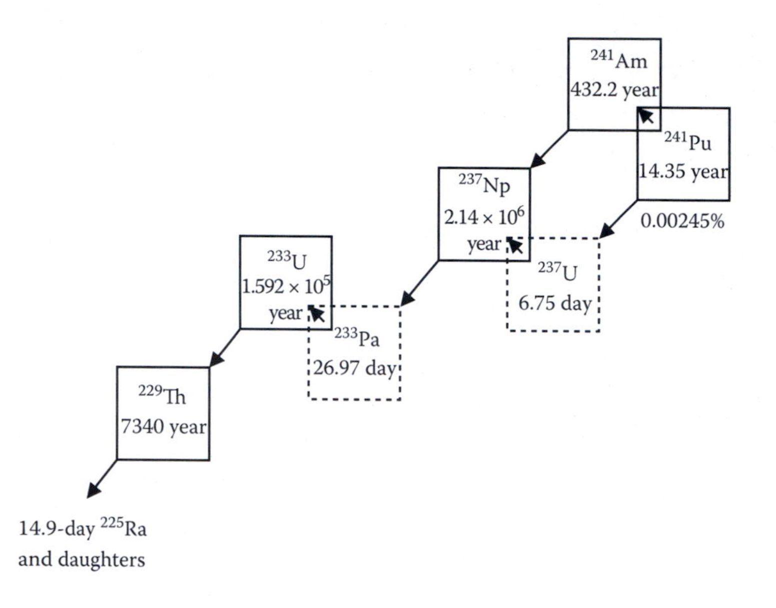

FIGURE 6.6 A radionuclide box diagram showing the chronometric relationships among the 4n + 1 nuclides. These data were used in the construction of Table 6.4. The presence of the small α-branch in the decay of ^{241}Pu has a large effect on chronometers of the second order ($^{237}Np/^{241}Pu$) and higher.

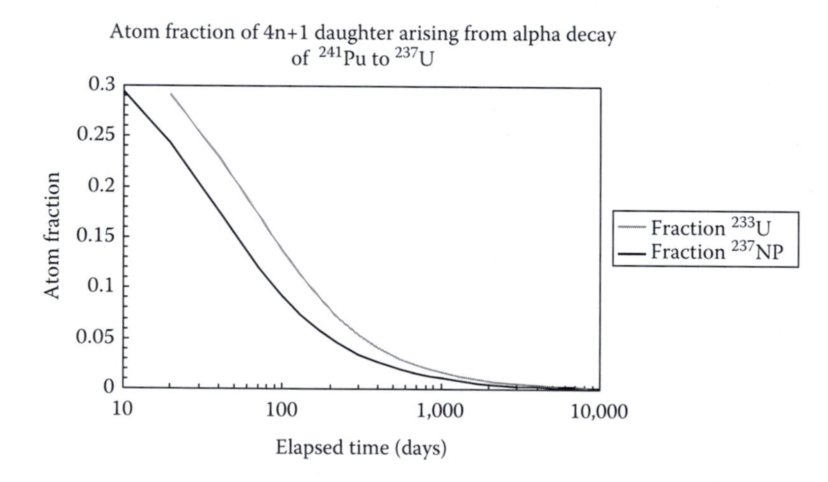

FIGURE 6.7 The fractions of the number of atoms of ^{237}Np and ^{233}U in a ^{241}Pu sample, produced by the minor decay branch through the 6.75-day ^{237}U intermediate.

the sample completely pure at the time of last separation, all of the chronometers should yield the same age (within measurement uncertainties). We usually find that, in a Pu sample, the $^{232}U/^{236}Pu$, $^{234}U/^{238}Pu$, $^{235}U/^{239}Pu$, and $^{236}U/^{240}Pu$ chronometers all yield the same age (as they should, because the purification of a Pu sample from one U isotope is as effective as the purification from all of them). When this age matches those determined from $^{241}Am/^{241}Pu$ and $^{230}Th/^{238}Pu$, which

TABLE 6.4
4n + 1 Chronometers

^{241}Pu Chronometers, to be used with ingrown analytes in Pu samples

Elapsed Days	^{241}Am/^{241}Pu	^{237}Np/^{241}Pu	^{233}U/^{241}Pu	^{229}Th/^{241}Pu
0	0.0000E+00	0.0000E+00	0.0000E+00	0.0000E+00
1	1.3226E–04	4.5118E–10	0.0000E+00	0.0000E+00
3	3.9682E–04	3.9687E–09	0.0000E+00	0.0000E+00
10	1.3233E–03	4.1217E–08	0.0000E+00	0.0000E+00
20	2.6483E–03	1.5369E–07	1.0930E–16	0.0000E+00
40	5.3034E–03	5.6503E–07	1.5116E–15	0.0000E+00
70	9.2987E–03	1.6276E–06	1.1696E–14	0.0000E+00
100	1.3310E–02	3.2232E–06	4.1669E–14	0.0000E+00
130	1.7336E–02	5.3557E–06	1.0432E–13	0.0000E+00
170	2.2728E–02	9.0410E–06	2.6260E–13	0.0000E+00
220	2.9507E–02	1.5015E–05	6.2858E–13	3.7372E–19
300	4.0445E–02	2.7780E–05	1.7648E–12	1.6013E–18
400	5.4274E–02	4.9396E–05	4.5292E–12	5.5584E–18
550	7.5354E–02	9.3947E–05	1.2689E–11	2.1186E–17
700	9.6842E–02	1.5354E–04	2.7510E–11	5.7921E–17
850	1.1874E–01	2.2877E–04	5.1177E–11	1.3114E–16
1,000	1.4108E–01	3.2022E–04	8.5996E–11	2.5921E–16
1,300	1.8703E–01	5.5421E–04	1.9895E–10	7.8048E–16
1,700	2.5112E–01	9.7978E–04	4.7104E–10	2.4216E–15
2,200	3.3599E–01	1.7126E–03	1.0869E–09	7.2489E–15
3,000	4.8356E–01	3.4153E–03	3.0207E–09	2.7598E–14
4,000	6.9060E–01	6.6365E–03	7.9801E–09	9.7743E–14
5,500	1.0553E+00	1.4371E–02	2.4302E–08	4.1289E–13
7,000	1.4970E+00	2.6723E–02	5.8591E–08	1.2795E–12
8,500	2.0321E+00	4.5319E–02	1.2267E–07	3.2743E–12
10,000	2.6804E+00	7.2279E–02	2.3367E–07	7.3927E–12
13,000	4.4171E+00	1.6305E–01	7.0413E–07	2.9396E–11
17,000	8.0569E+00	4.1388E–01	2.4123E–06	1.3401E–10
22,000	1.6195E+01	1.1520E+00	8.9873E–06	6.5900E–10
30,000	4.6880E+01	4.9668E+00	5.5213E–05	5.6703E–09
40,000	1.7105E+02	2.6270E+01	4.0618E–04	5.7096E–08
55,000	1.1702E+03	2.7048E+02	6.0237E–03	1.1993E–06
70,000	7.9709E+03	2.5089E+03	7.3564E–02	1.9064E–05

^{241}Am Chronometers, to be used with ingrown analytes in Am samples

Elapsed Days	^{237}Np/^{241}Am	^{233}U/^{241}Am	^{229}Th/^{241}Am
0	0.0000E+00	0.0000E+00	0.0000E+00
1	4.3908E–06	1.8739E–17	0.0000E+00
3	1.3172E–05	4.4627E–16	0.0000E+00
10	4.3910E–05	1.5653E–14	0.0000E+00
20	8.7821E–05	1.1781E–13	0.0000E+00

(Continued)

TABLE 6.4 (*Continued*)
4n + 1 Chronometers

^{241}Am Chronometers, to be used with ingrown analytes in Am samples

Elapsed Days	$^{237}Np/^{241}Am$	$^{233}U/^{241}Am$	$^{229}Th/^{241}Am$
40	1.7565E–04	8.4056E–13	0.0000E+00
70	3.0741E–04	3.8524E–12	8.4046E–19
100	4.3918E–04	9.7600E–12	3.1274E–18
130	5.7098E–04	1.8892E–11	7.9355E–18
170	7.4673E–04	3.6334E–11	2.0944E–17
220	9.6646E–04	6.6797E–11	5.1346E–17
300	1.3181E–03	1.3575E–10	1.4604E–16
400	1.7579E–03	2.5706E–10	3.7649E–16
550	2.4179E–03	5.1227E–10	1.0516E–15
700	3.0784E–03	8.5552E–10	2.2620E–15
850	3.7392E–03	1.2869E–09	4.1664E–15
1,000	4.4005E–03	1.8065E–09	6.9220E–15
1,300	5.7244E–03	3.1110E–09	1.5623E–14
1,700	7.4923E–03	5.4017E–09	3.5709E–14
2,200	9.7066E–03	9.1544E–09	7.8709E–14
3,000	1.3260E–02	1.7225E–08	2.0291E–13
4,000	1.7718E–02	3.0911E–08	4.8718E–13
5,500	2.4443E–02	5.9010E–08	1.2827E–12
7,000	3.1213E–02	9.6300E–08	2.6694E–12
8,500	3.8027E–02	1.4289E–07	4.8167E–12
10,000	4.4886E–02	1.9893E–07	7.8976E–12
13,000	5.8741E–02	3.3977E–07	1.7567E–11
17,000	7.7501E–02	5.8874E–07	3.9880E–11
22,000	1.0142E–01	1.0016E–06	8.7980E–11
30,000	1.4080E–01	1.9089E–06	2.2928E–10
40,000	1.9199E–01	3.4977E–06	5.6197E–10
55,000	2.7314E–01	6.9181E–06	1.5353E–09
70,000	3.5982E–01	1.1723E–05	3.3257E–09

^{237}Np Chronometers, to be used with ingrown analytes in Np samples

Elapsed Days	$^{233}U/^{237}Np$	$^{229}Th/^{237}Np$
0	0.0000E+00	0.0000E+00
1	1.1288E–11	1.5779E–19
3	9.9878E–11	1.4336E–18
10	1.0469E–09	4.2784E–17
20	3.8649E–09	3.1969E–16
40	1.3300E–08	2.2815E–15
70	3.3262E–08	1.0456E–14
100	5.6791E–08	2.6488E–14
130	8.1971E–08	5.1267E–14
170	1.1666E–07	9.8589E–14

(*Continued*)

TABLE 6.4 (*Continued*)
4n + 1 Chronometers

^{237}Np Chronometers, to be used with ingrown analytes in Np samples

Elapsed Days	$^{233}U/^{237}Np$	$^{229}Th/^{237}Np$
220	1.6068E–07	1.8122E–13
300	2.3151E–07	3.6820E–13
400	3.2018E–07	6.9700E–13
550	4.5320E–07	1.3884E–12
700	5.8622E–07	2.3176E–12
850	7.1923E–07	3.4846E–12
1,000	8.5225E–07	4.8894E–12
1,300	1.1183E–06	8.4123E–12
1,700	1.4730E–06	1.4589E–11
2,200	1.9164E–06	2.4687E–11
3,000	2.6258E–06	4.6338E–11
4,000	3.5126E–06	8.2908E–11
5,500	4.8427E–06	1.5756E–10
7,000	6.1729E–06	2.5596E–10
8,500	7.5029E-06	3.7810E–10
10,000	8.8329E–06	5.2399E–10
13,000	1.1493E–05	8.8688E–10
17,000	1.5039E–05	1.5182E–09
22,000	1.9472E–05	2.5442E–09
30,000	2.6565E–05	4.7319E–09
40,000	3.5429E–05	8.4102E–09
55,000	4.8724E–05	1.5888E–08
70,000	6.2017E–05	2.5709E–08

^{233}U Chronometer, to be used with ingrown analytes in U samples

Elapsed Days	$^{229}Th/^{233}U$
0	0.0000E+00
1	1.1920E–08
3	3.5761E–08
10	1.1920E–07
20	2.3841E–07
40	4.7682E–07
70	8.3442E–07
100	1.1920E–06
130	1.5496E–06
170	2.0264E–06
220	2.6224E–06
300	3.5760E–06
400	4.7679E–06
550	6.5559E–06
700	8.3436E–06

(*Continued*)

TABLE 6.4 (*Continued*)
4n + 1 Chronometers

^{233}U Chronometer, to be used with ingrown analytes in U samples

Elapsed Days	^{229}Th/^{233}U
850	1.0131E–05
1,000	1.1919E–05
1,300	1.5494E–05
1,700	2.0260E–05
2,200	2.6218E–05
3,000	3.5748E–05
4,000	4.7658E–05
5,500	6.5519E–05
7,000	8.3371E–05
8,500	1.0122E–04
10,000	1.1905E–04
13,000	1.5471E–04
17,000	2.0222E–04
22,000	2.6154E–04
30,000	3.5630E–04
40,000	4.7448E–04
55,000	6.5120E–04
70,000	8.2727E–04

^{241}Pu Chronometer (short), to be used with ingrown analytes in Pu samples; useful only for short elapsed times

Elapsed Days	^{237}U/^{241}Pu
0	0.0000E+00
0.08333	2.6885E–10
0.166667	5.3542E–10
0.291667	9.3102E–10
0.5	1.5792E–09
0.75	2.3389E–09
1	3.0794E–09
1.333333	4.0377E–09
1.666667	4.9638E–09
2	5.8587E–09
2.5	7.1451E–09
3	8.3671E–09
4	1.0631E–08
5	1.2674E–08
6.5	1.5372E–08
8	1.7685E–08
10	2.0264E–08
12	2.2365E–08
15	2.4809E–08

(*Continued*)

TABLE 6.4 (*Continued*)
4n + 1 Chronometers

^{241}Pu Chronometer (short), to be used with ingrown analytes in Pu samples; useful only for short elapsed times

Elapsed Days	$^{237}U/^{241}Pu$
19	2.7092E–08
24	2.8898E–08
30	3.0136E–08
36	3.0805E–08
43	3.1209E–08
51	3.1424E–08
59	3.1519E–08

^{237}Np chronometer (short), to be used with in-grown analytes in Np samples; useful only for short elapsed times

Elapsed Days	$^{233}Pa/^{237}Np$
0	0.0000E+00
0.08333	7.3821E–11
0.166667	1.4748E–10
0.291667	2.5768E–10
0.5	4.4056E–10
0.75	6.5873E–10
1	8.7551E–10
1.333333	1.1624E–09
1.666667	1.4468E–09
2	1.7288E–09
2.5	2.1473E–09
3	2.5605E–09
4	3.3711E–09
5	4.1612E–09
6.5	5.3089E–09
8	6.4133E–09
10	7.8211E–09
12	9.1585E–09
15	1.1040E–08
19	1.3334E–08
24	1.5889E–08
30	1.8552E–08
36	2.0835E–08
43	2.3089E–08
51	2.5216E–08
59	2.6948E–08
67	2.8358E–08
75	2.9506E–08

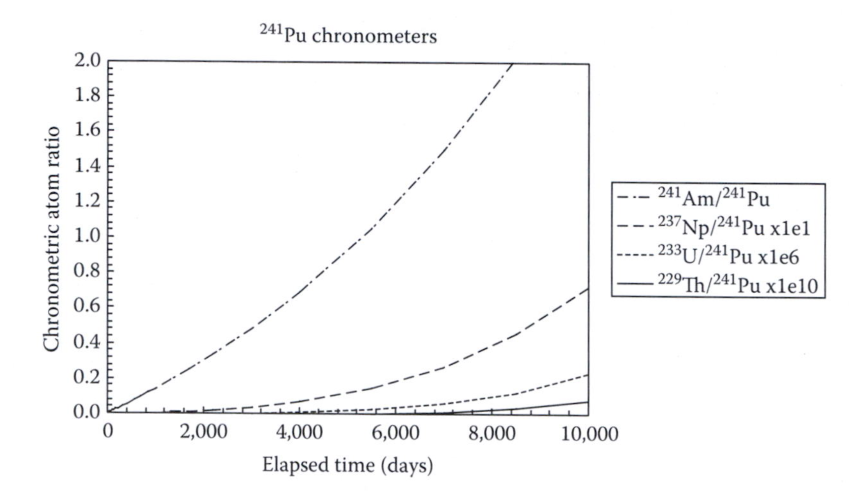

FIGURE 6.8 The concentration of each long-lived descendant nuclide arising from the decay of ^{241}Pu, expressed as an atom ratio relative to ^{241}Pu, as a function of time after separation/purification.

TABLE 6.5
Quantities of Selected Heavy-Element Progeny Nuclides in a Sample of 1 g of Pu Containing 6% ^{240}Pu, 0.91% ^{241}Pu, and 0.023% ^{242}Pu by Weight, after 1 Year of Ingrowth

Nuclide	Half-Life (Years)	Mass (ng)	Activity (dpm)
^{230}Th	7.54×10^4	0.0013	0.058
^{231}Pa	3.28×10^4	0.000013	0.0013
^{233}U	1.59×10^5	0.000056	0.0012
^{234}U	2.45×10^5	915	12,700.
^{235}U	7.04×10^8	26,300	126.
^{236}U	2.34×10^7	6250	897.
^{238}U	4.47×10^9	0.42	0.00032
^{237}Np	2.14×10^6	355	555.
^{241}Am	432.2	427,000	3.25×10^9

are chronometric pairs that can be measured with great precision and are based on the chemical properties of different elements, it is assumed that the sample was completely purified at the time of separation. However, for US weapons-grade Pu, ^{241}Am/^{241}Pu often gives a significantly larger value for the age than do the U isotopes. The only reasonable explanation for this observation is that, when the U isotopes were removed from the Pu sample for the last time, some Am was left in the material. This means that at any subsequent time, there will be more ^{241}Am in the Pu sample than can be explained by ingrowth, making it appear as though the decay process had been proceeding for a longer time and resulting in a value for the age that is too large. This is the basis for the use of chronometry as an indicator of incomplete fuel reprocessing, to be discussed later.

The following commentary applies to the errors associated with chronometry and the use of Tables 6.1 through 6.4. With very few exceptions, the decay properties used in the construction of the tables are known with greater accuracy than are the isotope ratios that the analyst is likely to measure. The error bars associated with the empirical chronometric atom ratios can then be taken as representative of nearly the entire error associated with the age determination. These uncertainties can be large or small, depending on the measurement technique and the age and size of the sample. However, these error bars do not translate directly to the same errors for the derived sample ages; that is, a 5% error for the value of a chronometric ratio does not necessarily result in a 5% error bar on the age, as the latter is dependent on the functional form relating the ratio to the age. For example, a sample of Pu was found to have a first-order chronometric atom ratio of $^{241}Am/^{241}Pu = 0.355 \pm 2\%$. Applying this error to the concentration ratio results in a range of values encompassed by an uncertainty between 0.348 and 0.362. Using the first section of Table 6.4, ages of 2265, 2303, and 2341 days are calculated, associated with the ratios of 0.348, 0.355, and 0.362, respectively. The error bar on the age is, therefore, 2303 – 2265 = 38 days, or 1.65%.

The chronometry of Pu samples is not limited to nuclides produced by the processes of α and β decay. The mass-240 Pu isotope decays almost completely by α emission; however, one in every 1.7×10^7 decays occurs by spontaneous fission (SF). In SF, the atom subdivides into two roughly equal parts, with the yield being distributed over products covering a wide range of atomic numbers and masses, similar to the yield distribution arising in the neutron-induced fission of ^{235}U or ^{239}Pu. Although the fraction of atoms in a Pu sample that undergoes SF decay does not appear significant, the neutrons that accompany the SF restrict the concentration of ^{240}Pu in materials used for weapons. One gram of weapons-grade Pu containing 6% ^{240}Pu by mass undergoes 1480 SF decays every minute.

Table 6.6 gives the quantities of several fission products that would grow into a 1-g sample of nominal weapons-grade Pu from the SF of ^{240}Pu in 1 year [6,7]. The ingrowth of ^{236}U in the same sample from α decay of ^{240}Pu is shown for comparison. The relative magnitudes of the SF- and α-decay branches, coupled with the distribution of the fission-product yield over a large number of nuclides, favor the formation of ^{236}U over any given fission product by a minimum of about 10^9. However, as the decay rate of a radionuclide is inversely proportional to its half-life, the activities of many of the shorter-lived fission products in the table (i.e., those with half-lives of the same order of magnitude as the ingrowth period) are comparable to that of long-lived ^{236}U. The variety of chemical properties represented by the range of fission products listed in the table makes chronometry involving the fission products in a Pu sample a valuable tool for determining the signatures of incomplete fuel reprocessing.

A superscript "a" next to the nuclide identification in column 1 of Table 6.6 denotes a short-lived nuclide that is essentially in equilibrium with ^{240}Pu after 1 or more years of ingrowth. This means that their activities are approximately equal to the SF activity of ^{240}Pu multiplied by the probability that the fission will result in that product (i.e., the cumulative fission yield, column 2). These species are not useful as

TABLE 6.6
Selected Spontaneous-Fission Products Present in 1 g of Pu Containing 6% ^{240}Pu after 1 Year of Ingrowth

Nuclide	Cumulative Yield (%)	Half-Life	Atoms	Activity (dpm)
^{85}Kr	0.07	10.8 years	5.3×10^5	0.06
^{86}Kr	0.47	Stable	3.7×10^6	
^{89}Sr[a]	1.1	50.5 days	1.7×10^6	16.2
^{90}Sr	1.4	28.5 years	1.1×10^7	0.5
^{93}Zr	3.0	1.5×10^6 years	2.3×10^7	2×10^{-5}
^{95}Zr[a]	4.8	64 days	9.3×10^6	70
^{99}Tc	6.9	2.1×10^5 years	5.4×10^7	3×10^{-4}
^{99m}Tc[a]	5.9	6 hours	4.5×10^4	87
^{103}Ru[a]	7.9	39.4 days	9.6×10^6	117
^{106}Ru	6.2	368 days	3.5×10^7	46
^{125}Sb	0.05	2.77 years	3.4×10^5	0.2
^{126}Sb[a]	0.03	12.4 days	1.1×10^4	0.4
^{127}Sb[a]	0.18	3.85 days	2.1×10^4	2.6
^{129}I	0.70	1.6×10^7 years	5.4×10^6	4×10^{-7}
^{131}I[a]	2.3	8.02 days	5.7×10^5	34
^{133}I[a]	8.2	20.8 hours	2.2×10^5	122
^{133}Xe[a]	8.2	5.25 days	1.3×10^6	119
^{136}Xe	7.5	Stable	5.8×10^7	
^{135}Cs	7.8	2.0×10^6 years	6.1×10^7	4×10^{-5}
^{136}Cs[a]	0.08	13.2 days	3.4×10^4	1.2
^{137}Cs	7.2	30.2 years	5.5×10^7	2.4
^{141}Ce[a]	5.7	32 days	5.6×10^6	84
^{144}Ce	4.0	285 days	2.1×10^7	35
^{143}Nd	4.7	Stable	3.7×10^7	
^{144}Nd	4.0	Stable	3.1×10^7	
^{145}Nd	3.3	Stable	2.6×10^7	
^{146}Nd	2.7	Stable	2.1×10^7	
^{147}Nd[a]	2.1	11 days	7.1×10^5	31
^{148}Nd	1.6	Stable	1.2×10^7	
^{150}Nd	0.75	Stable	5.8×10^6	
^{155}Eu	0.08	4.96 years	5.8×10^5	0.2
^{156}Eu[a]	0.05	15 days	2.3×10^4	0.7
^{236}U		2.3×10^7 years	1.6×10^{16}	900

Ingrowth of ^{236}U from α decay of ^{240}Pu in the same sample is included for comparison.
[a] Short-lived activity in secular equilibrium after 1 year of ingrowth.

process contamination indicators, even though they are also produced in high yields by a reactor irradiation. They decay away during the cooling period following reactor discharge and, even if the reprocessing chemistry is inefficient at removing them from the product Pu, continue to decay after the final chemical step until only the amounts that are continuously replenished by the SF decay of ^{240}Pu remain.

The short-lived fission products can be used as chemical-yield indicators for the longer-lived process-contaminant and age-determinant species. For example, a mass-spectrometric measurement of the number of atoms of long-lived ^{129}I in a chemical fraction separated from Pu must be corrected for losses in the separation procedure and sample preparation before it can be considered a signature. If the quantity of ^{240}Pu in a sample is measured, then the amount of 8-day ^{131}I in equilibrium with the sample can be calculated. Comparison with the amount of ^{131}I in the final iodine fraction, corrected for decay since it was separated from the ^{240}Pu parent activity, gives the efficiency of recovery of I from the Pu sample.

Krypton and xenon are gaseous elements that are chemically inert under most conditions. Table 6.6 contains entries for some of the isotopes of these elements that are produced in the SF of ^{240}Pu. The chemical properties of these nuclides provide an opportunity to determine a sample age associated more with metallurgy than with chemical reprocessing; plutonium is converted to the metal from a salt at some time after the last chemical purification step. Most of the fission products are chemically reactive species. Even fairly volatile elements will be retained to some extent in the melt, which provides no additional chemical separation from many species. However, the inert gases will be completely flushed out of the melt by the by-product process gases. The subsequent ingrowth of long-lived and stable Xe and Kr isotopes defines the casting time of a metal sample. Because the isotopic distributions of fission-product Kr and Xe are different from those of the naturally occurring gases, a minor admixture of the natural isotopes can be deconvoluted. However, the measurement becomes impossible in the presence of a significant quantity of air, which must be rigorously excluded during the early stages of the analytical procedure. One cubic centimeter of air at standard temperature and pressure contains 3×10^{13} atoms of Kr and 2×10^{12} atoms of Xe, introducing a background that would make it impossible to observe the radiogenic species by mass spectrometry.

The fission products in a Pu sample arise in decays of all of the even-mass isotopes, all of which have approximately the same partial half-life for SF. In a sample of weapons-grade material, therefore, contributions from the decays of ^{238}Pu and ^{242}Pu can be ignored, as both are present as relatively minor isotopes. However, in higher burn-up materials, the analyst may need to take these isotopes into account to obtain an accurate fission-product source term. The interaction of neutrons with the odd-mass isotopes gives rise to fission products through (n,f) reactions. The influence of a neutron flux will express itself in excess fission products, making the sample look older than it is. A discrepancy between the age determined through the ingrowth of the fission products and the age determined through the ingrowth of heavy-element nuclides could provide evidence of neutron exposure after chemical processing, a clue to the history of the material.

6.2 GRANDDAUGHTERS AND SPOOF DETECTION

Whenever possible, the analyst should measure multiple chronometers to guard against intentional misdirection ("spoofing"). Any individual signature that can be measured by the analyst can be introduced into a sample during fabrication with an intent to mislead. Although the tactic of adding, for example, γ-emitting signatures such as ^{241}Am or ^{226}Ra would deceive nondestructive field screening, it would be unsuccessful against full forensic analysis of multiple chronometers.

To successfully deceive the forensic analyst, the proliferators would first need to predict which chronometric signatures would be quantitated and then introduce specific materials to modify that particular signature. In the absence of such information, the proliferators would be required to modify all of them. To adjust the age of the U chronometers in a sample of weapons-grade Pu, for example, a peculiar mixture of U isotopes containing about 2% ^{234}U, 79% ^{235}U, and 19% ^{236}U (all atom-%) would need to be added. However, if a significant quantity of this material were added, the fabricator would need to address the distribution of the contamination throughout the Pu matrix. Uranium isotopes arising through decay processes should be distributed uniformly with the parent Pu isotopes. Uranium added to the process solutions before the oxalate is precipitated and fired to the oxide might mix inhomogeneously within the final product. Furthermore, if fabricated into metal, the U contaminant is more likely to concentrate in the slag than in the billet. Adding U to the host metal in the form of a melt is of limited value because U is not very soluble in molten Pu.

Imagine that a proliferant state has solved the daunting technical problem of distributing the Am and U isotopes, uniformly and in proper proportions, within a Pu sample so as to make it look older. But now consider the second- and third-order chronometers (granddaughters and great-granddaughters of radioactive decay): if the objective is to make the ^{238}Pu in a freshly separated Pu sample appear to be 8 years old, the values in Table 6.1 can be used to determine that ^{234}U must be added to adjust $^{234}U/^{238}Pu$ to a value of 0.0653. However, $^{230}Th/^{238}Pu$ must be adjusted simultaneously to a value of 7.54×10^{-7}. To tailor a sample of ^{238}Pu to appear to be 8 years older, it is necessary to add ^{234}U that is 4.07 years old to match the concentrations making up the ^{230}Th chronometer. Even were this done, though, the resultant $^{226}Ra/^{238}Pu$ signature would be too low by 24%.

The difficulty in uniformly dispersing the correct quantities of properly aged contaminants to achieve a consistent false age makes it essentially impossible to deceive a full forensic analysis. When questioned samples are sufficiently large, the determination of second- and third-order chronometers makes such spoofing prohibitively difficult. The determination of the distribution of first-order chronometers relative to the parent material in particulate specimens should also be a means to assess whether the daughters arose through decay processes or were introduced by mixing.

6.3 DETECTION OF INCOMPLETE FUEL REPROCESSING

If a daughter nuclide is incompletely removed from a sample containing its parent activity, the result for the age determined by that chronometric pair will always be too large. The daughter activity carried over will be in excess of what has grown into the sample, making it appear as if ingrowth had occurred over a longer time, until the sample becomes so old that the daughter activity approaches secular equilibrium with its parent. As a result, a chronometric analysis based on several pairs of genetically related nuclides can be a sensitive way of obtaining details about the chemical processes resulting in the questioned sample.

For example, U.S. weapons-grade Pu, which is produced by the PUREX process, always contains a small excess of Am along with a larger excess of Np, relative to the U chronometers that give a consistent, smaller age. It is perhaps possible that the Am was completely removed, but at an earlier time than was the U. However, we have observed trace amounts of ^{243}Am in Pu samples, and ^{243}Am is not a product of the radioactive decay of mixed Pu isotopes, cooled for at least 1 day after discharge from a reactor before being separated. The empirical ^{243}Am can be present only if it was residual from the original reactor fuel. Therefore, the Am was never completely isolated from Pu in the PUREX process. PUREX is much more effective at removing fission products, and we have performed good chronometric measurements of ^{137}Cs/^{240}Pu and ^{125}Sb/^{240}Pu ratios in several samples.

The relationship of chronometry to reprocessing signatures can be best illustrated by a pair of examples:

Example 6.1

We analyzed a sample of reactor-grade Pu, and, from the ^{234}U/^{238}Pu, ^{235}U/^{239}Pu, ^{236}U/^{240}Pu, ^{231}Pa/^{239}Pu, ^{230}Th/^{238}Pu, and ^{226}Ra/^{238}Pu chronometers, a consistent age for the sample of (5.6 ± 0.1) years was obtained. From the 4n + 1 chronometers (first section of Table 6.4), we observed the following:

$$^{241}\text{Am}/^{241}\text{Pu} = 0.3042 \quad \text{age} = 5.52 \text{ years}$$

$$^{237}\text{Np}/^{241}\text{Pu} = 3.52 \times 10^{-3} \quad \text{age} = 8.32 \text{ years}$$

and

$$^{233}\text{U}/^{241}\text{Pu} = 4.73 \times 10^{-9} \quad \text{age} = 9.42 \text{ years.}$$

It was clear that more ^{237}Np and ^{233}U were present in the sample than could be attributed to the decay of ^{241}Pu. We developed a hypothesis (prematurely) in which the Pu might have been used in conjunction with ^{233}U in some mixed-fuel reactor scenario. After all, the signatures of the other U isotopes were consistent with expectations from ingrowth, and we surmised that the only way that excess ^{233}U could be present was if it were added after reprocessing. However, a more careful analysis revealed our error: when the Pu was purified, a significant quantity of ^{237}Np remained as a contaminant, and the ^{237}Np/^{241}Pu chronometer made the sample appear to be too old. If we assume that ^{241}Am/^{241}Pu gives the true age of the sample, the ingrown quantities of ^{237}Np/^{241}Pu and ^{233}U/^{241}Pu were 1.42×10^{-3} and 8.15×10^{-10}, respectively. Therefore, the relative concentrations of ^{237}Np and ^{233}U that did not result from the decay of ^{241}Pu were ^{237}Np/^{241}Pu = $3.52 \times 10^{-3} - 1.42 \times 10^{-3} = 2.10 \times 10^{-3}$ and ^{233}U/^{241}Pu = $4.73 \times 10^{-9} - 8.15 \times 10^{-10} = 3.92 \times 10^{-9}$. Division to eliminate ^{241}Pu gives ^{233}U/^{237}Np = 1.86×10^{-6}. If we assume that ^{233}U was successfully removed from the sample at the reference time (as were the other U isotopes), then it could only have arisen by ingrowth from the excess ^{237}Np. Using the third section of Table 6.4, a ^{233}U/^{237}Np age of 5.82 years was found, which is in reasonable agreement with the ages obtained from the other chronometers. Of the heavy-element chronometric nuclides, only ^{237}Np was observed to be a nonradiogenic contaminant.

Example 6.2

A weapons-grade Pu sample was determined to be about 4 years old. Among the chronometers, $^{230}Th/^{238}Pu$ was found to be the largest outlier, giving an age of approximately 16 years. Interestingly, however, the $^{228}Th/^{236}Pu$ chronometer agreed fairly well with the consensus, giving an age of 3.2 years (but with a fairly large uncertainty). The analyst determined that this difference was an artifact of the short half-life of ^{228}Th (1.9 years) and was the result of the fact that two purification campaigns occurred in the history of the sample, well-separated in time, but neither of which was effective in removing Th. At the first separation, significant amounts of both ^{228}Th and ^{230}Th remained in the sample—daughters of the decays of the U isotopes residual in the reactor fuel. Several years passed, during which the ^{228}Th decayed significantly, whereas the longer-lived (75,000-year) ^{230}Th did not. When the second chemistry was performed, Th again was incompletely removed. Four years later, when the chronometric analysis was performed, the ^{230}Th concentration was anomalously high, yielding an age of 16 years. The ^{228}Th, however, had its concentration reduced by decay, as well as by the two chemical separations. Fractionally, the second chemical separation reduced the residual ^{228}Th much closer to zero than it did the ^{230}Th concentration, making the $^{228}Th/^{236}Pu$ age approximately correct. Had the analyst been able to measure the $^{229}Th/^{241}Pu$ chronometer, it would be expected to give an age more in accordance with 16 years than 4 years (because the ^{229}Th half-life is long). Unfortunately, ^{229}Th was not detectable in the final counting fractions.

At low burn-ups of U reactor fuels, such as those used for the production of weapons-grade Pu, the concentrations of most of the stable fission products are reasonably constant relative to each other and to the total quantity of Pu produced, independent of reactor power. Table 6.7 gives the results of an ORIGEN2 calculation for fission-produced stable Nd isotopes in a Pu production reactor, operated at a power level of 125 MW_t/t for an irradiation time necessary to produce weapons-grade Pu (6% ^{240}Pu). For this calculation, that interval was about

TABLE 6.7
Neodymium Isotope Ratios (by Mass), and the Production of ^{143}Nd *vs* Total Pu Production (Weapons-Grade, 6% ^{240}Pu), for a Production Reactor Operated at a Power Level of 125 MW_t/t

Mass Ratio	Reactor Production	Natural Nd
$^{142}Nd/^{143}Nd$[a]	0.00019	2.212
$^{144}Nd/^{143}Nd$[b]	0.949	1.973
$^{145}Nd/^{143}Nd$	0.699	0.692
$^{146}Nd/^{143}Nd$	0.572	1.478
$^{148}Nd/^{143}Nd$	0.351	0.487
$^{150}Nd/^{143}Nd$	0.160	0.484
$^{143}Nd/^{tot}Pu$	0.0194	—

Note: Natural Nd isotopics are included for comparison.
[a] The $^{142}Nd/^{143}Nd$ mass ratio is very sensitive to reactor flux.
[b] Assumes complete decay of 285-day ^{144}Ce to ^{144}Nd.

12 days. Values are given relative to the production of ^{143}Nd, which in turn is given relative to the production of total Pu. The isotopic composition of naturally occurring Nd is also listed for comparison. The distribution of isotopes in fission-product Nd is significantly different from that in natural Nd. If the distribution of isotopes in a Nd fraction separated from a Pu sample is measured, the set of isotope ratios is deconvoluted into three components: chronometric Nd, arising from the SF of ^{240}Pu after the sample was purified; radiogenic Nd, produced in the fission of the reactor fuel and present as a contaminant in the analytical sample; and natural Nd, introduced into the sample by the reagents used in its purification. Components 1 and 2 can be resolved if the age of the sample is known. However, the relative amounts of components 2 and 3 provide more information about the chemical procedure used for reprocessing.

Reagent-grade solvents contain approximately 100 parts per billion Nd (by mass) as impurities, and Nd concentrations in industrial solvents are likely higher. The solubility of reactor fuel (mostly uranium oxide) in nitric acid is certainly less than 500 g/L, and the weapons-grade Pu content of the fuel is probably no more than a few tenths of a percent of the U. As a consequence, the total chemical volume involved in the early steps of an aqueous fuel-reprocessing procedure must be on the order of 10 L/g of Pu. Therefore, impurities in the reagents introduce about 1 mg of natural Nd to the process for each gram of Pu, which is associated with 75 mg of fission-product Nd (Table 6.7). Even if decontamination factors of 10^8 are achieved for Nd in the reprocessing, there will still remain approximately 1 ng of the initial Nd content present in the Pu sample, which is adequate for mass-spectrometric analysis. Natural Nd introduced as contaminants in the lesser quantities of reagents used in later chemical steps, or in metallurgical processes, will also be present in the final sample and at higher chemical yield. The relative amounts of natural and fission-product Nd are determined by the purity and volumes of the reagents used in the chemistry and are likely unique for a given reprocessing plant.

REFERENCES

1. G. Faure, *Principles of Isotope Geology*, 2nd edn., Wiley, New York, 1986.
2. W.F. Libby, *Radiocarbon Dating*, 2nd edn., University of Chicago Press, Chicago, IL, 1955.
3. M. Stuiver and H.E. Suess, On the relationship between radiocarbon dates and the true sample ages, *Radiocarbon*, 8, 534, 1966.
4. M.J. Aitken, *Physics and Archaelogy*, Interscience, New York, 1961.
5. R.P. Keegan and R.J. Gehrke, A method to determine the time since last purification of weapons grade plutonium, *Int. J. Appl. Radiat. Isot.*, 59, 137, 2003.
6. J.B. Laidler and F. Brown, Mass distribution in the spontaneous fission of ^{240}Pu, *J. Inorg. Nucl. Chem.*, 24, 1485, 1962.
7. A. Prindle and K.J. Moody, unpublished data.

7 Techniques for Small Signatures

The last section discussed chronometry and how the concentrations of nuclides that are linked by decay can be used to determine the time elapsed since the last chemical purification. In many instances, particularly for short times or more distant genetic relationships, the atom ratios of the related species can be quite small, approaching zero. In those cases, particularly if sample sizes are limited, a radionuclide concentration needs to be determined from very few atoms. This can be done by atom counting or by decay counting and may require a chemical separation to remove an overwhelming background activity.

7.1 CHEMICAL SEPARATIONS AND REDUCTION OF BACKGROUND

Radiochemical forensic analysis is a labor-intensive activity. However, it is necessary because of the enormous dynamic range over which the chronometric pairs and radionuclides residual from incomplete fuel reprocessing must be measured. Following a radiochemical analysis, a typical table of radioanalyte concentrations can span 15–20 orders of magnitude, depending on the combined mass of the analytes in the original sample.

Several radiochemical techniques can be used to facilitate the detection of small numbers of analyte atoms. The first is concentration: if an analyte is dispersed throughout a large volume, the ability to detect it can be reduced. In the case of radiation counting, an extended source invites increased interference from the natural background radiation. In the case of atom counting, an expanded matrix volume increases the possibility of second-order effects producing false counts at a given value of charge-to-mass. With conservative techniques such as evaporation (when the analyte is not volatile), volumetric techniques apply, and a reduction in volume by a given factor results in a corresponding increase in concentration. With nonconservative techniques such as ion exchange (where analyte can be lost in liquid transfers or irretrievably on the resin), yield-measuring procedures using isotopic carriers or tracers must be employed.

Dilution is also a technique used by the radiochemist to reduce background. In radiation counting, high count rates produce second-order effects (such as summing) that degrade the performance of counters and make the determination of a radionuclide concentration less accurate. Similar effects occur with the introduction of too much analyte into the ion source of a mass spectrometer. The resultant formation of charged, multiatomic species can bias isotope ratios, and the ion detectors would be overwhelmed by the signals. Dilution is usually conservative (unless reactive reagents are mixed or large changes in solute concentration occur).

Dilution is a tactic for producing many calibration standards. For example, gas proportional counters can be calibrated with ^{241}Am sources that are dilutions of a dissolved specimen of electrorefined Am metal. The weight of the metal sample defines the number of atoms of ^{241}Am in the original sample, which is then quantitatively dissolved in dilute HCl. The original sample and its initial dissolution would be far too radioactive to be of any use for calibrating analytical instruments, so the original stock solution must be quantitatively diluted using volumetric glassware and HCl of similar concentration as the original solution. Aliquots of the dilution, quantitatively transferred to platinum counting plates, contain a known number of atoms of ^{241}Am, defined by the dilution factor and the original sample content. In this manner, the radioactive decay rate of the weightless calibration source can be related to a measured mass.

Of course, the principal way that a chemical separation reduces background is through the removal of interfering species. These techniques are intrinsically nonconservative, so tracer and carrier methods must be applied to determine a chemical yield. In radiation counting, removal of a bulk analyte, such as Pu or U, can reduce the count rate of a sample sufficiently for it to be counted closer to the detector, with greater efficiency. There are radionuclide-specific interferences that can be remediated by an appropriate radiochemical separation, and two examples of interfering contaminants that are routinely addressed by the forensic radiochemist follow.

First, a sample containing mixed Pu isotopes always contains a small quantity of ^{241}Pu, which decays with a short half-life (14.4 years) to ^{241}Am. The α decay of ^{241}Am (E_α = 5.486 MeV) occurs in the same portion of the spectrum as does the decay of ^{238}Pu (E_α = 5.499 MeV), which is also always present in a sample of mixed Pu isotopes. A source prepared without chemical purification will produce an α-particle spectrum with a peak at about 5.49 MeV that results from a combination of the decays of ^{238}Pu and ^{241}Am. They are unresolved by a standard α-counting system, which is characterized by an energy resolution of approximately 15–20 keV. Chemical purification before source preparation can resolve the contribution of each component of the mixture.

Second, a sample containing mixed U isotopes often contains a small quantity of ^{237}Np. The main peak in the α spectrum arising from the decay of ^{237}Np (E_α = 4.788 MeV) is at a similar energy as the peak arising from the decay of ^{234}U (E_α = 4.776 MeV). Compounding this effect is the granddaughter of ^{237}Np, the important chronometric nuclide ^{233}U (E_α = 4.824 MeV), which is also obscured. A radiochemical separation following appropriate tracer addition can result in a clean Np fraction. Separation of ^{233}U and ^{234}U cannot be accomplished radiochemically, but a clever milking scheme (discussed later) can resolve their relative concentrations.

Mass spectrometry (MS) benefits from such chemical separations in a similar way. An accurate MS determination of ^{241}Pu can only be accomplished if the isobaric ^{241}Am β-decay daughter is first removed. Similarly, the MS measurement of ^{238}Pu is always compromised by the presence of ^{238}U, which is found in sufficient concentration in many reagents to interfere with the quantitation of ^{238}Pu. Multiatomic ions can also interfere in some MS measurements. Thus, better success in determining the relative ^{239}Pu isotopic concentration in a sample occurs if U is removed, because the uranium hydride ion can add to the spectrometer signal at mass 239.

The nature of a sample of mixed radionuclides introduces a time dependence of the analyte concentrations that is unimportant in conventional analytical chemistry. Short-lived species decay, and radionuclides that are linked to one another by decay provide a pathway for the transfer of mass across element boundaries in a complicated journey toward stability, modeled by the Bateman equations (Section 2.3). As a result, all analyte concentrations should be reported for a specified reference time. Practically speaking, this time should be associated with an important event in the history of the sample—often the time of its interdiction or its receival at the radiochemistry facility. For convenience, our reference-time convention for calibration samples or exercise materials is the time when the sample (or an aliquot of it) undergoes its first nonconservative separation procedure.

The limitations of radioisotope chronometry arise from two sources: first, the accuracy to which nuclide concentrations can be determined, which includes the effects of uncertainties in nuclear decay data and the statistical uncertainties in counting data; and second, the accuracy with which chemical separation times are measured, in conjunction with the extent to which the elements are fractionated by the procedure. As an example, assume that the analyst must determine the concentration of ^{237}U (the daughter of the α-decay branch of ^{241}Pu decay) in a sample of mixed Pu nuclides. A reason could be to relate the chemical yield of a U fraction to that of the Pu fraction, or perhaps there was the possibility that the sample was recently purified (within the last month), such that a ^{241}Pu–^{237}U disequilibrium existed. Alternatively, the analyst might need a quick estimate of the ^{241}Pu concentration in the sample before MS data were available. To accomplish the separation, the analyst chooses to subject an aliquot of the analytical pot solution to an anion-exchange procedure. Both U and Pu are sorbed on the resin from 9 M HCl. After washing, Pu is then selectively eluted with HCl containing HI, leaving the U sorbed on the resin. During elution of the Pu, ^{241}Pu actively decays and leaves fresh ^{237}U behind on the column, and this deposition continues until the Pu is completely eluted, perhaps an hour later. At that time, secular equilibrium no longer exists between ^{237}U and ^{241}Pu for the nuclides remaining on the column. Unfortunately, Pu does not exit the column bed all at once, in a single drop, but rather elutes in a band that can encompass several mL of eluent if the column is large. Even if the analyst knows the beginning and end of the elution of Pu, the establishment of a Pu–U separation time is uncertain. In practice, the uncertainty in establishing this time is on the order of 30 min. A 30-min error in the separation time, applied to 6.75-day ^{237}U, results in an error of 0.2% in the concentration of the nuclide after decay corrections are made. This is effectively not very serious, as the uncertainty in the α-decay branch of ^{241}Pu is larger. However, this error becomes more important for shorter-lived species. For instance, a 30-min ambiguity in the Th–U separation time would introduce a 1.4% error in the concentration of 25.5-h ^{231}Th recovered from ^{235}U.

A lack of complete separation also introduces an error into the chronometry calculation. For the same Pu–U separation discussed earlier, assume that the analyst was able to establish a reasonable separation time, but that the separation was incomplete. That is, the kinetics of the column were slower than expected, the analyst began to elute U from the column before all of the Pu was removed, and the result was 1% of the Pu remaining in the U fraction. The analyst does not perform a second purification until 24 h later. If uncorrected, the surplus ingrowth of ^{237}U, from 1% of the ^{241}Pu for 1 day, will result in an error of 0.11% in the ^{237}U concentration (corrected for decay to the separation time). Again, this effect is more important for shorter-lived species. A similar 1% contamination of U in a separated Th fraction results in a 0.9% error in the reported ^{231}Th concentration at the reference time.

In summary, chemical separations are important in nuclear forensic analysis, particularly for the reduction of background to enable the measurement of small radionuclide signatures. However, the interdependence of chronometer signatures as a function of time causes the uncertainties in the performance of the chemical procedure, and the times at which separations take place, to affect the determination of the relative concentrations of the chronometers. The effect is most important for chronometric pairs in which one or both of the radionuclides are short lived, or chronometers that are approaching radioactive equilibrium because of the age of the sample. For most chronometers and most samples, the uncertainties caused by the timing and completeness of chemical steps are on the order of those resulting from counting statistics and nuclear-data inaccuracies. When more than one separation method is available, the analyst should choose the one that requires the shortest elapsed time and gives the greatest single-step purification factor.

7.2 RADIOCHEMICAL MILKING

The dynamic range in the determination of nuclide concentrations required for chronometry makes it impossible to perform most of the measurements without radiochemical separations. However, when the nuclide to be determined is present as a small admixture in an overwhelming quantity of isotopes of the same element, chemical separations are of limited use. For this case, the method of radiochemical milking has proven most valuable. In a milking experiment, a purified sample of the element incorporating the trace nuclide to be measured is prepared. This sample is then set aside for a predetermined length of time to allow the daughter activities to grow in. At an accurately determined time, the daughter activities are separated from the parent activities. The decay interval relates the isotope ratios among the daughter activities to those among the parent activities. If one or more concentrations of the parent isotopes are known, and the daughter of the nuclide to be quantified can be observed and measured, the concentration of the otherwise "invisible" isotope of the parent element can be determined.

An example is instructive: the ^{236}Pu content of a Pu sample is determined by the flux of the production reactor and the fraction of high-energy neutrons (hardness) of its neutron spectrum in that portion of the fuel assembly from which the Pu sample was recovered. ^{236}Pu arises primarily from high-energy (n,2n) reactions on ^{237}Np, followed by the decay of 22.5-h ^{236m}Np. The ^{237}Np concentration in the fuel is maximized at low flux, which allows time for its production from the β decay of ^{237}U or the α decay of ^{241}Am. Even though ^{236}Pu has a relatively short half-life (2.858 years), its characteristic decay γ-rays are weak. Its emitted α particles are of higher energy than those emitted by the other Pu isotopes, but its decay rate is low compared to those of the other Pu isotopes. It is difficult to fabricate a source that can be counted close enough to a detector to observe the ^{236}Pu α peak over random pileup events.

One of the analytical samples that we have processed contained 850 mg of weapons-grade Pu. After we created an initial analytical solution and aliquoted it for radiochemical analysis, approximately half of the residual material was used in a radiochemical milking experiment to determine ^{236}Pu. The fraction was purified by converting the sample to the nitrate form via repeated evaporations to a moist deposit after additions of nitric acid, following which the material was dissolved in 0.5 M HNO_3 (containing NO_2^-). Plutonium was then extracted into 0.02 M thenoyltrifluoroacetone (TTA) in benzene, leaving U (and Am) behind in the aqueous phase. The organic phase was washed several times with 0.5 M HNO_3, discarding the aqueous phase each time, and the time of the last phase separation was recorded as the Pu–U separation time. The Pu sample, still in the organic solvent, was capped and set aside. After 40 days, the sample was agitated with 0.5 M HNO_3, and the aqueous sample was retained as the U fraction. This fraction contained only the U and Am atoms that grew into the Pu sample during the decay period. The U fraction was further purified with several anion-exchange steps and was then volatilized onto a Pt counting plate.

In the α-particle pulse-height spectrum registered by the U sample, two activities were observed: ^{234}U (from the decay of ^{238}Pu) and ^{232}U (from the decay of ^{236}Pu), with an activity ratio of $^{232}U/^{234}U = 9.44 \times 10^{-3}$. From the various analytical measurements that were performed on the initial sample solution, the total number of ^{238}Pu atoms in the original sample was 3.09×10^{17}. From the tables in Chapter 6, 2.674×10^{14} atoms of ^{234}U grew into the sample during the 40 days of decay. The ^{234}U activity (from its 2.45×10^5-year half-life) was 1440 disintegrations/min (dpm). Therefore, the ^{232}U activity was 13.6 dpm, corresponding to 7.1×10^8 atoms. From the chronometry tables, $^{232}U/^{236}Pu$ at 40 days of ingrowth was 2.69×10^{-2}, so the number of atoms of ^{236}Pu present at the milking time was 2.64×10^{10}. This value should be corrected for 40 days of decay, giving 2.71×10^{10} atoms of ^{236}Pu at the separation time. The ratio of the α-decay rates of $^{236}Pu/^{238}Pu$ in the original sample was 2.6×10^{-6}. The milking technique takes advantage of the shorter half-lives of both ^{236}Pu and ^{232}U, relative to ^{238}Pu and ^{234}U, to produce an activity ratio much easier to measure.

The assay of ^{243}Am is another example. Americium-243 is an indicator of reactor flux and is present in a Pu sample only as a contaminant—an artifact of incomplete fuel reprocessing that provides a clue to the chemical process used to recover the sample from the spent fuel. Unfortunately, the residual signal from the decay of ^{243}Am is effectively overwhelmed by the decay of chronometric ^{241}Am, which grows into the sample from the decay of ^{241}Pu. Because ^{243}Am has a lower α-decay energy than ^{241}Am, its α particles are lost in the low-energy tail of the more intense activity. The α decay of ^{243}Am results in 2.35-day ^{239}Np, which eventually attains radioactive equilibrium with its parent activity. Neptunium-239 has easily observable γ-rays, but the intensity of the ^{241}Am γ activity overwhelms their signal.

The initial aqueous fraction from the extraction of Pu into TTA/benzene in the previous example (half of 850 mg of weapons-grade Pu) was retained, and the time of phase separation was recorded as the Pu–Am separation time. The aqueous phase was washed once again with TTA/benzene, and the organic phase was discarded. The aqueous phase was then evaporated to dryness, the activity dissolved in 10 M HCl, and the resulting solution passed through an anion-exchange column. Americium is not adsorbed by the resin, but Np is left behind on the column. The eluent was evaporated to dryness, the residue redissolved in 10 M HCl (+1 drop of HNO_3), and the anion-exchange step repeated with a fresh column. The temporal midpoint of this column was recorded as the Am–Np separation time. After 40 days, the solution was again passed through an anion-exchange column. Neptunium that had grown into the sample over the decay interval was sorbed by the resin. Again, the temporal midpoint of the column served as the reference time for the end of the decay interval. The Np fraction was recovered from the column, purified with several anion-exchange steps, and volatilized onto a Pt counting plate.

The final sample consisted of the radionuclides ^{237}Np (from the decay of ^{241}Am) and ^{239}Np (from the decay of ^{243}Am), and the absolute decay rate of the ^{239}Np was determined by γ-ray spectroscopy. Because of the nuclide's 2.35-day half-life, a decay correction was necessary to relate the empirical intensity of ^{239}Np emissions at the time of measurement to the end of the decay interval. The absolute decay rate of the ^{237}Np activity ($t_{1/2} = 2.14 \times 10^6$ years) was determined in two ways: a combination of α-particle proportional counting and α-particle spectrometry, and γ-ray spectroscopy of the 27-day ^{233}Pa daughter of ^{237}Np α decay, which grew into the sample over the next several months. Both methods yielded the same value for the decay rate of ^{237}Np, and the resultant ratio of activities at the end of the ingrowth interval (the milking time) was ^{239}Np/^{237}Np = 38.6. From the various analytical measurements that were performed on the pot solution, the total number of ^{241}Am atoms in the original sample at the Pu–Am separation time was 1.09×10^{18}. From the chronometry tables, 1.914×10^{14} atoms of ^{237}Np grew into the sample during 40 days of decay. The ^{237}Np activity (from the 2.14×10^6-year half-life) was 118 dpm, and the ^{239}Np activity at the time of separation from ^{243}Am was therefore 4550 dpm. Because the ingrowth period (40 days) was long compared with the half-life of ^{239}Np (2.35 days), ^{239}Np can be assumed to have been in secular equilibrium with ^{243}Am at the time of separation. The ^{243}Am in the original sample was therefore 4550 dpm, corresponding to 2.54×10^{13} atoms. The ratio of α radioactivities in the original sample at the separation time was thus ^{243}Am/^{241}Am = 1.4×10^{-6}. The milking technique takes advantage of the much shorter half-life of ^{239}Np relative to ^{237}Np to produce the readily measured activity ratio.

Milking techniques are not limited to a single-step process, and an example of a double-milking experiment follows. A questioned sample of high-enriched uranium (HEU), with a mass of about 1 g, was submitted for analysis. The concentrations of the nonnatural U isotopes, ^{232}U and ^{233}U, are indicative of prior use of enrichment-cascade feedstock in a reactor application, and they are characteristic of a particular batch of HEU. The α activity of the ^{232}U (higher in energy than the α particles emitted by the other U isotopes) was sufficiently intense that it could be observed in an α-particle spectrum over the continuum of pileup events, but the statistics of the result were poor. The concentration of ^{233}U (which emits α particles at the same energy as those emitted by the much more abundant ^{234}U) could be observed by MS, but the dynamic range of the measurement (approximately six orders of magnitude compared with ^{235}U) led to questionable validity of the result.

An aliquot of the analytical solution (HCl medium) containing 750 mg of U was analyzed. The volume of the solution was doubled by adding concentrated HCl, and it was passed through a large anion-exchange column. Uranium was sorbed by the resin, whereas Th was not. Uranium was then stripped from the column with dilute HCl, the solution was evaporated to a moist deposit, and it was dissolved in 9 M HCl and passed through a fresh anion-exchange column. The temporal midpoint of this column procedure was recorded as the U–Th separation time. The column that had sorbed uranium was stored for 40 days, after which a Th fraction was isolated by elution with additional 9 M HCl. The temporal midpoint of this elution was recorded as the end of the first milking interval. After several purification steps, the Th fraction was electroplated onto a Pt counting disk and lightly flamed.

The α spectrum of the Th fraction contained measurable activities of ^{230}Th (from ^{234}U decay) and ^{228}Th (from ^{232}U decay), and the ratio of their activities was ^{228}Th/^{230}Th = 3.27. From the various analytical measurements performed on the pot solution, the total number of ^{234}U atoms in the original sample at the U–Th separation time was 2.45×10^{19}. From the chronometry tables, 40 days of ingrowth produced 7.61×10^{12} ^{230}Th atoms, or an activity of 132 dpm. Thus, the ^{228}Th activity at the time of separation was 432 dpm, translating eventually into 5.82×10^{11} atoms of ^{232}U, in fair agreement with the direct measurement. Unfortunately, ^{229}Th ingrown from the decay of ^{233}U was not visible in the α spectrum of the Th fraction because of tailing from the higher-energy α peaks from the decays of ^{228}Th (and its daughters) into the spectral region where ^{229}Th was expected.

The Th fraction was recovered from the counting plate by partial dissolution of the Pt surface and was extensively purified by anion exchange. The temporal midpoint of the last column procedure was recorded as the Ra–Th separation time. After 40 days of ingrowth, the Ra fraction was isolated, purified, and volatilized onto Pt. Initially, the only activity observable in the α-particle spectrum of the Ra fraction was 3.66-day ^{224}Ra (from the decay of ^{228}Th) and its daughter activities. A second α-particle spectrum was then taken after 30 days of decay. The decay of ^{224}Ra had proceeded to the point that the α-emitting daughters of the β decay of 14.8-day ^{225}Ra (ingrown from the decay of ^{229}Th) were visible in the spectrum. At the time that the descendants of ^{225}Ra became observable, each was registering approximately 40 events/day in the α detector. The limited activity resulted in poor counting statistics, allowing a concentration calculation with an uncertainty of no better than ±4%, even after long counting intervals.

The ^{224}Ra and ^{225}Ra activities were corrected for decay to the end of the second milking period. These data were then used to determine the relative numbers of atoms of ^{228}Th and ^{229}Th in the first milking fraction (^{228}Th/^{229}Th = 0.384 atom ratio). This ratio was then used to relate the number of atoms of ^{232}U to the number of atoms of ^{233}U in the original sample (^{232}U/^{233}U = 1.70×10^{-4} atom ratio), in good agreement with the MS determination of the concentration of ^{233}U.

Radiochemical milking procedures for nuclear forensic applications are very labor intensive, require large quantities of sample, and necessitate ingrowth periods of a month or more, causing considerable delays in returning analytical data. The fundamental accuracy of the technique depends upon how well separation times and ingrowth periods can be established. When sequential milkings are performed (as in the HEU example given earlier), the activities in the second step are always less intense than those in the first step, which can result in counting fractions that emit too little decay intensity to measure accurately.

7.3 MASS SPECTROMETRY AND MICROANALYSIS

MS is one of the workhorses of nuclear forensic analysis, providing the capability to determine isotopic, elemental, and molecular abundances of an extraordinarily diverse array of materials with high sensitivity, precision, and accuracy. Samples can be solid, liquid, or gas, large (kilogram) or small (zeptomole), radioactive or stable, light (1 amu) or massive (kilodaltons). Mass spectrometers achieve their impressive performance by converting a portion of a sample into positively or negatively charged atoms or molecules, then separating and analyzing the elemental or molecular ions according to their respective mass-to-charge ratios.

In general, MS protocols are characterized by very high sensitivity, wide dynamic range, and great versatility. For example, bulk-sample MS techniques such as thermal-ionization (TIMS), inductively coupled plasma (ICP-MS), and multi-collector ICP-MS can routinely assay multielement signatures for elements as light as H ($Z = 1$) and as heavy as curium ($Z = 96$), with limits of detection in the attogram to picogram range. Bulk analysis techniques are often limited more by signal-to-noise issues, particularly associated with chemical processing, than by sensitivity. At the micro- to nanoscale, secondary-ion MS (SIMS) and laser-ablation MS can interrogate solid samples to determine elemental and isotopic abundances in situ for elements from H to Cm, at concentrations of ~1–100 ng/g, and with 10^9 dynamic range over spatial dimensions of 0.05–100 μm. All of these methods, and others, are discussed in more detail in Chapter 16.

7.4 RADIATION DETECTION

7.4.1 Interactions of Radiation with Matter

Some facets of the interaction of radiation with matter, mostly in the context of the discovery of radioactivity and the determination of its nature, have been discussed previously. This interaction will now be treated in more detail, as the operation of radiation detectors depends on the properties of the radiation being measured and on knowledge of the energy-loss mechanism of particles or photons as they interact with the detecting medium. More detailed information can be found in Siegbahn [1] for the interaction of radiation with matter, and in Knoll [2] and Price [3] for the operation of detectors and the quantitation of radioactivity.

Interactions of radiation with matter may result in attenuation (absorption), scattering, ionization, or excitation of the atomic electrons in the interacting medium, or the conversion of one type of radiation into another (e.g., positron annihilation). Interactions most often occur with the atomic electrons, but interactions with atomic nuclei cannot be neglected in some cases. Collisions between radiation and atoms can be elastic or inelastic. The products of an inelastic collision, which results in a change in the internal energy of either or both of the collision partners, provide the functional basis of detector systems.

The electrons orbiting the nucleus are not randomly arranged, but occupy certain energy states required by the restrictions of quantum mechanics. Electrons tend to fill these states from the most tightly bound to those with lesser binding energies. Electrons occupying an outer, partially unfilled electron shell (a collection of states with the same principal quantum numbers) are referred to as valence electrons. Energy can be transferred to any electron in the atom, raising it to a higher energy state. Such excitation may occur whenever an external process imparts sufficient energy to the atom to promote an electron to an unoccupied state. The specific quantity of required energy varies and depends on the particular electron under consideration, the atomic number of the atom to which it is bound, and whether the atom is already excited or ionized, but it is usually on the order of a few electron volts (eV). Excitation may be induced by collision with a moving particle, absorption of electromagnetic radiation, or energy transfer through the electrostatic interaction arising from the movement of a nearby charged particle. In bulk matter, excitations can be brought about by traditional chemical interactions, such as heat absorption, which interfere with the function of a material as a radiation detector.

If an atom absorbs sufficient energy to raise an electron to an energy at which its kinetic energy exceeds its electrostatic Coulomb attraction to the nucleus, the electron becomes an unbound particle. The residual atom is then ionized, and recombination will eventually occur, in which the positive ion is neutralized by capturing an electron. Such capture will initially populate a high-lying energy state in the resultant atom, followed by the emission of photons of discrete energies that accompany the transition of the electron to the most tightly bound empty state, thereby returning the atom to its ground state. This same process accompanies de-excitation of an atom when the energy imparted was sufficient to promote an electron to a higher state, but insufficient to remove it. If de-excitation and photon emission follow the excitation or recombination processes immediately, the emitted photons are referred to as fluorescence. If the excited atomic state exists for a sufficiently long time before de-excitation, the emitted photons are termed phosphorescence.

Occasionally, an atom will absorb enough energy that an electron is removed from one of the inner, more tightly bound electronic states, rather than from a valence state. The resultant hole state is filled by transfer of an electron from a higher-electron shell, which is often accompanied by the emission of a high-energy, x-ray photon. X-rays are emitted with energies characteristic of the energy spacings between the inner electron shells of the atoms, which are a function of the nuclear charge. Rearrangement of the atomic electrons following x-ray emission gives rise to second-order interactions that result in the emission of photons of lower energy and ejected valence electrons. The latter is referred to as the Auger process.

7.4.2 Decay Characteristics

Alpha decay involves the emission of a ^{4}He nucleus, usually with an energy between 4 and 9 MeV. Alpha particles are emitted at discrete energies, making spectral analysis possible. In the process of stopping in matter, these α particles interact principally with the atomic electrons of the stopping medium, leaving a dense trail of excited and ionized atoms in their path (high specific ionization). The energy loss occurs through the electrostatic interaction of the moving charge with the orbital electrons. At the beginning of its trajectory through matter, when the kinetic energy of the particle is still high, the incident α particle may penetrate the electron screen and interact directly with the atomic nucleus. The recoil energy imparted to the scattering nucleus can also result in excitation or ionization. With the exception of a rare nucleus–nucleus Rutherford interaction, very little energy is transferred in a given interaction, and α particles therefore tend to follow straight-line paths. The range of 5-MeV α particles in most gases at standard temperature and pressure is about 4–6 cm. In solids and liquids, the range is much shorter, merely 10–30 μm. Only minimal intervening material can be allowed to come between an α source and the detector material.

Occasionally, the nuclear forensic analyst may have cause to measure the heavy ions accompanying the fission process, either induced by neutron capture or arising from the spontaneous decay process. The interaction of fission fragments with matter is quite similar to that of α particles, except that the range is shorter and the specific ionization along the flight path is higher. The result is a potentially high probability of the local recombination of electrons and positive ions that complicates energy measurements.

Beta decay involves the emission of negatively or positively charged electrons, with decay energies most probably between 0.02 and 3.5 MeV. Unfortunately, β particles are not emitted at discrete energies because of the simultaneous emission of antineutrinos (neutrinos accompany β^+ emission), which carry off a portion of the decay energy. Spectroscopic measurements of individual β-particle energies are not performed in the forensic radiochemistry laboratory. The mass of the electron (1/1836 amu) is so small that very little momentum is transferred in collisions with atomic nuclei, and the main interactions of β particles with matter are through the excitation and ionization of orbital electrons. Most detector methods rely on ionization, but excitation is much more probable. At a comparable energy, a β particle moves much faster than an α particle. As a consequence, it has a lower probability of interaction induced electrostatically by the moving charge, a longer mean-free path between interactions, and a lower specific ionization. Because they are easily deflected, the path of β particles through matter can be convoluted. Although the analyst can measure a range, it is defined as the distance traveled in the initial forward direction and has nothing to do with the actual paths followed by the particles. The energy transfer per collision is high, resulting in an acceleration of charge that causes the emission of photons. The slowing down and stopping of electrons in matter is accompanied by Bremsstrahlung radiation that can complicate the measurement of a range.

Auger electrons and conversion electrons obey the same rules as β^- particles, although they are emitted at discrete energies. The range of a 20-keV electron is approximately the same as that of a 5-MeV α particle. For very low β energies, Auger-process electrons, and internal conversion electrons, no more than micrometers of material can intervene between a source and the detector medium.

Positive β particles (positrons) behave in the same way as β^- particles until they are nearly stopped. At low energies, β^+ radiation is captured by electrons, creating a short-lived orbital species (positronium) that eventually annihilates. Annihilation releases twice the rest mass of the electron, which is kinetically constrained to appear as two photons of 0.511-MeV energy each.

Gamma-rays and x-rays are electromagnetic radiation and have a much lower probability of interacting with matter than does charged-particle radiation. Although charged particles interact directly to produce excited atoms and ions, photons may interact in several different ways. Because both γ-rays and x-rays arise in transitions between discrete states, they are emitted at discrete energies. The interaction of a collimated, monochromatic beam of high-energy photons incident on an absorbing medium is not characterized by a range, but rather, by a characteristic half-thickness (T). The intensity (I) of the photon beam, after passing through a thickness X of absorber, is given by

$$I = I_0 \exp[-(\ln 2)\, X/T],$$

where I_0 is the incident photon intensity. The half-thickness is a constant whose value depends on the nature of the material and the energy of the photons in the beam. For the purposes of discussion relating to radiation detectors, T is assembled from contributions of three different processes.

First, the photoelectric interaction is the most important for low-energy photons (up to ~0.5 MeV) and materials with high atomic number (Z). The photon interacts with a single atom, and the entire quantum of energy is transferred to this atom, which subsequently ejects an orbital electron. The kinetic energy of the ejected electron (E_{kin}) is given by

$$E_{\text{kin}} = h\nu - B,$$

where

$h\nu$ is the energy of the incident photon
B is the binding energy of the ejected electron before the interaction

This results in a sawtooth discontinuity in the interaction probability near the energies of the electron shells.

Second, the Compton interaction process is most important for higher-energy photons (between 0.5 and 5 MeV). Rather than interacting with the atom as a whole, the photon scatters off an individual electron, imparting a portion of its energy to it. The energy of the incident photon is divided between the lower-energy outgoing photon and the kinetic energy imparted to the electron, very much as in a classical scattering (billiard-ball) collision. From relativistic kinematics, the energy of an outgoing photon, $h\nu'$, is

$$h\nu' = h\nu/[1 + h\nu\,(1 - \cos\phi)/m_0c^2],$$

where

$h\nu$ is the energy of the incident photon
m_0c^2 is the rest energy of the recoil electron
ϕ is the angle between the scattered and incident photons

Third, pair production occurs only with very high energy photons. In this process, the photon is converted into an electron–positron pair, usually in the vicinity of a nucleus (necessary three-body electromagnetic interaction). The threshold energy for the process is $2m_0c^2 = 1.022$ MeV, but it competes with Compton scattering only for photons with $h\nu \geq 3$ MeV. Any photon energy in excess of the rest masses of the two outgoing leptons is divided between them as kinetic energy. The electrons and positrons subsequently interact with matter in the manner described earlier.

The relative rates of the three modes of electromagnetic interaction can be expressed in terms of absorption (or attenuation) coefficients, μ_i. The various absorption coefficients are plotted for Al and Pb in Figures 7.1 and 7.2. The individual values of μ_i for the three modes sum to a total μ_0, which is related to the half-thickness by

$$T = (\ln 2)/\mu_0.$$

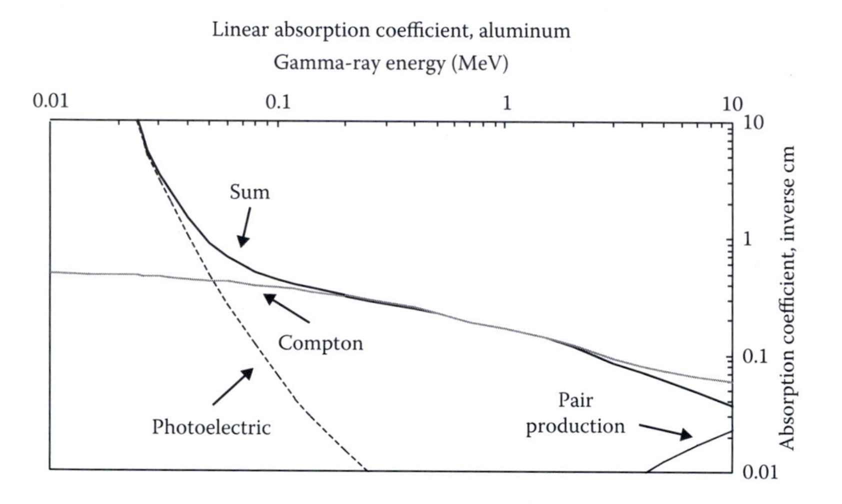

FIGURE 7.1 The components of the absorption (or attenuation) coefficient for photons in Al as a function of photon energy.

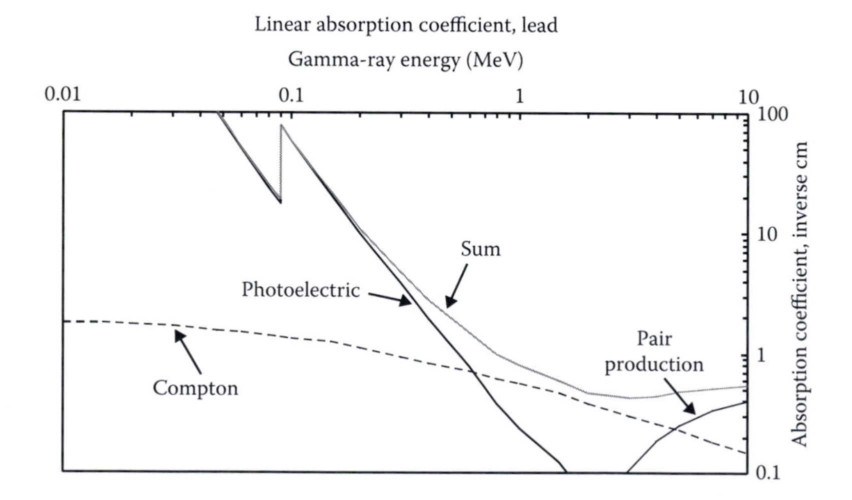

FIGURE 7.2 The components of the absorption (or attenuation) coefficient for photons in Pb as a function of photon energy. Note the discontinuity in the photoelectric coefficient at 90 keV, corresponding to the binding energy of the K-shell electrons of Pb.

On occasion, the radiochemical analyst will need to count neutrons, usually as a sensitive means of detecting the presence of heavy elements in a shipping container before discharge. Neutrons, being uncharged particles almost 2000× more massive than the electron, interact almost entirely with the nuclei of any stopping media. Because the dimensions of the nucleus are small compared to the extent of atoms, the probabilities of all types of neutron interactions are small and expressed in units of barns (10^{-24} cm^2). Neutrons do not induce primary ionization; they must therefore be detected by means of ionizing events caused by secondary radiation. Secondary radiation can arise from both elastic and inelastic neutron scattering on target nuclei, as well as following neutron capture. In both cases, the target nucleus is left in an excited state that de-excites through the emission of photons or charged particles, whose interactions with matter were described earlier.

7.4.3 Gas-Phase Detectors

The science of radioanalytical chemistry depends on the ability to accurately measure the intensities of the various radiations emitted by a source [4]. Early radiation detection instruments included the Geiger–Mueller (G–M) counter, introduced in 1928 [5]. The G–M tube is the prototype of most modern charge-collecting radiation detectors and is a sensing device that converts energy deposited by ionizing radiation into electrical impulses. The G–M tube consists of a volume of either He or Ar, with a front window that is sufficiently thin for β particles to penetrate. The inside of the container is plated with a conductive coating that serves as the cathode, and the anode is a wire that extends down the axis of the tube and is positively biased. Under the influence of the electric field, β electrons that penetrate the boundary of the tube, and the electrons released by primary interactions of the β particle with the counting gas, are accelerated toward the anode, acquiring sufficient energy that collisions with the counting gas create ion pairs. The ion pairs are then separated into electrons and positive ions by the potential of the tube and are similarly accelerated. This creates an avalanche of electrons, which discharges the tube and sends a signal to either a scaler or a ratemeter.

Discussion of the G–M tube introduces two interesting concepts that are common to many detectors. The first is quenching. Once fired, the discharge of an unquenched G–M tube is continuous because of the presence of delayed electrons. Some of these electrons arise from the decay of metastable electronic states in the counting gas and serves to delay the emission of ultraviolet radiation, with the subsequent production of photoelectrons and another avalanche. In addition, positive ions are collected by the cathode more slowly than are the electrons by the anode, as they are larger and thus less mobile. The process of neutralizing the positive ions may initiate the production of more photoelectrons. Quenching the G–M tube can be accomplished in one of two ways. In external quenching, the high voltage is removed from the tube once its discharge has been detected. In internal quenching, a small admixture of a polyatomic gas is added to the counting gas to absorb energy after an ionizing event. In internally quenched G–M counters, as positive ions move toward the cathode, there is a transfer of charge to the molecules of the quenching gas. When the charged quench molecules reach the cathode, they dissociate rather than produce electrons.

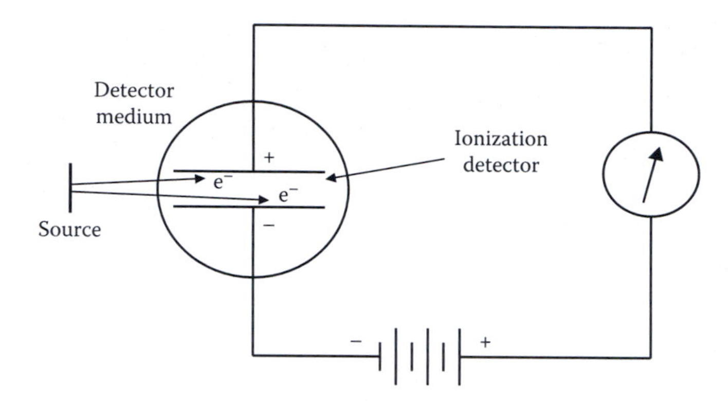

FIGURE 7.3 A schematic diagram of a general detector that functions via charge collection.

The second concept exemplified by the operation of a G–M tube is that of a region of operating bias. Clearly, without sufficient electrical potential, any β particles entering the G–M tube would not be accelerated sufficiently to create secondary ion pairs, and the primary ion pairs formed by direct interaction of the β particle with the counting gas would not be segregated into electrons and positive ions. Hence, no avalanche of electrons and no discharge of the G–M tube would occur. With too high an electrical potential, the counting gas will undergo charge-exchange reactions with the anode or cathode, and the tube would fire without any incident radiation. Between these two extremes are regions of bias that result in different operating regimes of the counter.

Figure 7.3 shows a general schematic diagram of a radiation counter that operates through charge collection. It consists of two parallel plates, or electrodes, across which a potential is applied by a battery. Current flow is read on the meter. As with the G–M tube, the electrode region comprises a material with which the incident radiation interacts. For the purposes of the discussion to follow, the assumption is that this material is a counting gas, but many of the concepts are also valid for solid materials. The interaction of ionizing radiation with the counting gas creates ion pairs, which induce small current flow between the plates and a resultant meter deflection.

The deflection of the meter depends on the extent of ionization produced by the radiation interaction and on the potential between the plates. In general, the current increases with voltage, but not in a linear fashion as might be inferred from Ohm's law. The ionized gas in a detector behaves in a more complicated fashion than does a simple resistor. A gas-filled counter, exposed to a radioactive source of constant strength, would produce a current *vs* voltage plot such as that in Figure 7.4. In gas-filled detectors, α and β radiations are classified as directly ionizing, whereas γ radiation and neutrons are indirectly ionizing. The interactions of γ-rays and neutrons are much more likely to occur in the dense walls of the detector than in the counting gas, making the response of the detector to these radiations principally the result of electrons ejected from the tube wall and into the gas volume.

When a radioactive emission enters the volume of gas, it creates ion pairs. The specific ionization along the path of an α particle is about 100–1000 times greater

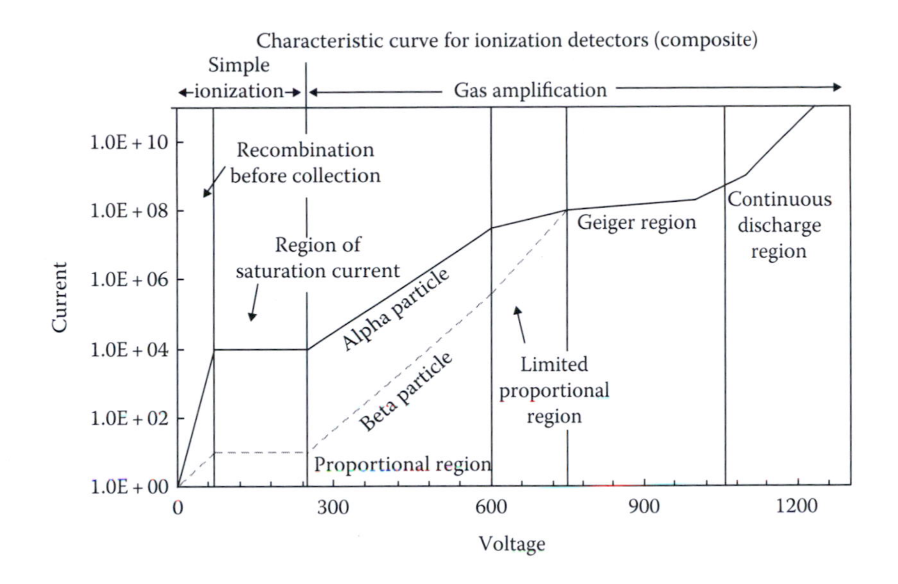

FIGURE 7.4 Gas-filled detector response as a function of applied voltage. This plot is sometimes referred to as the characteristic curve for a particular detector. This specific curve is schematic and can be considered a composite for various detectors of different designs.

than that for β particles, so an α particle traveling a certain distance through the gas volume produces 100–1000 times as many ion pairs as does the β particle in traversing the same distance. If potential is not applied to the electrodes, recombination of the electrons and positive ions occurs, and no current flows in the circuit. If, however, a small electrical bias is supplied, some of the electrons will be attracted to the positive electrode (the anode) and some of the positive ions will drift to the negative electrode (the cathode), causing current to flow. Those electrons and positive ions not collected will recombine to form neutral atoms. As the bias is increased, the fraction of ion pairs that recombines decreases, and the detected current increases. As the electrode bias is increased further, a state is eventually reached in which essentially all of the electrons and positive ions reach the electrodes. This results in a saturation current, and an increase in electrical potential beyond this point has little effect on the current until gas-amplification phenomena begin. The onset of the saturated-current mode is usually at an applied potential between 50 and 100 V, with the induced current between 10^{-15} and 10^{-12} amp.

A detector operating in the saturated-current mode is an ionization chamber. It is generally operated with an electrometer with amplification stages, which is required to detect such small currents. Because there is no gas amplification, there is no electron avalanche and no need for a quench gas. From Figure 7.4, it is clear that the saturated current derived from an ionization chamber is proportional to the number of ion pairs produced by the incident radiation. It is possible to accept or reject amplified signals in such a way that the detector records only α-particle events.

A continued increase in electrical bias applied to the electrodes would eventually induce an increase in the current produced by the incident radiation. This increase beyond the current supplied by collection of all primary ion pairs is explained by the formation of secondary ion pairs, mainly as a result of collisional processes generated by the accelerated electrons. Approximately 30–40 eV of energy must be supplied through impact to create a secondary ion pair in most counting gases. The number of ion pairs available for current conduction is increased, and this process is known as gas amplification. Because amplification applies to each ion pair individually, the magnitude of the induced current is still proportional to the number of primary ion pairs formed in the gas by the incident radiation. If an incident α particle induces 100 times more ion pairs than does an incident β particle, the current in the circuit caused by the passage of the α particle will still be 100 times greater than that induced by the β particle, although both currents will be many times larger than their saturation-current equivalents. This operating regime is known as the proportional region. Gas amplification varies from unity near the boundary with the saturated-current regime, to as much as 10^6 near the upper extreme of the proportional region.

Gas amplification has limits that are regulated by the total number of ion pairs that can be produced in a detector under the conditions at which that detector operates. The identity of the counting gas, its pressure and temperature, and the physical dimensions of the gas volume all serve to limit the number of ion pairs that can exist in the detector. This ceiling on charge density defines the limit of the proportional region. As an example, consider a gas-filled detector, with a physical limitation of 10^{10} on the possible number of ion pairs, being operated as a proportional counter with a gas-amplification factor of 10^6. During a certain operating interval, three events are detected, which deposit 10^2, 10^3, and 10^5 ion pairs in the detector. The total number of electrons detected at the anode should be 10^8, 10^9, and 10^{11}, respectively, but the third event generates only 10^{10} ions because of the limitations of the counter. The detector is operating proportionally in processing the signals from the first two events, but not when dealing with the third. At the upper end of the proportional region lies the region of limited proportionality, in which the restrictions imposed by the apparatus affect the counterresponse. Counters operating in the limited proportional region have found little utility for the quantitative measurement of radiation.

If the applied voltage is increased even further, all radiation, regardless of the number of primary ion pairs produced in the detector volume, will yield the same current flow. This is the avalanche condition that is familiar from the discussion of the operation of a G–M tube, and this portion of the characteristic curve of detector response is referred to as the Geiger region. At the upper end of the Geiger region, a potential is reached that results in multiple discharges, where the counting gas is interacting with the counter without being ionized by radiation. At these potentials, extremely high (and completely false) counting rates are reported by the detector, increasing rapidly with applied voltage.

From this discussion, the conclusion that a single detector could be designed to function in all operating regions might seem reasonable. This is not the case, however, as the physical dimensions of the electrodes, the gas composition, and the applied potential must be selected carefully for each region of the characteristic curve. Figure 7.4 is strictly schematic, a composite of several experiments. In each operating regime, a different method of current measurement must be employed. Each quantum of ionizing radiation causes a pulse of current to flow, with the pulse duration on the order of hundreds of ns to hundreds of μs. If current is measured with an integrating circuit, the pulses tend to be smoothed, generating a DC current that is averaged over several events if the count rate is sufficient. If current is measured with a differentiating circuit, the pulses are shaped, and no DC component of the current reaches the recording device. If the recording device is a scaler, the discrete pulses are counted to give the total number of ionizing particles triggering the detector. Ionization chambers may be operated as either differentiating or integrating devices, whereas proportional and Geiger counters are operated as differentiating devices.

In our counting laboratory, we use gas-filled proportional counters to measure the absolute count rate from sources of α-particle emitters. The counters are windowless, meaning that the sources are introduced into the counting gas (a mixture of 90% Ar and 10% methane, called "P-10") that fills the detector volume. The counter is operated at a voltage in the middle of the proportional region, where the gas-amplification factor is not limited by the physical constraints of the detector. Alpha particles, which induce specific ionizations much larger than those induced by β particles, produce large pulses from the differentiating circuit. A discriminator is included in the circuit to reject low-energy pulses (with the threshold established by counting a source of 3.2–MeV α-emitting ^{148}Gd and setting the threshold to reject <5% of the pulses). The efficiency of the counter for detecting 4–7 MeV α particles emitted from weightless sources mounted on Pt counting plates is 52%. The excess efficiency over 50% arises from α particles that elastically scatter from the substrate back into the counting gas. Addition of mass to the source, as well as fabrication on substrates with lower atomic number, serve to decrease the counting efficiency. The background count rate in these counters is determined by counting a blank Pt counting plate and is usually between 0.4 and 0.6 counts/min.

The resolving time of a detector is the time interval during which two or more ionizing particles that strike the detector will be counted as a single event. Some of this time arises from the physical interaction of radiation with the detector medium, and some from electronic signal processing. With the gas-filled counter, most of this time arises from the collection of charge. If a second ionizing event occurs while charge is being collected from a previous event, the multiple event will be integrated as a single one and produce an over-large pulse. There are electronic methods of rejecting some of these events based on the pulse shape. For the gas-filled proportional counter, the resolving time is typically about 1 μs. A dead-time correction must be made to the counting rate measured from sources that emit more than 10^4 α/s.

We also use proportional counters to detect and quantitate the rate of emission of β particles. In this case, sources are mounted external to the detector volume, and β particles enter the counting gas through a thin window. The window is thick enough that it stops α particles, which have the potential of producing current pulses that are difficult to discriminate against. The counters operate at higher bias potentials than do α counters, but they use the same P-10 counting gas because of its superior charge-collection properties. Beta particles are emitted from well-defined nuclear states with a broad range of kinetic energies, from zero to the decay Q-value (the end-point energy), as a consequence of co-emission of the antineutrino. Beta particles emitted from low-Q-value nuclides have a higher probability of possessing insufficient energy to penetrate the end-window and induce ionization in the counting gas than do β particles emitted from high-Q-value nuclides. Therefore, the efficiency of β counters is dependent on the end-point energy of the β emitter. This is the fundamental limitation of β counting. Alpha particles and γ-rays are emitted at discrete energies, and the spectroscopic determination of these energies provides the information necessary to calibrate the counter efficiency and energy response. With β particles, it is impossible to use the energies associated with a few events to determine an end-point. The analyst must know what nuclides are present in a β-emitting sample, what β end-points are represented by the decays of these nuclides, and the relevant counter efficiencies before the emission rate can be quantified. The main utility of the gas-proportional counting of β particles is measurement of the decay rates of those nuclides that do not emit other radiations, or the detection of radioactive emissions from very weak sources.

The best way to calibrate a β proportional counter is with a standard source containing a known quantity of the same mixture of nuclides as does the experimental sample, counted in the same configuration relative to the detector, and on the same substrate as the experimental sample. However, this approach is rarely possible, particularly as the analyst is usually working with unknowns, but β counters can be calibrated with commercial sources containing ^{90}Sr, ^{14}C, and ^{147}Pm, the decay of which do not include any γ-ray or α-particle intensity. The next best method of calibrating the efficiency of β counters is the use of a set of standards covering a wide range of end-point energies, followed by construction of an energy-dependent efficiency function. This method is less satisfactory because of changes in spectral shape caused by forbidden decay or multiple end-points, as well as by effects of conversion electrons accompanying any γ-ray emission.

The end-window of β proportional counters, which is thick enough to screen out α particles, also screens out very low energy β emissions of interest to the forensic analyst, such as ^{3}H (end-point energy = 18.6 keV) and ^{241}Pu (end-point energy = 20.8 keV). Other methods for measuring the β emission of these nuclides are discussed later, but the preferred modern methods for quantitating their concentrations are mass spectrometric.

Alpha-spectroscopic measurements have evolved toward the use of solid-state detectors. However, application of gas-filled ion chambers yet remains of value for these endeavors. The apparatus is referred to as a gridded ion chamber or as a Frisch-grid counter after the originator of the design [2]. A grid-type electrode is placed between the anode and the cathode and is maintained at a potential midway between the two electrodes. The source is mounted in the counting gas in such a way that all ion pairs produced by the interaction of the α particles with the counting gas (Ar) are initially confined between the grid and the cathode. This requirement often requires that the pressure of the counting gas exceed 1 atm to provide sufficient α stopping power. Positive ions drift to the cathode, while electrons are initially drawn to the grid, which is designed to be as permeable to them as possible. Once the electrons pass through the grid, the potential between the anode and the grid induces a current and, consequently, a drop in potential across a resistor. Now, however, the signal arises only from the drift of the electrons, rather than from both the electrons and the slow-moving positive ions. The result is an induced voltage signal, typically about 0.1 mV, with favorable time properties that can be amplified into a signal truly proportional to the energy of the α particle. This technique requires the use of relatively sophisticated preamplifiers and pulse-processing electronics to avoid losing the intrinsic energy and time resolution. The energy resolution attainable by a Frisch-grid counter depends on the number of primary electrons producing the signal in the grid–anode circuit, usually on the order of 0.2% (full width at half-maximum, FWHM) for α particles between 4 and 7 MeV. However, the actual resolution attained by a Frisch-grid counter is somewhat poorer because of the contribution of electronic noise to the processed signal. Quite often, range straggling caused by source thickness is the most important factor in the resolution of an α spectrometer system.

The insertion of the sample into the counting gas results in a high efficiency (~40%), which is very useful for samples formed from chemical fractions of limited activity. Frisch-grid counters can be designed to hold sources of large spatial extent, and the characteristic peak shape of the α spectrum from a Frisch-grid counter can be quite favorable for resolving closely spaced doublets. However, a drawback to the use of Frisch-grid counters is their sensitivity to electrons: any conversion electrons emitted during the α-decay process interact with the detector well within its resolving time and are summed with the α signal. The result is a high-energy "porch" on spectral peaks arising from decays of odd-atomic-number nuclides, such as ^{237}Np and ^{241}Am.

The last topic in the discussion of gas-filled detectors is the detection of neutrons. Neutrons are generally sensed through nuclear reactions that produce prompt, energetic, charged particles. Two pertinent reactions that proceed with high cross sections are

$$^{10}\mathrm{B} + \mathrm{n} \rightarrow {}^{7}\mathrm{Li} + \alpha$$

and

$$^{3}\mathrm{He} + \mathrm{n} \rightarrow {}^{3}\mathrm{H} + \mathrm{p}.$$

Both target materials can be loaded into gas-filled detectors, ^{3}He in its elemental state and B in the form of volatile $^{10}BF_3$. In both cases, the detection of a neutron is accomplished by the identification of an emitted charged particle, which creates ion pairs in the counter gas.

7.4.4 Solid-State Detectors

The basic operation of conduction-type, solid-state detectors is very similar to that of an ion chamber. An incident particle produces ion pairs. In the solid material, this results in a free electron and an electron hole in the structure of the bulk material. The positive ion is bound within the solid matrix and prevented from movement. When a potential is applied across opposite surfaces of the detector material, the electrons migrate to the anode. The holes may or may not migrate (through filling from neighboring electron states), depending on the nature of the material.

Solid-state detectors offer several advantages over gas-filled ion chambers. Their resolving time is much shorter, allowing higher count rates (semiconductor detectors have pulse rise times on the order of tens of ns). Most of the time associated with signal processing results from the electronics. Solid detectors are also intrinsically of much higher density, which results in greater stopping power and enables detection of photons. The number of ions produced in the interaction of radiation with a solid-state detector is greater than that produced in an equivalent interaction with a counting gas. Hence, improved energy resolution is attained.

The focus here is on detectors fabricated from the solid-state, intrinsic semiconductor materials, Si and Ge. In very pure samples of the crystalline solids of these materials, atomic electron energy levels delocalize, producing a band structure over the entire material. The highest occupied band of electrons, the valence band, contains its full complement of electrons allowed by the exclusion principle. The band above it, the conduction band, contains none. There is no mechanism for the movement of electrons through the solid unless electrons from the valence band can be promoted into the conduction band. Impurities can provide for conduction, but excessive quantities produce traps that delay charge collection and degrade detector performance. The energy gap between the two bands is 1.1 eV for Si and 0.75 eV for Ge.

The conductivity of Si and Ge can be enhanced by doping ultrapure material with very small amounts of elements such as P, As, or Al. For example, a small number of P atoms can be dispersed through a Si crystalline lattice without disrupting it. However, P has five valence electrons and Si only four, so every atom of P donates an electron that cannot be accommodated in the valence band. This electron is connected to the P^+ center, but its orbit is delocalized because of the permittivity of Si. The electron is effectively free and spends most of its time in the conduction band of the surrounding Si, acting as a conductor when electrical potential is applied across the crystal. Silicon doped with P is an n-type semiconductor, as conduction arises from an excess of negative charge over that which can reside in the valence band.

A small number of Al atoms (with three valence electrons each) added to a pure Si sample produces a p-type semiconductor, in which the valence band contains too few electrons to fill it. Such hole states make it possible for electrons to migrate through the valence band under the influence of an applied potential.

Early in the development of semiconductors, it was found that exposed materials were sensitive to radiation-induced ionization. An electrical potential applied to draw off any intrinsic charge carriers leaves behind a depleted layer that is particularly sensitive to the injection of new charge carriers by ionizing radiation. A further improvement is obtained by applying a reverse bias to a paired junction of n-type and p-type materials in close contact, drawing holes and electrons away from the junction and forming a diode. As the electric field increases, the fraction of charge carriers lost to trapping or recombination decreases. With a sufficiently high field, charge collection is complete, and this situation corresponds to the saturation region of an ion chamber. If only one type of particle is to be detected (e.g., α particles), surface-barrier detectors need not be operated at full saturation. The fraction of charge carriers lost for each event is nearly constant, so there is no loss of energy resolution.

For purposes of radiation detection, a semiconductor with a depleted volume of size comparable to the range of the radiation to be detected is required. Silicon has a larger band gap than Ge and is not as susceptible to the promotion of electrons to the conductance band by thermal interactions. Silicon detectors can therefore operate at room temperature, whereas Ge detectors must be cooled, usually with liquid nitrogen. In contrast, the atomic number of germanium ($Z = 32$) is significantly larger than that of silicon ($Z = 14$), making it a much more efficient material for γ-ray detection.

Silicon detectors are often used for the measurement of α spectra. The average energy needed to produce a charge–carrier pair in Si is 3.65 eV, compared with approximately 30 eV in the noble gases. Thus, the available charge that is proportional to the energy of the α particle is almost an order of magnitude greater and is distributed over a smaller area. The problem is to create a depleted region in the semiconductor that is thick enough to stop incident α particles, yet close enough to the surface to be within the range of the α particles impinging on that surface. In the commonly used surface-barrier detector, a slice of n-type Si is etched to clean the surface, which is then exposed to air to produce a thin film of silicon oxide, a p-type material. A thin film of Au, much thinner than the range of an α particle, is deposited on the oxide layer to provide an electrical connection. A reverse bias is applied across the Si crystal, between the Au layer and an ohmic contact on the back surface of the Si, which creates a charge-depleted region just behind the Au layer. The Au layer and the layer of silicon oxide are not detector materials and constitute a dead layer through which the α particles must penetrate before entering the semiconductor. If this dead layer is thin, the number of ion pairs deposited in the detector material is still proportional to the energy of the α particle, as the electrons are distributed along its path and only a small fraction is uncollected. Charge promoted to the conduction band is collected, and the signal is amplified to yield an energy measurement.

Surface-barrier detectors can also be produced by evaporating a thin layer of conductive Al onto the surface of p-type Si to form an n-type contact. These detectors, sometimes referred to as being "ruggedized" because they can be (carefully) cleaned with solvents, function in the same way as the Au-plated, n-type detectors, but with an operating bias of the opposite polarity. Surface-barrier detectors of both types are sensitive to light, as the thin entrance windows are optically transparent. Because of the nature of the interaction of α particles with matter, surface-barrier detectors are generally operated in a vacuum chamber that usually provides protection from ambient light.

The efficiency of a surface-barrier detector for measuring α particles is effectively determined by the solid angle subtended by the detector face relative to the extended source deposit and is independent of the energy of the α particle. Typical detector sizes are 1–5 cm^2 in active area, although larger detectors are available. With close spacing between the source and detector surface, efficiencies as high as 35% can be obtained. However, resolution of the detector suffers at these distances because of the number of α particles that transit the dead layer of Au and silicon oxide at oblique angles, adding a component of range straggling to the resolution of the peaks in an α spectrum.

Sources of nearly monoenergetic α particles (such as those emitted by ^{238}Pu or ^{241}Am) are used to test the performance of semiconductor detectors. These detectors are not very sensitive to interactions with electrons, so summing of α-particle energies with the energies of coincident conversion electrons is minimized. For these detectors, the limiting resolution resulting from the statistics of charge-carrier formation is approximately 3.5 keV for 5.5-MeV α particles. In practice, it is rarely possible to achieve a resolution better than 10 keV using commercial electronics, in part because of the combined effects of nuclear scattering and incomplete charge collection.

Germanium detectors have almost completely supplanted other systems for γ-ray spectrometry because of the energy resolution obtained and the relative convenience of use (despite the cooling requirement). A fundamental difference between the surface-barrier detectors used for α spectrometry and the Ge detectors used for photon spectrometry is the required depletion depth. Using semiconductor materials of normal purity, it is difficult to achieve depletion depths beyond 2 mm with applied bias voltages below those that would cause the semiconductor material to break down, much like the multiple-discharge regime in a gas-filled detector. Gamma-ray spectrometry requires much greater charge-depletion depths, necessitating great reductions in the concentrations of charge-carrying impurities in the materials.

There are two common ways of accomplishing the goal of greater charge-depletion depth. The first is compensation, in which the residual impurities are balanced by an equal concentration of dopant atoms of the opposite type. The most common method involves Li-ion drifting, which has been used to compensate both Si and Ge. The p-type detector crystal is warmed, and Li ions are diffused into the material, where they site themselves at the crystalline interstices to compensate the p-type contaminants. After completion of the Li drift, the detector must be maintained at liquid nitrogen temperatures to keep the Li^+ ions from continuing to drift, even in the absence of an applied bias potential.

Starting in the late 1970s, a process of zone refining was introduced to create very pure Ge material—so pure that compensation was no longer needed to produce a viable detector. In this process, a near-molten region is moved slowly down a long Ge billet. Impurities are more soluble in hot Ge than cold, so they tend to concentrate in the hot region. At the end of the process, the contaminated end of the billet is removed and discarded. This procedure may be repeated several times to achieve the desired purity. Impurity concentrations of $<10^{10}$ atoms/cm^3 have been achieved, where most of the residual impurities are of the p-type. This technique has been so successful that Li-drifted Ge detectors [Ge(Li)] are no longer produced, effectively supplanted by the high-purity Ge (HPGe) detectors. There is no equivalent method for Si, so Li drifting is still the only way to achieve high-depletion depths in Si crystals.

Germanium detectors for γ-spectrometric applications are usually in one of two geometrical configurations: planar or coaxial. A planar germanium detector is superficially similar in construction to a silicon surface-barrier detector. The electrical contacts are provided on the two flat surfaces of a Ge disk. The n-type contact can be created by evaporation of a small amount of Li on one surface of the detector, followed by a limited diffusion, or by ion implantation of donor atoms using an accelerator. The depletion region is formed by reverse-biasing the n–p junction, usually to encompass the entire detector cut to a thickness of 1–2 cm. Planar detectors are used principally for the spectrometry of low-energy photons;

high-energy γ-rays pass through the Ge wafer, depositing only a fraction of their energies and resulting in an increased continuum background in the detector.

Spectrometry of high-energy photons is accomplished with coaxial Ge detectors. The detector crystal is cylindrical in shape and has a cylindrical core removed along its long axis. The outer cylindrical surface is plated with one electrical contact (for p-type material, the n-type contact). The inner cylindrical surface makes up the opposite contact. The crystal can be fabricated to any arbitrary length, such that large, charge-depleted volumes can be obtained. The cylindrical hole can extend through the entire crystal or stop short of the front surface, in which case the front surface is also coated with the outer electrode. As bias is applied, the charge-depletion zone grows from the outer surface. Optimally, the entire crystal will be encompassed by charge depletion.

The average energy needed to produce a charge-carrier pair in Ge is 2.95 eV, which is slightly less than that required by Si. The basic limitation of the energy resolution of a Ge detector (aside from limitations resulting from noise in the signal processing electronics) is the statistical fluctuation in the number of ion pairs created for a given deposited energy. It is advantageous to have the smallest possible signal rise time, so the charge-collection time should be held to a minimum. In Ge, electron velocities continue to increase until the field reaches approximately 10^5 V/m, where they saturate at about 10^5 m/s. Hole states require a higher potential to reach their saturation velocity than do electrons. It requires 100 ns to collect charge in a detector having a critical dimension of 1 cm. Part of the energy of the interaction of radiation with the detector creates ion pairs, but part also heats the crystal lattice of the detector. For the 1332-keV γ-ray of ^{60}Co, the intrinsic resolution of Ge at liquid nitrogen temperature (77°K) is 1.6 keV (FWHM). In general, the width of the peak increases as the energy of the photon increases and the actual energy resolutions of the photopeaks in a Ge detector are determined via standard counting sources.

Like the surface-barrier detector, Ge photon detectors have a dead layer from which charge is incompletely collected. In addition, cooling requires that the detector be housed in a cryostat, which scatters low-energy photons. At low photon energies, the efficiency of the detector increases with energy. As the energy of the photon is increased, however, the probability of the photon undergoing a Compton interaction and scattering out of the crystal, or taking a shallow trajectory through the crystal in such a way that it is not completely absorbed, increases such that detector efficiency falls off with increasing photon energy (Figures 7.1 and 7.2). The energy at which the maximum efficiency occurs, and the rate at which that efficiency falls off to either side, is very much a function of the individual detector crystal. Both the efficiency response of the detector and its energy calibration are determined by counting primary radionuclide standards. Such standards may be fabricated from one nuclide with several γ lines of known intensity (e.g., ^{152}Eu) or from a mixture of multiple nuclides, each of which must be individually calibrated. As with the case of the conversion electrons accompanying α decay, care must be taken to place counting sources emitting coincident photons far enough from the detector that summing (two photons striking the detector within the resolving time) is unimportant.

Photon detectors suffer from background problems that α detectors do not. High-energy photons from the surrounding environment introduce a background spectrum that underlies all others. As a consequence, radioactivity counting rooms are generally shielded or constructed underground. Each detector is surrounded with several inches of Pb shielding and concentric metal liners (inside of which the counting sample can be placed at well-calibrated counting positions). Aged Pb is superior to freshly mined lead, which contains 22-year ^{210}Pb, a member of the U decay chain. The lowest backgrounds are realized with shielding materials that were mined long ago and that have been protected from cosmic rays and from the fallout from atmospheric nuclear testing.

The nature of the intrinsic interactions of photons with matter introduces features into the γ-ray spectrum. For high-energy photons, the probability that all of the photon energy will be deposited in the detector and add to the intensity at the energy of the photopeak can be quite small. There is always a significant probability that some of the energy delivered by the photon will leave the boundary of the charge-collection region. In that case, the photon will contribute to the spectrum at a lower energy, adding to the continuum of events underlying the low-energy portion of the spectrum. The kinematic constraints on Compton scattering cause photons scattered backward out of the detector crystal to transfer the maximum amount of energy to the forward-moving Compton electrons, which have energy somewhat less than that of the incident photon. This results in a step-like structure in the spectrum below each high-energy photopeak, called the Compton edge. Pair production caused by the interaction of high-energy photons with the detector material results in the creation of positron–electron pairs. The positron subsequently thermalizes and annihilates, producing two 511-keV photons. The 511-keV annihilation peak in a γ-ray spectrum is somewhat wider than surrounding photopeaks because of the motion of the positron relative to the detector medium upon annihilation (Doppler broadening). If one or both of the annihilation photons escape from the detector without absorbing, small peak artifacts appear in the resulting spectrum at energies 511 keV and 1.02 MeV below the principal γ-ray energy (single- and double-escape peaks, respectively).

Up to this point, discussion has focused on radiation detectors that require the formation of ions by the incident radiation and the subsequent collection and measurement of these ions. However, there are other ways in which radiation can be detected and quantitated. For the nuclear forensic analyst, the most important of these involves the production of photons by the incident radiation and their measurement though the use of scintillators.

7.4.5 Scintillation Detectors

As discussed earlier, the interaction of radiation with matter can result in the promotion of electrons in the atoms of the detecting medium into higher-lying bound states or can result in ionization. In the absence of an applied electric field, many of the resulting free electrons recombine with the positive ions to occupy upper excited states. De-excitation of these atoms results from the emission of photons, with wavelengths that span the electromagnetic spectrum from the infrared to the ultraviolet. The nuclear forensic analyst finds this phenomenon useful for the detection and quantitation of low-energy β emitters (e.g., ^{3}H and ^{241}Pu) with liquid scintillators, and for γ-ray detection using scintillation spectrometry. Instruments based on these principles are the descendants of the spinthariscope, just as charge-collection devices derive from the G–M counter and, before that, the electroscope.

There are a number of substances in which the atomic de-excitation process results in the emission of visible light. One of the earliest known fluors is ZnS, which scintillates upon being irradiated with α particles and which served as the basis for the spinthariscope. The fundamental principle on which this measurement works is that the number of photons produced in the fluorescent substance is proportional to the energy dissipated by the radiation in its passage through the fluor. Although it is possible to detect the flash of light by visual means in some instances, the scintillation method was not considered highly developed until the perfection of the photomultiplier (PM) tube, which converts flashes of light into a proportional number of electrons and amplifies the resultant pulse.

A basic scintillator detector system consists of a scintillator, which is intimately attached to a PM tube, perhaps through a light pipe or light guide of transparent material; a high-voltage supply, necessary for the operation of the PM tube; and a preamplifier, generally located at the base of the PM tube, which converts the high-impedance electrical pulses produced by the photomultiplier into strong low-impedance pulses that are capable of being conducted through a cable to an amplifier or scaler. The key to the operation of the PM tube is the photosensitive cathode. The PM tube end-window is usually silica or quartz to allow the passage of both visible and ultraviolet light.

The photocathode is a thin sheet of metal that, when struck by photons, emits electrons. The energy that can be transferred from photon to electron is equal to the energy of the photon. Many scintillators emit a blue light, with each photon quantum carrying approximately 3 eV of energy. The incident photons are absorbed, electrons are dislodged from the atoms of the photoemissive cathode material, and the free electrons migrate to the surface of the cathode. The potential barrier to emission of the electron from the surface of the cathode is higher than the residual energy of the electrons (after collisions) for most metals. However, if the cathode surface is coated with alkali metals, the potential barrier can be as low as 1.5 eV.

The operation of a PM tube is analogous to the operation of a gas-filled detector in the proportional region. Electrons are accelerated and undergo collisions to produce more electrons. Because the multiplication factor of the tube applies to each electron leaving the cathode, the output electric current signal of the PM tube is proportional to the initial electron production, which, in turn, is proportional to the number of photons striking the cathode and is roughly proportional to the energy deposited in the scintillator by the incident radiation. Electrons from the photocathode are emitted with energy <1 eV, and the body of the PM tube is evacuated to minimize collisional losses from electron scattering off gas molecules. The electrons are accelerated toward an electrode, called a dynode, which is held at a positive potential of several hundred volts relative to the cathode. The kinetic energies of the accelerated electrons that reach the dynode are almost entirely determined by the magnitude of the accelerating voltage, and the collision of a single electron with the dynode can produce as many as 30 free electrons per 100 V of potential. A fraction of these δ electrons will reach the appropriate surface of the dynode with sufficient energy to be emitted, but if the dynode is fabricated from the proper materials, the multiplication factor can be as much as 5–10. To achieve electron gains of 10^6 or more, all PM tubes employ multiple stages. Similar to emissions from the cathode, secondary electrons emitted from the surface of the first dynode have very low energies and must be accelerated toward a second dynode that is biased positively relative to the first one. This process can be repeated as many times as necessary. If the multiplication factor at each dynode is five, then 10 stages will yield an overall tube gain of 5^{10}, or about 10^7. The last dynode is actually the anode of the PM tube, and the current pulse in the anode is amplified to give the tube output signal.

The function of a scintillation detector relies on the transmission of light and the acceleration of electrons in vacuum. The width of the current pulse arriving at the anode is only a few tens of ns, but the resolving time of the system is limited by the electronics, rather than by the detector itself. Because the transmission of light to the PM tube depends on the spatial distribution of light produced in the scintillator, the efficiency of the detector is not only dependent on the distance and extent of the radioactive material, but also on its position relative to the phototube. This results in a detector efficiency that must be determined at each counting location, for each tube, and at each energy of incident radiation. The same problem is encountered with Ge semiconductor detectors.

For γ-ray scintillation detectors, single large crystals of sodium iodide containing a trace of thallium iodide as an activator are used almost exclusively. Thallium ions distribute themselves uniformly throughout the crystal and give rise to light at frequencies close to the maximum sensitivities of commercial PM tubes. The detector medium is designated a NaI(Tl) crystal. For high detector efficiency, the incident γ-rays must be highly absorbed by the fluor. Sodium iodide offers the advantage of both high density (3.7 g/cm^3) and high atomic number ($Z = 53$ for I). Large, transparent crystals of NaI can be grown in a dry, inert atmosphere, making it possible to create detectors with very high efficiencies for high-energy photons. Unfortunately, NaI(Tl) is very hygroscopic and quickly becomes opaque upon exposure to air. As a consequence, NaI(Tl) detectors are almost always fabricated in Al cans, with plastic or quartz windows for connection to the PM tube.

The yield of photons and the efficiency with which they pass to the phototube are much lower than the production of electrons and their subsequent collection in semiconductor detectors. As a result, the energy resolution of a NaI(Tl) detector for photon counting is very inferior to that of a Ge semiconductor detector. However, the larger efficiency of the NaI(Tl) detector makes it a particularly valuable tool in the forensic radiochemistry laboratory. NaI(Tl) crystals can be fabricated with a hole through part of their depth, called a well. The superior efficiency of the crystal for detection of radiation emitted from samples placed in the well makes it a valuable tool for monitoring the success of chemical separation procedures during isolation of fractions of very low activity.

Many organic substances have long been known to fluoresce when excited by x-rays or β particles. Anthracene and naphthalene are grown into clear, thin crystal plates that are excellent scintillators for β radiation. Many other organic substances, particularly aromatic compounds related to the simple phenols, scintillate in the blue and near-ultraviolet regions of the electromagnetic spectrum. Many organic liquids scintillate strongly if they are highly purified, and they can also dissolve small quantities of other aromatics that act as wavelength shifters. It is important that the resultant liquid be transparent to its fluorescent emissions. Many of these liquids are commercially available as scintillation cocktails, usually consisting of a matrix fluid (e.g., xylene or toluene), a principal scintillator (e.g., *p*-terphenyl), and a wavelength shifter (e.g., 2-(1-naphthyl)-5-phenyloxazole).

A liquid-scintillation counter consists of a PM tube that can be optically coupled to a transparent cell containing the scintillator fluid through a light guide. Dissolution of a radionuclide sample in the scintillator allows β particles (and low-energy photons) to be detected with nearly 4π geometry. In practice, the quantity of radioactive analyte that can be introduced into the scintillator is limited because its chemical presence interferes with the function of the liquid as a scintillator (i.e., its existence in solution may interfere with the collection of light through the introduction of turbidity or discoloration of the liquid). The correction for loss of detector response caused by these effects makes it very difficult to perform a truly quantitative measurement by liquid scintillation counting. Preparation of sources for scintillation counting is discussed elsewhere in this book.

Although it is being supplanted by MS techniques for accurate measurement, liquid scintillation counting is used by the nuclear forensic analyst to determine the concentrations of low-energy β emitters. The determination of the 21-keV β-emitter ^{241}Pu is problematic because of the high levels of α decay that occur in a typical Pu sample, as well as the ingrowth of ^{241}Am (which emits conversion electrons in its decay). By counting a freshly separated sample and adjusting the discrimination of the output signal of the preamplifier, the analyst may obtain a reasonable approximation of the concentration of ^{241}Pu. In addition, the determination of ^{3}H in environmental samples via liquid scintillation counting is a mature art.

7.4.6 Empirical Application and Spectra

The analysis of α-particle intensity data obtained with a gas-filled proportional counter is fairly straightforward. Because the efficiency of the detector is essentially independent of the energies of the emitted α particles, it is applied to the count rate as a whole to obtain the absolute disintegration rate. The only exception to this tactic involves samples that emit significant amounts of short-lived Rn isotopes. For example, samples containing 3.66-day ^{224}Ra will always contain 55-s ^{220}Rn in radioactive equilibrium with the α decay. Some of the Rn, which is a noble gas, migrates out of the necessarily thin α source and enters the counting gas. Decays of ^{220}Rn that occur in the counting gas are detected with nearly 100% efficiency, compared to the 52% efficiency for those that decay in the counting source. Some fraction of the decays of the daughter of ^{220}Rn decay, 0.14-s ^{216}Po, will also occur in the counting gas, as not all of these more refractory atoms have a chance to drift to the counter walls before decay. Our counters are flow-through devices, which are constantly purged with a flow of counting gas; this mitigates the effect of 3.8-day ^{222}Rn, which is largely swept from the gas volume before decay, but the effect of the diffusive loss of radon on the counting rate from the source itself is difficult to calculate. Thorium sources must be counted as soon as practical after separation from Ra in order to minimize the effect of Rn evolution. The decay rates of Ra-containing samples cannot be quantified with a gas-filled counter. The correction of decay rates for detector dead-time is accomplished with a set of decay-rate standards.

Analysis of β-particle data from gas-filled proportional counters is complicated by the response of the detector to β radiations of different end-point energies. The most common way of resolving the various components of β emission that make up the radioactivity characteristic of a source is through decay-curve analysis. A relevant example is as follows: a Sr source separated from a Pu sample will contain β activities from the decays of 50.5-day ^{89}Sr and 29-year ^{90}Sr, both of which are prominent fission products. Eventually, 2.7-day ^{90}Y will come into radioactive equilibrium with ^{90}Sr. If the activity measured from the counting source is tracked as a function of time, it will increase at first from the ingrowth of ^{90}Y and then decrease according to the half-life of ^{89}Sr until it approaches the level of activity supplied by the combination of ^{90}Sr and ^{90}Y. Resolution of decay curves into components of each contributing activity can be performed graphically or with commercial software. This determination results in a set of initial activities to which energy-dependent detector efficiencies can be applied. However, the method does require some patience.

In our experience, a mixture of Sr isotopes arising from the SF of mixed Pu isotopes must be counted for about 200 days before the initial activity of ^{90}Sr can be determined. Many analysts attempt insertion of degrader foils of varying thickness to alter the portion of the β spectrum reaching the detector. Again with mixed Sr isotopes as the example, the β end-point of ^{90}Sr is 546 keV, that of ^{89}Sr is 1488 keV, and that of ^{90}Y is 2281 keV. In principle, a foil could be interposed between the source and the counter to absorb most of the radiation associated with ^{90}Sr and ^{89}Sr (Bremsstrahlung, being largely photons, is characterized by a half-thickness rather than a range). It is still necessary to measure the fraction of decays of the sample that penetrate the foil as a function of time, but the admixture of ^{89}Sr can be reduced to the point that the decay (mainly ^{90}Y in equilibrium with ^{90}Sr) may only have to be followed for a few days. Counting efficiencies through an absorber are calibrated with radionuclide standards.

Alpha-particle and γ-ray spectrometry share many features of data acquisition. The detectors operate in pulse mode, and the output of the detector element (whether gaseous or solid) is converted from a low-amplitude current or voltage pulse to a higher-amplitude linear voltage pulse, with amplitude proportional to the energy deposited in the detector matrix. Preamplification of small current or voltage signals must take place at the detector to minimize the magnitude of signal noise caused by cable capacitance. The preamplifier (or preamp) is the avenue through which the power supply biases the detector volume. For proportional counters, where scalers are used to record the number of pulses as a function of time, the demands of signal processing are minimal. When energy information is required, however, an amplifier converts the preamp output pulse to a voltage signal, the amplitude of which is proportional to energy. This is the first step in pulse-height analysis and results in energy spectra. It is at the amplification stage that issues of peak shape/rise time, pileup rejection, continuum reduction, and timing are addressed. This step is particularly important for γ-ray spectrometry with Ge detectors. The complicated shape of the detector crystal results in pulse shapes that vary from event to event, depending on where the incident radiation enters the detector. In addition, conversion of

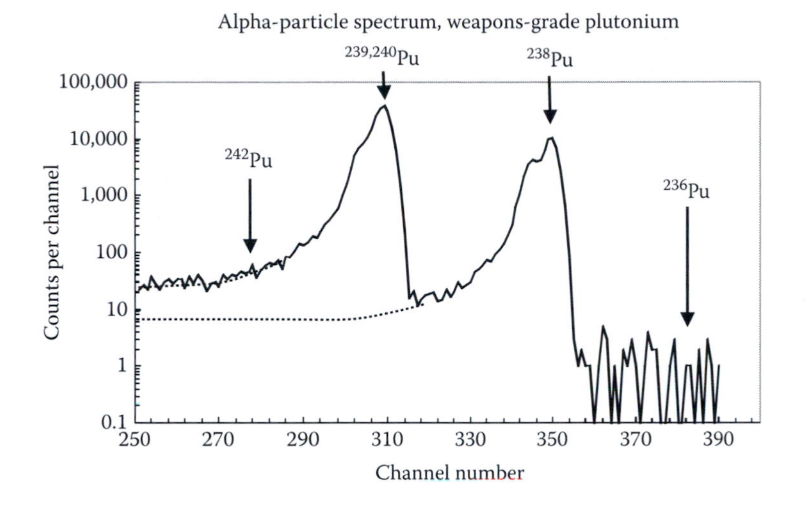

FIGURE 7.5 A representative α-particle spectrum of chemically purified weapons-grade Pu.

the preamplifier output pulse into a well-behaved, energy-proportional voltage pulse depends on where the voltage is picked off the wave form. The amplitude of the amplifier output pulse is measured and converted to a digital number with an analog-to-digital converter (ADC) and is then stored as an event in a multichannel analyzer (MCA). Detection of numerous events associated with a particular digital output from the ADC will result in that number of events stored in a channel of the MCA after the acquisition period is complete. The counts in each channel of the MCA represent the spectrum of radiation reported by the detector element.

Figure 7.5 shows a section of a representative α-particle spectrum, taken with a Si surface-barrier detector. The sample was freshly purified, weapons-grade Pu volatilized onto a stainless steel substrate. It was essentially weightless, and the energy resolutions of the peaks, as well as their shapes, were defined by the response of the detector. Any residual mass in the sample would result in a broadening of the peaks and an increase in the amount of the tailing of high-energy peaks into low-energy peaks. The shapes of the tops of the peaks are the result of fine structure in the decay. In the decay of even-mass Pu isotopes, there is some probability of α decay to the first excited state of the daughter, giving rise to a low-energy satellite peak. The conversion of MCA channel number to energy was accomplished through a calibration function, which was obtained by counting standard sources. The efficiency of the detector was not precisely known, but was effectively independent of energy so that ratios of activities could be measured. The pileup events induced by more than one α particle entering the detector during the resolving time precluded observation of the decay of ^{236}Pu at 5.77 MeV. The detector also responded to rare fission events arising from the SF decay of ^{240}Pu, which had pulse heights that far exceeded the range of the MCA and were tracked separately with a scaler.

The characteristic peak shape of each α spectrum is somewhat different, being a complicated function of the detector response and the thickness and extent of the deposited radionuclides making up the counting source. As a result, analysis of α spectra and calculation of activity ratios are more primitive than for analyses of γ-ray spectra. Alpha peaks are treated as histograms, summed between low and high limits defined by a fraction of the peak height (usually 1% or 2%, depending on the separation of peaks). The intensity underlying lower-energy peaks as a result of tailing from high-energy peaks is subtracted from the histogram graphically. This correction can be relatively small, as in the tail intensity under the $^{239,240}Pu$ peak, or can be large, as for the tail correction to ^{242}Pu. To convert an activity ratio to an atom ratio requires not only the half-lives of the two nuclides, but also the decay branch resulting in the observed radiation. For the Pu isotopes giving rise to the spectrum of Figure 7.5, this branch is effectively 1.0. However, the decay of ^{212}Bi generates α particles 36% of the time, so any α intensity resulting from ^{212}Bi decay must be divided by 0.36 to give the total decay rate. The situation with ^{241}Am is more complicated still, because of conversion-electron summing. The α decay of ^{241}Am populates an excited state in ^{237}Np that promptly emits conversion electrons, some fraction of which adds to the energy signal produced by the α particles and results in a high-energy shoulder on the peak in the α spectrum. This intensity has been lost from the main peak and must be added back into the sum for an accurate intensity assessment.

Figure 7.6 shows a section of a representative γ-ray spectrum taken with a coaxial Ge semiconductor detector. The source was a ^{152}Eu calibration standard, counted 0.5 cm from the end-window of the detector. Data were collected for photons with energies between 60 and 2000 keV and stored in 4096 MCA channels, but only the data between 500 and 1500 keV are displayed in the figure. This figure shows many of the features common to γ-ray spectra. Gamma-ray photopeaks arising from the complete absorption of photons emitted in the decay of ^{152}Eu are labeled with their energies in keV (only for the more intense peaks, however). For the high-energy photons in the spectrum, there is a greater probability that the photon will not be

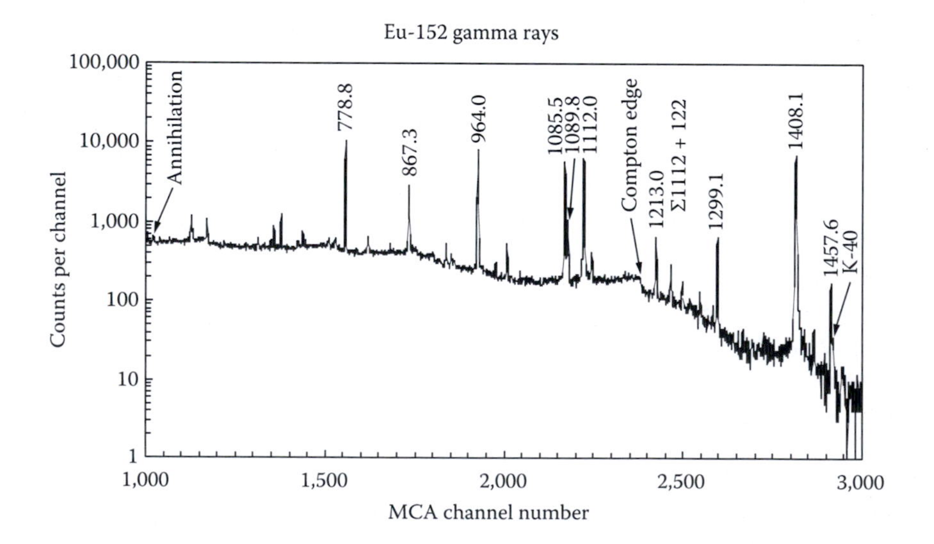

FIGURE 7.6 A section of a representative HPGe γ-ray spectrum: a ^{152}Eu calibration source counted close to the detector.

completely absorbed and will contribute to the intensity of the background continuum at energies lower than the photopeak. One component of this continuum is caused by Compton interactions, in which a photon is scattered out of the detector crystal in the direction from which it came, transferring a quantity of energy to the Compton electron (which is absorbed). The electron is kinematically constrained to a maximum value that is less than that of the incident photon, and it gives rise to a sudden increase in the intensity of the continuum at energy values less than the maximum electron value, termed a Compton edge. The Compton edge associated with the 1408-keV photopeak is located near channel 2400 and labeled in the figure. There is another Compton feature in the spectrum, arising from photons associated with the photopeaks at 1085.8 and 1112.0 keV. It is more difficult to observe because of the higher photoelectric continuum, the lower intensities of the decay branches, and the smearing out caused by the two transitions being near one another. Nevertheless, there is a Compton feature in the continuum at an energy just greater than the 867.3-keV photopeak. It can interfere with an accurate determination of the magnitude of the background, which is necessary to reliably determine the peak area.

There is a small branch for ^{152}Eu to decay by β^+ emission, hence the small annihilation peak at 511 keV. Also observed are spectral doublets (two incompletely resolved peaks of similar energy and intensity observed in the same spectrum), one at 1085.8 and 1089.8 keV, arising from two different transitions in the decay of ^{152}Eu, and one at 1457.6 and 1460.8 keV, resulting from a transition in ^{152}Eu and the decay of ^{40}K (detector background). Doublets and higher-order multiplets are more common at lower energies in complicated spectra. Samples emitting a variety of low-energy (<100 keV) photons should be counted with a planar Ge detector. The planar detector often offers lower continuum backgrounds and better energy resolution, both of which aid in the accurate analysis of multiplets.

Some of the lower-intensity peaks in Figure 7.6 are a result of summing, resulting from the close proximity of the source to the detector. One of these is indicated with the Σ symbol. There is a $J^\pi = 3^+$ state in ^{152}Sm at an excitation energy of 1234 keV, populated in the decay of ^{152}Eu. Because of the change in quantum numbers, this state cannot decay directly to the $J^\pi = 0^+$ ground state through the emission of a single 1234-keV photon. Instead, its preferred decay path is the emission of a 1112-keV photon (whose photopeak is visible in the figure), populating the $J^\pi = 2^+$ first excited state at 122 keV, followed by the emission of a 122-keV photon to arrive at the ground state. These two photons are emitted in cascade. The lifetime of the 122-keV state is only approximately 1.5 ns, so the two photons are emitted effectively simultaneously for Ge detection.

If both photons strike the detector and are completely absorbed, the sum peak is observed. In this case, not only does summing create a feature in the γ-ray spectrum that does not truly exist, it also decreases the intensities of both the 1112- and 122-keV photopeaks. Cascade summing can be reduced by increasing the source-to-detector distance, thereby reducing the chance that both photons will strike the detector simultaneously. However, the forensic analyst often has limited quantities of sample to work with. When the radionuclides being quantitated have photons in cascade, the analyst must weigh the effect of counting the sample at a distance (and reducing the counting efficiency) against counting the sample close to the detector (and compromising the intensities of photons from summing effects). Summing is discussed in more detail later.

The quantitation of the intensities of photopeaks measured in a γ-ray spectrum is a mature field of study [6–8]. Computer software exists that can input efficiency functions, energy calibrations, and background information, and then integrate the areas of peaks in spectra acquired from experimental counting samples. This software can then subtract a component for the intensity of the continuum under the peak (the quantitation of which remains a matter of some disagreement) and generate a table of photon energies and associated intensities in absolute photons per minute. Energy and efficiency calibrations are benchmarked against counting standards, but are largely theoretical constructs that can be modified for attenuation (by both the source matrix and any intervening foils) and finite-source extent. A functional fit to the peak shape, whose width is a function of energy, is used for the resolution of multiplets in complex spectra. Analogous to α-particle spectra, photon intensities are converted to a number of atoms using the half-life of the nuclide and the intensity of the decay branch giving rise to the photon [9]. For example, 30.1-year ^{137}Cs, a fission product that can be used for chronometry of Pu samples, emits a single photon in its decay with a probability of 85%. A popular calibration nuclide, 13.23-year ^{152}Eu, has more than 100 known γ-ray lines in its decay scheme, with high-intensity lines spanning the energy interval between 120 and 1408 keV. Its strong 344.1-keV line is emitted with a probability of 27.2%. Thorium-231 (25.5 h) has a high probability of emitting an 84-keV photon, so the preferred detector for counting ^{231}Th is likely planar rather than coaxial.

Because photons of several different energies can arise in the decay of certain nuclides, if the efficiency of the detector has been modeled correctly, they should all give the same value for the number of parent atoms when corrected for the appropriate decay branches. The best value for the number of atoms is obtained by taking a weighted average of the values determined from all of the photons emitted by the radionuclide. However, some fraction of the uncertainty in both the nuclear intensity data and the efficiency response of the detector is systematic in origin (rather than random) and should be applied to the results of the averaging procedure instead of during calculation of the average.

As discussed previously, establishing a common time to which all radionuclide concentrations are referenced is required by the radioactive nature of the chronometric species. As a consequence, measurements of emitted radioactivity must be made at known times, although this requirement is less important in MS applications because the half-lives of the analytes tend to be long. The time attributed to the data of a particular count is usually the midpoint of the counting interval. As long as the length of the counting interval is short relative to the half-life of the analyte species being measured, the midpoint time can be used in decay-curve analysis and in simple extrapolations back to the reference time. However, radioactive decay is exponential rather than linear, so for long counting intervals, a finite-counting-time correction must be made. For a single-component, simple radioactive decay, the count time associated with a counting interval of duration Δt is calculated to be a time δ after the start of the interval:

$$\delta = (1/\lambda)\ln\{\lambda\Delta t/[1-\exp(-\lambda\Delta t)]\}.$$

In the limit of long half-lives or short counting intervals, $\delta \rightarrow \Delta t/2$, as expected.

7.4.7 Coincidence and Anticoincidence Counting

As mentioned previously, radioactive decay often results in the emission of more than one energetic particle and/or photon. For instance, the α decay of ^{241}Am is frequently (36%) accompanied by the emission of a 60-keV photon and more frequently by the emission of conversion electrons [9]. A gas-filled Frisch-grid counter is often used to measure the energy spectrum of α particles emitted by an Am source. Frisch-grid counters are insensitive to interactions with photons, but low-energy electrons deposit some of their energy in the counting gas. Since the nuclear state produced by the α decay is not isomeric, the emissions of the α particles and electrons are effectively simultaneous, within the resolving time of the detector. If more than one particle enters the detector at a given time, the total energy deposited in the detector is interpreted by the electronics as a single energy signal. This is the reason that the ^{241}Am α-particle peak observed in a Frisch-grid spectrum has a high-energy shoulder.

In the earlier example, conversion-electron summing has taken intensity from the main α-particle peak and placed it elsewhere in the spectrum. For true quantitative measurements, the analyst must work to avoid the effects of summing, and there are two types of summing [2]: chance coincidence and true coincidence. The preceding example is a true coincidence; both radiations striking the detector are emitted by the same atom at the same time, and the probability of summing is related to the product of the efficiencies of the detector for observing the two emitted radiations. If the activity of the sample is increased, the probability of the conversion electrons emitted in the decay of one atom and the α particle emitted by a different atom both entering the detector at the same time increases; this is an example of chance coincidence. The probability of chance coincidence can be reduced by preparing a less intense source, but this tactic has no effect on the probability of a true coincidence.

One method of reducing the relative rate of true coincidences is to decrease the sensitivity of the counter for the detection of one of the emitted radiations. For instance, when performing a spectrometric measurement on an Am source, surface-barrier detectors are preferred over gas-filled detectors because they are less sensitive to the detection of conversion electrons. Further, when two photons are emitted simultaneously, it is often possible to selectively reduce the detection efficiency for the lower-energy photon by covering the source with a layer of cadmium and/or lead. Measurement of the intensity of a low-energy photon emitted in coincidence with a high-energy photon is better accomplished with a thin low-energy photon spectrometer (LEPS) detector than with a thicker coaxial detector.

A second method of reducing the effect of true coincidences is to back the sample off, away from the detector. If the detection efficiency for both radiations is reduced, then the coincidence probability (proportional to the product of the individual efficiencies) is reduced relative to the single-event rate (proportional to each efficiency). For most applications, and particularly in γ-ray spectrometry, increasing the distance between the source and the detector is the preferred method to reduce coincidences of both types. However, if the radioactive emissions from the source are weak, this approach also increases the length of time required to obtain a statistically meaningful spectrum. This is especially important when the low-intensity analyte is intermingled with a much more intense emitter, or when the decay rate becomes comparable to, or less than, the detector background.

One way to circumvent the increased distance required to minimize the effect of true coincidences is to perform a "near-far" measurement with a stronger calibration source of the limiting nuclide. The source is counted sufficiently far from the detector that summing is not a significant factor, after which the source is moved closer to the detector face and counted again. Based on these measurements, a summing correction factor is calculated. This factor applies only to close-in counts of a source of the same nuclide, with the same geometry, at the same location, relative to the same detector. Once the "near-far" calibration factor has been calculated, the analyst can report a quantitative concentration derived from a close-in count, even for nuclides whose radioactive emissions are effectively simultaneous.

The coincident emission of more than one radiation signature in a single decay has given rise to the use of multiple detectors to enhance the detection of one nuclide over another in a mixture, to increase the efficiency of detection of photons, and to reduce the prominent spectral continuum caused by Compton scattering between high-energy photons and the detector material. In most cases, however, these techniques enable the detection of rare processes to the detriment of the analyst's ability to quantify them. Coincidence is the enemy of quantitation.

A "clover" detector consists of four coaxial, HPGe photon detectors in a close-proximity clover configuration, sharing a common cryostat. A γ-ray that deposits its full energy in one petal of the clover will contribute to the photopeak intensity, while those undergoing Compton scattering out of the detector volume will deposit only part of their energy, adding to the low-energy continuum in that detector. The scattered γ-ray may encounter one (or more) of the other detectors in the clover, resulting in coincident energy signals in more than one element of the clover. If treated individually, each of the signals adds to the continuum in each detector spectrum. If coincident energy signals are summed, the continuum is reduced and the photopeak intensity is strengthened, a process referred to as "add-back." Like the calculation of near-far corrections, add-back results in a nuclide- and counting-geometry-specific efficiency function and a broadened peak shape, both of which are hard to model because the Compton scattering process is influenced by the electric or magnetic multipolarity of the incident photons [10]. If operated as four individual detectors, a clover detector provides additional efficiency for weak sources counted at a distance. If the add-back feature is used, the impact on the accuracy of quantitation can be severe, and measurements quickly become spectroscopic rather than spectrometric. The application of clover detectors to nuclear forensic measurements is limited.

The application of β–γ coincidence counting is a common measurement technique. If a sample is mounted in the detector volume of a gas-filled proportional counter, the efficiency for detecting β particles approaches 100% (a "4π" counter). If a γ-ray detector is mounted nearby, the coincidence rate between β particles and simultaneously emitted photons is determined solely by the efficiency of the photon detector. The β–γ coincidence method is invaluable in the production of calibration sources, and accuracies in emitted photon intensities of as little as ±1% can be achieved [2]. The method is less useful in the quantitation of the radioactive emissions from a mixed radionuclide source.

Compton rejection by anticoincidence is a valuable technique for increasing the analyst's ability to detect weak γ-ray emissions from a minor component in complicated mixtures of radionuclides. The method is similar to that employed by a clover detector, but Compton events are rejected, rather than added back into the photopeak [11,12]. The apparatus usually consists of a cylindrical Ge detector surrounded by an annular detector whose efficiency is more important than is its energy resolution, quite often a scintillator such as bismuth germanate or NaI(Tl) [2]. Complete absorption of a photon by the central detector results in no signal in the annulus. For photons that do not deposit their full energies in the central detector, escaping scattered photons have a high probability of depositing all or part of their energy in the annular scintillator. If photons are detected simultaneously in the central detector and the annulus, the event in the central detector is rejected, resulting in a reduction of the height of the Compton continuum in the spectrum relative to the intensities of the photopeaks. Escape peaks resulting from pair production are also reduced in magnitude, although the lifetime of the positron in the detector matrix can delay the detection of escaping radiation by the annular detector.

Compton rejection by anticoincidence is a tool for reducing the continuum under weak, low-energy photopeaks, resulting in an improved sensitivity for detection. However, if the nuclide to be quantified has a complicated decay scheme, the annular scintillator must be shielded from direct interaction with the true-coincident photons emitted by the source. The detection in the annulus of photons emitted by the source causes the rejection of fully absorbed, in-cascade photons from the central detector spectrum, adversely impacting quantitation.

REFERENCES

1. K. Siegbahn, ed., *Alpha-, Beta- and Gamma-Ray Spectroscopy*, Vols. 1 and 2, North-Holland, Amsterdam, the Netherlands, 1964.
2. G.F. Knoll, *Radiation Detection and Measurement*, 3rd edn., Wiley, New York, 2000.
3. W.J. Price, *Nuclear Radiation Detection*, McGraw-Hill, New York, 1964.
4. D.H. Wilkinson, *Ionization Chambers and Counters*, Cambridge University Press, London, U.K., 1950.
5. H. Geiger and W. Mueller, Das Electronenzaehlrohr, *Phys. Z*, 29, 839, 1928.
6. J.T. Routti and S.G. Prussin, Photopeak method for the computer analysis of gamma-ray spectra from semiconductor detectors, *Nucl. Instrum. Methods*, 72, 125, 1969.
7. R. Gunnink, R.A. Meyer, J.B. Niday, and R.P. Anderson, Precise determinations of high-energy gamma rays and errors in the pair-peak method, *Nucl. Instrum. Methods*, 65, 26, 1968.
8. R. Gunnink and J.B. Niday, *Computerized Quantitative Analysis by Gamma-Ray Spectrometry*, Lawrence Livermore National Laboratory Report UCRL-51061, Vol. 1, Livermore, CA, 1972.
9. R.B. Firestone and V.S. Shirley, eds., *Table of Isotopes*, 8th edn., Wiley, New York, 1996.

10. P.M. Jones, L. Wei, F.A. Beck, P.A. Butler, T. Byrski, G. Duchene, G. deFrance, F. Hannachi, G.D. Jones, and B. Kharraja, Calibration of the new composite "clover" detector as a Compton polarimeter for the EUROGAM array, *Nucl. Instrum. Methods Phys. Res.*, A362, 556, 1995.
11. P.J. Nolan, D.W. Gifford, and P.J. Twin, The performance of a bismuth germanate escape suppressed spectrometer, *Nucl. Instrum. Methods Phys. Res.*, A236, 95, 1985.
12. J. Verplancke, W. Schoenmaekers, W.H.A. Hesselink, A. Hacquebord, J. Penninga, and A. Stolk, A HP Ge-telescope as Compton suppression spectrometer, *IEEE Trans. Nucl. Sci.*, NS-33(1), 340, 1986.

8 Collateral Forensic Indicators

8.1 STABLE ISOTOPES

8.1.1 LEAD

Atmospheric pollution from fossil fuel combustion has increased dramatically over the last century. The ubiquity of particulate contamination worldwide has, however, created a powerful new tool for nuclear forensic analysis—Pb isotope profiling (or "fingerprinting"). Lead alkyls, added to gasoline for their antiknock properties, are emitted as micrometer-sized particulates from automobile exhaust and, as the major source of atmospheric Pb, are a well-recognized imprint of modern society on the Earth's ecosystem [1].

Lead is composed of four stable isotopes, only one of which (^{204}Pb) is nonradiogenic. The abundances of the other isotopes have increased through time by amounts controlled by the time-averaged content of Pb, Th, and U in different geologic reservoirs on Earth. The variation of U/Th/Pb ratios in natural materials, together with the large differences in half-lives of the three parent isotopes (^{235}U, ^{238}U, and ^{232}Th), have produced large variations in the Pb isotope compositions of terrestrial Pb deposits [2,3]. The stable isotopes of Pb provide a powerful isotopic fingerprint for nuclear forensic science because Pb emitted to the atmosphere from industrial sources generally has an isotopic composition distinct from the Pb present in rocks at the earth's surface. In addition, in most urban areas, industrial Pb overwhelms the natural Pb component [3]. Systematic variations in the Pb isotopic composition of gasoline have, for example, been calibrated and used to estimate the age of episodic gasoline releases [4].

Lead is also often associated with the handling of special nuclear materials (e.g., as shielding), providing an added opportunity to use Pb isotope analysis to provide clues about the location of previous handling and storage of the SNM, or about the provenance of the Pb itself. An example of the use of Pb isotopes for geolocation is provided in the discussion of the Bulgarian HEU investigation (Chapter 20).

8.1.2 OXYGEN

Nonnuclear materials adhering to, or incorporated into the surfaces of, radioactive samples may provide valuable clues to the postproduction handling of the nuclear material. Oxygen (and H) isotopes may provide insights about the location at which surface oxide layers formed, as the isotopic composition of water vapor in the atmosphere exhibits variations according to geographic location and season in a well-documented and predictable manner [5]. The $^{18}O/^{16}O$ ratio of rainwater in the United States, for example, varies by about 2% from an ^{18}O-rich composition in northernmost Montana to an ^{18}O-depleted composition in the Florida keys [3]. Although a 2% variation may not seem large, it is roughly 100× greater than the analytical uncertainty in O-isotope measurements by gas-source isotope ratio MS (see Chapter 16).

Oxygen isotope abundances in U and Pu oxides may also be useful for delineating their origins. The utility of this geolocation indicator depends on the original source of oxygen in the nuclear material, the degree to which the O-isotopic composition of nuclear material varies according to production location, and the rate at which oxygen isotopes in surficial oxide layers exchange with the environment.

Initial studies revealed significant differences in O-isotope ratios of uranium oxides manufactured in different geographic locations, suggesting that variations in $^{18}O/^{16}O$ ratios in UO_2 fuel pellets correlated with respective values in meteoric water [6]. However, a recent investigation was unable to confirm those results [7]. Hence, the application of oxygen isotope measurements as investigative evidence in nuclear forensic science is presently unconfirmed.

8.2 INORGANIC ELEMENTS

The forensic interpretation of trace inorganic constituents in materials like paint, alloys, and gunshot residues is a mature science [8,9]. Here, the forensic significance of inorganic elemental contaminants in heavy-element samples, independent of the stable isotopic makeup of these elements, is considered. The meaning of residual fission products in a sample of U or Pu as indicators of incomplete fuel reprocessing, and that of residual decay daughters as clues to the chemical processes employed in the last stages of purification or metallurgy, has been previously discussed. Although most of these radiogenic indicators are assessed via their radioactive emissions, the stable isotopes that are included in the sample also provide information. As shown earlier, deviations in the isotope ratios of elements such as O and Pb provide signatures that can be used for geolocation.

Elemental signatures in samples of heavy elements can be difficult to interpret. In general, these samples arise from an industrial environment in which inexpensive, practical- or reagent-grade chemicals are used. Even most reagent-grade chemicals contain inorganic contaminants beyond those found in the ultrapure reagents that are often used in radioforensic analysis. Therefore, any heavy-element sample, be it a chemical compound, a pure metal, or an alloy, will contain a positive trace-analytical signature for virtually every stable element in the periodic table. Elemental contaminants of natural isotopic composition are impossible to interpret in the absence of a comprehensive database if they are in concentrations of parts-per-million or less. On occasion, this limitation can be circumvented if the natural isotopic mixture from the reagents has been blended in a measurable ratio with the radiogenic isotopic mixture from incomplete reprocessing or decay processes. We previously discussed the interpretation of the mixture of Nd isotopes residual in a Pu sample as being controlled by the specific volumes of the reprocessing reagents, using the stable radiogenic isotopes as a pseudo-tracer.

In weapons applications, important elemental contaminants arise from processes involved in metallurgy and weaponization (the actual use of the material in a nuclear device). The chemical processes resulting in the product that is feed for the production of metals and alloys are usually free of inorganic contaminants beyond those present in the reagents. This is particularly true for U, whose production can involve a protocol in which volatile UF_6 is formed. Given modern efforts to reduce many of the nuclear stockpiles, interdicted samples of heavy-element oxides or other salts may arise from the destruction of metal weapons parts via burning or hydriding. In this case, the salts may retain many of the signatures of the production of the original metal sample.

There are three types of inorganic metallurgical signatures: residual reductant used in the production of the metal, contaminant species incorporated in the metal from the metallurgical environment (from initial reduction through forming and machining operations), and small amounts of metals deliberately added to the matrix materials to moderate allotropic phase or chemical reactivity.

In general, samples of actinide fluorides, oxides, or chlorides are reduced to the metal through the action of Na, C, Mg, or Ca and may require I or S to provide the extra heat necessary to produce a liquid metal in the reaction. The process actually used is determined by batch size, chemical affinity, and the desired purity of the final product. Different reductants are soluble to varying degrees in the various actinide metals. Either the metallic product pools at the bottom of the reaction vessel and the salt cake of residual reactants, reaction products, and flux materials floats on top, or the metal is sequestered as beads suspended in the salt cake. After cooling, the salt cake is discarded, and the metallic sample is further refined to remove contaminants. The vessel in which the reduction is carried out can also contribute to contamination of the sample. Excess Be may be introduced through the use of a beryllia crucible for the reaction. Molds made of other metals or graphite are often coated with CaF_2 before casting is performed. In addition, tool bits from machining operations can leave a signature on the surface of the metal, particularly if it is in a particularly intractable allotropic form.

The alloying of U and Pu is often used to stabilize the crystalline form of a more workable polymorphic state. For instance, pure Pu metal consists of the monoclinic α phase at room temperature; α-Pu is hard, brittle, and difficult to machine. However, by the addition of a small quantity of Ga to form a Pu–Ga alloy, the fcc δ phase (found in pure Pu between 310°C and 450°C) can be stabilized at room temperature. The alloy is soft, ductile, and easy to machine. For Pu, Ga and Al alloys are the most likely encountered by the radioforensic analyst. Alloys of U are extensive and varied, with the minor alloy components strongly determined by the intended use. Uranium alloys with Nb or Zr are particularly common in nuclear applications. The U–Mo alloys have good chemical resistance and metallurgical properties, but the large neutron-capture cross section of Mo makes it unsuitable for weapons applications.

The distributions of these products in the sample material may be quite different, depending on their source. Residual reductant and alloying metals tend to be distributed uniformly through a metallic sample, whereas contaminants introduced by refining and machining may be concentrated on surfaces. Contaminant elements that make simple intermetallic compounds with the actinide matrix may be concentrated on metallic grain boundaries; for example, Fe forms a compound with Pu, Pu_6Fe, which participates in the definition of the grain structure of the wrought metal [10]. Inhomogeneities in the parent metal sample may be retained by a product salt generated by dismantlement activities.

Weaponization of an actinide part can leave inorganic signatures in materials derived from that part. In addition to the presence of other heavy elements, contamination by elements at the other extreme of the periodic table can also indicate weaponization. Beryllium metal has a small capture cross section and is an excellent neutron reflector that is relatively transparent to other radiations. The stable compound, $PuBe_{13}$, can be formed by heating a mixture of the elements. Thus, measurement of $PuBe_{13}$ on the surface of plutonium oxide granules could indicate the surface contamination of a metallic Pu part with Be and its subsequent destruction. Chapter 5 discussed the use of ^6LiD as thermonuclear fuel. Salts of other low-Z materials may have relatively low capture cross sections and be reasonably transparent to x-rays emitted by the nuclear mechanisms of an explosive device. The presence of Li or B in an actinide specimen (beyond residual reagent levels) may therefore be an indicator of weaponization.

In samples arising from reactor applications, the isotopic signature of the heavy-element matrix is probably more useful for forensic interpretation than are the minor constituents. The weapon environment is considerably more forgiving of incompatible materials than is the reactor environment, as the weapon must only function for microseconds, whereas the reactor must remain safe and stable for years. Nevertheless, the requirements of good dimensional stability in the fuel under irradiation, combined with a cladding material that is compatible with both the hot fuel and the coolant, can result in inorganic signatures in reactor fuels that are characteristic of reactor type.

The fuel itself generally consists of fairly pure metal, an oxide, or a carbide. In pressurized water (PWR) and boiling water reactor (BWR) fuels, a burnable neutron poison is often added to produce a more uniform power output over the lifetime of the fuel element (Gd or B are often used for this purpose). Fuel element cladding is usually stainless steel or a Zr alloy, and the Zr used in reactor applications is much lower in Hf (a neutron poison) than is normal industrial Zr. In natural-U-fueled, gas-cooled reactors, the cladding is often Mg and Zr. Pebble-bed reactor cladding often incorporates silicon carbide.

Much work remains to be done to develop inorganic trace concentrations in nuclear materials as forensic signatures. A large fraction of this development will involve the construction of a database of characteristic contaminants against which a sample of questioned material can be compared.

8.3 ORGANIC ANALYSES

Nonradiologic items, such as glass fragments, soil, pollens, paper, DNA, or dust, are often present in samples acquired for nuclear forensic analysis. Such collateral evidence can furnish valuable information that can aid the nuclear investigation, such as the route traversed by the sample, the individuals involved, sources of packaging materials, and so forth. These nonradiologic specimens are frequently of the same types analyzed by mainstream forensic science for criminalistics applications. As a consequence, although nuclear forensic science is a discipline heavily oriented toward radioisotopic and inorganic analyses, it can at times also make productive use of the organic chemical techniques developed in crime labs over many years. It is beyond the scope of this book to provide in-depth treatment of all such effort areas, but comprehensive reviews are available in the forensic literature. What follows here are highlights of organic forensic activities that are most germane to nuclear investigations.

8.3.1 High Explosives

High-explosive (HE) forensic analyses can be quite valuable during inquiries into nuclear smuggling or nuclear terrorism. In the case of an exploded radiologic dispersal device, for example, identification of residual traces of the HE is an important lead in the ensuing investigation. For an RDD interdicted before detonation, detailed analyses of the explosive (after render-safe) will measure empirical signature species, such as synthesis by-products, degradation compounds, and process impurities. These indicators provide a material characterization that may be valuable for intelligence purposes, comparison to other questioned specimens, and perhaps geolocation information. Such predetonation forensic considerations also hold for the interrogation of an improvised nuclear device, while postdetonation HE analyses could be applicable to nuclear scenarios for investigations of hydrodynamic testing by proliferant or terrorist groups.

The most widely used HEs are HMX, RDX, TNT, PETN, tetryl, nitroglycerine, and EGDN, and the more notorious plastic explosives (e.g., C-4, Composition B, Semtex, and Detasheet) are composite formulations of these HEs and other chemical compounds. Moreover, in the specialty area of nuclear weapons development, insensitive HEs such as TATB are also important concerns.

Forensic methods used for the identification and quantitation of HE species include gas chromatography–mass spectrometry (GC-MS), Fourier-transform infrared (FTIR) spectrometry, thin-layer chromatography (TLC), gas chromatography-thermal energy analyzer (chemiluminescence), and high-performance liquid chromatography–mass spectrometry (LC-MS). Recent literature references providing detailed overviews and primary source material are Refs. [11,12].

8.3.2 Hairs and Fibers

These items of trace organic evidence are routinely analyzed in criminalistics labs. Microscopic examination of individual hair specimens can provide a wealth of valuable class-characteristic information, and this method is also the traditional basis for intercomparison of samples. Such hair identifications include the species of origin, a racial indication if human, and the somatic derivation (location on the body) of the sample. Hair structure, color, cosmetic treatments (e.g., bleaching, dyeing, and permanent wave), length, the presence of drugs or disease, or other abnormalities may provide more distinctive data about the individual or individuals involved. Highly specific information is also available from hair specimens through DNA methods comprising mtDNA analysis. A comprehensive resource for the forensic assessment of hair is Ref. [13].

Similarly, textile fibers are identified and compared primarily through optical and FTIR microscopies. However, other forensic methods employed for fiber analysis have included hot-stage microscopy (for melting-point determinations), solvent solubility tests, pyrolysis GC, TGA, and TLC or HPLC (for dye analyses).

Fibers are principally classified as natural or man-made. Natural fibers are further divided into vegetable (e.g., cotton, linen), animal (wool, silk), and mineral (asbestos) types. Anthropogenic fibers are diverse and many and encompass generic categories such as rayon, polyester, nylon, fluorocarbon (Teflon), acrylic, aramid (Kevlar), olefin (polyethylene, polypropylene), and spandex. In addition to fiber identification, characteristics such as color, diameter, and cross section provide further information relevant to an investigation. Extensive textile compendia and databases of optical and infrared properties of fibers are available, and a complete treatment of forensic fiber analysis can be found in Ref. [14].

8.3.3 Inks and Papers

Forensic analyses of inks and papers contribute to the broader discipline of questioned document examination. Many modern inks consist of organic dyes in solvents (often glycols) and insoluble pigments in liquid suspension. The forensic examiner compares different exemplars of inks for similarities and differences, with the identification or characterization of an ink corresponding to a chemical determination of its formulation.

A variety of dyes, such as methyl violet, rhodamine red, and Victoria green, impart color to inks. Other organic components include preservatives, resins, viscosity adjusters, and fatty acids, and consequently, ink compositions are most often interrogated by TLC, HPLC, and GC-MS analyses. In addition, forensic testing of the extent of ink dryness can provide the age of ink on paper. The US Secret Service maintains a historic ink library for identification and comparative age-dating studies.

Traditional inks include India, fountain pen, ballpoint pen, and typewriter ribbon varieties. Today, however, the modern document examiner is also confronted with outputs from copy machines, ink-jet printers, fax machines, and laser printers. To potentially differentiate among these various toners and other similar media, advanced forensic techniques, such as laser-desorption mass spectrometry and microRaman spectroscopy, are being explored.

Comparative paper analyses are conducted to determine similarities and the likelihood of common origin. These assessments typically consist of measurements of physical characteristics, watermark examinations, fiber analyses, chemical analyses, and trace-element distributions. With enough points of comparison, the latter determinations can provide particularly individualized data. Trace elements in paper originate predominantly from processing instrumentation and raw materials—chiefly the impurities in paper additives. It is improbable that different paper manufacturers could make equivalent products with respect to the relative concentrations of the identical trace elements.

An aspect of paper analysis that has proven particularly useful in some nuclear smuggling investigations is fiber examination. In addition to its straightforward comparative value, identification of the fiber content of paper can provide insights on pulping processes, as well as geolocation information. For example, forensic measurement of the identities and percentages of various hardwood and softwood species within a paper can narrow the place of manufacture to selective regions of the world (see Chapter 20).

The chemical components of paper include fluorescent whiteners, coatings, sizings, fillers, waxes, and plasticizers. Organic analyses can provide material profiles of different combinations of starch, hydrocarbons, glues, phthalates, and other species.

Finally, counterfeit currency is an integral component of worldwide terrorist activities in general, and all aspects of questioned document examination are applied to counterfeit bill analyses. Recent reviews of this general activity area are Refs. [15,16].

8.3.4 Fingermarks

Organic deposits that are important evidence in any forensic investigation are fingermark residues. They consist of lipid compounds on the skin surface from sebaceous glands and the epidermis, and particularly from eccrine glands on the fingertips and palms of the hands. The predominant chemical classes of fingermark compounds are amino acids, fatty acids, long-chain fatty-acid esters, and alcohols, whereas specific, identified analytes include palmitic acid, oleic acid, dodecanoic acid, myristic acid, stearic acid, and cholesterol.

Although current forensic research is exploring the implications of the chemical variability of fingermark residues, the major activities of fingermark examination are processing a scene to develop latent prints (including palm prints), photographing or lifting the visible traces, image processing, comparison with known fingermarks in one or more databases (e.g., AFIS—the Automated Fingerprint Identification System), and expert interpretation. Numerous protocols exist for visualizing latent prints on various surfaces, and Ref. [17] is a comprehensive reference for fingermark technology.

8.3.5 Other

There are also a number of other organic techniques that could provide important collateral evidence in a nuclear forensic investigation, depending on the specific nature of the samples. They include DNA analyses [18,19], the characterization of plastics and polymers [20], adhesives and tapes [21], and hydrocarbon/lubricant measurements (e.g., to chemically profile the fuel oil in ANFO low-explosive).

REFERENCES

1. C.C. Patterson and D.M. Settle, Review of data on eolian fluxes of industrial and natural lead to the lands and seas in remote regions on a global scale, *Mar. Chem.*, 22, 137, 1987.
2. B.G. Dalrymple, *Ancient Earth, Ancient Skies*, Stanford University Press, Stanford, CA, 2004.
3. W.T. Sturges and L.A. Barrie, The use of stable lead 207/206 isotope ratios and elemental composition to discriminate the origins of lead in aerosols at a rural site in eastern Canada, *Atmos. Environ.*, 23, 1645, 1989.
4. R.W. Hurst, Age dating of gasoline releases using stable isotopes of lead, in *Geochemical Fingerprinting in Environmental Geology* (Amer. Assoc. Petrol. Geol. Short Course #9), R.W. Hurst and J.B. Fisher, eds., San Diego, CA, 1996.

5. C. Kendall and T.B. Coplen, Distribution of oxygen-18 and deuterium in river waters across the United States, *Hydrol. Proc.*, 15, 1363, 2001.
6. L. Pajo, K. Mayer, and L. Koch, Investigation of the oxygen isotopic composition in oxidic uranium compounds as a new property in nuclear forensic science, *Fresenius J. Anal. Chem.*, 371, 348, 2001.
7. J. Plaue, Forensic signatures of chemical process history in uranium oxides, PhD thesis, University of Nevada, Las Vegas, NV, 2013.
8. S.H. James and J.J. Norby, eds., *Forensic Science*, CRC Press, Boca Raton, FL, 2003.
9. R. Saferstein, *Criminalistics*, Prentice Hall, Upper Saddle River, NJ, 1998.
10. O.J. Wick, ed., *Plutonium Handbook*, Vol. 1, American Nuclear Society, LaGrange Park, IL, 1980.
11. A. Beveridge, ed., *Forensic Investigation of Explosions*, 2nd ed., CRC Press, Boca Raton, FL, 2012.
12. C.R. Midkiff, Arson and explosive investigation, in *Forensic Science Handbook*, Vol. I, 2nd ed., R. Saferstein, ed., Prentice-Hall, Upper Saddle River, NJ, 2002, Chap. 9.
13. J.R. Robertson, ed., *Forensic Examination of Hair*, CRC Press, Boca Raton, FL, 1999.
14. J.R. Robertson, C. Roux, and K. Wiggins, *Forensic Examination of Fibres*, 3rd ed., CRC Press, Boca Raton, FL, 2012.
15. R.L. Brunelle, Questioned document examination, in *Forensic Science Handbook*, Vol. I, 2nd ed., R. Saferstein, ed., Prentice-Hall, Upper Saddle River, NJ, 2002, Chap. 13.
16. J.S. Kelly and B.S. Lindblom, eds., *Examination of Questioned Documents*, CRC Press, Boca Raton, FL, 2006.
17. R. Ramotowski, ed., *Lee and Gaensslen's Advances in Fingerprint Technology*, 3rd ed., CRC Press, Boca Raton, FL, 2012.
18. R.C. Shaler, Modern forensic biology, in *Forensic Science Handbook*, Vol. I, 2nd ed., R. Saferstein, ed., Prentice-Hall, Upper Saddle River, NJ, 2002, Chap. 10.
19. J.M. Butler, *Fundamentals of Forensic DNA Typing*, Academic Press, London, U.K., 2010.
20. S. Palenik, Microscopy and microchemistry of physical evidence, in *Forensic Science Handbook*, Vol. II, R. Saferstein, ed., Prentice-Hall, Englewood Cliffs, NJ, 1988, Chap. 4.
21. P. Maynard, K. Gates, C. Roux et al. Adhesive tape analysis: Establishing the evidential value of specific techniques, *J. Forensic Sci.*, 46(2), 280, 2001.

9 Sample Matrices and Collection

Specimens investigated within the diverse facets of nuclear forensic analysis have encompassed many types of sample matrix. The handling, collection, preservation, and storage of samples are important to the quality and ultimate interpretation of the final data. As in any forensic investigation, specimens must be obtained that are representative of the questioned material. Inhomogeneities in sample matrices must be considered, and for nuclear forensic specimens, such heterogeneity may be important clues to the origin and prior manipulations of the sample.

The most significant analytes for nuclear source information (radioactive nuclides and isotopic ratios) are unaffected by any chemical perturbations of a system, and the molecular speciation of radioactive indicators (other than U and Pu) is generally unimportant in the final analysis. The most serious uncertainties in these analytes would be competing levels of background fallout in the specimen or any potential cross-contamination between different samples.

The same is not true for route forensic evidence, however, and sampling protocols and storage conditions can be more important for valid data and interpretation of results from those analyses. Forensic analysts do not always perform primary evidence collection, so it is imperative that fire department, National Guard, Hazmat, law enforcement, and other first-responder personnel understand the requirements of adequate sample acquisition.

9.1 SOIL/SEDIMENT MATRICES

Soils and sediments are collectors that integrate signature species over time and that can indicate the presence of undeclared nuclear facilities. Such specimens are readily collected, manipulated, and stored, and they serve as proficient sorbents of actinide, lanthanide, and transition-metal analytes. These chemical classes correspond to nuclear fuels, fission products, and activation products, respectively. The radiochemical indicators are easily isolated from soils and sediments by extraction or ion exchange, with subsequent measurement by an assortment of instrumental techniques.

Nuclear analytes sorb on soils by both wet and dry atmospheric deposition processes, as well as by accidental releases and transport activities, and they have their highest concentrations near the release points. A monitoring program to measure concentrations in discrete samples, along with the application of the relevant meteorological conditions, can provide radioactivity contours that indicate the location of an undeclared facility. The forensic specimens should be collected over a wide area and from minimal soil thickness.

Soils are generally less sensitive accumulators than vegetation, as they are more prone to interferences (often of a heterogeneous nature) from the higher background levels of various signature nuclides. These backgrounds derive from the history of nuclear technology dating from the time of the Manhattan Project, and they are particularly exigent for long-lived species, such as ^{129}I, ^{137}Cs, ^{235}U, and ^{239}Pu.

Silts and sediments of fine particle size are efficient collectors of the same radionuclear species as soil. However, for other than near-shore sampling, effective collection equipment can be relatively complicated and unwieldy. Rather than atmospheric deposition, sediments gain their signatures from localized discharge sources (e.g., process cooling waters), and the retained signals are quite dependent on their individual geochemistries. Thus, for example, transition metals and Pu are quantitatively retained by fine-grained sediments in nearly all environments, the sorption of trace ^{137}Cs depends inversely on the ionic strength of the aqueous medium, and ambient redox conditions determine the mobility of U. Some tracers, such as the fission products ^{99}Tc and $^{103,106}Ru$, fractionate appreciably between aqueous and solid phases and can therefore be relatively ubiquitous among collected samples. Sediments are less effective collection media for such analytes, but when signature species do efficiently sorb on sediments, a well-planned collection strategy can provide time-history information about the operations of a questioned nuclear facility.

9.2 VEGETATION MATRICES

Terrestrial vegetation can collect and concentrate, often by several orders of magnitude, fugitive radioactive species released into the ambient environment. Vegetation is an integrating collector that is straightforward and fast to sample and that requires no specialized equipment. Two collection modes are possible with plants: the exposed portion can serve as an air collector, while the root system samples ambient soil and groundwater.

Plant foliage collects small particles from the atmosphere, and leaf texture, morphology, and composition are all factors in the overall efficiency of the process. Thus, leaves with waxy or sticky exteriors are typically better collectors than those with dry surfaces. Specific plants with large surface areas also optimize such collections, so that grasses, pine needles, and mosses are superior samplers for nuclear analytes. Although fresh grass clippings would likely be indicative of recent releases, trees such as pines and firs drop needles only every few years; time-history intelligence from an individual tree is therefore possible. The rate of foliage loss by a tree depends on local environmental stressors, and younger growth is generally lighter in appearance with less physical damage to needle structure. In some instances, therefore, multiple timeline data on integrated emissions at a given area can be determined through analyses of distinct layers of fallen needles. The concentration of airborne species at an off-site collector will normally vary inversely with distance from the source, and vegetation has provided effective signatures of large US nuclear facilities over distances of many kilometers.

For groundwater sampling, however, rooting strength and time within the growing cycle (i.e., during more active plant transpiration) are the critical factors. The more deeply rooted plants are typically more efficient in sequestering signature species, but in this regard, analyte bioactivity is also essential. Whereas vegetation tends to reject accumulations of nonbiogenic U or Pu, nuclides such as ^{3}H, ^{90}Sr, and ^{131}I may be readily concentrated from both groundwater and soil.

9.3 WATER MATRIX

Many mobile radioactive signatures can be found in aqueous effluents from nuclear activities. After discharge into surface or subsurface waters, sampling is achieved in the former via several techniques. In one, a portable, impeller-type water pump separates analytes from their aqueous medium by pumping through a filter cartridge consisting of a mixed-bed ion-exchange resin for dissolved species and a tandem filter of appropriate pore size for particulates. Another tactic is the collection of water grab-samples in proper containers. Each bottle is initially cleaned, thoroughly rinsed several times with the medium to be sampled, filled nearly to the top, and tightly capped. These specimens are then transported back to a laboratory for radiochemical analyses.

9.4 FAUNA MATRICES

Many aquatic and terrestrial biota can remove and sequester ultratrace quantities of radioelements from their environment. Such bioaccumulators can concentrate nuclear signatures from an ambient matrix (often water) by several orders of magnitude. The specific bioconcentration factors are organism dependent and are also effectively determined by the radiochemistry of the analyte. For example, bivalve filter-feeders, such as oysters and clams, collect transition-metal species, as well as certain organic contaminants, from dilute aqueous sources. Similarly, some mollusks accumulate fission-product ^{90}Sr within their shells. Thus, a regional nonproliferation surveillance program might investigate the time-history of thyroid specimens from local slaughterhouses because mammalian thyroids are excellent bioconcentrators of radioiodine.

Living organisms do not typically accumulate a comprehensive suite of signature species, as most are quite selective about which elements are retained. Thus, fauna do not concentrate diagnostic actinides. However, such collectors are ubiquitous and can be effective long-term integrators of various environmental radioactive species over their lifetimes. Although special handling and storage considerations may be necessary for such specimens, and higher backgrounds from local fallout can be a factor, their collection and forensic analyses could supply indicative signals of nuclear activities.

9.5 OTHER MATRICES

Air sampling has been effectively employed in nuclear nonproliferation surveillance for many years. Typically performed with high-volume air collectors at throughput rates of tens of cubic meters per minute, diagnostic releases from a source can be acquired over distances of many kilometers. Air filters will sequester particulate radioactivity, whereas activated carbon or other solid sorbent traps will retain aerosolized radioiodine and ^{3}H. Specialized systems could also be deployed to collect the nonreactive radioactive gases, such as ^{41}Ar, ^{85}Kr, and fission-product radioxenon. However, these efforts are resource-intensive and nontrivial, acquire species identical to those emitted by the activities of worldwide nuclear operations (thereby making attribution complex), and rely more fundamentally on accurate background corrections. In the assay of ^{85}Kr, global background continues to rise as the nuclear power and production industries reprocess more irradiated fuels. Nevertheless, air sampling, followed by proper laboratory analyses and interpretation, can identify signatures representative of reactor operations and nuclear reprocessing. Real-time monitoring of volatile species in air, such as by sensitive mass spectrometry with total analysis cycles of a few hundred milliseconds [1], has become more feasible with continuing advances in science and technology.

Various other collection media have been explored for nuclear analytes to lesser extents, and they are only listed here. These include mammalian hair, the feathers of birds, and biofilms that coat solid substrates under water surfaces. Conventional forensic laboratories analyze human hair for evidence of illicit drugs, and other analytic protocols allow similar interrogations for diagnostic radiochemical species [2]. Similarly, the algae and bacteria in biofilms can remove and concentrate some elements (e.g., phosphorus and iodine) from river water.

9.6 COLLECTION TACTICS

The tools and techniques favored by forensic analysts to collect questioned samples can be diverse [3,4], and many are those commonly encountered in good laboratory practice. They include items such as powder-free, disposable gloves (e.g., latex, nitrile, or PVC vinyl) for personnel and intersample contamination control, spatulas, forceps, sticky tape, wipe papers/cloths, razor blades, bulk-cutting utensils, and transfer pipettes. The ensuing analyses are likely to be multidisciplinary. They are also likely to focus on diverse exemplars of trace and ultratrace evidence, as a very large fraction of contemporary criminal investigations has come to rely on such efforts.

A variety of forms of unique samples will be obtainable, potentially admissible in court, and likely the source of important intelligence information. Such evidentiary specimens might include Special Nuclear Material and other radioactive species, along with transport indicators such as hair, fibers, glass, metal, plastics, soil, dust, pollen, paper, ink, diverse particulates, and biological material. However, trace analytes are particularly sensitive and can be readily compromised through careless or inappropriate collection techniques.

For nuclear incident scenarios, additional apparatus for radiation monitoring and the protection of collection personnel are often desired. These items would include α and β–γ survey meters, film-badge and finger-ring dosimeters, and possibly extensive personal protective equipment, such as full anti-contamination suits, respirators, and lead aprons.

The primary containers used for specimen storage and transport after initial collection can be important considerations in nuclear forensic investigations. The most inert, and therefore optimum, material for these purposes is Teflon; however, it is relatively expensive. Solid samples intended only for radioactivity or isotopic analyses for nuclear source information are often sealed in multiple layers of zip-lock polyethylene bags. High-purity, screw-cap, polyethylene vials (polyvials), of approximately 20 mL capacity, are outstanding primary containers for such source-data solid specimens, as well as for aqueous liquid samples. These polyvials (the ones without the metal liner in their caps) have been measured to generate very low blank values of leachable inorganic species, even when contacted by strong acid media.

In specimens collected for organic species, however, even high-purity polyethylene is a source of significant chemical phthalates (plasticizers) that will contaminate a questioned sample and adversely affect ensuing instrumental analyses. For those samples, an ultraclean glass vial is the primary container of choice. They are available in convenient capacities of 20, 40, and 60 mL, as well as other volumes, with caps lined with Teflon. Septum tops with composite silicone–Teflon liners also exist in order to directly sample vials, without removing their caps, via syringe or solid-phase microextraction fibers. However, glass vials should not be used for questioned inorganic specimens, particularly if liquid and acidified, because glass contributes significant concentrations of elemental species to experimental blanks.

Many of the protocols of good laboratory practice apply likewise to field sample collection, and they are designed to optimize the PARCC parameters (precision, accuracy, representativeness, completeness, and comparability). The acquisition of replicate samples is one such tactic. Collection of suitable blank specimens is another—and one that cannot be overemphasized. Background field blanks and a trip blank are very important for samples acquired on-site and transported off-site for analysis. Valid interpretation of experimental results, particularly for chemical or molecular species, usually cannot be made outside the context of suitable blank determinations. Similarly, collectors must be mindful of the need for specimen integrity and minimize both external contamination and cross-contamination between collected samples (e.g., by disposing of gloves or other personal protective equipment that come in contact with a sample before collecting another).

Once questioned and blank specimens are acquired, they must be adequately stabilized and preserved until preinstrumental sample preparation can be performed. Many factors can alter the chemical speciation of a collected sample, including photolysis by sunlight, elevated temperature, hydrolysis, redox reaction, and an inappropriate container. The proper specimen container is a very important consideration. Material composition guidelines were discussed earlier, and, for example, a dark-glass (amber) vial will additionally minimize photolytic decomposition of sensitive organic compounds. Sample preservation tactics depend on the target analytes and specimen matrix. Thus, preservation of solid samples can be effected by storage at approximately 4°C in an ice-packed cooler. Some water collections may be stabilized in the same way, whereas others should be acidified to pH ≤ 2 with analytic reagent-grade (or better) acid.

Unfortunately, a valuable reference for such evidence collection has not yet been released into the public domain. A scientific working group (SWG) under FBI auspices completed a best practices document in 2009 that addressed topics such as scene assessment; the reconnaissance entry; collection teams and methods; field screening and tools; specimen containers, packaging, and preservation;

decontamination; sample transport; and other considerations [5]. The report was authored by representatives of the FBI Laboratory Division (Hazardous Materials and CBRN units), three federal US national laboratories, the US EPA, US FDA, US CDC&P, and Royal Canadian Mounted Police and Defence Groups. The work was refereed and was in-press for open publication by the FBI journal, *Forensic Science Communications*, but has since become a casualty of the apparent demise of both the journal and the SWG.

For nuclear-focused forensic investigations, the eventuality of a criminal prosecution must be appreciated throughout the course of all conducted activities, and important legal consequences arising from specimen collection, preparation, analysis, and reporting are virtually assured. Forensic procedures must result in legally defensible data, the quality of which can influence the "general acceptability" of the information in court. Sample chain-of-custody is a critical consideration in this regard: it must exist in totality to confirm specimen identity and ensure that it was not modified from its original condition. Chain-of-custody criteria are essential for all evidence without unique markings or morphology, so that conclusive identification may be made at any future time. They are particularly necessary for items with the potential for loss or tampering, and for specimens that encompass a change in chemical or physical nature, such as aqueous rinses, swipes, or sticky-tape samples.

A chain-of-custody sequence must have a real (not hearsay) beginning and a real end, usually at final analysis (if completely destructive) or at archival storage. Absolute accountability must be traced and documented throughout all sample transfers, and authorized signatures are required at each reassignment. Only sanctioned personnel must handle a specimen at any time. Assurance of an unbroken evidence chain can be promoted by mechanisms such as locked containers, sealed plastic bags, and tamper-proof or tamper-indicating seals.

However, readily identifiable articles introduced at trial may not require customary chain-of-evidence documentation. Such evidence could include items with stamped or imprinted serial numbers, exceptional and acknowledged shapes, or unique marks made by investigators in the field.

REFERENCES

1. D.M. Chambers, L.I. Grace, and B.D. Andresen, Development of an ion store/time-of-flight mass spectrometer for the analysis of volatile compounds in air, *Anal. Chem.*, 69(18), 3780, 1997.
2. P.M. Grant, A.M. Volpe, and K.J. Moody, Forensic analysis of hair for inorganic and actinide signature species indicative of nuclear proliferation, *J. Intel. Com. R&D*, Intelink, 2009.
3. M.R. Byrnes, *Field Sampling Methods for Remedial Investigations*, Lewis Publishers, Boca Raton, FL, 1994.
4. S.C. Drielak, *Hot Zone Forensics*, Charles C. Thomas, Springfield, IL, 2004.
5. The Scientific Working Group on Forensic Analysis on Chemical, Biological, Radiological, and Nuclear Terrorism (SWGCBRN), Best practices for the collection of CBRN evidence, FBI Laboratory, Quantico, VA, 2009.

10 Radiochemical Procedures

Because of space limitations, a complete treatment of the isolation of chemical fractions from heavy-element samples cannot be presented here. The literature of the chemistry of the actinide elements is rich and varied. Unfortunately, many of the published procedures do not work at all, or work only under limited conditions, because of the complex nature of actinide elements in solution. The focus here is on a few relevant areas in which some success has been achieved in our laboratory. However, an excellent resource is the National Academy of Science/National Research Council Series on the Radiochemistry of the Elements, published by the US Atomic Energy Commission. Excellent reviews on the chemistries of U [1] and Pu [2] are also available.

Radioanalytical chemistry is a destructive technique; as a consequence, it is always advisable to save some portion of the sample if there is sufficient material to do so. Before radiochemical analysis, traditional forensic measurements such as fingermark development and hair and pollen recovery should be performed, and particles for morphology assessments should be isolated. During these procedures, it must be remembered that the sample may not be uniform, and fractionation of bulk material should be minimized. The radiochemical sample itself can be subjected to nondestructive assay by spectrometry and neutron counting, and in the past, we have also heated this specimen before dissolution to take an organic headspace gas sample.

10.1 DISSOLUTION

A radiochemical analysis begins with the creation of an analytic solution from the sample. The quantitative dissolution of a solid sample to create a stable solution that is the original radiochemical stock or "pot" solution can be the most difficult part of the procedure. Although samples of Pu and U metal dissolve readily in hydrochloric acid, a residue often remains even after heating. If chronometry beyond that involving the heavy elements is an experimental goal, collection of noble gases and volatile fission products must be accomplished during the dissolution of Pu metal samples. Therefore, the reaction vessel in which the dissolution takes place must be incorporated within a gas-tight manifold. Reagents used to dissolve the sample must be purged with He to remove traces of Xe and Kr dissolved via contact with air. The specimen is then dissolved under a flow of He gas (supplied through a cryogenic trap to remove impurities) that carries the released gas-phase products into a collection bottle. Separation of Xe and Kr from the He carrier gas is accomplished cryogenically. If the volatile fission products I and Ru are to be collected, carriers for both should be added to the reaction vessel before the dissolution. Following the collection of noble gases, I_2 and RuO_4 can be produced by the addition of Cl_2 to the reaction vessel, followed by boiling the solution. RuO_4 and I_2 are trapped by bubbling the carrier gas through a solution of dilute HCl containing hydroxylamine hydrochloride. The chemical apparatus between the reaction vessel and the trap must be wrapped and heated with heating tape to prevent premature condensation. The trap solution is subsequently processed for I and Ru, with their chemical yields determined gravimetrically.

Addition of nitric acid to U residuum, followed by digestion, can result in an effective solution, as uranium oxide dissolves readily in nitric acid. Plutonium oxide will not dissolve completely in nitric acid, however, particularly if it has been high-fired. Any sample containing a significant quantity of Pu must be treated with hydrofluoric acid, irrespective of the initial chemical form. Invisibly small plutonium oxide particles, even colloidal ones, are stable against reaction with most mineral acids, and their presence interferes with the chemistry. Many an unfortunate analyst has received Pu in the form of the $(–PuO_2–)_n$ polymer ("poly-plut"), which can be indiscernible in aqueous suspension even at high concentrations.

Poly-plut can form spontaneously in otherwise-stable pot solutions unless the acid concentration is maintained at greater than 1 M. The only known methods to destroy poly-plut are long-term boiling in concentrated nitric acid, a modified CEPOD (catalyzed electrolytic plutonium oxide dissolution) process [3], or addition of HF or H_2SiF_6 with heating. Unfortunately, the presence of fluoride in the pot solution is undesirable, as it interferes with the subsequent equilibration of added tracer activities with sample analytes, particularly Np, Pa, and Th. Before constituting the final pot solution, it is desirable to reduce the volume of the sample by evaporation to a moist deposit, addition of a small amount of perchloric acid, and evaporation to dense white fumes. This procedure ensures that excess fluoride has been expelled from the sample. Tc_2O_7 can distill from the specimen in $HClO_4$, although the efficiency is poor. Either ^{99m}Tc can be used as a chemical-yield indicator, or an added Re carrier can be used for an approximate gravimetric yield. An alternative to the perchloric acid step (which can be problematic to execute in the current regulatory environment of the United States), the sample can be successively evaporated to a moist residue several times after additions of concentrated nitric acid. For most actinide samples, the preferred final pot solution matrix is hydrochloric acid, with a concentration between 2 and 6 M.

If the matrix of a questioned sample is other than metal or a mixture of salts of the heavy elements, preparation of a stable analytical pot solution may pose a formidable challenge. See the case study of the counterforensic investigation of U enrichment plants (Chapter 21) for examples of particularly intractable samples. The principal working rule for the preparation of the final pot is avoidance of more than trace quantities of fluoride or phosphate in the stock solution.

10.2 TRACER EXCHANGE BY REDOX

Typically, each analytic solution is sampled three times for radiochemistry. One sample (labeled U) is processed without added tracers. The second sample (labeled S) is traced with ^{246}Cm, ^{243}Am (with 2.35-day ^{239}Np in secular equilibrium), and ^{236}Pu

(in which the relative amounts of ^{236}Pu and the ^{232}U and ^{228}Th decay daughters are known). The third sample (labeled N) is traced with ^{237}Np with 27-day ^{233}Pa in secular equilibrium. Typically, a few microliters of the pot solution are analyzed by inductively coupled plasma mass spectrometry for metal composition. If sufficient pot solution exists, a 10-mL aliquot is drawn for γ spectrometry; otherwise, the U fraction is counted before chemistry.

Each specimen is heated under a stream of air to reduce the volume to a moist residue; then 9 M HCl containing HI is added to reduce Pu, Pa, and Np. (Today it can be difficult to procure HI in the United States without a Drug Enforcement Agency license, as it can be used in the synthesis of illicit methamphetamine.) However, only unstabilized (PO_3^{3-} free) HI should be used in the procedure. Each sample is again evaporated to a moist deposit. If there is more than 10 mg of U or Pu in the analytical aliquots, this step is repeated. Each moist deposit is then dissolved in 9 M HCl and evaporated to expel excess HI and I_2 (reaction product). It is then carefully redissolved in 8 M HNO_3 and reevaporated to effect oxidation of Pu, Pa, and Np and equilibration of the tracers with analyte. On occasion, upon reducing the nitric acid volume, an intractable precipitate forms. It is most likely rutile (TiO_2), which can only be dissolved via alkali fusion or prolonged treatment with sulfuric acid. Rather than do this, however, we centrifuge the samples and discard the solid after a nitric acid wash. Very little heavy-element analyte is carried by rutile, and tracer–analyte equilibration is near completion when the precipitation occurs, so these losses are effectively negligible. If there are time pressures for the analysis, the separation procedures can begin at this point. If not, a second oxidation/reduction cycle is recommended, particularly if precipitation occurred during the nitric acid step. Independent of how many redox cycles are performed, however, the last evaporation from nitric acid must be carried out immediately before the first separation step to minimize disequilibrium of short-lived Np and Pa tracers from analyte nuclides as a result of hot-atom chemistry.

10.3 CHEMICAL SEPARATIONS

For environmental or site-inspection samples, the transition-metal content of the pot solution can limit the success of the ensuing ion-exchange chemistry. If the sample matrix contains a significant quantity of iron, it can be removed without depleting the desired analytes by solvent extraction with methyl isobutyl ketone before the final dissolution in nitric acid.

The final, moist nitrate deposits are dissolved in 10 M HCl and heated. After cooling, the solutions are loaded onto anion-exchange columns packed with Dowex-1X8 resin. The columns should be large enough that no more than the upper 25% of the resin bed is loaded with analyte ions. It is preferable that the resin bed be supported within the column by something other than a porous plug of glass wool, as there are elution steps with solutions containing dilute HF. Saran wool can often be used for this purpose.

The eluent solution from the column load, and from subsequent 9 M HCl washes, is collected from the U and S columns, but can be discarded from the N column. This eluent solution contains the heavy elements Ra, Ac, Th, Am, and Cm; the fission products Cs, Sr, and the lanthanides; and undesirable matrix materials such as Ni, Na, K, and Ca. The times that these, and all subsequent fractions, are collected should be recorded as separation times, and they are used for decay corrections of short-lived activities. Both collected fractions should be evaporated to dryness as quickly as possible for the next separation step.

Following the last 9 M HCl elution, Pu is eluted from the columns with a warm solution of 10 M HCl containing HI. Care must be taken to not heat this solution too long, as Np begins to migrate when the acid concentration decreases to less than 9 M. The Pu fractions from all three columns should be retained. Protactinium is then eluted with a solution 9 M in HCl and 0.02 M in HF; fractions from the N and U samples should be retained. Fission-product Zr is also eluted in the Pa fraction. Neptunium is eluted next, with a solution 4 M in HCl and 0.1 M in HF. Fractions from U and S samples should be retained. Finally, U is eluted from the columns with 0.1 M HCl, and all three fractions are collected and retained. Fission-product Sb is eluted with the U fractions.

The ion-exchange separation just described involves an elution column. Analytes are either fixed to the resin or are relatively unbound and are washed through the column. As the elution solution is changed, so are the adsorbed analytes. This column provides a good initial separation, as each fraction is decontaminated from the other analytes by factors of from 10^2 to 10^4. Particularly for the case of samples isolated from a Pu matrix, each collected fraction must be further purified, and these purification steps are often based on the same Dowex-1X8 anion-exchange column technique. Although any tracer activity may have some presence in a heavy-element sample, the activities specially selected as tracers are unlikely to be more than minor constituents.

The initial tracing and fractions can be collected and combined to perform a spiked/unspiked analysis. The Pu and U fractions are collected from the N column for another purpose (mass spectrometry). Among the samples collected, the Np samples and the Ra/Ac/Th/Am/Cm samples have the highest analytical priority because of half-life considerations. Once separation from Am is effected, the ^{239}Np tracer decays with a 2.35-day half-life; similarly, ^{231}Th decays with a 25 h half-life after it is separated from ^{235}U.

As a check on chemical yields calculated from added tracer activities, the yields can also be determined from short-lived activities that are intrinsic to the sample. For example, the chemical yield of U can be measured from the recovery of added ^{232}U tracer activity, but in Pu samples, it can also be determined from ^{237}U. There is a 0.00245% branch for the decay of ^{241}Pu by α-particle emission. In Figure 10.1, the decay rate of ^{237}U is plotted, relative to the decay rate of ^{241}Pu, as a function of time after the last chemical separation. If it is known that the Pu specimen was last purified more than 1 month previously, it can be assumed that 6.75-day ^{237}U in the analytic samples is in secular equilibrium with ^{241}Pu, near the saturation value set by the branching ratio of [activity(^{237}U)/activity(^{241}Pu)] = 2.45×10^{-5}. If the quantity of ^{241}Pu in the sample is known (through Pu tracing or the application of gravimetry to MS-measured isotope ratios), the chemical yield of U can be determined from the amount of recovered ^{237}U (after decay correction to the time that U and Pu were separated). Similarly, the quantity of ^{231}Th in a Th fraction separated from U can be used as a check on the chemical yield determined from an added ^{228}Th tracer.

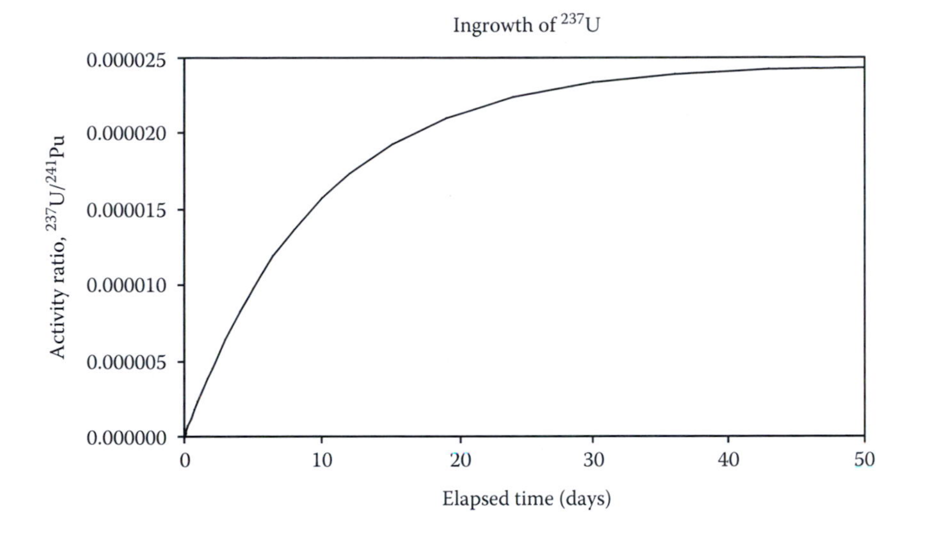

FIGURE 10.1 Ratio of the activities of ^{237}U and ^{241}Pu as a function of time following separation.

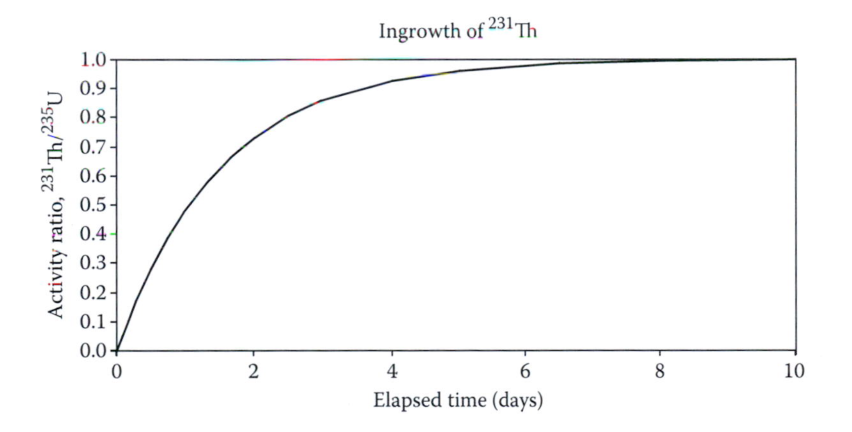

FIGURE 10.2 Ratio of the activities of ^{231}Th and ^{235}U as a function of time following separation.

From Figure 10.2, a U sample that is older than a few days must have an equilibrium amount of ^{231}Th relative to ^{235}U; that is, [activity(^{231}Th)/activity(^{235}U)] = 1.0.

The Ra/Ac/Th/Am/Cm fractions must be processed quickly for two reasons. First, the half-lives of the short-lived Th isotopes make it important that there be minimal delay. Second, although all five elements are essentially unbound during the anion-exchange elution with 9 M HCl, they may be slightly fractionated with respect to one another. Radium is traced with 3.66-day ^{224}Ra, the daughter of 1.9-year ^{228}Th, and equivalent chemical yields of both Th and Ra must be assumed or the ^{228}Th concentration will affect the concentration of ^{224}Ra through disequilibrium. If the second step of the chemistry on this fraction is performed quickly, the equivalence assumption is less important. If the time between the first separation and the subsequent separation of Th from Ra and Ac is kept to a minimum, uncertainties in the relevant decay corrections are minimized.

The dry Ra/Ac/Th/Am/Cm fractions are dissolved in 8 M HNO_3, evaporated to dryness, dissolved again in 8 M HNO_3, and passed through fresh Dowex-1X8 anion-exchange columns. These columns can be fairly small in capacity, as Th is likely to be the only analyte to be adsorbed on the resin. The other elements are collected as new Ra/Ac/Am/Cm fractions. Thorium is eluted from the columns with 9 M HCl and is another example of elution ion exchange.

The Ra/Ac/Am/Cm fractions are evaporated to dryness, dissolved in 1 M HCl, again evaporated to dryness, and redissolved in 0.2 M HCl. If a large quantity of Ni is present in the samples, the dry material may be treated with 1 M NH_4OH, after which dimethylglyoxime is added. The bright-red, Ni–ammonia–dimethylglyoxime complex is extracted into chloroform, after which the aqueous phase is evaporated to dryness and dissolved in 0.2 M HCl.

The separation of the Ra/Ac/Am/Cm fractions into components of individual elements is accomplished by a combination of elution and breakthrough ion-exchange. The new solutions are loaded onto cation-exchange columns filled with Dowex-50X8 resin beads, with the elution solution collected as waste. All of the analytes of interest are bound to the columns under these conditions. Dowex-50X4 resin can also be used, but the analytes would then be less tightly bound at all concentrations of HCl. The columns must be large enough to accommodate all of the cations in the samples. A decent assumption is that the fraction before dissolution in 0.2 M HCl contains about 0.5 eq/g, and the column size should be large enough that exchanger loading only occurs in the top 25% of the resin bed. The best column performance is obtained when large volumes of water, 2 M HCl, 6 M HCl, and water again are used to condition the resin beds before use. It is a good idea to prepare the columns and start this process the day before the columns are to be used.

The acid concentration of the eluting solution is gradually increased to 2 M HCl. This step results in the removal of much of the alkali-metal and alkaline-earth contamination from the columns [4–6] (see Table 10.1). Fission-product Cs can be eluted from the columns with 1 M HCl, and fission-product Sr with 2 M HCl. The 2 M HCl elutions are collected in open glass planchets, which are then dried under a lamp before α counting. The 2 M HCl elution is continued until all visible mass is gone or Am activity begins to appear in the eluent. At acid concentrations greater than 2 M, all four analyte elements become mobile, exchanging with the resin and moving down the column under solvent flow. Americium and Cm are least bound, followed by Ra and then Ac. However, at this acid concentration, analyte movement is quite slow. When the acid concentration is gradually increased from 2 to 12 M over a span of several free-column volumes, it becomes possible to collect fractions of Am + Cm, Ra, and Ac, in that order.

The narrow spread in K_d values across the rare-earths, Am, and Cm does not allow their separation from each other on the previous cation-exchange column. The separation of the trivalent actinides from the lanthanides is accomplished with a saturated-HCl column. The resin used is AG MP-50 (cation exchange), and the column eluent is

TABLE 10.1
Distribution Coefficients for the Sorption and Elution of Selected Cations from Dowex-50 Columns with Various Concentrations of HCl

Cation	0.5 M HCl	1 M HCl	2 M HCl	4 M HCl	6 M HCl
Na^+	12	5.6	3.6		
K^+	29	14	7.4		
Cs^+	44	19	10.4		
Ni^{2+}	70	22	7.2		
Ca^{2+}	150	42	12.2		
Al^{3+}	320	60	12.5		
Sr^{2+}	220	60	18		
Am^{3+}		150	24	6	
Cm^{3+}		150	24	6	
Eu^{3+}		150	24	6	
La^{3+}		260	48	10	
Ra^{2+}		250	70	24	16
Ac^{3+}			220	50	29

concentrated HCl that was saturated with HCl gas to increase the acid concentration to 13 M. During saturation, the liquid HCl container should be immersed in a room-temperature water bath to dissipate heat and increase the solubility of the HCl. The analytic sample is transferred to the top of the resin bed with minimum drops of acid, creating as narrow a band as possible at the top of the breakthrough column. Elution with 13 M HCl results in a trivalent actinide fraction that exits the column first, well-separated from all but the heaviest rare-earths (which are not produced in fission). To improve separation from the heavier lanthanides, ethanol can be added to the HCl solution, before saturation with HCl gas, to the level of 10% by volume.

Americium and Cm are separated from each other with a breakthrough procedure in which the activities are eluted from an anion-exchange column with methanolic nitric acid [7,8], with Cm preceding Am from the column. Addition of methanol to the column eluent favors the formation of polynitrate anionic species. Thus, the dry Am + Cm samples are dissolved in a small volume of a solution of 90 mL of methanol and 10 mL of concentrated (70%) HNO_3. Americium and Cm are adsorbed at the top of the resin bed under these conditions. The activities are then eluted from the column with a solution composed of 30 mL methanol, 39 mL water, and 31 mL concentrated HNO_3 (the final solution volume is <100 mL because of the mixing contraction). With the decrease in the methanol concentration, Cm becomes nearly unbound and elutes from the column in a few free-column volumes. Americium is significantly delayed, but also ultimately elutes from the column under these conditions. When the Am + Cm fraction from a Pu sample is processed, there is generally sufficient ^{241}Am in the sample that the Cm fraction is cut just before the first detectable β–γ activity from Am. This step may need to be repeated several times on the Cm fraction to yield a sample completely free of residual ^{241}Am.

The chemical properties of the rare-earths change only slightly from element to element, and for this reason, their separation by classical methods can be extremely difficult. The light lanthanides can be separated from one another with the anion-exchange procedure outlined earlier for the separation of Am and Cm. However, as the *Z* of the rare-earth increases, the effectiveness of the separation is reduced and becomes increasingly dependent on the methanol concentration of the eluent. The usual method for separating the rare-earths is by elution from a cation-exchange column with α-hydroxyisobutyrate (α-but) solutions. The procedure is more strongly dependent on the particle size of the resin, the temperature of the column, and the flow rate of the eluent than are the other procedures previously discussed. Optimum columns are fabricated from a commercial Dowex-50X 12 resins that has been separated into specific size fractions by letting the resin beads settle through upwelling water and obtaining an appropriate sample at a height defined by the water flow rate. This tactic results in a sized resin fraction for which all beads have nearly the same diameter. The column is mounted in a heating block maintained at 76°C. Caution must be exercised to prevent the column from going dry under these conditions. Further, any new aliquot of solution added to the top of the resin bed must reach thermal equilibrium with the column before eluent flow begins. The eluent flow rate should remain constant throughout the bulk of the separation (other than during loading and conditioning operations) and should be less than 1 $mL/cm^2/min$ [9].

The column can be run in either of two modes: pH controlled or concentration controlled. Under pH-controlled conditions, the concentration of the α-but ion is maintained at a constant value (usually between 0.2 and 0.3 M), and ammonium hydroxide is added to adjust the pH to a value of 3.4–4.2 (depending on the analyte to be separated). The advantage of pH control is that it allows the analyst to load the column with dilute HCl. After loading, the column is conditioned with solutions of ammonium chloride and water. The buffer allows the initial eluent solution to maintain constant pH through the elution, making the column performance reproducible. In the concentration-controlled mode, the elution of the lanthanides is determined by the concentration of the α-but in eluent solutions produced by dissolution of the reagent acid. Care must be taken that the eluent solutions do not come into contact with mineral acids or their fumes before use.

Thus, dry analytic fractions are dissolved in 0.05 M α-but to create the load solutions. Elution of consecutive rare-earths is accomplished with solutions of concentrations between 0.22 and 0.38 M. Under either control method, the lanthanides are eluted in reverse order, from heaviest to lightest. Typically, our laboratory performs the concentration-control procedure, collecting small fractions in 2 mL tubes that are subsequently counted for indications of short-lived fission products (e.g., ^{147}Nd) with a NaI(Tl) well-counter. Neodymium fractions collected from this column are sufficiently free from neighboring rare-earths that a MS analysis can be performed without further purification.

Radiochemical milking procedures are often performed with large volumes of analyte solution, looking for small signatures. In general, these procedures are not used for environmental or site-inspection samples, but only when bulk quantities of U or Pu are available for analysis. Milking procedures were treated previously.

10.4 MS ANALYSIS AND REAGENT PURITY

Uranium, Pu, and Nd fractions from the N and U samples are analyzed by MS and are usually submitted for analysis as dry deposits within Teflon containers. All other counting samples are prepared in vacuo by volatilization from W filaments onto Pt counting plates. The activity is dissolved in HCl and transferred to the filaments, which are warmed under an air stream to expedite evaporation. However, the samples to be volatilized must not contain nitrate, or the solution will attack the filaments.

Chemicals of normal reagent-grade purity suffice for the separation of analytes that do not occur in nature. There should be no significant effect on the analyses of samples of Np, Pu, Am, or Cm, provided reagent volumes are kept small so that mass is not introduced into the final fractions. For all other analytes, reagents that have been doubly distilled in Teflon should be used, even in the early stages of the chemistry, and contact with glass should be minimized. Reagent-grade chemicals and glass both contain significant levels of U and Th. The α-but reagent that is commercially available often contains a significant quantity of lanthanide mass; it should always be purified by vacuum distillation before use, regardless of its source or advertised purity.

REFERENCES

1. J.J. Katz, *Chemistry of Uranium*, McGraw-Hill, New York, 1951.
2. J.M. Cleveland, *Chemistry of Plutonium*, American Nuclear Society, LaGrange Park, IL, 1979.
3. L.A. Bray and J.L. Ryan, Catalyzed electrolytic dissolution of plutonium dioxide, *Radioact. Waste Manag.*, 6, 129, 1982.
4. F.W.E. Strelow, An ion exchange selectivity scale of cations based on equilibrium distribution coefficients, *Anal. Chem.*, 32, 1185, 1960.
5. F.W.E. Strelow, R. Rethemeyer, and C.J. Bothma, Ion exchange scales for cations in nitric acid and sulfuric acid media with a sulfonated polystyrene resin, *Anal. Chem.*, 37, 106, 1965.
6. R.M. Diamond, K. Street Jr., and G.T. Seaborg, An ion-exchange study of possible hybridized 5f bonding in the actinides, *J. Am. Chem. Soc.*, 76, 1461, 1954.
7. V.A. Bochkarev and E.N. Voevodin, Separation of americium and curium by anion exchange using solutions containing a mixture of methanol and nitric acid as the eluent, *Sov. Radiochem.*, 7, 459, 1965.
8. I.A. Lebedev, B.F. Myasoedov, and L.I. Guseva, Use of alcoholic solutions for the isolation and purification of americium and curium with anion-exchangers, *J. Radioanal. Nucl. Chem.*, 21, 259, 1974.
9. G.R. Choppin and R.J. Silva, Separation of the lanthanides by ion exchange with alpha-hydroxy isobutyric acid, *J. Inorg. Nucl. Chem.*, 3, 153, 1956.

11 Inorganic/Isotopic Sample Preparation

The conditioning and preparation of samples for radiation counting have also been treated to an extent in other parts of this book.

11.1 ALPHA COUNTING

Samples for α counting must be thin enough that range straggling does not prevent α particles from entering the sensitive area of the detector or degrade the peak structure of the spectrum. This is accomplished by purifying α-emitting chemical fractions from all inert contaminants. In the case of long-lived α emitters, such as isotopes of U and Th, care must be taken to balance the decay rate of the source against the quantity of analyte mass that accompanies it. Spreading an α-emitting sample over a larger area helps alleviate this problem until the dimensions of the source became comparable to the area subtended by the detector, at which point counting geometry causes a loss of counting efficiency. Final samples for α counting are usually limited to lateral dimensions of 2.5 cm. Alpha samples for absolute emission measurements must be mounted on the same substrate as the calibration standard (usually Pt) so that the degree of backscatter (α particles emitted into the sample backing that are subsequently reflected backward into the detector volume) is the same.

Samples for α counting are made by stippling, by vacuum volatilization, or by electrodeposition. In stippling, the substrate is warmed, and small drops of a solution containing the analyte are delivered from a transfer pipette, taking care to distribute the activity over the substrate in a quasi-uniform fashion. In vacuum volatilization, the analyte solution is dried on a W filament that is then mounted in vacuum between two electrodes. A current is passed through the filament, heating it white-hot, resulting in the analyte volatilizing from the filament and onto a substrate mounted above it. In electrodeposition, the analyte is dissolved in a plating solution that is placed in a cell, the bottom surface of which incorporates the conductive sample substrate. An electrode is suspended in the solution, a current passed through the cell, and the analyte electrodeposited onto the substrate.

In all three cases, activity can be fixed to the substrate, and extraneous mass removed, by passing the final sample through a Bunsen-burner flame (except in instances of elements such as Po, which are volatile at low temperatures). Stippling can be quantitative, with 100% of the analyte transferred from solution to substrate. The technique is also fast, but results in samples of inferior quality for spectrometric measurements. Vacuum volatilization results in much more uniform, thin deposits that are quite satisfactory for α spectrometry. If executed properly, the technique takes only slightly more time and effort than does stippling. Unfortunately, the technique is not quantitative, with significant losses (~50%) caused by a fraction of the volatilized analyte missing the substrate foil. Electrodeposition (or electroplating) can be nearly quantitative and results in a thin, uniform counting source appropriate for spectrometry. However, the technique is labor intensive and time consuming. Electrodeposition can initiate from an aqueous solution of near-neutral, pH-buffered ammonium chloride containing the analyte as an impurity and run at low voltage (~6 V) and high current (~100 mA). Alternatively, it may be conducted in an isopropanol solution of the nitrate salt of the analyte, to which enough 0.1 M HNO_3 has been added to produce a current (at an applied bias of 100 V) on the order of a few mA. In both instances, the optimum plating current at a given voltage is a function of the area of the source and the substrate-electrode spacing. It should be determined for each plating cell and plating solution.

11.2 BETA COUNTING

Samples for β counting need not be as thin as α samples, as even low-energy electrons have long ranges relative to α particles. The main problem in quantitative β counting is the fact that β particles are not emitted with discrete energies, which is a consequence of the simultaneous emission of the antineutrino. As a consequence, although the β source need not be thin, it must be uniform in thickness and of approximately the same thickness as the analytical standard used for calibration of the counters. Beta-particle sources can be assayed through attenuating media, but the same materials must then also be associated with the calibration count. The activity deposited for the experimental fraction should be of the same lateral dimensions as those of the calibration standard (unless the detector is much larger in area than either source, or the sources are counted at a significant distance from the detector).

Beta-particle sources can be made using the same techniques as for the preparation of α sources: stippling, vacuum volatilization, and electrodeposition. In addition, β-emitting radionuclides can be coprecipitated with small amounts of stable carrier and filtered into well-defined spots on porous paper. The weight of the deposit can be used for the determination of a chemical yield; this technique is rapid and can be nearly quantitative. Scintillation counting for β emitters (as well as α emitters, for that matter) requires the preparation of a liquid sample that can be added to a scintillation cocktail. This technique is discussed in some detail elsewhere in this book and is otherwise outside the scope of this section.

11.3 GAMMA COUNTING

Gamma- and x-ray emitting samples can be prepared in a variety of ways, including those described earlier for α and β counting. The penetrating nature of γ radiation allows the analyst to prepare sources in a variety of different media and geometrical configurations. As with β counting, if a γ-ray source is counted close to a detector, the solid angle that it subtends relative to the detector should be benchmarked against a radionuclide standard of the same physical dimensions. Low-energy photons are attenuated by matter more effectively than high-energy photons, with high-Z materials being more effective attenuators than low-Z materials. Thus, the analyst should not attempt quantitation of low-energy photon emission through a Pb container, but a plastic one is permitted for this application. Gamma-emitting nuclides in solution can be readily quantitated if the dimensions of the container are well defined and the quantity of attenuating material in the container wall facing the detector is known. Solutions of radionuclides that are interrogated for γ-rays must be stable. Any precipitation (or adsorption on the container walls) that might occur during a count would change the source geometry and affect the accuracy of radionuclide concentration determinations.

11.4 INORGANIC TECHNIQUES

Inorganic analysis can be accomplished by several methods. One of these is x-ray fluorescence (XRF), commonly used in the radioanalytical laboratory to identify the elemental components of questioned materials nondestructively. Typically, either solid or liquid samples are surveyed for the presence of elements with $Z > 13$ using this technique. Contiguous solid samples (such as metal foils) are affixed to Al counting plates. Subsamples of powdered or granular solids are finely ground and distributed on a piece of double-sided plastic tape affixed to the center of a counting plate. Liquid samples are adsorbed on 2-cm diameter Whatman 40 filter papers and dried under a heat lamp before mounting. If necessary, a thin layer of Mylar can be placed over the XRF sample to isolate it from the analytical staff. However, this tactic significantly degrades the quality of data obtained with the technique. Organic substrates should be selected that are free of incorporated chlorine ($Z = 17$). Counting plates should be fabricated from relatively pure Al, incorporating few transition metals, to decrease the magnitude of the blank measurement.

Another inorganic technique is neutron activation analysis, followed by radiation counting. A dry analytical sample is weighed and packaged into a container of polyethylene, quartz, or Be. A liquid sample can be dried in quartz under a heat lamp. Polyethylene or quartz containers are sealed with heat, whereas Be containers are threaded. The sample is then irradiated with neutrons in a nuclear reactor. A γ-ray assay of the activated package can be used to identify and quantitate the materials present in the sample. Although neutron activation analysis is far more sensitive for many elements than is XRF, allowing interrogation of smaller samples, reactors that are available for this purpose are becoming quite rare, and the induced activities can interfere with radionuclide signatures associated with unirradiated material. Both XRF and neutron activation analysis are nondestructive, however. A specimen is therefore available for additional forensic analyses that might ensue, as well as for display. Moreover, the ability of a prosecuting attorney to present the actual evidentiary item at trial can be a significant factor in the choreography of legal proceedings.

A more regular way of measuring inorganic components in a nuclear forensic sample is via mass spectrometry (MS). The most common uses of MS for such specimens involve the characterization of the inorganic constituents in the analytical pot solution and the determination of element isotope ratios in a purified analyte fraction, either for an isotope dilution measurement or for a direct determination of the enrichment of a heavy element. In most of these cases, the sample should be submitted for MS analysis as a dry deposit in the bottom of a Teflon container. This tactic allows the spectrometrist to dilute the sample to an appropriate concentration using completely characterized reagents. A purified fraction containing microgram levels of analyte is usually more than adequate for the relevant MS measurements.

Teflon containers must be scrupulously cleaned before use for MS techniques. A successful method is to boil Teflon apparatus twice in 6 M HCl for 1 h each, then once in aqua regia for an additional hour. The Teflon parts should be rinsed thoroughly, after which they should be boiled in 8 M HNO_3 for 4 h, and then for 2 h in distilled water that had been slightly acidified with HNO_3. After the final cleaning step, the containers are thoroughly rinsed with deionized water and allowed to dry upright on absorbent wipes. However, even with this procedure, Teflonware has a memory of past samples, particularly U, and we suggest that new Teflonware (cleaned as described earlier) be used for important nuclear forensic samples.

There is frequently some dynamic tension between the mass spectrometrist and the radiochemist when it comes to the transfer of samples. For MS applications, the standard glassware employed by the radiochemist is considered too dirty, as mineral acids leach alkali metals, boron, and silica from the glass. Unfortunately, Teflon is not appropriate for many radiochemical steps, particularly those involving volumetric manipulations (such as aliquoting and tracing). The reasons are that Teflon does not wet properly, is not sufficiently transparent, is semiporous to many inorganic species, and contains nontrivial interstitial fluoride. Thus, a tracer-level Pu sample manipulated in Teflon can suffer significant losses through sorption in the container walls. Reagent purity is not as great a factor in radiochemical applications as it is for MS. Finally, the concentration of acid in the analytical sample, and the identity of the anionic species comprising the acid, both affect the MS analysis. In general, a protocol that results in the submission of dry samples in Teflon for MS, with a rough estimate of the microgram-level of analyte in a specimen, yields the best analytical results and the least confusion.

In instances where a ratio is inadequate and an absolute concentration is required, an analytical pot solution whose matrix is 0.1 M HCl or HNO_3 is preferred over other mineral acids and other concentrations. However, for samples containing significant quantities of Pu, the acid concentration should be at least 0.5 M to avoid polymerization. Unusual sample matrices may result in pot solutions based on different acids (see, e.g., the case study on the counterforensic investigation of the U.S. enrichment facilities in Chapter 21). In these instances, the radioanalytical chemist and the mass spectrometrist must work together to determine the analytes whose measurement would be affected by the solution matrix.

The MS samples discussed here derive from a homogeneous solution, usually an aliquot of a much larger sample. Problems involving microscale samples are divided into two groups, isolated particles and localized inhomogeneities in bulk materials, and techniques for making these measurements are discussed elsewhere in this book. Small, isolated particles, such as tiny crystals, hairs, fibers, and filings, present difficulties in handling and mounting. We have had success embedding or adhering these species on a graphite matrix, which reduces effects of electric charge on the samples. However, accurate MS analyses of bulk samples quite often depend not so much on the mounting technique but, rather, on whether an adequate standard can be found to mount in the same fashion as a questioned specimen.

12 Organic Sample Preparation

Organic analyses are often employed for perpetrator, route, or explosives information in nuclear forensic investigations. The most versatile and effective instrumental technique for these applications is gas chromatography–mass spectrometry (GC–MS), and consequently, sample pretreatment and preparation for chromatography are central to the experimental protocols. Standard analytical tactics, such as weighing, pH adjustment, sonication, dilution, centrifugation, evaporation, and filtration, are extensively applied for organic sample prep. A result of many of these techniques is the concentration of diagnostic analytes from the questioned specimens before GC injection, and indeed, the enrichment of trace organic species from large-volume matrices is often an objective. Because each sample-prep step requires additional time and is a potential source of error, it is prudent to minimize such manipulations. Optimizing laboratory techniques is particularly important for dilute–concentration samples.

12.1 EXTRACTION

Chemical extraction methods are regularly employed as an essential practice in the work-up of organic specimens. Although other variations on this theme include ultrasonic extraction, accelerated solvent extraction, and microwave-assisted extraction, classical liquid–liquid solvent extraction remains an indispensable procedure for organic sample prep. Methylene chloride is an excellent reagent for removing nonpolar organic species from aqueous and solid samples, whereas a 3:1 mixture of methylene chloride/isopropanol will extract a more inclusive suite of general polar and nonpolar organic compounds. This broad approach is often modified by the specific considerations of an individual investigation, however. Thus, for example, acetone is a solvent of choice for the recovery of a broad spectrum of high-explosive analytes.

Whenever feasible, identical subsamples of a given specimen should be subjected to all of the related types of forensic interrogation necessitated by an investigation. In that way, the likelihood is greatest that the different analyses are performed on effectively the same material. The final consolidation of all forensic data into a self-consistent interpretation of the evidence is then as straightforward as possible.

12.2 SOLID-PHASE MICROEXTRACTION

Relatively new systems for organic sample preparation are solid-phase microextraction (SPME) fibers. SPME comprises a chemical sorbent coated on the surface of a glass fiber that is sequestered within the protective sheath of a syringe needle. The fiber is withdrawn from the barrel and exposed to the headspace over a questioned specimen (or briefly dipped into a liquid sample) to concentrate analytes on the SPME substrate. The fiber is then retracted into the assembly, a Teflon cap placed over the needle opening to protect the collected specimen, and the apparatus transported to a GC–MS instrument for direct sample introduction into the injection port. SPME can therefore reduce hours or days of tedious sample preparation to a few minutes. Moreover, a fiber can be reused many times after cleansing by thermal desorption in a hot GC injection port. The process generates no chemical waste streams, and it reduces bulk quantities of potentially dangerous materials to relatively innocuous nanogram to microgram levels adequate for GC–MS analysis. However, a weakness of SPME is the fragility of its fused silica core, though more flexible fibers have recently been developed to reduce breakage.

A variety of SPME fibers with different chemical coatings are commercially available. They include carboxen, polydimethylsiloxane (PDMS), divinylbenzene (DVB), and carbowax, and they are deployed singly or as composite substrates targeted at different classes of compounds. Carboxen/PDMS collects and concentrates low-molecular-weight species, PDMS/DVB sequesters amines and nitroaromatic compounds, and carbowax/DVB is optimum for alcohols and polar compounds. Several comprehensive textbooks have been written on the development and practice of SPME [1–3].

12.3 DERIVATIZATION

Another valuable sample-prep tactic before organic analysis by GC–MS is compound derivatization to improve the detectability or chromatographic separation of eluted species. Derivatization entails chemical reaction of analytes with an added reagent to form products clearly identified as having originated from the reactants. Motivation for this tactic arises in investigations in which the instability or large polarity of suspected signature compounds precludes their successful analysis by GC–MS.

Organic functional groups subject to derivatization methods are active hydrogens, carboxylic acids, alcohols, phenols, amines, amides, nitrosamines, and several others. Even inorganic anions may be thus functionalized for GC–MS assay. The major chemical reactions employed for sample derivatization are silylation, acylation, and alkylation, while silyl reagents, particularly BSTFA [bis(trimethylsilyl)trifluoroacetamide], are very versatile for GC analyses. The essence of BSTFA is the trimethylsilyl (TMS) group: $-Si-C-(CH_3)_3$. Substitution of TMS for an active hydrogen in a molecule, for example, lessens the compound polarity and reduces hydrogen bonding. The resulting TMS derivative is therefore more volatile, as well as more stable, because of the inactivation of reactive sites. Silylated species are less polar, easier to detect, and more thermally robust, and many molecular signatures that are nonvolatile or unstable at 200°C–300°C can be effectively chromatographed following silylation.

A specimen is generally derivatized after purification chemistry has been performed to isolate a clean experimental fraction before instrumental analysis. The refined sample is often split, with one fraction subjected directly to GC–MS analysis and the other derivatized before injection. A greater range of polar and nonpolar chemical species can be successfully analyzed as a result. In general, derivatization enhances compound stability by protecting unstable groups, and it also reduces the polarity and increases the volatility of organic signatures that might otherwise decompose on heating. In effect, it allows GC separations that are impossible with underivatized compounds.

Derivatization in tandem with SPME is also a useful protocol for forensic analysis. The derivatization can be accomplished on-fiber and can be executed either after or simultaneous with the extraction.

REFERENCES

1. J. Pawliszyn, *Solid-Phase Microextraction: Theory and Practice*, Wiley-VCH, New York, 1997.
2. S.A. Scheppers-Wercinski, ed., *Solid-Phase Microextraction: A Practical Guide*, Marcel Dekker, New York, 1999.
3. J. Pawliszyn, *Applications of Solid-Phase Microextraction*, Royal Society of Chemistry, Cambridge, U.K., 1999.

13 Extraordinary Sample Issues

The possible admixture of radioactive contamination with the evidentiary items of conventional forensic analyses poses an unusual problem for traditional criminalistics. Such samples of "mixed" evidence require special handling and treatment to perhaps decontaminate a questioned specimen, generate separate waste streams of radioactive and combined radioactive/hazardous wastes, or even conduct standard examinations within the confines of a controlled-environment glove box or hot cell.

Specialty sample receival facilities must be designed and assembled for this purpose. For the most part today, laboratories able to process large quantities of Special Nuclear Material (SNM) or other radioactivities are incapable of effectively isolating low-level forensic indicators (such as trace evidence, DNA, pollen, or even radionuclear signatures) from a specimen. Conversely, the large majority of labs competent to analyze the environmental signatures are not permitted to handle useful quantities of SNM. In the first instance, the facility would compromise the sample; in the latter, the sample would contaminate the facility. This dilemma is especially significant for the forensic analyses of smuggled specimens because the background indicators can be vital for intelligence or criminal prosecution. They will often provide valuable information about the location where the sample was manipulated and packaged, the individuals involved, and the route over which it was transported before interdiction. In addition, trace amounts of various radioisotopic species are often essential for source attribution.

13.1 THE RDD

In the category of extraordinary predet nuclear forensic evidence, the unexploded improvised nuclear device (IND) and radiologic dispersal device (RDD) are paramount. For considerations of personnel radiation dose and exposure safety during forensic sampling, a worthy RDD should be the more dangerous situation since ionizing radiation emissions from a credible IND would likely be much less hazardous than those from an RDD. Should detonation occur, of course, personnel adjacent to either device would be sorely distressed. To first approximation, the successful explosion of an IND would effectively dictate that subsequent analytic work is directed toward postdet efforts focused on source characterization. Following a successful RDD event, however, evidence collections for both radioactive and diverse collateral forensic signatures would remain possible and would indeed be likely.

The RDD is basically a radioactive IED (improvised explosive device). Although often placed in venues dealing with weapons of mass destruction, as but an instrument of mass disruption, the RDD is designed for psychological impact and mass panic, rather than devastation and extensive fatalities. However, a recent study has suggested that detonation of an RDD within a large urban environment would result in much higher economic burdens than previously estimated [1]. Although initial consequences would likely result in >$1 billion in lost regional gross domestic product, negative psychological residuals on the general public were forecast to contribute an additional $16 billion in costs over the ensuing 10 years. Contributing factors included a general loss of economic transactions within the affected zone (not solely food products) and demands by workers for higher wages to return to their jobs.

13.1.1 Isotopes and Commercial Uses

Industrial isotopes are the primary concern for the radioactive component of an RDD. The activity regime of such sources spans the range from <μCi used for smoke detectors to multi-MCi levels used in gamma irradiators, thereby making RDD analysis an order-of-magnitude science. Throughout this section, and in this book in general, the Curie is used as the practical unit of radioactivity, rather than the SI-defined Becquerel. Most sensible nuclear scientists not bound exclusively by ultra-trace analysis retain the Curie for effective terminology because of the irrelevance of formal SI radioactivity units for real-world applications. Unfortunately, authors are often forced to convert artificially to Bq by journals and other publications for reporting, yet these same publications will print data entries with constructs such as hundreds of thousands of terabecquerels for their measure. The only real function for SI-forced units, such as hundreds of kilo-teraBq and their ilk, is as input fodder for lay reporting of radioactivity issues in sensationalist media. Huge numbers must certainly reflect huge problems, so, for example, 10,000 nCi rather than 10 μCi for data from Three Mile Island (pre-Bq era). Bq units for much of radiochemistry (and all of NFA, where atoms present at specific times are calculated) are equivalent to posting distance measurements on highway signs in Angstroms.

Although terrorist application might well incorporate whatever radioactivity happened to be conveniently available, the entries in Table 13.1 summarize the relevant

TABLE 13.1
RDD Radionuclides

^{A}Z	Half-Life	A_{SP} (Ci/g)	α (MeV)	β_{max} (keV)	γ (keV)	R/hr/Ci at 0.3 m	Other
^{60}Co	5.27 years	1100	—	318	1173, 1332	16	
^{90}Sr-	28.9 years	140	—	546	—	—	Bremss.
^{90}Y	64.1 h	550k	—	2280	—		
^{137}Cs-	30.1 years	88	—	512	—	4	
^{137m}Ba	2.55 minutes	540M	—	—	662		
^{192}Ir	73.8 days	>450	—	672	296–468	6	
^{226}Ra	1600 years	1	4.8	—	186	0.1	Daughters
^{238}Pu	87.7 years	17	5.5	—	—	0.9	SF
^{241}Am	432 years	3.4	5.5	—	60	0.2	+ Be/Li → n
^{244}Cm	18.1 years	81	5.8	—	—	—	SF; 6570-year ^{240}Pu
^{252}Cf	2.64 years	540	6.1	—	—	—	SF

nuclear parameters of those nuclides considered most effective for an RDD. Since any radionuclide could serve to induce the panic factor in public reaction to an RDD, some analysts have included other commercial isotopes in their lists of favorites (e.g., 10.6-year ^{133}Ba and 120-day ^{75}Se radiography and calibration sources). Thus, although the listing in Table 13.1 is most worthy, it should not be considered inclusively comprehensive.

Probably the most extensively utilized industrial radioisotopes, ^{60}Co and ^{137}Cs are often found in high-dose applications that require source strengths ranging from kCi to tens of MCi. These consist of industrial irradiators for food and medical sterilization, research irradiators, and medical therapy (such as the γ-knife, blood irradiators, and teletherapy). Other uses of ^{60}Co and ^{137}Cs include commercial radiography devices, fluid processing, level gauges, and calibration sources.

The primary uses for ^{90}Sr are in thickness gauges and as the long-lived heat supply in terrestrial radioisotope thermoelectric generators (RTGs). The latter application entails kCi–MCi activities and is popular for remote power sources in lighthouses and for military use on the north coast of Russia.

Ir-192 is a familiar source in industrial radiography, incorporating activity levels of approximately 0.1–10 Ci. It is also used for nondestructive testing devices. ^{226}Ra is found in luminescent dials and has the longest physical half-life for environmental persistence of any of the species in Table 13.1. It also decays to a multiplicity of diverse α-, β-, and γ-emitting radioactive daughters (including volatile 3.8-day ^{222}Rn), whose varied chemistries could require complex cleanup and decontamination considerations.

A component of smoke detectors, density-thickness gauges, static eliminators, lightning preventers, and other devices, ^{241}Am has myriad commercial uses. In addition to potential direct radiation for an RDD (432 years, 60-keV γ, and 5.5-MeV α), it is also a prolific source of neutrons when $^{241}Am_2O_3$ is admixed with low-Z elements such as Be or Li. Indeed, the AmBe (α,n) sealed source is a common industrial tool for neutron radiography, activation analysis, well logging, and nuclear reactor start-up. These high-activity applications use up to 20 Ci of ^{241}Am to generate 10^7–10^8 n/s, and there are several thousand such sources in use in the United States. There also exist more than 20,000 lower-activity AmBe sources in the United States for moisture-density gauges (<50 mCi; 10^5 n/s). Of course, device dissemination by the HE of an RDD could well separate the two integral components such that neutron emission in debris might be nonexistent or of lesser concern, but the primary radiation emissions of the Am would still require remediation.

Relatively small quantities of ^{238}Pu are used in smoke detectors and cardiac pacemaker batteries, while intermediate-level activities are necessary for radioisotope heater units and extraterrestrial RTGs. Both ^{244}Cm and ^{252}Cf are sources of spontaneous-fission neutrons. The former is utilized in mining operations for analyses of drilling slurries and materials excavated from pits. The latter is used for well logging, moisture gauges, and explosives surveillance of airline luggage. Principal decay daughters of both nuclides are 6560-year ^{240}Pu along with other long-lived species, thereby increasing the environmental persistency of these systems.

13.1.2 Radiation Devices and Threat Potential

There are limited isotope production facilities around the world that have viable capacities to synthesize the vast quantities of radionuclides required for commercially practical operations. For example, the Northern Hemisphere includes REVISS in the United Kingdom, MDS Nordion in Canada, Mayak Production Association (PO Mayak) in Russia, and the Department of Atomic Energy in India; in the Southern Hemisphere, such facilities exist at the Comision Nacional de Energia Atomica in Argentina and the South African Nuclear Energy Corporation.

In 2009, the US Nuclear Regulatory Commission (NRC) deployed the National Source Tracking System (NSTS) mandated by the 2005 Energy Policy Act. The NSTS is a centralized national registry that provides cradle-to-grave accountability for specific radioactive materials, including sources used in irradiators, teletherapy devices, radiography, and well logging. It contains data on source ownership; location; contact information; license number; make, model, serial number, identity, and activity of the source; and all transfers between licensees from initial procurement to final disposal. Anyone holding and making use of a controlled source must now possess an NRC license, or one from one of 35 authorized agreement states. These recent US protocols for source tracking are not universally employed, however.

The largest commercial radiation installation is the industrial irradiator. Typically the size of a small building or large room, it deploys tens of MCi of ^{60}Co distributed over hundreds of individual source pencils. Primary applications are in the areas of medical instrument sterilization and food irradiation. A few hundred of these units exist worldwide, and since they are produced and sold by corporations in developed countries, they possess general industrial safety/security systems and are self-protecting to a reasonable degree. However, with the source half-life of only 5.3 years, they must be reloaded, in whole or in part, relatively frequently. The possibility of theft during such operations is a concern of the counterterrorism community.

A typical research irradiator utilizes a few tens of kCi of 30-year ^{137}Cs. The lesser activity and lower γ energy translate to considerably less shielding required than for the industrial irradiator, and the longer half-life makes frequency of reloading operations a virtual nonissue. Several hundred units are in use worldwide, often within universities and medical schools. However, because of the itinerant nature of students, the variable scientific interests among students and faculty alike, and fluctuous programmatic funding cycles, research environments are more prone to source inattention, uncertain stewardship, and periods of neglect. The chemical form of ^{137}Cs in such irradiators is CsCl, an alkali-metal salt that is easily dispersed.

A notable variant of the research irradiator for RDD concerns was Project Gamma Kolos in the former Soviet Union (FSU). It was an applied agricultural research program during the 1970s and 1980s to determine the results of ionizing radiation on seeds and plants. A shielded, 3.5-kCi source of ^{137}Cs was housed in a small truck and driven between different farms and fields to study radiation effects on various crops and their growing cycles. Up to 1000 of these vehicles were operational during the program peak, and groups of orphaned radiation sources were ultimately accumulated in unprotected, open-storage depots. In 2003, only nine were accounted for and securely impounded. It is worrisome that such a large supply of long-lived ^{137}CsCl was so easily accessible in readily mobile form.

Radiographic and well-logging sources have widespread use internationally for diverse applications. In addition to ^{60}Co and ^{137}Cs, ^{192}Ir is also very popular for radiography, and activity levels of 10 Ci and lower are typical. On the order of 10,000 new radiography sources are sold annually, with opportunities for falsified purchases relatively high, and lost or stolen sources familiar occurrences. Significant quantities of radioactivities are commercially available in extremely compact packaging, making potential diversion for an RDD reasonably unencumbered by obstacles of bulk (if protective shielding is of little or no concern, i.e., suicide missions). For example, one vendor sells 1–10 Ci sources of ^{60}Co and ^{137}Cs, doubly encapsulated in welded stainless-steel capsules, having diameters between 0.5 and 2 cm. Their lengths vary from 3 to 6 cm, as a function of source strength.

Around 10,000 well-logging sources are currently in use, with activity levels ranging up to a few tens of Ci. ^{137}Cs is the most popular γ-emitter, while AmBe, AmBe analogs (e.g., PuBe, RaBe), and ^{252}Cf are favorite neutron sources. A tandem γ + n device is used to evaluate geologic oil formations near exploratory drill holes. A 10–20 Ci AmBe induces neutron activation for elemental analysis, as well as performs compensated density measurements, while 30 Ci of ^{137}Cs measures compensated dual resistivity. The instrument is 16 cm diameter × 16 m long, contains packed powders, and is very portable. Theft is a concern. Despite the 432-year half-life of ^{241}Am, sealed AmBe of some design embodied more than 50% of the 11,000 unwanted sources reclaimed in the United States from 2001 to 2010.

Orphaned and abandoned ^{90}Sr RTGs (typically 5–500 kCi each) from the 1000 or so lighthouses in northern Russia are scattered around the FSU. They are unwieldy, 360–3600 kg in weight, and can radiate significant residual heat. However, ^{90}Sr–Y does not emit primary penetrating radiation and is therefore somewhat easier to secret than either ^{60}Co or ^{137}Cs.

Much of the commercial source information presented in this section was taken from post-9/11/01 investigations by Van Tuyle and Mullen [2], and more details may be found in this primary reference. However, no attempt here was made to update their facts in light of developments that may have occurred over the past 10 years (e.g., advent of electron linear accelerators for sterilization or potential displacement of AmBe sources by small D–T accelerators).

13.1.3 Maximum-Credible Source and RDD Aftermath

Despite the commercial availability of profuse quantities of radionuclides with penetrating radiation as industrial sources, they are not the RDD analyst's worst scenario. The max-cred nightmare is irradiated fuel elements from a nuclear reactor. In addition to residual fuel and a multiplicity of actinide elements, irradiated fuel can also contain many MCi of mixed fission products. Comprised of diverse chemical species and redox states, the fission products span a practical range of roughly one-third of the periodic table, from transition metals to the mid-lanthanides [$Z = 30$ (Zn) $\rightarrow$ $Z = 65$ (Tb)]. In contrast to single-nuclide radioactive sources, RDD dissemination of burned reactor fuel into the environment would yield a complex and heterogeneous milieu of radiochemical species that would present extraordinary challenges for decontamination efforts.

The magnitude of the radiation background of a spent-fuel source may be appreciated from the following discussion. Consider a single 3.6-m fuel assembly from a pressurized-water reactor (PWR) that had been irradiated for 30–35 GW-d/MT. Even after 5 years in a cooling pond, this fuel burn-up would result in residual radiation that provided a whole-body dose of ~5000 rem/h at 1 m (i.e., a human LD_{50} in ~5 min). Doubling the cooling time to 10 years would decrease the exposure rate to only ~3 kR/h at 1 m.

The potential target pool for theft or terrorist destruction of irradiated fuel rods is international and large. At the end of 2011, 543 total power reactors existed worldwide: 435 in operation, and the remainder either under construction or ordered. In the United States alone, more than 75 locations house >65,000 metric tons of spent fuel, and >2,000 metric tons are added to the inventory each year. In 2012, Khalid Ali-M Aldawsari, a Saudi national, was convicted by a Texas court for plotting bomb strikes against former US military personnel, the home of George W. Bush, dams, and nuclear power facilities.

The catastrophic 1986 disaster at the Chernobyl nuclear power reactor in Ukraine released 3%–4% of the active fuel inventory across national boundaries and contaminated 10^5 km^2 of land. This emission produced 400 times more environmental radioactivity than the Little Boy bomb dropped over Hiroshima to end World War II. While the Chernobyl incident might appear analogous to a super-RDD device containing irradiated reactor fuel, important differences between the two situations make this an unworthy comparison. First is the disparity in sheer magnitude of the source terms. Absent complete destruction of a nuclear plant by hostile force, a credible terrorist RDD would not approach Chernobyl in either explosive energy or quantity of radioactivity, even given the PWR fuel assembly mentioned earlier. Moreover, an RDD would be a violent pulse of short duration. Chernobyl, however, was a steady-state continuous release that required many hours to even extinguish the fires (one enclosed in Reactor 4 burned for days), let alone contain all of the radioactive release.

Association of Chernobyl with Little Boy is indeed more appropriate. From solely fallout considerations (i.e., without the factors of blast radius and prompt γ–n radiation release), this reactor accident was more similar to an IND than to an RDD. For example, the ground-burst detonation of a 1 Mt nuclear weapon results in a downwind fallout pattern with dose contours inversely proportional to distance: 4.5 kR extends to ~35 km from ground-zero; 450 R to ~100 km; and 50 R to ~240 km. For comparison, if one models a plausible RDD based on the worst radioactive source incident to date (Goiania, Brazil, 1987; 1.4-kCi ^{137}Cs radiotherapy source stolen from a hospital for scrap metal; 4 fatalities; >100 contaminated), an illustrative RDD aftermath can be computed. Distribution of 1.5 kCi of readily dispersed ^{137}CsCl by 45 g of TNT in an urban environment is predicted to contaminate an area of ~1 km^2 (~5 city blocks). The dose contours are correspondingly smaller and of shorter distance than from a nuclear blast: 5 R to ~80 m; 1 R to ~250 m; and 500 mR to ~350 m. Thus, the impacts of the two devices are extremely different and quite dissimilar in geographic extent; an RDD is not equivalent to an IND for credible reduction to practice; and the RDD is not an instrument of widespread mass destruction.

The good news is that no terrorist RDD has yet been exploded. The bad news is that an RDD has relatively high potential to eventually occur, likely very much higher than an IND.

13.1.4 Historic Nuance

Prior to 9/11, a dirty radioactive bomb meant something very different to the nuclear weapons community. Since then, however, various media have associated the term "dirty bomb" as being synonymous with the RDD. Historically, nuclear weapons have been characterized as having fundamental, "clean," and dirty properties. A "clean" warhead is, more accurately, merely cleaner than otherwise, as it minimizes fission reactions (and residual fission products) relative to fusion reactions. Examples of such designs are the enhanced-radiation devices and the neutron bomb.

To explain the original meaning of the historic dirty bomb, a two-stage thermonuclear explosive device must be considered. In this design, x-rays from a fission primary induce thermonuclear fusion in a secondary, resulting in the emission of high-energy neutrons. These thermonuclear neutrons then react with the device radiation case, often cheap (i.e., unenriched) U metal, to produce more yield via high-energy fission in the case (inducing more fission products). This is the classic fission–fusion–fission design.

Along with lesser amounts of activation products and residual actinides, the predominant residual radioactivities present after such detonation are copious quantities of mixed fission products. From zero-time to one-half year later, the collective half-life of these mixed fission products is approximately 8 days. However, were it desirable to appreciably increase postblast fallout effects and environmental decontamination efforts, an element with suitable neutron-activation properties could be alloyed into the radiation case. One such element, Co, would capture device neutrons to form ^{60}Co and thus present site remediation with an additional fallout component having extremely penetrating γ radiation and a 5.3-year half-life. This "cobalt bomb" was proposed in the early days of weapons development by Leo Szilard and was the original concept of a "dirty bomb."

13.2 MIXED EVIDENCE

Suitable sample acquisition and handling protocols for radioactive evidence require developments for the preservation of all diverse forensic signatures that could be obtained from a specimen or its container and packaging materials. Safety considerations for such activities are also important. Analytes such as hair, DNA, and pollen require exceptionally clean sampling protocols, whereas fingermark investigations need the conscientious protection of surface areas. Acceptable methods must also specify a prioritized sampling order, such that forensic information is not lost to the later analyses; the sequence in which specimens are processed can have a considerable influence on the quality of data obtained. Thus, for example, latent fingermark processing with physical developer solution may have degrading effects on subsequent DNA recovery.

A suggested order of examinations, absent any radioactive contamination considerations, has been outlined in a proposed strategy for processing traditional evidence from a rendered-safe RDD or IND at a forensic laboratory [Special Agent, US Federal Bureau of Investigation, personal communication]:

1. Visible trace evidence (hairs, fibers, etc.) is removed from external surfaces.
2. The condition of the evidence as received is photographed and documented.
3. Latent fingermark examination is conducted, including stripping off and investigating any tape; additional trace evidence, if present, is isolated.
4. Tape is examined for polymer content and end-match assessment.
5. Material residues are collected from the high-explosive containment vessel.
6. Tool-mark examination is conducted.
7. Metallurgical exams of unusual or significant metals are conducted.
8. Device components and functions are assessed.

Other evidence-specific factors could alter the order, however. For example, if paper were present, its isolation for questioned-document examination would take place before any contact fingerprinting methods were implemented. Similarly, should the case have a DNA component (e.g., a sealed envelope or gummed labels), noncontact fingermark screening by laser would occur first. Of course, any required radioactivity decontamination procedures might also necessitate changing an initial forensic sequence as a consequence of overall method development.

More fundamental research investigations, exploring core foundations of established techniques, may also be necessary for reliable nuclear forensic analyses. For example, if DNA evidence is exposed to significant levels of ionizing radiation for some period of time, would the conventional methods of bioscience (such as the polymerase chain reaction or restriction fragment length polymorphism typing) still give expected results? If not, what are the empirical bounds on critical parameters that may be required for successful analyses? The answers to these and other similar questions are likely essential for the effective practice of various aspects of the discipline.

Most of the applied research in this area over the past 10–15 years has explored the effects of ionizing radiation on the quality of conventional forensic evidence. An early study, stimulated by the decontamination of mail from viable biologic pathogens after 9/11, determined that 30–50 Mrads of electron-beam irradiation somewhat degraded DNA on envelopes. However, it did not hinder the ability to measure full nuclear STR profiles or mitochondrial HVI sequence haplotypes [3]. Similar studies were conducted on the impacts of electron beams on latent-print recovery [4] and inks [5], as well as of γ radiation on fiber evidence [6]. Although various evidentiary decontamination procedures have been explored within two US national laboratories, very little work has been published on radioactive decon methods for the pretreatment of evidence prior to criminalistics examinations (Ref. [7] is a notable exception). More research efforts are needed in this area.

REFERENCES

1. J.A. Giesecke, W.J. Burns, A. Barrett, E. Bayrak, A. Rose, P. Slovic, and M. Suher, Assessment of the regional economic impacts of catastrophic events: CGE analysis of resource loss and behavioral effects of an RDD attack scenario, *Risk Anal.*, 32(4), 583, 2012.
2. G.J. Van Tuyle and E. Mullen, Large radiological source applications: RDD implications and proposed alternative technologies, LA-UR-03-6281, Los Alamos, NM, 2003.
3. A.G. Withrow, J. Sikorsky, J.C.U. Downs, and T. Fenger, Extraction and analysis of human nuclear and mitochondrial DNA from electron beam irradiated envelopes, *J. Forensic Sci.*, 48(6), 1302, 2003.
4. R.S. Ramotowski and E.M. Regen, The effect of electron beam irradiation on forensic evidence. 1. Latent print recovery on porous and non-porous surfaces, *J. Forensic Sci.*, 50(2), 298, 2005.
5. R.S. Ramotowski and E.M. Regen, Effect of electron beam irradiation on forensic evidence. 2. Analysis of writing inks on porous surfaces, *J. Forensic Sci.*, 52(3), 604, 2007.
6. M. Colella, A. Parkinson, T. Evans, J. Robertson, and C. Roux, The effect of ionizing gamma radiation on natural and synthetic fibers and its implications for the forensic examination of fiber evidence, *J. Forensic Sci.*, 56(3), 591, 2011.
7. A. Parkinson, M. Colella, and T. Evans, The development and evaluation of radiological decontamination procedures for documents, document inks, and latent fingermarks on porous surfaces, *J. Forensic Sci.*, 55(3), 728, 2010.

14 Field Collection Kits

The section of this book on sampling and collection techniques emphasized the fundamental importance of effective protocols for those activities. The reason is that good science must begin at the incident or crime scene, not at a laboratory workbench. The choice of sample collection apparatus is key to worthy specimen acquisition, and the contents of a nuclear forensic field kit should be tailored to the anticipated needs of a specific investigation. It is impossible to cover all contingencies here, but the example of a specific case study is illustrative.

In the mid-1990s, the Livermore Forensic Science Center was tasked to perform "counterforensic" sampling at the US gaseous diffusion facilities conducting U isotopic enrichment. The plants were under consideration for inspection by the International Atomic Energy Agency (IAEA) for the purposes of international safeguards and accountability. The concern of US authorities was whether an inspector with an alternate agenda could learn secret information through analyses of environmental samples of opportunity collected surreptitiously at the sites. The LLNL Forensic Science Center (FSC) therefore conducted a modest sampling campaign to evaluate the magnitude of potential risk to sensitive information and technology from such accessibility. Although a general objective of the exercise was the broad assessment of plant vulnerabilities, a collateral nuclear forensic investigation of the US Special Nuclear Material production complex was also successfully executed. More detail may be found in Chapter 21.

The field collection hardware expressly chosen for the anticipated requirements of sampling the gaseous diffusion facilities was readily contained within two hard-shell, one-man-portable suitcases. A detailed inventory of the two cases is given in Table 14.1. A wide range of possible sampling tools was assembled, from remote tongs and a hacksaw for bulk specimens to hand-held vacuums fitted with tandem coarse and cellulose filters for the collection of fine particulates.

Although some swipe samples were obtained in the collection, the most useful samplers for this particular investigation were spatulas (for diverse solid debris) and horseshoe magnets (for opportunistic ferromagnetic materials). The latter were covered with plastic wrap during the accumulation of magnetic specimens, which were subsequently transferred into zip-locked polyethylene bags by physically detaching the plastic wraps from the magnets.

For this campaign, the collected samples were placed in either high-purity polyethylene, liquid-scintillation vials, or zip-locked polyethylene bags for primary containment. Another zip-locked bag served as a secondary sample container, and the doubly enclosed specimens (all solids in this case) were placed in individual plastic boxes for transport back to FSC labs. The boxes were sealed with vinyl

TABLE 14.1
Contents of Nuclear Field-Sampling Kit for Counterforensic Collections at Gaseous Diffusion Facilities

Applicators, cotton-tipped	1 Package
Backpack/duffel bag	1
Bags, garbage, polyethylene	>6
Bags, sample, polyethylene	25
Bags, zip-lock, 10 and 20 cm	>60
Batteries, AAA, AA, D	>4 each
Blades, razor	15
Bottles, polyethylene, 20, 30, 60, and 125 mL	15 each
Bottles, Teflon, 250 mL	4
Bottles, Teflon, wash	4
Boxes, plastic, 4 × 4 cm and 6.5 × 9 cm	20 each
Brushes, fine paint	5
Clamps, hose (various sizes)	5
Clipboard	1
Clips, alligator	5
Clips, paper, heavy-duty	10
Compass	1
Cutters, wire	1 pair
Disks, high-purity graphite, SEM	4
Files, rectangular, 15 cm	3
Flashlights, small and large	2 each
Foil, aluminum	1 roll
Forceps, curved, 14 cm	4
Forceps, straight, 14 cm	4
Glasses, safety	3 pair
Gloves, latex, XL	1 box
Hacksaw, with spare blades	1
Kimwipes, small	2 boxes
Labels	>50
Magnets, bar and horseshoe	10 each
Mirrors, dental	3
Papers, filter	
Millipore, cellulose, 6 cm × 0.45 μm	>50
S&S, #604, 12.5 cm	>50
Whatman, cellulose, 2.5 cm × 0.45 μm	>50
Whatman, #4, 9 cm	>50
Pens, writing and marking	5 each
Pipettes, transfer, with 10-mL syringe	10
Pliers, needle-nose	1
Pole, telescoping, 1.2–3.6 m	1
Post-it pads, various sizes	1 each
Sandpaper, 600 grit	6 sheets

(*Continued*)

TABLE 14.1 (*Continued*)
Contents of Nuclear Field-Sampling Kit for Counterforensic Collections at Gaseous Diffusion Facilities

Scissors	3
Scrapers, razor	3
Screwdriver, slot	1
Seals, tamper-proof	>50
Spatulas, stainless steel	5
Swipes, absorbent	>30
Swipe kits	3
Forceps, filter, 13 and 26 cm	1 each
Forceps, straight, 32 cm	1
Hemostat	1
Tape, double-stick	2 rolls
Tape, measuring, 7.5 m	1
Tape, Teflon	1 roll
Tape, vinyl	2 rolls
Tie-wraps, plastic	10
Tongs, 65 cm	1
Vacuums, hand-held	2
Vials, glass, with Teflon cap liners, 20, 40, and 60 mL	10 each
Wire	1 roll
Wrap, plastic	1 roll

tape, and tamper-indicating seals were placed at various locations on the exteriors. Sample-identification labels were positioned on both the primary containers and the outer boxes.

A more general treatment of radiologic evidence collection (as well as chemical and biologic), along with examples of other sampling tools, may be found in Ref. [1].

REFERENCE

1. S.C. Drielak, *Hot Zone Forensics*, Charles C. Thomas, Springfield, IL, 2004.

15 NDA Field Radioactivity Detection

The first steps in the interdiction of a sample of radioactive material take place in the field, far from the controlled environment of a radiochemical forensic facility. The field analyst will have but limited instrumentation to help make a material identification, discover the level of hazard associated with it, and package it for safe and reliable delivery to the laboratory or other disposition site. Instruments for the determination of radiation in the field must be easy to use, rugged, and portable. These instruments have two primary purposes: protection of personnel and initial identification and rough characterization of the material.

This book has purposely avoided explicit discussion of the rules associated with exposure to radiation and the steps needed for protection from that hazard. In part, it is because the rules are complicated and evolve with time and depend on whether the individual is a radiation worker, on the types of radiation involved in the exposure, and on the degree to which the situation is deemed an emergency. Most of the rules involving permissible dose are based on an incomplete knowledge of the levels of ionizing radiation that cause appreciable injury to a person at any time during their lifetime [1].

Field first-response personnel are taught that radiation exposure is a function of time, distance, and shielding, and that less exposure is better than more. However, practical and effective shielding may not be available in the field. Dose is minimized by spending as little time as possible in the vicinity of the source and realizing that the intensity of exposure to penetrating radiation decreases with the square of the distance. Apparatuses as simple as a pair of tongs can reduce dose to the fingers by more than an order of magnitude. A radiation counter capable of unattended function is preferable to one that requires an operator.

The field responder must appreciate the fundamental difference between exposure resulting from radioactive contamination and that from a radiation field. The dose associated with a radiation field derives from those penetrating emissions that accompany the decay of a radioactive sample, and it is mitigated by time, distance, and shielding. Radioactive contamination is a surface phenomenon that is an entirely different hazard. The transfer of exposed radionuclides from the sample to nominally uncontaminated areas (including the interior of the body) can take place through physical contact or, in the case of small particles, through the air. Thus, the radiation field associated with a milligram of Pu is a negligible hazard at any distance, as the principal decay modes of the Pu isotopes do not involve emission of significant photons or high-energy electrons. However, that same sample is a severe contamination hazard. If exposed, the radioactivity can be transferred from surface to surface and can even become airborne. Alpha radioactivity is of very short range.

If deposited on the skin, α-emitting radionuclides do not cause direct injury, but if deposited in the lungs, they can do an enormous amount of damage. Field operations involving removal of layers of packaging material should always be performed wearing full personal protective equipment, including gloves, respirator, and disposable outer garments.

The field analyst should become familiar with the rules intended to govern behavior in the presence of a poorly characterized radiation hazard. He or she should also become familiar with the modes by which radioactivity interacts with the body and with the principles of radiation dosimetry. Although somewhat dated, reference [2] is a thorough description of the fundamentals.

The field analyst should also be equipped with instrumentation useful for both personal protection and the characterization of unknown material. Instruments for the detection of α and β–γ contaminations are recommended, as well as a monitor for the assessment of radiation field and exposure. In situations in which a sample of Pu might be interdicted, the field analyst would be well served by a neutron meter. However, neutron instruments are bulky and cannot be carried everywhere. The analyst should also possess small, radioactive sources to check instrument performance because, although an instrument might appear intact and possess a fresh battery, it is no guarantee that it is functioning correctly. The Th-impregnated mantle from a Coleman lantern provides an excellent check-source for most instruments. These lanterns are exempt from most shipping regulations dealing with the transportation of radioactivity, and, if lost or forgotten, can often be obtained at a hardware store in many locations. A neutron detector would require a small ^{252}Cf or ^{248}Cm source for performance verification and calibration. The analyst should never be deployed to the field with survey instrumentation without the means of checking its function.

The fundamental properties of radiation detectors are given elsewhere in this book, and a good overview of greater depth is presented in reference [3]. There are two common types of detector used for monitoring a β–γ radiation field: the direct-current ion chamber and the portable Geiger tube. In the former, ions arising from the interaction of penetrating radiation with the gas in the chamber (usually air) are collected [4]. The size and shape of the chamber are designed to capture all secondary products of the incident ionization in the interior gas, collecting essentially the entire charge produced in the detector volume. Because radiation exposure is defined in terms of the ionization of air, the instrumentation is particularly appropriate for the measurement of absorbed dose. The most common of these instruments measures the current produced in the air volume. An electric potential remains continuously across what is effectively a gas-filled capacitor, which is continuously recharged by the flow of current in an external circuit that compensates for the rate of discharge induced by the radiation field. The current in the circuit, equal to the rate of ionization in the chamber, is measured with a meter. If ionizing events are frequent in comparison to the detector response time, a constant current is observed. In low-radiation fields, the meter pulses in response to individually detected events.

In Geiger-tube-based survey instruments, a ratemeter counts discharges of the tube within a given time interval. In the Geiger region, all detected pulses, regardless of initial energy, produce the same output energy. Therefore, although a low-energy photon produces little dose, it generates the same signal as a high-energy photon.

Even though these detectors are often calibrated in units of dose rate, the discharge rate is dependent on the incident β or γ energy, particularly at low energy. The dose scale is calibrated only for photons of a given energy. Nevertheless, because the detector count rate scales linearly with the number of incident monochromatic photons striking the detector in a given time interval, the actual dose is proportional to the measured dose. Some efforts have been made to compensate for the extra sensitivity of the tube to low-energy photons by incorporating a thin external layer of Pb or Sn [5]. By their nature, Geiger-tube instruments are more effective at lower count rates than are ion chambers, and the Geiger-tube instrument is usually more compact and facilitates surveys in limited areas.

Both types of dose-measurement instrument can be equipped with a shield that provides rough discrimination between β particles and γ-rays. With the ion chamber, it usually takes the form of an end-cap that fits over the entrance window to the air volume. For the Geiger-tube detector, it is often a cylindrical shield that can be rotated to expose a thin area of the tube wall. With the shields off or open, the detectors measure both β and γ radiation; with the shield in place, β particles are excluded from the active volume of the detector, although it remains sensitive to interactions with γ-rays. However, energetic β particles can still interact through the closed shield via Bremsstrahlung radiation.

Detection of contamination on surfaces is a different problem than measurement of dose. Although some contamination-monitoring instrumentation that works on the principle of scintillation counting is available, most monitoring instruments are based on the ion chamber. In these instruments, the detector is usually large in area but relatively shallow in the direction from which the radiation is incident. In this way, penetrating radiations that result in dose are discriminated against, and less penetrating radiations associated with surface contamination are favored. Of course, the entrance window to the detector volume must be sufficiently thin that the radioactivity can penetrate to the sensitive volume, a requirement particularly important for the detection of α particles. An α-survey instrument includes a discrimination circuit to reject low-ionization events caused by interactions of β particles. The detection of surface contamination is more art than that necessary for the detection of dose. The analyst must move the probe over the contaminated surface without touching it, yet without being so far away that the detector efficiency is affected. Similar to the ion-chamber instrument used for measurement of dose, the contamination-survey instrument tends to pulse if the level of radioactivity is low. The analyst must therefore move slowly enough to allow the instrument to respond. This tactic is particularly important in performing exit surveys of personnel leaving a potentially contaminated interdiction site.

Samples of the heaviest elements emit neutrons, and a neutron measurement in the field can provide a valuable clue as to the presence of these nuclides. Of particular interest are the even-mass isotopes of Pu, which are present in almost any mixture of Pu materials diverted from either reactor or weapons applications. Neutron survey instruments, often called rem-meters, incorporate a neutron detector at the center of a ball of paraffin or polyethylene, which moderates incident neutrons to the thermal energies most efficiently absorbed by nuclear reactions. A ball of polyethylene approximately 30 cm in diameter gives the detector an equivalent response to all incident neutrons of energies from thermal to 14 MeV [6]. The neutron detector itself is usually a sealed-tube BF_3 proportional counter, which is relatively insensitive to γ-rays but provides a means by which low-ionizing neutrons produce ionization in the detector. The reaction $^{10}B(n,\alpha)^{7}Li$ has a thermal-neutron cross section of 3800 b. The energy release of 2.78 MeV is shared among the α particle, the recoiling ^{7}Li nucleus, and internal excitation energy of ^{7}Li. The particles stop in the counting gas through collisions with atomic electrons, thereby producing ionization.

The earlier discussion focused on the use of survey instruments for the protection of personnel in the field by detecting both radioactive contamination and penetrating radiation dose. The same instrumentation can also provide valuable information on the nature of materials in an interdicted package containing radionuclides. The Geiger β–γ detector can be used to determine the relative amounts of penetrating γ radiation, compared to less penetrating β radiation, through the use of a shield. Medical sources containing ^{60}Co or ^{137}Cs can thereby be discriminated from aged, mixed fission products with such detectors. The β–γ contamination detector can be used in conjunction with a set of Al foils of various thicknesses to determine the distribution of ranges of β particles emitted from a mixed radionuclide source, as well as, consequently, a rough discrimination between the quantities of low- and medium-energy β emitters. This technique can be used to infer the presence or absence of pure β emitters, such as $^{90}Sr(^{90}Y)$ ($Q_\beta = 2.28$ MeV), ^{247}Pm ($Q_\beta = 0.225$ MeV), and ^{3}H ($Q_\beta = 0.019$ MeV). Because every commercial instrument has slightly different construction, this practice must be thoroughly calibrated in the laboratory before being deployed in the field.

Similarly, an α survey instrument can be used as an enrichment meter. The short range of the α particles means that only a top, thin layer of any material contributes to the α-particle emission rate of that material. Therefore, as the energies of most α decays are approximately equivalent (4–8 MeV), the detected rate from an "infinitely thick" sample is mostly dependent on the specific activity of the material. Given a sample that can be visually identified as U metal (which has a characteristic color and oxidation behavior), its enrichment can be determined with some accuracy using an α survey instrument. Because of the presence of ^{234}U, a high-enriched uranium sample emits 100 times more α particles per cm^3 in a given time interval than does an equivalent specimen of U of natural isotopic composition.

A sample that emits a significant number of neutrons may consist only of limited species: a Pu/Be-type source, in which powdered Be or another low-Z element is intimately mixed with an α emitter to produce neutrons via the (α,n) reaction; a spontaneously fissioning heavy element, the most likely candidate being ^{252}Cf; or a large quantity of Pu. Generally speaking, a Pu sample can be distinguished from the other two possibilities through its decay heat. The α-particle decay rate of any Pu sample is so much higher than its fission rate that a sample emitting a significant number of neutrons must also be significantly warmer than ambient temperature, which can be determined with a gloved hand. An important precaution for the field analyst is that a sample that contains a near-critical quantity of Pu should not be placed next to a neutron detector, which is encapsulated within an excellent neutron moderator/reflector (i.e., polyethylene). The act of measuring the neutron flux could actually cause a criticality excursion in some instances. Thus, the field analyst should first assess the temperature of a sample before approaching it with a neutron survey instrument. If the sample is warm to the touch, neutrons should be measured at a conservative source-detector distance.

During the last several years, there has been a concerted effort to produce and market nuclear spectroscopy instrumentation that is appropriate for use in the field [3,7,8]. Most of these instruments are based on scintillation or semiconductor technology, are intended to give positive radionuclide identifications based on γ-ray emission, and are designed to be operated by response staff having only limited training and deployed in well-defined scenarios. Although fairly good at identifying common γ-emitting isotopes characterized by simple spectra, these instruments do not identify all possible isotopes of concern and will often misidentify nuclides. Our experience with these instruments is not good. An experienced analyst is necessary to interpret the data in any complicated scenario, and the software is not sufficiently sophisticated to know when trained help is needed. It is often far superior to place faith in clever analysts than in clever gadgets. However, this area of research in field instrumentation is currently receiving much attention and should not be ignored. Some of the instruments now available are quite sophisticated and could be worthy for certain nuclear forensic applications.

There is a final consideration for field surveying and the transportation of seized material: an interdicted sample must be returned from the field, where it exists in an unregulated environment, to the laboratory, which is a regulated environment. This disconnect can cause severe problems in the timeliness of shipping and receipt, as well as the ability of the analyst to begin work on the sample after it arrives in the laboratory. The sample must be transferred from the interdiction point, where it had been in possession of individuals operating outside the rules governing the handling of nuclear materials (and who may not have understood, or merely ignored, the safety concerns associated with those materials), to the analytical laboratory, where all rules and regulations apply. In this case, the advantage is to the criminals. The first step in the transportation process is field characterization. Although the regulations that apply to transportation of radioactive or hazardous materials evolve over time, a good overview, although now somewhat dated, is reference [9]. After an appropriate shipping container is identified on the basis of the field survey, shielding must be added as necessary to reduce the radiation dose rate from the surface of the container to within regulatory limits. The additional time and maneuvering necessary for such requirements invariably delay the start of laboratory forensic analyses and could result in the loss of important data in the ensuing investigation.

REFERENCES

1. H.F. Henry, *Fundamentals of Radiation Protection*, Wiley-Interscience, New York, 1969.
2. F.W. Spiers, Radiation dosimetry, Course 30, in *Proceedings of the International School of Physics 'Enrico Fermi'*, G.W. Reed, ed., Academic Press, New York, 1964.
3. G.F. Knoll, *Radiation Detection and Measurement*, 3rd ed., Wiley, New York, 2000.
4. J.S. Handloser, *Health Physics Instrumentation*, Pergamon Press, New York, 1959.

5. D. Barclay, Improved response of Geiger-Muller detectors, *IEEE Trans. Nucl. Sci.*, NS-33, 613, 1986.
6. G.G. Eichholz and J.W. Poston, *Principles of Nuclear Radiation Detection*, Ann Arbor Science, Ann Arbor, MI, 1979.
7. EG&G Ortec Corp., Safeguards/NDA measurement, available at: http://www.ortec-online.com/Solutions/nuclear-safeguards-non-destructive-assay.aspx, accessed June 29, 2014.
8. Leidos homepage, available at: https://www.leidos.com/products/security, accessed June 29, 2014.
9. R.C. Ricks, W.L. Beck, and J.D. Berger, Radiation Emergency Assistance Center/Training Site, Oak Ridge Associated Universities, Radioactive materials transportation information and incident guidance, report DOT/RSPA/MTB-81/4, Department of Transportation, Washington, DC, 1981.

16 Laboratory Analyses

This chapter addresses the laboratory analytical techniques most commonly applied in nuclear forensic investigations and describes the strengths and limitations of the various techniques, as well as provides illustrative examples of the types of samples to which each technique is best suited. More specific information on the use of these techniques can be found in the chapters containing examples of case studies (Chapters 20 through 27).

16.1 RADIATION COUNTING SYSTEMS

Radiation counters are discussed in general terms elsewhere in this book. This section focuses on instruments that a nuclear forensic analyst deploys in a well-equipped laboratory, as well as on issues pertaining to the successful operation of a counting lab.

16.1.1 Counting Lab

The counting lab should be proximate to the radiochemical forensic lab, yet isolated from it to avoid contamination and changes in background induced by fluctuations in the sample load in the analytical facility. The best way to accomplish this isolation is vertically: The optimum location for a counting lab is underground, under a thick layer of dirt or magnetite to attenuate the cosmic-ray background. Lab coats, booties, gloves, and other personal protective equipment should never exit the analytical lab and enter the counting lab. If personal protective equipment is needed in the counting lab, it should be donned at the entrance to the facility. Hands and feet should be monitored for contamination before entering the counting lab. Sticky, step-off, entry mats can be of some value if they are changed frequently.

Shielding and some detectors can be quite heavy, making elevator access to the counting lab a necessity. A good, stable air conditioning system is also required, both for the removal of cryogenic and counting-gas effluents and for the control of temperature. The electronic gains of gas-filled counters for α spectrometry are affected by small shifts in temperature, and many older electronic modules also suffer temperature instabilities. The counting facility must have a reliable supply of filtered and stable electric power, as well as a separate circuit for pumps and other noncounting apparatus. If Ge detectors are power-cycled down and up quickly, they can suffer damage. A counting laboratory should have sufficient battery backup power to smooth short power transients, as well as an auxiliary electric generator for longer outages. The power distribution to the detectors should also contain a cut-off circuit such that, in the event of a power fluctuation, power will only be resupplied to the electronics in an approved fashion. Some electronic noise degrades the performance of radiation detectors used for spectroscopic measurements. Such detectors should be supplied with filtered power, and transient use of power tools should be confined to a different power circuit.

The counting laboratory also requires lines for vacuum and for the delivery of counting gases. The pumps for producing vacuum should be located at some distance from the lab to maintain both audible and electronic noises at a minimum. Vacuum lines should be supplied with cold traps to help prevent pump oil from back-streaming and depositing on exposed detector parts. The cold traps can be operated with either a regular supply of liquid nitrogen or via cryomechanical cooling. If a mechanical cooler is used, it should plug into a separate power circuit from that used for the detectors to minimize microphonic noise from the compressors. Counting gases (principally, Ar, P10, and methane) are supplied at the pressures required by the counters (slightly above atmospheric pressure for flow-through proportional counters, and at ~1.5 atmospheres for gridded ion chambers), with flows that are controlled by needle valves on the counters themselves. Counting gases should be delivered through metal tubes with in-line ceramic electrical insulators. In many labs, gases are delivered to the facility in metal tubes but are then delivered from a wall supply to the counters via plastic tubes. Although this tactic provides electrical isolation of the counters and minimizes ground loops, the plastic tubes degrade with time, and at some point, the counters may suffer a catastrophic loss of counting gas. Gases that are supplied in pressurized cylinders, particularly those containing Ar, suffer from a particular background problem. When delivered to a facility, they often contain a measurable quantity of 3.8-day ^{222}Rn, which introduces an interference in α spectra at the energy of ^{238}Pu and ^{241}Am decays. This problem can be alleviated by maintaining a supply of these gases in-house at all times and not using any particular cylinder until it has been stored for a minimum of 2 weeks. In general, counting gases are supplied from a gas manifold connected to several bottles of the gas. Regulators are set such that when one bottle (or bank of bottles) is emptied, the next bottle comes online. When an empty bottle is replaced by a new one, a small quantity of air is introduced into the manifold between the bottle and regulator. This volume should be evacuated or purged before opening it to the system that delivers the counting gas to the counters.

16.1.2 Counter Shielding and Systems

In general, α counters do not require shielding in excess of that provided by the underground environment. The lower limitations on α counts imposed by background events are controlled more by contamination than by background radiation. It is necessary to shield γ- and x-ray counters, however. Photon shielding materials must exclude cosmic rays and environmental radiation from the detector volume (i.e., photons from ^{40}K and the natural U and Th decay chains). A layer of 20-cm thick Pb is sufficient for this purpose, particularly if the counting lab is underground. However, Bremsstrahlung in the Pb has a long range, as do neutrons from cosmic-ray interactions with the intervening materials. It is usual to line the inside of the Pb shield with layers of Cd, Cu, and Al. A large volume of air around the source (~30 cm in any direction) is effective in screening photons and electrons that have been down-scattered into the 150–200 keV range [1].

Another purpose of a layered shield is to reduce the background arising from fluorescence of the shielding materials induced by radiation emitted from the counting sample itself. Lead x-rays are often visible in photon spectra taken from samples counted in unlined Pb shielding caves. The attenuation of these x-rays via use of the layered shield is effective.

In selecting detector shielding, care must be taken to obtain low-background materials. Modern steels can contain 5.3-year ^{60}Co, incorporated either from fallout from atmospheric testing in the 1950s and 1960s or by improper disposal of ^{60}Co γ irradiators, which have been known to become incorporated in scrap metal [2,3]. Natural sources of Pb contain 22-year ^{210}Pb, which is a member of the ^{238}U decay series. Any Pb ore will contain U at some level, and although chemical recovery of the Pb removes the U, it cannot remove ^{210}Pb. The decay of ^{210}Pb produces 5-day ^{210}Bi (which emits 1.1-MeV β particles and, consequently, Bremsstrahlung) and 138-day ^{210}Po (which emits α particles); these nuclides grow into purified Pb and achieve equilibrium. Different batches of Pb suffer from ^{210}Pb contamination to varying extents [4]. In general, older materials are preferred over new material in fabricating shielding for low-background counting applications, as ^{210}Pb, ^{60}Co, and such are reduced in these materials by decay. Steel from World War I battleships and Pb ballast from sunken eighteenth-century Spanish vessels are in demand for shielding applications.

Samples should be received at the counting laboratory in appropriately shielded containers, both to prevent transfer of loose contamination and to avoid exposing the counters (which may be busy with other samples) to penetrating radiation. Gamma-spectroscopy samples should be stored in a shielded container near the laboratory door. Samples for α counting, which intrinsically involve fixed, but unencapsulated, radionuclides, are usually contained in plastic or paper folders until delivered to the vicinity of the counters.

Quite often, the counting samples generated in a nuclear forensic analysis outnumber the available radiation counters. All counters should optimally be equipped with automated sample changers, such that when a given count is finished, the sample is removed from the counter and replaced with another sample, which is then counted. In this way, the analyst is relieved of double duty in the middle of a prolonged radiochemical campaign, reducing the probability that loose contamination will be introduced into the counting facility. Sample changers for α and β counters consist of a rotary turret. The detector itself is surrounded by a collimating shield so that radiations emitted by nearby samples do not interfere with the current count. Sample changers for γ counters usually consist of a turret or a belt coupled to an elevator. Following the completion of a γ count, the elevator lowers the sample to the turret, which rotates the next sample into a position to be raised into the proximity of the detector. In this way, the detector is shielded from the samples in the turret, both by intervening shielding material and by distance. One issue that all radiation counters share is the fact that the efficiency of the counter is usually strongly dependent on the position of the source relative to the detector during the count. With counters for which efficiency is an issue, the turret or turret/elevator must deliver the sample to a well-defined location before the count begins. In our laboratory, both α and γ counters are available in which the source-to-detector distance is adjustable to several predetermined values that span distances from <1 cm to ~1 m. Close-in counting should be avoided for all but the least intensely radioactive samples to avoid pileup and summing effects.

Beta counting and some types of α counting result in a scalar number of events devoid of spectral information. In this case, the counting data should be associated with the start time of the count, the live time of the count, and the detector number, position, and counting geometry. These data should be stored electronically for future retrieval. Alpha and γ pulse-height analysis (PHA) results in spectral data: α spectra can be stored as 512- or 1024-channel spectra, whereas γ spectra are usually stored in 4096 or 8192 channels. Similar to scalar data, each spectrum should be associated with a start time, a live time, and the particulars of the count before electronic storage. Electronic data can be stored locally or transported across a local network to a central location. However, there should be a local display of the data while spectra are being accumulated. Another way of taking spectral data, called list mode, involves recording individual, time-tagged energy events. In the forensic analysis of list-mode data, the first action on receiving the data is to sum them into spectra, so there is little added value in that data-collection mode (except for specific cases of sources of mixed radionuclides with disparate half-lives and decay intensities).

16.1.3 Particle and Photon Detection

Upon considering an actual configuration for the counting laboratory, with the number of radiation counters of each type needed for radiochemical analyses of a suite of nuclear forensic samples, input parameters include the number of analytes the radiochemists will separate, the likely sample load, and the time pressures that the analysts are likely to encounter from the political arena. As a model, the following is based on our personal experience.

16.1.3.1 Beta-Particle Counters

Beta counting is rarely employed because most of the short-lived β-emitting analytes also emit γ-rays, which are characteristic of the decaying nuclides. However, long-lived fission products such as ^{93}Zr and ^{99}Tc, and other potential analytes like ^{14}C and ^{3}H, are best determined via their β activities. A set of attenuating Al foils of various thicknesses can assist in the identification of nuclides in samples counted with a gas-filled proportional counter. A single, gas-filled β counter with an eight-position turret is recommended. The need for liquid scintillation counting (LSC) is too rare to support such an instrument, but extramural environmental organizations for scintillation-count analyses of chemical fractions containing ^{3}H or ^{14}C can be contracted.

16.1.3.2 Alpha-Particle Counters

Alpha counting is used extensively in the nuclear forensic analysis of samples of U, Pu, and other heavy elements. Alpha 2π-counting, with decay-energy-insensitive, windowless, gas-proportional detectors, generates a scalar value that is proportional to the absolute α-decay rate of the sample. Alpha PHA counting, in which spectral data are taken with Si surface-barrier detectors or gridded ionization (Frisch-grid) chambers, is used to divide the total decay rate obtained from 2π-counting into its isotopic components. Frisch-grid detectors, by virtue of their high efficiencies, are very useful for counting low-activity sources. However, conversion-electron summing makes them of dubious utility for counting Am or Np isotopes. Silicon detectors suffer less from summing effects, and high-activity sources can be counted if the turret is designed so that it can be backed away from the detector. Both types of PHA counter should be configured so that the preamplifier output signal is split, and higher-energy events arising from spontaneous-fission decays can be stored as a scalar quantity, either in a separate data storage location or incorporated into the α spectrum. One tactic is to reserve channel 1023 in each α spectrum for spontaneous-fission events. Counting labs should have at least two 2π counters, with four- or six-position turrets, four Si surface-barrier detectors with six-position turrets, and four gridded ionization counters with six-position turrets.

16.1.3.3 Gamma-Ray Counters

A wide variety of intrinsic (or high-purity) Ge (HPGe) counters are available from commercial vendors. All nuclear forensic photon-counting needs are met via two different types: an Al-windowed, coaxial detector with an efficiency relative to NaI(Tl) of 25%–40%, and the low-energy photon spectrometer (LEPS) planar detector with a Be end-window. They are available in the downward-looking geometry so that an elevator/turret sample changer could be implemented. Although high-efficiency, well-type detectors can be valuable in the radiochemistry laboratory to check the performance of chemistry procedures, the difficulty in deconvoluting summing effects makes them of little use for quantitative γ-ray spectrometry. Many detectors are available with Compton suppression. A high-efficiency, low-resolution photon detector is placed annularly around an HPGe detector. Any events that take place in the Compton annulus at the same time as an event in the HPGe detector are used as a criterion for rejecting the Ge event, excluding it from the spectrum. In this way, events in which only a fraction of the energy of the photon is deposited in the Ge crystal are discriminated against, and the intensity of the continuum background under low-energy photopeaks is reduced. Compton suppression is of some value for increasing the precision with which low-energy photons can be measured in the presence of more intense high-energy photons. A well-equipped nuclear forensic counting laboratory will contain at least three coaxial detectors, one with Compton suppression, and at least one LEPS-type detector. All four detectors should be equipped with multiposition sample turret/elevators.

16.1.3.4 Neutron Counters

Most of the utility of neutron counting is within a sample receival facility, and little application is found in our analytic counting laboratory. Any unknown sample should be interrogated for neutron emission (signaling the possible presence of Pu) before the shipping container is opened. Analytical fractions containing a significant quantity of spontaneously fissioning nuclides should not be introduced into the counting facility, as neutrons are difficult to shield against and damage the function of HPGe detectors.

A counting laboratory containing the equipment described earlier, used in conjunction with a mass spectrometry (MS) laboratory, should be sufficient to generate all needed radioforensic data from a suite of samples of limited size in a reasonable time. Required radiation-counting capacity for a more extensive analysis is larger (e.g., the gaseous-diffusion facility counterforensic investigation described in Chapter 21).

16.1.4 Chemistry Lab Application

Some radiation counting support is required directly in the radioanalytical laboratory to monitor the progress of separations as they take place. For example, the operation of the rare-earth/trivalent actinide separation on a saturated-HCl cation-exchange column involves the generation of sample tubes, each of which contains a few drops of eluent. The actinides are eluted from the column before the lanthanides, many of which emit intense γ radiation in their decays. The isolation of those tubes containing actinides from those containing rare earths is made with the aid of a radiation counter in the laboratory.

The utilities necessary for radiation counting in a chemistry laboratory are similar to those required in the counting laboratory, but requirements of continuity and background are less stringent. Instruments in the counting laboratory that become obsolete or contaminated are often transitioned into the radiochemistry laboratory. However, because of the proximity to radiochemical work, some of which involves large amounts of primary sample, shielding requirements in the laboratory are at least as restrictive as those in the counting facility. The radiochemistry laboratory environment also includes reagent fumes that are absent in the counting laboratory, so that delicate electronic modules should be wrapped in plastic for protection. Similarly, any portions of the counters that come into contact with the samples themselves should be plastic coated or disposable. In this way, potential contamination from the outside of the labware can be isolated from the detector, thereby reducing the buildup of background contaminants. Similarly, contact between protective rubber gloves and system electronics should be avoided to minimize opportunities for spreading contamination.

In addition to survey instrumentation, the performance of radioforensic chemical separations can be successfully monitored with only three counters: a flow-through, P-10 gas proportional counter for α particles; a flow-through, methane-gas proportional counter for β particles; and a NaI(Tl) well-counter for γ-rays. The outputs of the first two counters go to scalers; the third counter yields spectral information that requires either a multichannel analyzer that permits peaks to be integrated from the screen or a set of single-channel analyzer/scaler units. The principal laboratory delay encountered in many forensic analyses results from a backlog of counting fractions at the α counter. When possible, more than one of these units would be a productive luxury.

16.2 TRITIUM ANALYSIS

The hydrogen isotope tritium (^3H or T) is a nuclide used in the function of nuclear weapons, a component of the fuel that provides a small thermonuclear yield to cause the fission fuel to be consumed with greater efficiency (boosting). Tritium has also found wide use in industry and medicine, both as the agent that drives the luminescence of some wristwatches and building exit signs and as a label in thousands of radiopharmaceuticals [5,6]. As a result, T is the weapons material most likely to be diverted into proliferant channels and, along with ^{241}Am in smoke detectors, is one of the most widely and openly distributed radionuclides.

The use of heavy water (D_2O) as both a moderator and medium of heat exchange in CANDU-type reactors results in the continuous exposure of large quantities of deuterium (^2H or D) to high neutron fluxes [7]:

$$^{2}H\ (n,\gamma)\ ^{3}H \quad \sigma = 0.52\ \text{mb}.$$

Even though the capture cross section is small, approximately 2400 Ci of T are produced per electric megawatt in a CANDU reactor. Although this is the source of a serious contamination nuisance, the dilution of the radioactivity in the other isotopes of hydrogen makes it of little use in the weapons world without expensive isotope enrichment. Even so, T-bearing material from this source could appear on the black market. Practically speaking, T can also be produced by neutron irradiation of the minor ^3He component ($1.4 \times 10^{-4}\%$) of natural He [8]:

$$^{3}He\ (n,p)\ ^{3}H \quad \sigma = 5300\ \text{b}$$

or, more likely, through the irradiation of enriched ^6Li targets with neutrons:

$$^{6}Li\ (n,\alpha)\ ^{3}H \quad \sigma = 940\ \text{b}.$$

Fortunately, the radiotoxicity of T is very low as a consequence of the short range of the β particles emitted in its decay, about 0.5 μm in tissue [5].

$$^{3}H \rightarrow {}^{3}He + \beta^{-} \quad t_{1/2} = 12.3\ \text{years} \quad Q = 18.6\ \text{keV}.$$

Unfortunately, this same property makes T difficult to detect and to quantify in the radioanalytical laboratory. No characteristic γ-rays are emitted in T decay, and as most radionuclides decay with the generation of more ionization than does T, the presence of other radionuclides can act to mask a low-level tritium presence. As a result, all analytic techniques used for the determination of bulk T involve a chemical step: either purification of the T (along with the other isotopes of hydrogen) and its preparation in a standard chemical form (usually water) for direct radiation counting [9], or the isolation of its decay daughter, ^3He, after a suitable ingrowth period, for a mass-spectrometric measurement [10]. These techniques and others [11] are also applicable for environmental monitoring, but they are not a focus of this work. As for most investigations, the collection of a representative analytical sample from a questioned specimen is the first issue for an analyst. The most common forms of T encountered in nuclear forensic applications are molecular (gaseous), aqueous (incorporated in either the water molecule or, in the case of radiopharmaceuticals, the solute), or organic (usually incorporated in a liquid, but possibly as either a volatile or semivolatile compound).

A gaseous sample may be counted by mixing a known aliquot into a counting gas and using a gas proportional counter or current-ionization chamber. An interdicted gas cylinder containing a sample suspected of having a T component must be sampled through manipulation of the gases. A gas manifold with valves, bottles, and a vacuum pump can be used to collect a sample containing a known fraction of the original, if volumes and pressures are known with high accuracy. A liquid-phase T sample can be aliquoted much like any liquid, provided the analyte is in true solution. Whether liquid or gas, however, care must be taken during sampling operations to exclude atmospheric moisture from the sample and ensure that the sample does not evaporate or leak from its container.

An elemental sample is converted to water by mixing it with an excess of oxygen or air and passing the mixture over a Pd catalyst on a molecular sieve (a combustion trap) [12]. Hydrogen gas reacts at ambient temperature (or slightly above), forming water that is adsorbed in situ on the sieve. A known quantity of T-free hydrogen can be added to a second aliquot of the gaseous sample and treated in the same way, in preparation for an isotope dilution analysis via MS. Simple gas-phase organic molecules can be burned in this same apparatus at more elevated temperatures [13,14]. The trap can be made to release water by evacuating it at high temperature (~500°C), with the water vapor then collected in a cold trap.

Organically bound T can be released in the form of water through a variety of oxidation methods, either wet or dry. An example of wet oxidation entails the destruction of the organic material with H_2O_2 in the presence of a ferrous-ion catalyst [15]; other methods involve digestion of mixtures containing $HClO_4$, but these must be used with caution. The evolved water fraction must be corrected for addition of hydrogen atoms from the reagents. Dry oxidation (combustion) is accomplished by heating the sample in the presence of oxygen, with the combustion products passing over heated copper oxide and the resulting water collected in a cold trap.

The usual technique for measuring T in an analytical sample is LSC. Appropriate solid or liquid samples are introduced into a scintillation cocktail, either directly or after conversion to water. Because unambiguous detection and measurement of T is complicated by the presence of other radionuclides in the sample, conversion to water and subsequent distillation to purify the sample is the preferred forensic method unless the analyst is certain that no other radionuclides are present. Instruments for LSC of T samples are available from a wide range of commercial vendors. The art of LSC involves the synthesis of a scintillation solution with an appropriate emulsifier to enable handling the T-bearing aqueous specimen without an unpredictable degradation of performance. Reference [16] provides a good discussion of this topic. Standards for counter calibration are available from the National Institute of Standards and Technology, and treatment of the necessary compensation resulting from the efficiency change inherent in the introduction of different amounts of standard solution into the scintillator can be found in reference [17].

A more modern technique used for T analysis is the detection and quantitation of the T decay daughter, ^3He [10]. An aqueous sample is initially degassed by cycling it between 90°C and liquid-nitrogen temperature, evacuating and discarding the headspace gas each time. The last such cycle defines the beginning of the ingrowth period for the daughter, the necessary time depending on the quantity of T in the specimen and the sensitivity of the mass spectrometer. Samples containing as few as 10^5 atoms of ^3He can be determined to a precision of a few percentage points. After the ingrowth period, the sample is heated and cooled, and the headspace gas is collected and introduced into a mass spectrometer. The ingrowth time and the number of daughter atoms produced in that interval delineate the number of T atoms in the analytical sample. The specificity of the technique is such that very little sample preparation is required for aqueous samples. The technique is also nondestructive. For situations in which there are few samples to be analyzed and the number of atoms of T in each sample is fairly large, this technique is superior to LSC. However, in situations in which large sample throughput is required, LSC is the method of choice. LSC, ^3He MS, and accelerator MS of hydrogen isotopes all offer their own advantages and disadvantages for the analysis of very dilute samples [10,11], but such discussion is beyond the scope of this book.

16.3 IMAGING AND MICROSCOPY

The role of microscopy is to provide a magnified image of a sample, allowing the observation of features beyond the resolution of the unaided human eye (roughly 50–100 μm). The ability to identify and characterize diverse suites of samples rapidly, accurately, and without compromising the integrity of the sample is an essential starting point of most forensic investigations. A variety of microscopy techniques are applied in nuclear forensic science, using photons, electrons, and x-rays to probe the physical, chemical, and structural makeup of samples at resolutions ranging from 1 Å to tens of micrometers. Highlighted here are the most commonly used microscopies.

16.3.1 Optical Microscopy

The practice of light microscopy dates back more than 300 years, and optical microscopy remains one of the most basic and fundamental characterization techniques in nuclear forensic analysis. The optical microscope is often the first instrument used to examine a questioned sample in fine detail, and it allows the forensic scientist to answer the simple, yet vital, question, "What does the sample look like?" before proceeding with more extensive (and often destructive) analyses. Ranging from simple hand lenses to sophisticated compound microscopes, optical microscopy reveals details of color, surface morphology and texture, shape and size, tool marks, wear patterns, surficial coatings, corrosion, and mineralogy [18,19].

Two basic types of compound microscopes find widespread use in forensic science: the stereomicroscope and the polarized-light microscope. Both use a dual array of lenses (objective and eyepiece) to form a magnified image of the sample. The stereomicroscope uses separate optical paths for each eye to produce a three-dimensional image at relatively low magnification (~2–80×). Stereomicroscopes are very useful for examining samples with complex shapes, as well as for dissecting or aliquoting samples for additional analyses. The polarizing microscope passes light through a set of polarized filters to gain additional information about the nature of the sample from optical properties, such as crystallinity, anisotropy, pleochroism, and birefringence. Polarizing microscopes use either transmitted or reflected polarized light, and can be used to study both transparent and opaque materials at magnifications of up to 1000×. The limit of resolution is set by the wavelength of the light used; the maximum theoretical resolution of conventional microscopes is 200 nm, but values closer to 1 μm are more commonly achieved. Both stereo and polarizing microscopes can be equipped with digital cameras for photomicroscopy. An excellent description of the uses of optical microscopy in forensic science, with many illuminating case studies, is given in reference [20].

16.3.2 Scanning Electron Microscopy

The attractive feature of optical microscopy is that it is relatively simple to use—samples can be analyzed in air with little or no preparation, and the images are usually in natural color and at useful magnifications of 2–1000×. Despite these advantages, the most commonly applied imaging tool in nuclear forensic analysis is not optical microscopy, but scanning electron microscopy (SEM). SEM offers superior resolution, much greater depth of focus, and when combined with x-ray analysis (described later), direct imaging of elemental composition. In the SEM, a finely focused electron beam with an energy of ~1–25 keV is rastered over a sample. The interaction of the incident electron beam with the sample produces backscattered electrons, secondary electrons, Auger electrons, x-rays, and photons. By measuring the intensity of one or more of these types of particles as a function of raster position, an image of the sample is constructed and displayed. Each type of emitted particle conveys different information about the sample, and, by choosing the appropriate detection mode, either topographic or compositional contrast can be revealed in the image [21–23]. Secondary electrons, for example, arise from inelastic collisions between incident electrons and atomic electrons within the outer few nanometers of the surface and carry information about sample topography. Backscattered electrons have energies more nearly comparable to that of the incident electron beam and carry information about the mean atomic number of the area being imaged. Backscattered electrons can be used to map spatially resolved phases of disparate chemical composition [24]. However, more quantitative information about chemical composition is carried by the characteristic x-rays [22,23].

The quality of SEM images is highly dependent on sample preparation. Specimens must be compatible with the high vacuum of the SEM, and insulating materials must be coated with a thin, 5–10 nm conductive layer (usually Au, Pd, or C). With thermionic, W filament sources, image resolution is limited to about 10 nm, with a corresponding maximum magnification of 100,000. With field-emission electron sources, the resolution can be better than 1 nm, with a corresponding maximum magnification of 1,000,000. Sample size is limited by the physical dimensions of the SEM specimen chamber and is typically on the order of 5–10 cm, although SEMs capable of handling samples up to 20 cm in diameter are available.

The environmental SEM (ESEM) is a recent addition to the forensic scientist's toolbox. The ESEM retains most of the features of a conventional SEM, but removes the requirement for a high-vacuum environment. Wet and nonconductive samples may be examined in their as-received states without modification or preparation. The ESEM uses a gas-ionization detector to produce high-resolution, secondary-electron images in a gaseous environment (at pressures as high as ~50 Torr and temperatures >1000°C).

For conventional SEM analysis, samples must be nonvolatile in order to survive the high-vacuum environment and conductive to prevent charging by the incident electron beam. Nonconductive samples are typically coated with a thin (2–10 nm) layer of C, Au, or Ir. These requirements are eased with an ESEM, which can operate under various sample-chamber atmospheres. Powdered samples are typically adhered to a specimen stub using conductive tape or adhesive, or set into epoxy resin and polished. Analysts must take particular care to ensure that the sample is adequately dispersed. Common methods of dispersion include sonication using various solvents and repeated contact transfers. Dispersion will depend heavily on the characteristics of the sample and often requires scoping efforts to ensure success.

16.3.3 Transmission Electron Microscopy

Transmission electron microscopy (TEM) lies at the opposite end of the spectrum from optical microscopy: it is difficult to use and requires elaborate (and often expensive) sample preparation. Nevertheless, its unique capabilities for ultrahigh, 0.1-nm spatial resolution (2000× that of the best optical microscope) and for revealing microstructure information via electron diffraction make TEM an important tool in many nuclear forensic investigations. In TEM, a high-energy (100–1000 keV), focused electron beam is transmitted through a very thin sample foil (typically tens to hundred nanometers) [25–28] in a high-vacuum column (10^{-4} to 10^{-7} Pa). Elastic and inelastic scattering from interactions between the electron beam and the specimen give rise to a variety of signals that provide information on the microstructural and material properties of the specimen, including morphology, crystal structure and crystallographic orientation, interfaces and crystal defects, chemical composition, valence states and chemical bonding, electronic band structure, and magnetic structure. TEM is unique among materials characterization techniques used in nuclear forensics in that it enables near-simultaneous examination of microstructural, chemical, and crystallographic features, from very small regions (down to nm) and over a wide range of magnification (100–1,000,000×).

TEM offers two modes of operation: imaging and diffraction. The imaging mode produces a magnified image of the region of the sample illuminated by the incident electrons. The diffraction mode provides an electron diffraction pattern, analogous to an x-ray diffraction (XRD) pattern, from the illuminated region of the sample. Electron diffraction patterns may be indexed by the same procedures used in XRD, and they can be used to identify phases on an extremely fine spatial scale. Just as in SEM, characteristic x-rays are generated by the interaction of the incident electron beam with the sample. X-ray analysis can be combined with TEM imaging and diffraction to provide comprehensive information on the internal microstructure of a questioned specimen with nm spatial resolution [27].

Sample preparation is one of the most difficult and important aspects of TEM analysis. High-quality samples have a thickness comparable to the mean-free-path of the electrons that travel through the sample, typically ~tens to a few hundreds of nm. Depending on the material and the desired information, different techniques are used for the preparation of thin sections, including mechanical milling, ultramicrotome sectioning, chemical etching, ion etching, and focused ion beam (FIB).

FIB makes use of energetic Ga ions to prepare thin samples for TEM and can produce very thin foils from specific areas of interest. However, FIB may also alter the composition or structure of the material through Ga implantation. Additional information on TEM sample preparation is available in Refs. [29–31].

16.3.4 Electron Microprobe Analysis

Electron microprobe microanalysis (EMPA) provides spatially resolved, quantitative chemical analysis of small, selected areas of samples based on the generation of characteristic x-rays excited by the focused beam of energetic electrons. EMPA has many similarities to SEM, including the basic design of the electron column and the detectors for secondary and backscattered electrons, but with a shift in emphasis from imaging to accurate determination of major- and minor-element concentrations [23,32]. In EMPA, characteristic x-rays are generated when the beam of energetic electrons (5–25 keV) strikes a sample. Compared with SEM, EMPA uses much higher beam currents (tens to hundreds of nA) and achieves correspondingly poorer spatial resolution. For thick samples, the spatial resolution of chemical analyses is limited to ~1 μm by the spreading of the electron beam after entering the sample. For labile materials, the electron beam is often defocused to a diameter of 10–50 μm to reduce damage caused by sample heating and loss of volatile species (e.g., Na or bound water).

EMPA uses both the energy-dispersive x-ray spectrometer (EDS) and the wavelength-dispersive spectrometer (WDS) to measure the intensity of the characteristic x-rays. EMPA can detect all elements from Be to Pu, with accuracies approaching 1% and detection limits typically in the range of 50–200 μg/g. Modern EMPA systems incorporate computer-controlled x–y stages, which allow the analyst to map the distributions of specific elements of interest onto a two-dimensional array with μm spatial resolution.

The accuracy of EMPA is strongly affected by sample preparation. Accurate quantitation requires flat, well-polished samples, devoid of topographic features, and polishing procedures developed for optical microscopy are adequate for most materials [33]. Special techniques are required for small grains, which are often encountered in nuclear forensic investigations. Grains larger than ~10 μm can be mixed with epoxy and set in a mold (care is required to not reduce the density of grains below a usable limit by adding excess epoxy). Once cured, the epoxy can be removed from the mold and polished to expose the grains. Grains smaller than ~5 μm are very difficult to polish and are best analyzed by dispersion on a flat, polished, electrically conducting substrate that contains none of the elements of interest [34]; pyrolytic graphite and metallic Si are commonly used for this purpose. The grains can then be analyzed by EMPA using relatively low beam current. As with SEM, insulating materials must be coated with a conductive coating, and C is normally used in order to minimize absorption of the lower-energy characteristic x-rays.

The major application of EMPA in nuclear forensic science is quantitative analysis of the elemental composition of solid materials. Concentrations can be measured with an accuracy of 1%, as long as suitable standards are available and the sample is well prepared. Backscattered-electron imaging in combination with EDS allows rapid characterization of the various phases in a specimen and can reveal elemental zoning and other phase relations that are otherwise undetectable by other techniques.

16.3.5 X-Ray Microanalysis

The characteristic x-rays generated by interactions between energetic (keV) electrons and the sample in SEM, TEM, and EMPA carry information on chemical composition and provide a convenient and effective way of determining elemental concentrations in almost any solid sample, including small particles. The characteristic x-rays are emitted when electrons collide with atoms in the sample and cause inner-shell electrons to be ejected. An inner-shell vacancy leaves the atom in an excited state, and the atom loses this excess energy, either by the movement of an outer-shell electron to fill the inner-shell vacancy (accompanied by the emission of an x-ray photon) or by the ejection of an outer-shell (Auger) electron. An x-ray produced through interaction with an inner-shell electron has energy equal to the difference in energy between the two shells involved in the transition. For a given transition between specific inner and outer shells, the energy of the x-ray increases uniformly with increasing atomic number. As a result, each element produces x-rays with unique and characteristic wavelength (energy) [23,32,35].

The characteristic x-rays can be analyzed by either of two methods. An EDS uses the photoelectric absorption of x-rays in a semiconductor detector, usually Si(Li), to simultaneously measure the energy and intensity of incident x-rays. EDS systems provide an easy-to-use method of measuring x-ray spectra over a broad energy range (0.2–20 keV) and can detect elements of the periodic table from B to U. Detection limits are typically 0.1% for routine EDS analysis of silicate and oxide materials. A WDS operates on the principle of Bragg diffraction of the x-rays incident on an analyzing crystal. Only x-rays satisfying the relation ($n\lambda = 2d \sin \theta$) are constructively reflected into a detector, and the x-rays are dispersed according to wavelength rather than energy. WDS provides much higher energy resolution (~1 eV compared to ~100 eV) and a much higher signal-to-noise ratio (~10×) compared to EDS. WDS will detect elements from Be to Pu, with typical detection limits of 0.01% [23,35]. However, two noteworthy disadvantages of WDS are lower x-ray intensity, requiring much higher electron-beam current and significantly greater mechanical and electronic complexity.

X-ray microanalysis is particularly valuable in nuclear forensic investigations for the relative ease with which x-ray intensities can be accurately quantitated to yield elemental concentrations in questioned samples. The cornerstone of x-ray microanalysis is the proportional relationship between x-ray intensity and concentration. In practice, quantitative analysis requires careful comparison to matrix-matched standards and the use of algorithms such as CITZAF [32,34] to correct for the effects of backscattered electron emission, fluorescence, and absorption. In favorable cases (e.g., ferrous alloys, aluminosilicates, and simple oxides), the uncertainties in concentration can be <0.5% absolute.

16.3.6 Optical Spectroscopy

Optical spectroscopy is used to obtain molecular information on bulk samples by measuring incident light transmitted or reflected by the sample. The light is separated into discrete wavelengths and measured as absorptions, and different molecules have unique absorption bands based on their chemical composition [36]. A number of different optical spectroscopy techniques can be used, including visible (Vis), near-infrared (NIR), infrared (IR), Raman, and fluorescence. We focus on IR and Vis/NIR in reflectance mode as the optical wavelength bands with the greatest value for nuclear forensic analysis [37].

The Vis range spans 400–700 nm, NIR covers 700–2,500 nm, and IR from 2,500 to 20,000 nm. Molecular bonds vibrate at characteristic frequencies in the IR region. If a particular molecular vibration results in a change in dipole moment, the molecule can absorb IR radiation of that specific frequency, exciting that vibration. Absorption at specific frequencies is characteristic of certain bonds, and the IR spectrum thus identifies the bonds and functional groups within a molecule. In general, IR measures molecular motions associated with bending, twisting, and stretching of molecular bonds, and it is typified by discrete bands that are characteristic of different chemical entities. NIR measures overtones and combination bands of the motions detected in the IR; these absorption bands tend to be weaker and broader, yet are still a source of important chemical information. More detail on Vis/NIR reflectance spectroscopy is given in Section 16.6.2.

The current embodiment of IR analysis is Fourier-transform infrared (FTIR) spectroscopy, and it is used for structure determination and identification of both organic and inorganic compounds. It is particularly useful for nondestructive screening of suspected explosive and other extraordinary compounds, and for initial investigation of various route evidentiary specimens in general.

An FTIR spectrometer consists of a broadband IR source (2.5–50 μm wavelength), an interferometer that produces a temporally dispersive signal, and a signal processing unit that Fourier-transforms the output and converts it into an intensity *vs* wavelength spectrum. These instruments irradiate a sample with a wide range of IR frequencies and subsequently measure the intensity of the reflected or transmitted radiation as a function of frequency. Extensive libraries of spectra help identify unknown compounds, but unambiguous chemical identification generally requires the use of an orthogonal analytic technique, such as MS or nuclear magnetic resonance spectrometry. Through the use of an IR microscope, FTIR can be performed on samples as small as 10 μm in diameter.

While IR and Vis/NIR spectroscopy can unambiguously identify pure compounds, many processed nuclear materials (e.g., uranium ore concentrate) contain a mixture of compounds that result in complex IR spectra. Chemometric algorithms, such as principal component analysis (PCA) or partial-least-squares derivative analysis (PLSDA), are often used to rapidly classify unknown materials via reference to spectral libraries. Vis/NIR spectra can also provide chemical information, but the combination and overtone bands are challenging to interpret. Consequently, Vis/NIR spectroscopy is heavily reliant on techniques such as PCA and PLSDA to compare and assign spectra. However, Vis/NIR spectroscopy is entirely nondestructive, and samples may be analyzed as received, without special preparation. Thus, different veins in ore specimens can be directly assayed with a noncontact measurement. Grinding and homogenizing the sample facilitates a uniform response from a bulk specimen, while rotating the sample during analysis helps minimize effects from specular reflection and inhomogeneities.

16.4 MASS SPECTROMETRY

MS is one of the workhorses of nuclear forensic science, providing the capability to measure isotopic, elemental, and molecular abundances of an extraordinarily diverse array of materials with high precision and accuracy. Samples can be solid, liquid, or gaseous, large (kg) or small (zeptomoles), radioactive or stable, light (1 amu) or massive (kDa). Mass spectrometers achieve their high sensitivity, precision, and accuracy by converting a portion of a sample into positively or negatively charged ions and then separating and analyzing the ions according to their respective mass-to-charge ratios. Mass spectrometers are generally distinguished from each other according to the method by which ions are generated (e.g., thermal ionization, electron-impact ionization, or inductively coupled plasma ionization) and the type of mass analyzer (e.g., magnetic sector or radiofrequency [rf] quadrupole). They range in size from somewhat smaller than a credit card to large research instruments filling small buildings [37–39]. Modern mass spectrometers are highly specialized, and nuclear forensic investigations can employ as many as nine different types of instruments, depending on the nature and complexity of a sample. Described here are the operating principles and basic characteristics of the mass spectrometers most commonly applied in nuclear forensic analysis. For clarity, the discussion is organized according to the type of analysis: isotope ratio, elemental or molecular abundance, and spatial resolution.

16.4.1 Isotope-Ratio MS

Isotope-ratio MS (IRMS) is used to determine the isotopic abundances of specific elements in a sample. Mass spectrometric methods are able to determine concentrations of both radioactive and stable isotopes with a precision and accuracy of better than 10 ppm (0.001%). To achieve such high precision, IRMS uses magnetic-sector instruments almost exclusively, operated in environmentally controlled laboratories [40,41].

Thermal ionization mass spectrometry (TIMS) is used to measure isotope ratios of bulk samples and particles. TIMS has been the traditional mainstay of isotopic-ratio measurements for nuclear forensic applications, but it has been augmented, and in some cases replaced, over the last 10 years by multicollector, inductively coupled plasma mass spectrometry (MC-ICPMS). TIMS is the preferred technique for elements with low ionization potentials, including Sr, Pb, the lanthanides, and the actinides. Concentration determinations may also be made using artificially enriched tracers and isotope dilution analysis (Section 4.1).

In TIMS, a small volume (typically 1–10 μL), containing fg–ng quantities of a chemically purified and isolated analyte, is deposited on a metal filament (e.g., ultra-high purity Re or W). The filament is then heated to temperatures of 1000°C–2500°C in the high vacuum of the ion source by passing an electric current through the filament. If the ionization potential of a given element is sufficiently low compared to the work function of the filament, a fraction of the analyte atoms will be ionized via interaction with the high-temperature filament surface. TIMS ion sources commonly incorporate two filaments arranged opposite each other. One filament evaporates the sample via thermal heating, while the other ionizes evaporated atoms and molecules from the hot filament surface. Ions formed by thermal ionization possess uniformly low kinetic energy (0.1–0.2 eV), and nearly all TIMS employs single-focusing, magnetic-sector mass analyzers in which ions are separated according to their respective mass-to-charge ratios in an electromagnet. Modern TIMS instruments incorporate a combination of Faraday-cup and electron-multiplier detectors (as many as 12) for simultaneous collection of all isotopes of interest over a large range of isotope abundance. TIMS is capable of measuring ion currents as large as 10 nA and as small as 0.1 aA (10^{-10} nA), achieving a precision on isotope abundance ratios approaching 2 parts in 10^6 [42–44].

One drawback to TIMS is the requirement for time-consuming and labor-intensive sample preparation and filament-loading techniques, specific to each element, which can take weeks to months to develop and perfect. In many laboratories, TIMS has been replaced by MC-ICPMS. In MC-ICPMS, a sample is dissolved in acid and chemically processed to produce a purified solution, which is then nebulized in a spray chamber and aspirated into an Ar plasma. The high temperature of the plasma (5000°K–8000°K) dissociates the sample into its atomic constituents and ionizes these neutral species with very high efficiency (>90% for elements with a first ionization potential of <8 eV). Positively charged ions extracted from the plasma have a wide dispersion in energy (up to several hundred eV), and MC-ICPMS instruments use either double-focusing mass spectrometers or a gas-filled collision cell to reduce the energy spread and achieve the desired mass resolution and abundance sensitivity. Although very efficient, a plasma source is inherently much less stable than a thermal ionization source, and all MC-ICPMS instruments rely on simultaneous collection of the isotopes of interest to achieve high precision and accuracy. The multicollector detection systems are essentially identical to those used in TIMS. The concentration detection limits of MC-ICPMS are generally <1 pg/g (10^{-12} g/g) and reduce to the fg range (10^{-15} g/g) for favorable elements. A single analysis typically consumes 100 fgto 100 ng of sample, depending on the desired level of precision, with precision and accuracy generally comparable to TIMS [41,44]. The high efficiency of the plasma source for a very wide range of elements is an important advantage of plasma-source MS, and it is expanding the scope and importance of isotope-abundance measurements for nuclear forensic analysis [41,45,48].

Neither TIMS nor MC-ICPMS provide high efficiency for the electronegative elements (C, N, O, and S) or for the noble gases (He, Ar, Kr, and Xe). These groups of elements are best analyzed with dedicated instruments known collectively as (gas-source) IRMSs. For both sets of elements, samples are chemically processed, and the elements are extracted and purified as gases [38,39,46,47]. Gaseous analyte molecules are introduced into the ion source, where they are bombarded with an intense beam of energetic electrons (typically 50–70 eV) produced by thermionic emission from a W or Re filament. Ionization occurs primarily by electron impact (EI; i.e., direct collision between an electron and a gas molecule). Positive ions are swept from the source by a (positive) ion-repeller voltage and are subsequently mass-analyzed in a single magnetic sector. Many stable-isotope mass spectrometers incorporate a dual-inlet system in which a gas of known isotopic composition is fed into the ion source at regular intervals during analysis. The isotopic composition of the unknown analyte is then always calculated relative to that of the known standard, and not as an absolute ratio. For example, oxygen isotope ratios are always represented as deviations relative to standard mean ocean water. Stable-isotope MS generally requires mg or larger samples, from which the C, N, O, or S must be liberated by chemical reactions, generating ratios with a precision and accuracy of 0.01%–0.1%. Noble-gas mass spectrometers often use specially modified ion sources to improve the ionization efficiency for elements with very high first-ionization potentials, and they utilize very small internal volumes to minimize analytical blanks. Noble-gas spectrometers can achieve zeptomole (10^{-20} mol) detection limits [47].

16.4.2 Trace-Element MS

Trace-element MS is used to determine the abundances of trace inorganic constituents in a questioned sample. Compared to IRMS, greater emphasis is placed on the ability to analyze a wide variety of disparate materials than on extremely high precision and accuracy. For trace-element MS, accuracies of 1%–10% are considered acceptable, compared to 0.001%–0.1% with IRMS.

A diverse array of mass spectrometers is employed to measure elemental concentrations rapidly and with minimal sample preparation. By virtue of its ability to analyze nearly all elements exclusive of the gases, ICP-MS has become the preferred technique for minor- and trace-element analysis in nuclear forensics. Samples are generally dissolved in acid and the resulting aqueous solution nebulized in a spray chamber and aspirated into an Ar plasma, as described in the previous section for MC-ICPMS. Solids may also be analyzed by coupling ICP-MS to laser ablation or other commercially available solid-sourcing systems. With minimum detection limits at the parts-per-trillion (ppt; 10^{-12}) level, ICP-MS is generally the most sensitive analytical technique for rapid, multielement analysis of both solids and liquids [40,41,45]. Most commercial ICP-MS instruments use an rf quadrupole mass analyzer. The quadrupole can be tuned through a wide mass range very quickly (a scan from 1 to 300 amu taking <1 s), while providing adequate mass resolution and allowing much easier maintenance than magnetic-sector analyzers. The lower precision and decreased sensitivity relative to a magnetic-sector instrument is not a significant drawback for most nuclear forensic applications. If needed, higher precision may be obtained by time-averaging multiple scans to achieve an increased signal-to-noise ratio.

Detection limits for ICP-MS depend on several factors. Dilution of the sample before it can be aspirated into the spray chamber has a major effect. The quantity of sample held in solution, in turn, is governed by ion-suppression effects and by the level of dissolved solids that can be tolerated by the spray chamber. Ion suppression caused by high concentrations of U in solution is a particular problem for many analyses in nuclear forensic applications. The response curve of the mass analyzer also plays an important role, as most rf quadrupole analyzers exhibit much greater sensitivity for elements in the middle of the mass range (i.e., Sr to Lu). In general terms, detection limits are 0.1–10 pg/L, with analyte requirements in the fg–ng range. The practical limits of detection for elements abundant in nature are often determined by necessary corrections for the analytical blank (i.e., analyte concentrations in the acidic solution to which no contributions have been intentionally added).

The extraordinary sensitivity of ICP-MS can be particularly important in a nuclear forensic investigation for detecting ultratrace amounts of actinide elements. The sensitivity of MS is, for example, at least 1000× greater than α spectroscopy. Trace-element analyses by ICP-MS have typical accuracies of 2%–5%, although higher accuracy may be obtained with isotope dilution [44]. Simple sample-introduction techniques at near-atmospheric pressure have enabled ICP-MS to become a particularly versatile tool for many types of nuclear forensic investigation, and many easy-to-apply sample-introduction systems for liquid, solid, and gaseous materials have been successfully adapted to ICP-MS. These include laser ablation, high-performance liquid chromatography (HPLC), gas and ion chromatography (IC), capillary electrophoresis, and electrothermal vaporization, vastly broadening the scope of materials to which ICP-MS may be applied [40,45].

In glow-discharge mass spectrometry (GDMS), the sample serves as the cathode of a glow discharge (Ar is usually the support gas). The specimen is bombarded by Ar ions, with sputtered, neutral species from the sample ejected into the plasma. In the plasma, the neutrals are ionized by either EI or, more typically, by collision with metastable Ar atoms (Penning ionization). GDMS is an effective technique for measuring trace-element concentrations in solid samples with minimal sample preparation. GDMS suffers from very few matrix effects, is quantitative (with typical accuracy ~20%), and can be used as a sensitive survey tool over a very wide dynamic range spanning >11 orders of magnitude (detection limits ranging from <10 pg/g to a few μg/g, depending on the element) [49,50]. GDMS has become an industry standard for the analysis of trace elements in bulk semiconductor and metal samples. However, GDMS lacks the precision and accuracy associated with radiochemistry, TIMS, or ICP-MS. Care is required for GDMS analyses of heterogeneous samples, as the sampled volume is small and there is little opportunity for homogenization by dissolution or mixing.

16.4.3 Accelerator Mass Spectrometry

Accelerator mass spectrometry (AMS) measures long-lived radioisotopes by counting atoms directly, rather than via α, β, or γ emissions from radioactive decay. Such direct-atom counting becomes advantageous when the number of radioactive atoms, divided by their mean lifetime in months, becomes <~1. Thus, AMS provides an increase of a factor of ~10^6 compared to β counting for the detection of ^{14}C, and the gain in sensitivity further increases for nuclides with longer half-lives (e.g., ^{233}U).

AMS is predicated on the removal of interfering molecular species and high differentiation between isobars. If both objectives can be attained with good efficiency, then the MS of many natural radioisotopes using mg (or smaller) samples is possible. AMS discriminates against molecular species by accelerating ions to high energy (>0.5 MeV), such that an average of three or more electrons are removed by passage through a thin foil or an equivalent mass of gas. It then separates isobars using particle-identification techniques, such as time-of-flight and energy-loss measurements [51]. Originally developed for ultrasensitive analyses of rare environmental isotopes like ^{14}C and ^{129}I, AMS has also proven effective for nuclear forensic analysis of the trace uranium isotopes, ^{233}U and ^{236}U, with a detection limit for $^{233}U/^{238}U < 10^{-13}$ [52].

16.4.4 MS and Microanalysis

The MS techniques discussed in the preceding paragraphs are extremely sensitive, but pertain largely to the analysis of homogeneous bulk samples. Insofar as many nuclear forensic samples are physically and chemically heterogeneous, complementary techniques capable of isotopic and trace-element analyses on a microscale also play a central role in many nuclear forensic investigations.

Secondary-ion mass spectrometry (SIMS) is the most widely used technique for spatially resolved analyses of isotope and trace-element abundances. SIMS can be applied to any solid sample, from golf-ball-size pieces of metal, to counterfeit bills, to dust grains only a few tens of nm in size [53–56]. SIMS uses a high-energy (keV), finely focused, primary ion beam (typically O_2^+, Cs^+, or O^-) to sputter a specimen surface. The primary ions penetrate a few nm into the sample and, in transferring their kinetic energy via collision cascades, generate positive and negative secondary ions characteristic of the sample. SIMS may be used to measure any element, from H to Pu, over a range in secondary-ion intensities of $>10^9$. The sputtering process is very matrix dependent, and accurate isotopic and trace-element quantitation requires matrix-matched standards with sputtering behavior similar to that of the unknown analytes. Secondary ions have a very broad energy distribution that peaks at a maximum of a few eV, but with the distribution extending out to hundreds of eV and requiring the use of double-focusing mass analyzers. Sputtering of chemically complex samples, such as those frequently encountered in nuclear forensic examinations, generates copious numbers of molecular ions and creates the additional requirement of high mass-resolving power ($m/\Delta m > 3000$). This is necessary in order to separate singly charged elemental ions from molecular species that occur at the same nominal mass (e.g., $^{24}Mg^{16}O^+$ and $^{40}Ca^+$ at mass 40). The mass-resolving power of commercial SIMS instruments is variable from ~300 to 20,000, depending on instrument settings.

SIMS can be operated in any of three fundamental modes, based upon the nature of the sample and on the type of information required by the forensic investigation. In the depth-profiling mode (used to reveal changes in chemical or isotopic composition as a function of penetration beneath a surface) and the microprobe mode (used to map changes in elemental and isotopic distributions across a sample surface), a finely focused primary-ion beam is rastered across the sample in a manner similar to the electron beam in an SEM. The secondary-ion signal is correlated with the position of the primary-ion beam to generate a mass-resolved, secondary-ion image. Sputtering of the sample by the focused primary-ion beam produces a sensitive record of isotopic and chemical variations in three dimensions [53–56]. This capacity for three-dimensional analyses of isotopes and trace elements is unique to SIMS. The lateral resolution is determined by the primary-beam diameter and can be as good as 50 nm with the Cameca NanoSIMS (although μm resolution is more typical [57]). The depth resolution is limited by atomic mixing and is typically 1–2 nm. Bulk-analysis mode uses a defocused, static primary beam to produce maximum sensitivity while sacrificing depth and lateral resolution. Most commercial SIMS instruments are stigmatically focusing, such that the spatial position of ions leaving the sample surface is maintained and magnified through the instrument. Stigmatic focusing allows the real-time visualization of mass-resolved images through the use of a position-sensitive detector, such as a phosphor screen or resistive anode encoder.

The ion-yields of different elements vary by many orders of magnitude, with corresponding detection limits generally in the range of 1–500 ng/g. The accuracy of trace-element analyses is approximately 2%–10%, depending on the availability of suitable standards [55,56]. Isotope-ratio measurements are conducted in analogous fashion, but with special attention to sample charging and alignment of the secondary-ion beam in the mass spectrometer. The accuracy of isotope-ratio measurements is typically 0.01%–0.2%, depending on ion-yield and relative isotopic abundances [54,57]. The adoption of large geometry, multicollector mass spectrometers has significantly improved SIMS capabilities for isotope-ratio measurements, particularly for low-abundance isotopes such as ^{236}U [58].

SIMS is the technique of choice to measure isotope ratios and impurity concentrations in particulate samples. Using a finely focused primary-ion beam, SIMS can analyze particles in the μm and smaller size-range, weighing <1 pg, with a precision and accuracy of better than 0.5% [59,60]. In many nuclear forensic applications, a few U- or Pu-bearing particles of interest may be immersed in a sea of environmental detritus containing little in the way of forensic information. SIMS can be applied in a particle-search mode to locate and analyze these rare, but highly prized, particles. Particle searches may be carried out in both the ion-microscope mode, using a defocused primary beam (diameter >100 μm), or in the microprobe mode as described earlier. These particle searches can be conducted on fully automated systems, requiring no operator intervention except to load samples, and can be programmed to locate and identify a handful of "interesting" grains out of a population of thousands of grains of normal composition [60]. To execute particle searches with SIMS, particulate samples are dispersed on a conducting substrate (e.g., polished, high-purity graphite disks) or adhered to double-sided adhesive supports. Depending on the type of particle, a carbon coating may be applied to minimize charging effects.

Laser-ablation inductively coupled plasma mass spectrometry (LA-ICPMS) is another technique that provides spatially resolved isotopic and trace-element information. In LA-ICPMS, the spray chamber and nebulizer used to introduce solutions into the plasma torch in conventional ICP-MS analyses are replaced with an optical microscope and a laser-ablation cell. A high-powered laser, usually a ns-pulse-length Nd-YAG laser tuned to a 266-, 213-, or 193-nm wavelength, is used to ablate material from a solid sample [39,41,42,45]. Recent studies have shown that the use of ultrashort, fs laser pulses offers significant advantages for many applications of LA-ICPMS [61,63]. The aerosols created by ablation are entrained in an Ar gas stream and conveyed to the plasma torch; ionization and mass analysis then occur as for solution ICP-MS. Because ablation occurs at atmospheric pressure, sample preparation is minimal, and nearly any solid material can be analyzed. The spatial resolution is typically 10–100 μm, although laser-spot sizes down to a few μm may be obtained with some loss in analytic sensitivity.

LA-ICPMS is able to rapidly characterize questioned samples for minor- and trace-element compositions, much faster than either GDMS or SIMS, and chemical analysis via laser ablation requires smaller samples (μg) than does solution ICP-MS (mg).

As with SIMS, accurate quantitation of isotopic or trace-element abundances requires matrix-matched standards. Detection limits approach 1 ng/g, depending on the laser beam diameter, with accuracies ranging from 5% to 25% and dependent on the availability of suitable standards. Laser ablation coupled to MC-ICPMS is a powerful technique for in situ isotope-ratio measurements of a wide variety of elements, including the actinides [41,42,62,63]. The precision and accuracy are generally in the ‰-range for elements with concentrations >0.1 mg/g and in the %-range for elements at lower concentrations [41,42,63].

The principal techniques used to measure isotopic, elemental, and image information in nuclear forensic analysis are summarized in Table 16.1.

TABLE 16.1
Analytical Tools for Nuclear Forensic Analysis

Bulk Samples

Technique	Type of Information	Typical Detection Limit
Radiochemistry	Isotopic	10^4 atoms
Thermal-ionization mass spectrometry	Isotopic	Picograms to nanograms per gram
Plasma-source mass spectrometry	Isotopic, elemental	Picograms to nanograms per gram
Glow-discharge mass spectrometry	Isotopic, elemental	Picograms to micrograms per gram
X-ray fluorescence	Elemental	0.01 wt.%
X-ray diffraction	Molecular	~5 at.%

Imaging

Technique	Type of Information	Spatial Resolution
Optical microscopy	Imaging	0.5 μm
Scanning electron microscopy	Imaging	1 nm
Transmission electron microscopy	Imaging	0.1 nm

Microanalysis

Technique	Type of Information	Typical Detection Limit	Spatial Resolution (μm)
Laser-ablation, inductively coupled plasma mass spectrometry	Elemental, isotopic	Picograms to nanograms per gram	20
Secondary-ion mass spectrometry	Elemental, isotopic	1–100 ng/g	0.05–10
Scanning electron microscopy/ energy-dispersive x-ray spectrometry	Elemental	0.2–2 at.%	1
Electron microprobe microanalysis	Elemental	0.01 at.%	1

Source: Adapted from IAEA Nuclear Security Series No. 2, Nuclear forensics (sic) support: A reference manual, International Atomic Energy Agency, Vienna, Austria, 2006.

16.5 GAS CHROMATOGRAPHY–MASS SPECTROMETRY

Sophisticated instrumentation exists for the molecular identification and speciation determination of chemical compounds of interest in nuclear forensic analysis. Some, such as FTIR spectroscopy and nuclear magnetic resonance, are quite powerful under certain circumstances, but they require relatively pure samples for successful application. Questioned forensic specimens often will not satisfy this criterion without extensive chemical manipulation before instrumental interrogation, and consequently, methods that incorporate some form of chromatographic or electrophoretic separation are generally more versatile and applicable. Although gas chromatography (GC)–FTIR has been developed as a valuable analytic technique, the most widely utilized instrumental system for diverse and effective organic analysis is GC–MS.

GC–MS is a technique for detecting and measuring the concentration of trace organic constituents in a bulk sample [38]. In GC–MS, the components of a mixture are separated in the gas chromatograph and then identified in the mass spectrometer. The primary component of a GC is a narrow-bore tube maintained inside an oven. In the simplest arrangement, the analyte mixture is flash-vaporized in a heated injection port. The various components are swept onto, and through, the column by a He (or other suitable) carrier gas and are separated on the column based on their relative absorption affinities. Columns are usually coated with a special material to enhance separation of the components of interest. In the ideal case, all components elute from the column separated in time and are introduced into the MS as a time series. The MS, usually an rf quadrupole mass analyzer, detects and quantitates the concentration of each component as it elutes from the column.

Many different ionization methods can be used, but the most common is EI, in which a 70-eV beam of electrons bombards the sample molecules. Electrons hit a sample molecule and fragment it, ionize it, and impart kinetic energy. The mass-spectral fragmentation pattern of the resultant positive ions is characteristic of the molecule's structure. The relative abundance of ions of various masses (the mass spectrum) is characteristic of the parent molecule, and extensive libraries of EI mass spectra exist to help identify unknown compounds separated and detected by GC–MS.

GC–MS analysis results in a large output file containing MS fragment-ion data ($m/z+$ values and associated intensities) measured at increasing retention times of analytes on the GC column. If compounds with available and reliable certified standards are the only analytes of interest, absolute identifications reliant on matched retention times and corresponding ion-fragmentation patterns are obtained. Accurate quantitation may be obtained in this situation via measurement of a calibration curve (typically chromatogram peak area as a function of known species concentration), with interpolation of questioned-sample experimental data within the fitted calibration function of the primary standard. However, the CAS Registry contains >50 million (and counting) chemical substances, while the NIST/EPA/NIH (NIST08) MS database (for example) includes only 192,000 compounds. Dependable calibration standards are therefore generally unavailable for all analytes of potential interest in a forensic investigation. Thus, an analyst must often rely heavily on an evaluated match of empirical mass spectra to those in suitable comparison libraries.

GC–MS spectral analyses and species identification by manual analyst methods are often tiresome, inefficient, and seriously challenged by complex chromatograms. However, productive application may be made of computer software developed by the U.S. Department of Defense and NIST for the verification of international treaties. The Automated Mass Spectral Deconvolution & Identification System (AMDIS) applies suitable algorithms to experimental MS data to extract pure component spectra and related data from complicated chromatograms, then utilizes the information to evaluate whether the component can be ascribed to an established compound in a reference database.

The Livermore Forensic Science Center (FSC) uses two fundamental AMDIS outputs to evaluate the confidence in a possible identification. One is the net match factor: a final match quality value for association of the deconvoluted empirical component with the library spectra, where a perfect match ≡ 100. The other is the purity parameter: that portion of the total ion signal at component maximum scan that is credited to the deconvoluted constituent. For reasonable forensic results, minimum acceptance criteria are generally match factor ≥ ~80 and purity ≥ ~60%–70%. However, an identification with lower purity is occasionally accepted if, in the judgment of an experienced analyst, the corresponding match factor is suitably high and justification is afforded by individual interpretation of the experimental spectrum. (Similar to computer-assisted fingermark matching, final conclusions on a GC–MS spectrum by an expert mass spectroscopist are essential for high-quality results.) But the absence of a standard reference material for corroborating experimental measurements, of both GC retention time and MS ion-fragmentation profile, necessitates compound identification largely via database match. Consequently, despite any possibly formed opinion that identification of a chemical species by AMDIS alone was a virtual certainty, a conservative forensic scientist would nevertheless declare that identification to be "tentative" only.

While AMDIS successfully deconvolutes overlapping spectra, we have found that it often incorrectly reports the presence of several components when only a single compound is present, and it frequently fails to properly identify alkanes because of the very similar spectral features of this class of compounds. Despite its limitations, however, proper use of AMDIS is a valuable analytic tool that can provide a common protocol for GC–MS data reduction among independent laboratories.

GC–MS analyses provide very high analyte specificity, allowing accurate separation of extremely complex mixtures and exact identification of their individual species. Unique signature compounds within very complicated sample matrices can thus be analyzed. The protocols are also highly sensitive. Limits of detection for scanning GC–MS are on the order of nanograms of material, which translate to sensitivities of approximately 1 part in 10^{13} for cleaner samples and 1 part in 10^{11} for complex mixtures. However, in selected-ion-monitoring analysis mode, these limits improve to parts per 10^{15} to 10^{14}. Indeed, it is likely that no other analytic method affords as much diverse and diagnostic data per gram of sample as does GC–MS.

It is clearly beyond the scope of this book to attempt to cover all of the various aspects of the science of GC–MS, and the interested reader is referred to textbooks that afford more comprehensive treatments of the subject [65–67]. However, examples of GC–MS tactics from actual nuclear investigations are instructive and representative of effective forensic protocols for organic analytes.

In the matter of the smuggled high-enriched uranium (HEU) interdicted in Bulgaria (see Chapter 20), the Special Nuclear Material was sealed in a glass ampoule for primary containment. A Pb containment shield masked the radioactive emissions from external detection, and a yellow wax cushioned the ampoule from the Pb wall. A material "fingerprint" of this wax was measured by GC–MS. More than 45 discrete chemical compounds were identified, with 5 of them being distinctive naphthalenic species.

Volatile and semivolatile organic compounds were analyzed by GC–MS after sampling by headspace solid-phase microextraction (SPME). For each analysis, an approximate 0.5-g subsample of the questioned wax was placed in a 6-mL glass vial within either an Ar- or N_2-filled glovebag. The vial was then capped with a Teflon-lined silicone septum and heated with an exposed SPME fiber in the sample headspace. Trace species were thus preconcentrated in the fiber according to their solid–gas partition coefficients and their affinities for the SPME sorbent. Each wax aliquot was sampled twice. The first SPME exposure collected volatile compounds at 50°C for approximately 1 h on a 75-μm Carboxen fiber. The second collected semivolatile species for an hour over 55°C–70°C on a 65-μm Carbowax/DVB fiber.

The wax was also extracted with methylene chloride for GC–MS analysis of nonvolatile compounds via direct liquid injection. It was necessary to separate the dissolved wax in CH_2Cl_2 from an inorganic additive ($BaCrO_4$), and this was performed by centrifugation. The wax–CH_2Cl_2 supernatant liquid was isolated and evaporated to near-dryness at ambient temperature. The wax residue was then reconstituted in CH_2Cl_2 at a concentration of 100 ng/μL, and 1-μL aliquots were injected for analysis.

The GC–MS instrument used by this investigation was a Hewlett-Packard (now Agilent) 6890 GC interfaced to a 5973 quadrupole mass-selective detector. The MS was operated using standard EI ionization at 70 eV and calibrated with perfluorotributylamine tuning compound [PFTBA, $(C_4F_9)_3N$]. The default ions for the tuning procedure were $m/z = 69$, 210, and 502 amu, with their standardized intensity ratios approximately 100%, 45%, and 2.5%, respectively. The relative peak widths (FWHM) of both $m/z = 69$ and 502 were equal within ±0.1 amu. A standard chemical mixture was also analyzed to ensure column performance and the sensitivity of the MS.

The GC–MS system operating conditions for the analysis of the semivolatile and nonvolatile analytes are given in Table 16.2. The parameter settings for the volatile analytes were similar, with the principal differences being a 60-m J&W DB-624 column and an oven-temperature program of 40°C for 2 min, ramped at 8°C/min to 260°C, and held for 19.5 min.

However, the questioned wax component in the Bulgarian HEU inquiry was a bulk organic specimen. A more general GC–MS protocol implemented for a nuclear forensic investigation was exemplified by analyses of select samples from the gaseous diffusion plants (Chapter 21). In it, metals and other miscellaneous solid specimens were interrogated for trace organic species. These were isolated by immersing a questioned component in 3:1 methylene chloride/isopropanol in a glass vial and equilibrating it in an ultrasonic bath for 30 min. After sonication, the solvent was separated and evaporated to a final volume of approximately 100 μL. A 1- or 2-μL aliquot was then analyzed by GC–MS.

TABLE 16.2
HP 6890/5973 Gas Chromatography–Mass Spectrometry Operating Conditions for Semivolatile Analytes (Collected by 65 μm Carbowax/DVB SPME Fiber) and Nonvolatile Analytes (Extracted into CH_2Cl_2)

Instrument Parameter	Condition or Setting
Column	30 m J&W DB-17ms, 0.25 mm i.d., 0.25 μm film thickness
Carrier gas	He
Column flow	1.0 mL/min
Temperature program	40°C, 10°C/min, 300°C for 23 min
Solvent delay	3 min
Injection mode	Splitless
SPME injection time	>1 min
Injector temperature	250°C
Mass spectrometry transfer-line temperature	250°C
Mass spectrometry source temperature	230°C
Mass spectrometry quadrupole temperature	150°C
Mass range	45–800 amu
Scan rate	2 scans/s

Note: For the solvent extraction aliquots, the injection volume was 1 μL, injector flow was split at a 16:1 ratio, and the initial oven temperature was 50°C.

Because polar organic compounds are typically not amenable to GC analysis directly on methylsilicone columns, a fraction of the 3:1 solvent extract was derivatized with BSTFA (Chapter 12). Fifty microliters of the reduced-volume methylene chloride/isopropanol solution was transferred to a 4-mL glass vial fitted with a Teflon-lined screw-cap. The vial was then positioned in a laboratory heating/evaporation unit, and the liquid was taken to near-dryness under a stream of ultrapure N_2 or He at ambient temperature. Adequate BSTFA (~50–100 μL) was added to cover the solid residue, after which the sample was heated (with the vial cap on) in the thermal unit for 15–30 min at 60°C. A 1- or 2-μL aliquot from this treatment was then analyzed by GC–MS.

16.6 OTHER TECHNIQUES

16.6.1 Capillary Electrophoresis

Like GC–MS, capillary electrophoresis (CE) is also a versatile analytic technique employed for nuclear forensic analyses. With respect to inorganic chemical separations, CE can be viewed as an advanced variation of progressive ion-exchange technology via HPLC (5,000–15,000 theoretical plates, 5–500 μL samples, parts-per-billion to parts-per-million sensitivities, and 10–30 min separation times) and IC (1,000–5,000 theoretical plates, 100-μL samples, 0.1–1 ppm sensitivities, and 10–30 min separation times). The operational figures of merit for CE are on the order of 100,000–500,000 theoretical plates, 10-nL sample sizes, 0.1 ppm detection limits, and 5-min separation times.

Although effective protocols have been developed for nuclear fission and activation products [68,69], the chief applications of CE in nuclear forensic investigations are analyses of questioned high (HE)- and low-explosive (LE) samples. As with GC–MS, entire textbooks have been written about CE, and the laboratory specialist is referred to them for in-depth treatments [70–72]. Representative examples of CE protocols for explosives analyses follow.

Anionic LE species can be separated, identified, and quantitated on an uncoated, 50 cm × 50 μm i.d., fused-silica capillary column at ambient temperature. The applied potential = 25 kV (negative to positive), and the nominal expected current is approximately 10 μA. The electrolyte solution is Dionex IonPhor PMA anion buffer, containing pyromellitic acid, NaOH, hexamethonium hydroxide, and triethanolamine. Experimental preconditioning consists of a 3 min flush with the run buffer. A questioned sample, after passing through a 0.2-μm filter, is injected by hydrodynamic pressure at 50 mbar for a duration of 5 s. Analytes are detected with a diode array at ultraviolet wavelengths of 254, 230, and 200 nm (NO_3^- is inverted). With this protocol, the species $SO_4^=$, NO_2^-, NO_3^-, ClO_4^-, and ClO_3^- are eluted, in that order, in about 5 min.

For the LE cations, the same column and voltage conditions (except positive to negative) are employed with Dionex IonPhor DDP cation buffer (dimethyldiphenylphosphonium hydroxide, 18-crown-6, and 2-hydroxyisobutyric acid). Following the 0.2-μm filtration and sample injection at 30 mbar for 2 s, detection is performed at 220 nm. The analytes NH_4^+, Ba^{2+}, Sr^{2+}, Ca^{2+}, Mg^{2+}, and Li^+ are baseline-resolved and eluted within little more than 2 min (again, in that order).

CE is also valuable for HE analyses, particularly when HMX or RDX are suspected, because these compounds are thermally unstable under GC operating conditions. An effective CE protocol for HE species is analogous to the LE methods provided earlier, but with an operating electrolyte of approximately 10 mM sodium tetraborate and 60 mM sodium dodecyl sulfate, at pH ~8.5–9.

16.6.2 Vis/NIR Reflectance Spectroscopy

Since 2006, the FSC has regularly applied reflectance spectroscopy in nuclear forensic examinations. The early work was entirely classified, and open-literature reports on related applications appeared only somewhat later [73]. During the course of casework analyses, an attribution analyst from another agency inquired about the specific color(s) of interdicted HEU oxide specimens. The issue of sample color was also raised by traffickers during trial and was alleged to be an important consideration in their defense.

Assessment of uranium oxides by perceived color has been a common tactic, as it is intuitive, rapid, nondestructive, and requires no special apparatus. However, such estimations can be quite subjective and within the eye of the beholder. Factors such as lighting, color acuity of the observer, his/her mood at the moment, and other such considerations may influence an evaluation. Consequently, we have placed such appraisals on a technical basis, via standard light source and spectrometer, to effect an objective measurement of color.

In addition, by also measuring NIR wavelengths (700–2500 nm) along with the visible (400–700 nm), rapid, nondestructive interrogation of specimens can provide more information than mere color differences. NIR spectral analysis can measure C–H, O–H, and N–H vibrational overtones and combination bands, and may provide chemical speciation information about the composition of a sample. In contrast to infrared spectrometry (FTIR), Vis/NIR reflectance measurements are performed without contacting the specimen or otherwise potentially compromising the sample.

In fact, the color of uranium oxide species has been shown to correlate with process information for some production protocols. For example, the thermal decomposition of ammonium diuranate [$(NH_4)_2U_2O_7$] to evolve volatile NH_3, H_2O, and N_2 produces U_3O_8. If the calcining temperature is below ~100°C, the resultant oxide is pale yellow; between 100°C and 400°C, dark yellow to red-orange; between 400°C and 650°C, orange or red to a green tinge; and above ~650°C, bottle-green to black [74]. Analogous characteristics have been noted in the syntheses of UO_2 and UO_3 [75].

Vis/NIR reflectance spectroscopy is a noncontact, nondestructive analytical technique that can provide chemical information on questioned specimens without prior sample preparation. We currently employ a backpack-portable, Vis/NIR spectrometer equipped with three separate detectors that span consecutive spectral regions: 350–1000, 1000–1800, and 1800–2500 nm. The light source is a 20 W tungsten–halogen lamp. A bifurcated fiber-optic bundle transmits light to the sample surface, collects the reflected light, and returns it to the spectrometer. Measurements are typically made from outside of the container, using optics to focus the light onto the specimen surface, and each analysis consists of an average of 10 scans over the complete range of the spectrometer, 350–2500 nm. Five analyses are normally performed on each sample, with each analysis interrogating a different surface location of the same specimen (some samples may be spun with a small rotating stage during measurement).

In this manner, visible spectra could be evaluated for, for example, hints of green (500–550 nm) in an ostensibly black specimen (see Figure 16.1). Data reduction and chemometric analyses are then performed using commercially available software packages. One incorporates discriminant analysis with cross-validation and mean-centering, and provides three-dimensional scores plots that can display spectral clustering into discrete groups for the various samples. Postacquisition pattern recognition algorithms, with analyses of spectral residuals, can be applied to empirical data to achieve direct comparisons or to search spectral libraries to determine the most probable match(es). Alternatively, we have also used data pretreatment with a standard–normal–variate transformation to eliminate interferences from scattering and particle-size effects (as well as to center relative data around zero absorbance), followed by PCA to assess sample clustering via scores plots. (Absorbance = $\log(\text{Reflectance})^{-1}$ and is directly proportional to concentration.)

Thus, although visual inspection can be a useful tool for assessing samples, instrumental spectroscopic techniques provide a more objective result with greater resolution. And while spectra from different specimens may appear very similar or even equivalent, chemometric analyses often differentiate them via fine-resolution data.

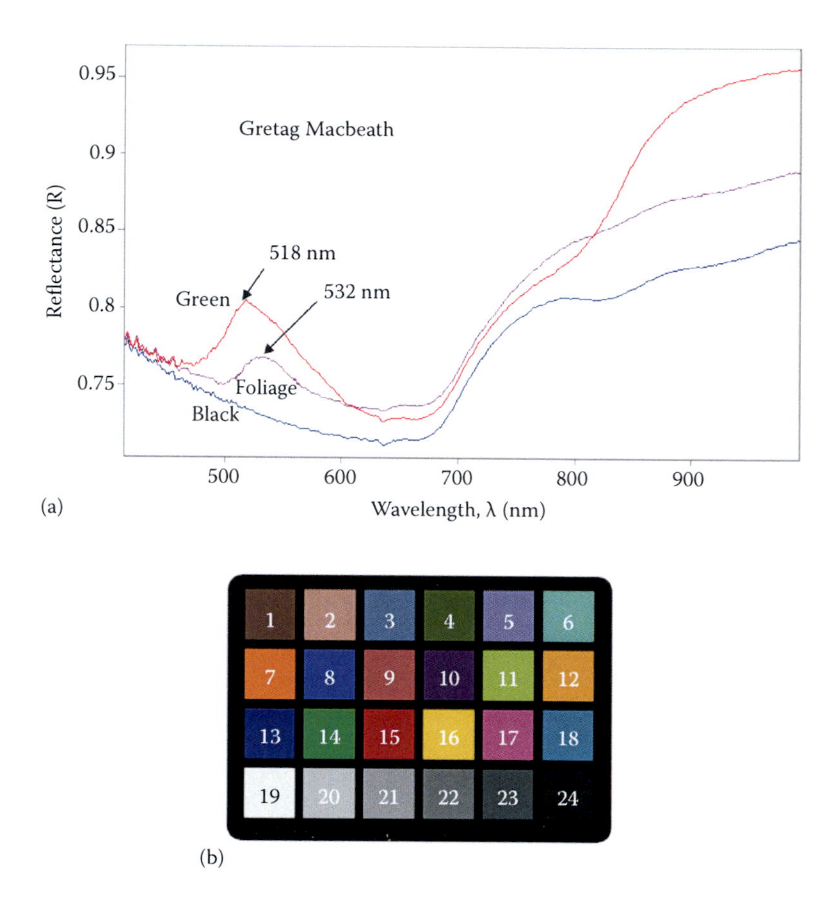

FIGURE 16.1 Reflectance spectra (a) for three defined colors in the Gretag Macbeath standard color-checker chart; (b) foliage = #4, green = #14. (Courtesy of Greg Klunder, LLNL Forensic Science Center, Livermore, CA.)

16.6.3 X-Ray Diffraction

XRD is a nondestructive technique used to conduct compositional and structural analyses of crystalline and semicrystalline materials [76,77]. XRD provides information on the constitution, concentration, and purity of multicomponent crystalline systems, the crystallographic nature of materials, and the crystallite (grain) size.

X-ray diffractometers consist of three basic elements: an x-ray tube, a sample holder (goniometer), and a detector. X-rays are generated in a cathode-ray tube by bombarding a target with high-energy electrons. The electrons dislodge inner-shell electrons of the target, producing characteristic x-rays specific to that target; Cu, Fe, Mo, and Cr are commonly used. Subsequent filtering, by foils or crystal monochrometers, produces the monochromatic x-ray beam needed for diffraction, and these x-rays are then collimated and directed onto the sample. As the sample and detector are rotated, the intensity of elastic, reflected x-rays is recorded. When the geometry of the incident x-rays impinging the specimen satisfies the Bragg equation, constructive interference occurs and a peak in intensity ensues. A detector records and processes the signal and converts it to a count rate.

The geometry of an x-ray diffractometer is such that the sample rotates in the path of the incident x-ray beam at an angle θ, while the detector is mounted on an arm to collect the diffracted x-rays rotating at an angle of 2θ. For typical powder patterns, data are collected at preset 2θ angles from ~5° to 70°.

All commercially available XRD instruments are equipped with data analysis software that utilizes databases compiled and organized by the International Centre for Diffraction Data. These databases contain hundreds of thousands of patterns to aid in phase and structure identifications, and many open-source software packages are also available.

16.6.4 X-Ray Fluorescence

X-ray fluorescence (XRF) is a nondestructive analytical method used to determine the chemical composition of a solid or liquid sample [78]. Specimens can be analyzed as bulk solids, pressed pellets, loose powders, or liquids. Although XRF spectrometry lacks the ultralow detection limits and high precision/accuracy of ICP-MS, modern XRF instruments are capable of measuring the concentration of elements from Na to U in a matter of minutes with minimal sample preparation. XRF spectrometry is particularly useful when sample alteration must be kept to a minimum, and it is often used as a screening tool in nuclear forensic investigations to guide additional analyses by MS. XRF spectrometers are often used in tandem with microbeam microscopies (such as SEM and TEM) to provide an in situ element quantitation capability. In addition, most XRF instruments can be customized for specific applications, such as the measurement of elements lighter than Na.

In XRF, a primary x-ray beam is generated by an x-ray tube having a Rh, Mo, or W anode target, and is used to bombard a sample with high-energy x-rays. The characteristic secondary x-rays induced in a sample are collimated and directed toward either a WDS or EDS for detection. With EDS, a Si detector is used to discriminate the different energies of the characteristic x-rays; with WDS, the characteristic x-rays are directed to a crystal with a given d-spacing. For the latter, only photons of the correct energy to satisfy Bragg's equation will be constructively diffracted and directed toward a detector. More information on the generation of characteristic x-rays and the detection of x-rays from different elements by WDS and EDS may be found in Section 16.3.5.

Elemental concentrations are quantified relative to a calibration curve for each element. Ideally, a matrix-matched functional relationship is determined for the elements of interest using a series of standards. However, in cases where a matrix-matched standard calibration cannot be obtained, XRF software may be used to calculate semiquantitative concentrations via the standardless fundamental-parameters method. This technique calculates expected x-ray intensities from the well-determined x-ray energies, the optical geometry, sample information, and other basic properties. In most cases, comparison of these theoretical intensities to empirical measurements allows a determination of chemical composition of the specimen within an accuracy of ~10%.

XRF is widely used in many industrial and research settings, and is particularly useful for applications where chemical compositions are required but ICP-MS data are unnecessary (e.g., quality control). XRF is less expensive, more user-friendly, and more versatile than ICP-MS, and it requires but minimal sample preparation. XRF is also useful in nuclear forensic analyses to quantitate elements that are frequently not measured by ICP-MS (e.g., Si, Cl, Br, S, and P).

REFERENCES

1. D.E. Watt and D. Ramsden, *High Sensitivity Counting Techniques*, Macmillan, New York, 1964.
2. J.O. Lubenau and J.G. Yusko, Radioactive materials in recycled materials, *Health Phys.*, 68, 440, 1995.
3. J.O. Lubenau and J.G. Yusko, Radioactive materials in recycled metals—An update, *Health Phys.*, 74, 293, 1998.
4. A. Alessandrello, F. Allegretti, C. Brofferio et al., Measurement of low radioactive contaminations in lead using bolometric detectors, *Nucl. Inst. Meth. Phys. Res.*, B83, 539, 1993.
5. Yu.I. Moskalev, ed., *Tritium Oxide*, Atomizdat, Moscow, Russia, 1968.
6. E.P. Cronkite, *Tritium and Other Labeled Organic Compounds Incorporated in Genetic Material*, National Council on Radiation Protection and Measurements, Washington, DC, 1979.
7. K.Y. Wong, T.A. Khan, and F. Guglielmi, *Canadian Tritium Experience*, Ontario Hydro, Toronto, Ontario, Canada, 1984.
8. R.B. Firestone and V.S. Shirley, eds., *Table of Isotopes*, 8th ed., Wiley, New York, 1996.
9. W.C. Reinig, J.M.R. Hutchinson, J.J. Koranda, A.A. Moghissi, R.V. Osbourne, and H.G. Ostlund, Committee 36 on tritium measurement techniques for laboratory and environmental use, Tritium measurement techniques, NCRP report 47, National Council on Radiation Protection and Measurements, Washington, DC, 1976.
10. K.A. Surano, G.B. Hudson, R.A. Failor, J.M. Sims, R.C. Holland, S.C. MacLean, and J.C. Garrison, Helium-3 mass spectrometry for low-level tritium analysis of environmental samples, *J. Radioanal. Nucl. Chem.*, 161, 443, 1992.
11. M.L. Chiarappa-Zucca, K.H. Dingley, M.L. Roberts, C.A. Velsko, and A.H. Love, Sample preparation for quantitation of tritium by accelerator mass spectrometry, *Anal. Chem.*, 74, 6285, 2002.
12. H.G. Ostlund and A.S. Mason, Atmospheric HT and THO. I. Experimental procedures and tropospheric data 1968–1972, *Tellus*, 26, 91, 1974.
13. F. Begemann and I. Friedman, Tritium and deuterium in atmospheric methane, *J. Geophys. Res.*, 73, 1149, 1968.
14. A. Haines and B.C. Musgrave, Tritium content of atmospheric methane and ethane, *J. Geophys. Res.*, 73, 1167, 1968.
15. B. Sansoni and W. Kracke, Rapid determination of low-level alpha and beta activities in biological material using wet ashing by OH radicals, in *Rapid Methods for Measuring Radioactivity in the Environment*, International Atomic Energy Agency, Vienna, Austria, 1971.
16. D.L. Horrocks, *Applications of Liquid Scintillation Counting*, Academic Press, New York, 1974, Chap. 6.
17. J.D. Davidson and P. Feigelson, Practical aspects of internal-sample liquid-scintillation counting, *Int. J. Appl. Radiat. Isot.*, 2, 1, 1957.
18. M. Pluta, *Advanced Light Microscopy*, Elsevier Science, Amsterdam, the Netherlands, 1988.

19. W.C. McCrone and J.G. Delly, *The Particle Atlas*, Vols. 1–4, Ann Arbor Science, Ann Arbor, MI, 1973.
20. N. Peteraco and T. Kubic, *Color Atlas and Manual of Microscopy for Criminalists, Chemists and Conservators*, CRC Press, Boca Raton, FL, 2004.
21. D.E. Newbury and D.B. Williams, The electron microscope: the materials characterization tool of the millennium, *Acta Mater.*, 48, 323, 2000.
22. J.I. Goldstein, D.E. Newbury, D.C. Joy, C.E. Lyman, P. Echlin, E. Lifshin, L. Sawyer, and J.R. Michael, *Scanning Electron Microscopy and X-Ray Microanalysis*, 3rd ed., Springer-Science, New York, 2003.
23. D.C. Joy, Scanning electron microscopy for materials characterization, *Curr. Opin. Solid State Mater. Sci.*, 2, 465, 1997.
24. D.H. Krinsley, K. Pye, S. Boggs Jr., and N.K. Tovey, *Backscattered Scanning Electron Microscopy and Image Analysis of Sediments and Sedimentary Rocks*, Cambridge University Press, Cambridge, U.K., 1998.
25. M. de Graef, *Introduction to Conventional Transmission Electron Microscopy*, Cambridge University Press, Cambridge, U.K., 2003.
26. D.B. Williams and C.B. Carter, *Transmission Electron Microscopy: A Textbook for Materials Science*, Springer-Science, New York, 2009.
27. D.C. Joy, A.D. Romig, and J.L. Goldstein, eds., *Principles of Analytical Electron Microscopy*, Plenum, New York, 1986.
28. L. Reimer, *Transmission Electron Microscopy: Physics of Image Formation*, Springer, Berlin, Germany, 1984.
29. J. Ayache, L. Beaunier, J. Boumendil, G. Ehret, and D. Laub, *Sample Preparation Handbook for Transmission Electron Microscopy: Techniques*, Springer-Science, New York, 2010.
30. R.M. Anderson, ed., *Specimen Preparation for Transmission Electron Microscopy II*, MRS Symp. Proc. Vol. 199, Materials Research Society, Pittsburgh, PA, 1990.
31. J.C. Brauman, R.M. Anderson, and M.L. McDonald, *Specimen Preparation for Transmission Electron Microscopy of Materials*, MRS Symp. Proc. Vol. 115, Materials Research Society, Pittsburgh, PA, 1988.
32. S.J.B. Reed, *Electron Microprobe Analysis and Scanning Electron Microscopy in Mineralogy*, Cambridge University Press, Cambridge, U.K., 1996.
33. B. Lister, The preparation of polished sections, Inst. Geol. Sci. Rep. No. 78/27, HMSO, London, U.K., 1978.
34. J.T. Armstrong, Quantitative elemental analysis of individual microparticles with electron beam instruments, in *Electron Probe Quantitation*, K.F.J. Heinrich and D.E. Newbury, eds., Plenum, New York, 1991.
35. L.C. Feldman and J.W. Mayer, *Fundamentals of Surface and Thin Film Analysis*, Elsevier Science, New York, 1986.
36. N. Tkachenko, *Optical Spectroscopy: Methods and Instrumentation*, Elsevier Press, Amsterdam, the Netherlands, 2006.
37. J.W. Plaue, G.L. Klunder, I.D. Hutcheon, and K.R. Czerwinski, Near-infrared reflectance spectroscopy as a process signature in uranium oxides, *J. Radioanal. Nucl. Chem.*, 296, 551, 2013.
38. J.H.E. Duckworth, R.C. Barber, and V.S. Venkatasubramanian, *Mass Spectrometry*, Cambridge University Press, Cambridge, U.K., 1986.
39. J.S. Becker, *Inorganic Mass Spectrometry: Principles and Applications*, John Wiley & Sons, Sussex, U.K., 2008.
40. J.S. Becker and H.-J. Dietze, Inorganic mass spectrometric methods for trace, ultratrace, isotope and surface analysis, *Int. J. Mass Spectrom.*, 197, 1, 2000.
41. T.R. Ireland, Recent developments in isotope-ratio mass spectrometry for geochemistry and cosmochemistry, *Rev. Sci. Instr.*, 84, 011101, 2013.

42. J.S. Becker, Inorganic mass spectrometry of radionuclides, in *Handbook of Radioactivity Analysis*, 3rd ed., M.F. L'Annunziata, ed., Academic Press, Oxford, U.K., 2012, p. 833.
43. S. Bürger, L.R. Riciputi, D.A. Bostick, S. Turgeon, E.H. McBay, and M. Lavelle, Isotope-ratio analysis of actinides, fission products, and geolocators by high-efficiency, multicollector, thermal-ionization mass spectrometry, *Int. J. Mass Spectrom.*, 286, 70, 2009.
44. S. Richter and S.A. Goldberg, Improved techniques for high accuracy isotope-ratio measurements of nuclear materials using thermal ionization mass spectrometry, *Int. J. Mass Spectrom.*, 229, 181, 2003.
45. H.E. Taylor, *Inductively Coupled Plasma Source Mass Spectrometry*, Academic Press, London, U.K., 2001.
46. R.A.W. Johnstone and C.G. Herbert, *Mass Spectrometry Basics*, CRC Press, Boca Raton, FL, 2002.
47. D. Porcelli, C.J. Ballentine, and R. Weiler, Noble gases in geochemistry and cosmochemistry, in *Reviews in Mineralogy and Geochemistry*, Vol. 47, Mineralogical Society of America, Washington, DC, 2011.
48. M. Wallenius, A. Morgenstern, C. Apostolidis, and K. Mayer, Determination of the age of highly enriched uranium, *Anal. Bioanal. Chem.*, 374, 379, 2002.
49. V. Hoffmann, M. Kasik, P.K. Robinson, and C. Venzago, Glow-discharge mass spectrometry, *Anal. Bioanal. Chem.*, 381, 173, 2005.
50. A. Bogaerts and R. Gijbels, New developments and applications in GDMS, *Fresenius J. Anal. Chem.*, 364, 367, 1999.
51. A.E. Litherland, X.-L. Zhao, and W.E. Kieser, Mass spectrometry with accelerators, *Mass Spectrom. Rev.*, 30, 1037, 2011.
52. S.J. Tumey, T.A. Brown, B.A. Buchholz, T.F. Hamilton, I.D. Hutcheon, and R.W. Williams, Ultrasensitive measurements of ^{233}U by accelerator mass spectrometry for national security applications, *J. Radioanal. Nucl. Chem.*, 282, 721, 2009.
53. R.G. Wilson, F.A. Stevie, and C.A. Magee, *Secondary-Ion Mass Spectrometry: A Practical Handbook for Depth-Profiling and Bulk Impurity Analysis*, Wiley-Interscience, New York, 1989.
54. A. Benninghoven, F.G. Rudenauer, and H.W. Werner, *Secondary-Ion Mass Spectrometry: Basic Concepts, Instrumental Aspects, Applications, and Trends*, John Wiley & Sons, New York, 1987.
55. G. Gillen and D. Bright, Tools and procedures for quantitative microbeam isotope-ratio imaging by secondary-ion mass spectrometry, *Scanning*, 25, 165, 2003.
56. R. Odom, Secondary-ion mass spectrometry, *Appl. Spectrosc. Rev.*, 29, 67, 1994.
57. P. Hoppe, S. Cohen, and A. Meibom, NanoSIMS: Technical aspects and applications in cosmochemistry and biological geochemistry, *Geostand. Geoanal. Res.*, 37, 111, 2013.
58. Y. Ranebo, P.M.L. Hedberg, M.J. Whitehouse, K. Ingeneri, and S. Littmann, Improved isotopic SIMS measurements of uranium particles for nuclear safeguard purposes, *J. Anal. Atomic Spectrom.*, 24, 277, 2009.
59. G. Tamborini and M. Betti, Characterization of radioactive particles by SIMS, *Mikrochim. Acta*, 132, 411, 2000.
60. L.R. Nittler and C.M.O'D. Alexander, Automated isotopic measurements of micron-sized dust: Application to meteoritic presolar silicon carbide, *Geochim. Cosmochim. Acta*, 67, 4961, 2003.
61. R. Hergenroder, O. Samek, and V. Hommes, Femtosecond laser-ablation elemental mass spectrometry, *Mass Spectrom. Rev.*, 25, 551, 2006.
62. D.C. Baxter, I. Rodushkin, and E. Engstrom, Isotope-abundance ratio measurements by inductively coupled magnetic-sector field mass spectrometry, *J. Anal. At. Spectrom.*, 27, 1355, 2012.

63. A.M. Duffin, G.L. Hart, R.C. Hanlen, and G.C. Eiden, Isotopic analysis of uranium in NIST SRM glass by femtosecond laser ablation MC-ICPMS, *J. Radioanal. Nucl. Chem.*, 296, 1031, 2013.
64. IAEA Nuclear Security Series No. 2, Nuclear Forensics (sic) Support: A Reference Manual, International Atomic Energy Agency, Vienna, Austria, 2006.
65. W. Jennings, E. Mittlefehldt, and P. Stremple, *Analytical Gas Chromatography*, 2nd ed., Academic Press, New York, 1997.
66. W.L. Budde, *Analytical Mass Spectrometry: Strategies for Environmental and Related Applications*, American Chemical Society, Washington, DC/Oxford University Press, New York, 2001.
67. J. Yinon, ed., *Forensic Mass Spectrometry*, CRC Press, Boca Raton, FL, 1987.
68. G.L. Klunder, J.E. Andrews, P.M. Grant et al., Analysis of fission products using capillary electrophoresis with on-line radioactivity detection, *Anal. Chem.*, 69, 2988, 1997.
69. M.N. Church, J.D. Spear, R.E. Russo et al., Transient isotachophoretic-electrophoretic separations of lanthanides with indirect laser-induced fluorescence detection, *Anal. Chem.*, 70, 2475, 1998.
70. S.F.Y. Li, *Capillary Electrophoresis: Principles, Practice, and Applications*, Elsevier, Amsterdam, the Netherlands, 1993.
71. M.G. Khaledi, ed., *High-Performance Capillary Electrophoresis: Theory, Techniques, and Applications*, Wiley, New York, 1998.
72. J.R. Petersen and A.A. Mohammad, ed., *Clinical and Forensic Applications of Capillary Electrophoresis*, Humana Press, Totowa, NJ, 2001.
73. G.L. Klunder, J.W. Plaue, P.E. Spackman et al., Application of visible/near-infrared reflectance spectroscopy to uranium ore concentrates for nuclear forensic analysis and attribution, *Appl. Spectrosc.*, 67, 1049, 2013.
74. R.C. Merritt, *The Extractive Metallurgy of Uranium*, U.S. Atomic Energy Commission, Washington, DC, 1971.
75. E.P. Hastings, C. Lewis, J. Fitzpatrick et al., Characterization of depleted uranium oxides fabricated using different processing methods, *J. Radioanal. Nucl. Chem.*, 276, 475, 2008.
76. D.L. Bish and J.E. Post, ed., Modern powder diffraction, in *Reviews in Mineralogy*, Vol. 20, Mineralogical Society of America, Washington, DC, 1989.
77. D.B. Cullity, *Elements of X-Ray Diffraction*, Addison-Wesley, Reading, MA, 1978.
78. G. Fitton, X-ray fluorescence spectrometry, in *Modern Analytical Geochemistry: An Introduction to Quantitative Chemical Analysis for Earth, Environmental, and Material Scientists*, R. Gill, ed., Addison Wesley, Longman, U.K., 1997.

17 Inferred Production Estimates

Fissile fuels are required for the operation of nuclear weapons. Although there are some relatively exotic materials that can be used for this purpose (e.g., ^{233}U), the most likely fuels for weapons applications are Pu and U highly enriched in ^{235}U (HEU or Oralloy). Plutonium and HEU were first produced in the 1940s by the United States and the Soviet Union at the inception of their nuclear weapons programs (see Chapter 1). Since that time, large quantities of both have been produced in many countries—including some without acknowledged weapons programs. Plutonium and HEU are difficult to produce, requiring a large initial investment in facilities and the mastery of dozens of engineering and scientific disciplines. Both fuels require extensive physical protection and safeguarding, in part to prevent criticality accidents and (in the case of Pu) the accidental release of a highly toxic substance.

17.1 URANIUM

As discussed previously, HEU is produced by enriching U of normal, or near-normal (e.g., reactor-modified), isotopic composition so that it contains equal to or more than 20% fissile ^{235}U (i.e., the International Atomic Energy Agency (IAEA) definition of HEU). This enrichment is accomplished by one of several methods, most of which take advantage of the subtle chemical difference between ^{235}U and ^{238}U that is induced by their different average inertias in the same thermodynamic environment. The production of HEU is necessarily accompanied by production of vastly more depleted U (in which the ^{235}U content is reduced below the 0.71% level found in most natural sources). Even countries without nuclear weapons programs produce or acquire HEU for use as fuel in high-flux, civilian research reactors. An even larger application of HEU, only indirectly involved in weapons applications, is as fuel in high-power-density naval propulsion reactors.

The conventions of Ref. [1] allow for five grades of U based on the enrichment (or depletion) of natural material: depleted U, which contains less than 0.71% ^{235}U; natural U, which contains about 0.71% ^{235}U; low-enriched U (LEU), which contains more than 0.71% but less than 20% ^{235}U; HEU, which contains from 20% or more, but less than 90% ^{235}U; and weapons-grade U (Oralloy), which contains 90% or more ^{235}U. In the United States, early production of HEU and Oralloy was accomplished by electromagnetic isotope separation. However, the total amount produced by this technique was only about 1 t. The vast bulk of the weapons inventory (~650 t) and highly enriched material for reactor applications (~350 t) was produced by gaseous diffusion at either Oak Ridge, Tennessee, or Portsmouth, Ohio. With reserves of Oralloy beyond that necessary to maintain the stockpile, HEU for weapons production ended in the United States in 1964 with the cessation of production at the Oak Ridge K-25 plant. The Portsmouth plant continued to produce HEU for reactor applications into the early 1990s.

Consumption of HEU in the United States (where extensive records were released in the 1990s [1–3]) is hard to estimate because of recycling. Material that is still nominally HEU, and is recovered from spent naval and research-reactor fuel, is often blended with depleted U to make LEU reactor fuel or is recycled back through the reactors as HEU. Before 1968, HEU was used in reactors at Savannah River, South Carolina, to irradiate external targets of ^{6}Li to make ^{3}H. After 1968, HEU fuel was also used to irradiate external targets of depleted U to produce Pu for military applications. Including the consumption of HEU and Oralloy in weapon tests, it is estimated that the total expenditure of the US stock of weapons-grade material is about 350 t.

In the Soviet Union, production of enriched U mirrored the US program, with electromagnetic isotope separation followed by gaseous diffusion. In the 1960s, the Soviets began replacing gaseous diffusion with the gas centrifuge, the replacement being complete by 1992. Production of HEU in the Soviet Union was probably at least as extensive as in the United States, resulting in significant stockpiles. Through a Russian–US agreement (Megatons-to-Megawatts; see Chapter 1), the former Soviet Union's (FSU's) stock of HEU is being blended with 1.5%-enriched LEU and sold to the United States as fuel for civilian reactors (the US domestic specifications for the ^{234}U and ^{236}U content of civil fuel precluded blending with depleted or natural U). In the United Kingdom, France, and China, production of weapons-grade U has been via gaseous diffusion. The United Kingdom has also received HEU from the United States under a barter agreement in exchange for Pu; in addition, the United States has provided enrichment services for the United Kingdom. Like the United States and FSU, France has stopped producing HEU for weapons applications, relying on reuse of a stockpile. Global research in U enrichment is extensive, and many techniques are being explored that could result in small analytical samples of HEU trading on the black market [4].

17.2 PLUTONIUM

Production of Pu isotopes unavoidably accompanies the production of fission products in the U fuel of a nuclear reactor. Plutonium is produced by neutron capture of the fertile ^{238}U present in mixed U isotopes, followed by two relatively short-lived β decays. Although Pu produced for weapons applications is preferentially made in reactor fuel of minimal ^{235}U enrichment over natural U, any civil power reactor can be a source of weapons-usable Pu. Plutonium is extracted from spent nuclear fuel by a set of chemical separations known as reprocessing. Although many tons of Pu have been produced for use in nuclear weapons, even larger quantities have been produced by civil power reactors. It is estimated that more than 200,000 metric tons of U and Pu, in the form of irradiated spent-fuel assemblies, have been discharged from civil power plants as of 2004, and only a small fraction of this material has been subjected to reprocessing [5]. Even the most conservative projections estimate that this quantity will increase to more than 1 million metric tons by 2020. The potential energy sequestered in this material has driven civilian programs in Europe and

Japan to reprocessing, with the intent of using the recovered Pu (mixed with fertile ^{238}U or ^{232}Th) as light-water reactor fuel to generate further electricity [6].

There are three grades of Pu [1], based on the level of ^{240}Pu (produced by the $^{239}Pu(n,\gamma)$ reaction) in the reprocessed material: reactor-grade Pu, which contains more than 18% ^{240}Pu; fuel-grade Pu, which contains more than 7% but less than 18% ^{240}Pu; and weapons-grade Pu, which contains less than 7% ^{240}Pu. As treated earlier, the concentration of spontaneously fissioning ^{240}Pu in a sample of fissile ^{239}Pu directly limits its use in weapons applications because of the possibility of preinitiation during assembly of the critical mass. Research has been conducted in Pu enrichment (or depletion of ^{240}Pu in a sample of ^{239}Pu). During the 1960s, it was explored with Calutrons and gas centrifuges; laser isotope separation then showed promise in the 1980s, but ultimately failed because of decreased demand for weapons-grade Pu [7].

Weapons-grade Pu isotopic content is controlled by limiting the extent of the irradiation of the parent U material in the reactor fuel, and the ^{240}Pu limitation imposes a constraint on the fuel burn-up of ~400 MW-day/t. This value is much smaller than when optimizing economic performance of a natural-U-fueled reactor (at least 4,000 MW-day/t) or a reactor running on LEU (at least 30,000 MW-day/t) [8]. These disparate reactor conditions lead to a dichotomy in reactor operations. In the military fuel cycle, production reactors are designed and operated not for power generation, but for the production of weapons-grade Pu. The low burn-up fuel does not require an extensive cooling period before reprocessing, unlike higher burn-up fuel arising in the civilian fuel cycle, where the goal is power production. The majority of spent civil fuel, containing vast quantities of reactor-grade and fuel-grade Pu, is unreprocessed and held in storage. Most spent-fuel assemblies arising from the civil fuel cycle are stored at the generating reactor facility for the short term and then sometimes moved to longer-term offsite storage after a few years [5].

Most of the weapons-grade stockpiles of the United States, FSU, and China were derived by reprocessing fuel discharged from light-water-cooled, graphite-moderated reactors. The United Kingdom, France, and North Korea have used gas-cooled, graphite-moderated reactors (Magnox type) to generate most of their weapons-usable Pu inventories. The United States, FSU, France, India, and Israel have also recovered Pu from heavy-water-moderated and cooled reactors as a by-product of ^{3}H production, and the United States and France have obtained weapons-grade Pu from external, depleted-U targets. The United States and FSU reprocessed Pu in mass-production operations performed in criticality-safe chemical apparatus. The British, French, and Chinese preferred smaller-scale batch operations in which limitations imposed by criticality safety were not as stringent. The United States reduced Pu production operations in the 1960s because sufficient supplies recovered from retired weapons were available to support the stockpile. Production operations for weapons applications were halted entirely in 1988, and the total US inventory of weapon-grade Pu stands at approximately 90 t.

17.3 SNM STOCKS

Nuclear production in both civil and military domains expanded rapidly in the decades following World War II. As a result, knowledge of the scale and locations of Pu and HEU inventories has become increasingly important. Spent nuclear fuel accumulates steadily with the operation of civil and military nuclear reactors. Production of HEU for nonweapons use has decreased, but stockpiles of HEU have expanded enormously as a consequence of arms-reduction treaties between weapons states. Most spent-fuel assemblies will eventually be stored or buried. However, international agreements involving the reprocessing of spent reactor fuel, and the use of recovered Pu for new fuel elements for other reactors, have led to the legitimate circulation of Pu between countries [9]. If Pu from dismantled warheads is eventually burned in reactors, this circulation will increase.

In a perfect world, these activities would require that the quantities and chemical states of fissile materials held by all countries, whether weapons states or not, be closely monitored and controlled. Of course, such is not the case. In weapons states, the inventories of HEU and Pu are indicators of military capability, and consequently, exact amounts and dispositions are usually classified. Information on civilian inventories is also not widely available. The IAEA and other international monitoring bodies hold information closely to protect the identities of the countries and commercial companies that provide them with data. Although this information is not open to scrutiny [1], high-level summaries are available [10].

The United States has recently declassified information about stocks of military HEU and Pu [1–3], but for most other countries, this information must be inferred. Projections of civil U requirements are routinely published by the power industry [11,12], and consumption of U over these estimates may be attributed to military use. The capacity for producing HEU can be estimated using published spreadsheets [13,14], and overall production estimates for Pu are performed by studying the design and operating histories of production reactors. HEU manufacture is estimated from power consumption and the production of cascade tails. Arms-control activities provide the (classified) opportunity to count nuclear warheads, each of which can be estimated to contain some quantity of Pu and HEU.

Samples of HEU and Pu have appeared on the black market [15,16], and many of these materials can be traced to the FSU. During its breakup into several independent republics, the procedures by which legitimate control of these materials was maintained were not always followed. The problem appears to have been short-lived, and major seizures of nuclear materials (such as those listed in Table 17.1) have decreased. However, the need for greater transparency in the knowledge and control of fissile assets remains a serious concern.

TABLE 17.1
Large-Scale Seizures of Weapons-Grade Materials in the 1990s

Date	Description
October 1992	1.5 kg of HEU seized near Moscow
March 1994	3.0 kg of HEU seized in St. Petersburg
August 1994	0.4 kg of Pu seized at the Munich airport
December 1994	2.7 kg of HEU seized in Prague

Note: HEU, highly enriched Uranium.

17.4 ANALYSIS

The radiochemical analyst, when presented with a sample of HEU or Pu, is faced with the daunting prospect of determining the source of the material from the results of the forensic analyses, without a comprehensive database listing the properties of legitimately owned and controlled nuclear materials. In all probability, the analyst will not have access to information beyond that described earlier but must still be able to interpret the analytical data sufficiently to guide the progress of the investigation. Is the interdicted specimen a sample of material that was legitimately produced by an acknowledged weapons state or an industrial entity operating under international safeguards that later fell into the wrong hands, or is it the result of nuclear proliferation activities? If the latter, are there nuclear signatures in the sample that would indicate the extent of the proliferators' production capabilities and the quantities of materials that had been produced or acquired? If the former, are there signatures that might indicate whether the interdicting agency now possessed all of the diverted material, or only an aliquot serving as a sales exemplar? Without access to a comprehensive database, the analyst must use the clues inherent in the sample itself to attempt to infer the capabilities of the originators of the material.

Thus, the analysts gain access to any information within the sample that they are clever enough to extract. These data include the isotopic composition of the host material, the relative concentrations of ingrown daughter and granddaughter activities, radioactive contaminants and their distributions throughout the matrix, and stable inorganic contaminants and their distributions.

The isotopic composition of the host material is a strong indicator of whether it was obtained by diversion from a legitimate owner or was the consequence of proliferation activities. In a sample of HEU, the presence of long-lived U isotopes that do not occur in nature (^{232}U, ^{233}U, and ^{236}U) is an indicator that the originating agency had operating nuclear reactors, was engaged in fuel reprocessing, and had an insufficient supply of U to operate separate U enrichment and Pu production facilities at some point in the past. The bulk of the US and FSU supplies of HEU contain easily measurable quantities of ^{232}U and ^{236}U. The absence of these nuclides in an HEU sample, termed "virgin" HEU, is an indication of either a small-scale enrichment program or a historically well-supplied one. The ^{236}U content of US and FSU stocks of HEU is usually measured in tenths of a percent; higher levels of ^{236}U in HEU imply that the source of the material is engaged in the reprocessing of research-reactor or naval-propulsion fuel.

Interpretation of the isotopic content of a Pu sample is best accomplished by the application of a reactor-inventory computer code. The operation of one such code (ORIGEN2) was described in Chapter 3 on engineering issues. The utility of this calculation is directly dependent on the accuracy and availability of cross-section libraries for each reactor type. The relative quantities of ^{238}Pu, ^{239}Pu, and ^{240}Pu are strongly influenced by the reactor type, the total burn-up, and the enrichment of the parent-U reactor fuel. Once the reactor type has been established, the ^{242}Pu content can be used to establish the flux at which the reactor operated. Comparison of measured ^{241}Pu ($t_{1/2}$ = 14.35 years) with that predicted by the code gives an estimate of the time when the fuel was discharged from the reactor. The older the Pu, the more likely that it was produced by an acknowledged weapon state. Plutonium of weapons-grade composition is unlikely to have derived from any application related to the civilian fuel cycle, and it is consequently an indicator of military reprocessing activities. A power reactor undergoing intermittent operation required by frequent refueling for weapons-grade Pu production would certainly be noticed by counterproliferation and intelligence assets.

Chronometry (the interpretation of the concentrations of ingrown daughter radionuclides in a sample as a set of ages) establishes the date at which a sample of mixed U or Pu isotopes was last purified. Within a set of ages measured for a given sample, an age that appears anomalously long is likely an indicator of a chemical process that was not specific for the daughter activity. For instance, the PUREX process is not very effective at removing Np from the Pu product and often leaves a small amount of Am contaminant as well. The age derived from the relative concentrations of ^{241}Am and ^{241}Pu is often slightly larger than that derived from the relative quantities of U and Pu, and the ^{237}Np/^{241}Pu age is usually older still. This disequilibrium pattern is a key to identifying the chemical operation used in the reprocessing. On the basis of the published literature and other considerations, it is most likely that any modern production facility constructed to reprocess spent fuel on a large scale would be designed around the PUREX process. A Pu sample that was derived by other means is either very old (globally, PUREX replaced other reprocessing technologies in the 1950s) or is the result of small-scale, experimental batch operations. In this case, the age of the sample estimated from the comparison of the measured ^{241}Pu content and a reactor calculation becomes critical.

Chronometry based on granddaughter activities is a means of detecting whether the material has intentionally been altered to look older through the addition of daughter activities. This spoofing tactic is very difficult to implement and is almost guaranteed to generate an inhomogeneous product characterized by an array of inconsistent ages. The detection of any but the clumsiest attempt at "age-tailoring" (usually through the addition of ^{241}Am to a Pu sample) is an indication of a high level of sophistication of the originating agency.

Another way to determine the chemical processes by which a sample was produced is through the measurement of nuclides that are residual from the parent environment, rather than those arising from decay processes of the matrix material itself. These can be remnants from the parent ore, from the action of a nuclear reactor (e.g., fission products), or from metallurgical processes. A recovered radioelement that is composed of both fission-product isotopes and the isotopic mixture found in nature can be deconvoluted to determine the scale of the reprocessing operations giving rise to the sample, under the assumption that reagent-grade chemicals were used.

The relatively high level of fission-product activity in the Bulgarian sample of HEU (described in Chapter 20) may have resulted from scaling down an industrial process designed for the recovery of LEU. The scaling could have been required by the constraints of criticality safety, indicating both large-scale and intermediate-scale reprocessing operations taking place in the country of origin. Residual ^{242m}Am and ^{243}Am in Pu samples, remaining from incomplete fuel reprocessing, are strong indicators of reactor flux when used in conjunction with a reactor-inventory code. Plutonium produced from near-natural U via high-flux irradiation most likely derives from a weapons state, as does that produced by very low flux irradiation. The latter situation probably indicates Pu recovered from a depleted-U blanket mounted around a high-flux reactor core.

The production of metal and the forming of metal shapes is another indication of the sophistication of the fabricating agency. Examination of the microstructure of metals and alloys, and the determination of their metallic phases, provide a variety of clues. Rolling and extruding operations often introduce contaminants onto the surfaces of fabricated parts. For example, Oralloy foils obtained from Oak Ridge often have a surface contamination of weapons-grade Pu at the ppb level. Even without knowledge of where the material came from, this juxtaposition of radionuclides proves that the plant producing the HEU metal is engaged in nuclear weapon activities and not solely in the fabrication of propulsion-reactor fuel.

The United States and FSU produced Pu in streams, the isotopic contents of which evolved slowly with time. Plutonium weapon parts were likely produced from a single sample of the stream, taken at a given production time. The recycling of retired stockpile components provides the means by which disparate samples are combined. We assume that small-scale proliferant operations, performed on a batch scale with a limited source of input material, will result in parts with an identifiable source (e.g., a given reactor signature), even if more than one batch is included in the fabrication of the part. Recycled weapon material from an established weapon state may contain Pu from widely different sources (e.g., a Savannah River reactor blanket and a light-water-cooled, graphite-moderated fuel assembly). The analyst should always interrogate any weapons-grade metal sample for inhomogeneities.

The focus here has been on signatures of weapons-grade materials, but the same arguments also apply to other matrices. Bulk U and Pu samples have the highest concentrations of signature species and are, consequently, the easiest subjects for a nuclear forensic analysis. However, one of the case studies at the end of this book (Chapter 21) describes the analyses of a set of site-inspection debris samples that allowed the reverse-engineering of the US uranium enrichment complex. In terms of inferring scale of operations from the signatures in a sample, this case study is an excellent example. In addition, the Bulgarian HEU investigation was a good example of the productive forensic examination of nonnuclear, collateral evidence.

REFERENCES

1. D. Albright, F. Berkhout, and W. Walker, *Plutonium and Highly Enriched Uranium 1996: World Inventories, Capabilities and Policies*, Oxford University Press, New York, 1997.
2. Department of Energy, *Plutonium: The First 50 Years*, Department of Energy, Washington, DC, 1996.
3. D. Albright and K. O'Neill, eds., *The Challenges of Fissile Material Control*, Institute for Science and International Security Press, Washington, DC, 1999.
4. A.S. Krass, P. Boskma, B. Elzen, and W.A. Smit, *Uranium Enrichment and Nuclear Weapons Proliferation*, Taylor & Francis, New York, 1983.
5. International Atomic Energy Agency, *Guidebook on Spent Fuel Storage*, 2nd ed., Technical Reports Series No. 240, International Atomic Energy Agency, Vienna, Austria, 1991.

6. European Atomic Energy Community, *A Plutonium Recycling Scenario in Light Water Reactors*, Commission of the European Communities, European Atomic Energy Community, Harwood Academic, London, U.K., 1982.
7. S. Blumkin and E. von Halle, *A Method for Estimating the Inventory of an Isotope Separation Cascade by the Use of Minor Isotope Transient Concentration Data*, Oak Ridge National Laboratory Report K-1892, Oak Ridge, TN, 1978.
8. International Atomic Energy Agency, *Reactor Burn-Up Physics*, Panel Proceedings Series, International Atomic Energy Agency, Vienna, Austria, 1973.
9. A. Chayes and W.B. Lewis, eds., *International Arrangements for Nuclear Fuel Reprocessing*, Part III, Ballinger Publishing, Cambridge, MA, 1977.
10. International Atomic Energy Agency, *Country Nuclear Fuel Cycle Profiles*, Technical Reports Series No. 404, International Atomic Energy Agency, Vienna, Austria, 2001.
11. Energy Information Administration, *World Nuclear Fuel Cycle Requirements 1990*, Department of Energy Report DOE/EIA-0436 (90), Government Printing Office, Washington, DC, 1990.
12. Organization for Economic Cooperation and Development (OECD), *Nuclear Energy and Its Fuel Cycle, Prospects to 2025*, Nuclear Energy Agency, OECD, Paris, France, 1982.
13. International Atomic Energy Agency, *Manual on the Projection of Uranium Production Capability, General Guidelines*, Technical Reports Series No. 238, International Atomic Energy Agency, Vienna, Austria, 1984.
14. Supply and Demand Committee, Uranium Institute, *The Balance of Supply and Demand, 1978–1990*, Supply and Demand Committee, Uranium Institute, Mining Journal Books, East Sussex, England, 1979, Appendix E.
15. W.J. Broad, Preparing to meet terrorists bearing plutonium, *New York Times*, Sunday, August 1, 1993.
16. C.R. Whitney, Germany seizes nuclear material, *San Francisco Examiner*, Sunday, August 14, 1994, p. A-2.

18 Materials Profiling

18.1 CRIMINALISTICS COMPARISONS

In conventional forensic science for criminalistics applications, the assessment of latent human (dermatoglyphic) fingermarks and palm prints is a long-standing and valuable investigative technique [1]. It is fundamentally an exercise in pattern recognition, where the print/mark ridge characteristics of a questioned specimen are compared to those of one or more suspects or to those of a larger subject population within a searchable database. A computer is often used to narrow the number of comparison possibilities, but the ultimate determination of match quality is made by a professional print examiner. For criminal prosecution, the minimum number of equivalent ridge points required to declare a match between questioned sample and known print can vary among different jurisdictions. Nonetheless, print visualization and examination have been productive forensic techniques for more than a century, and the underlying tenet remains that no two individuals have exactly the same ridge attributes.

Over the last 30 years, another individualization method based on genetic variation has developed with extraordinary success. It was termed DNA "fingerprinting" by its pioneers [2]. Several comprehensive overviews of forensic DNA typing, both nuclear and mitochondrial, are available in the literature (e.g., [3,4]), and more in-depth information can be found in them. The fundamental principle of DNA fingerprinting is that the specific sequence of A, T, G, and C bases in DNA establishes all of the genetic characteristics of an individual life-form. An important historic technique of this science is restriction fragment length polymorphism (RFLP) analysis, and it has vast discriminating power between different samples. However, increasingly popular today is the PCR-STR method, which entails an initial multiplex polymerase chain reaction (PCR) procedure. Then, individualizing lengths of short tandem repeats (STRs) in noncoding regions of DNA are explored. As little as 1 ng of DNA is extracted from a sample, and a region containing each STR is PCR-amplified and separated by capillary electrophoresis on the basis of the number of repeats of a short DNA sequence (allele). The result is an overall profile of STR alleles (genotype), and the statistical strength of such DNA comparisons increases with the number of measured STRs. For the 13 core STR loci used in the United States for a 13-STR profile, in conjunction with the statistics of FBI-specified allele frequencies across ethnic populations, the probability is 10^{-15} that two unrelated Caucasians, for example, would have indistinguishable STR profiles. Individual assessment against a larger populace is performed via national databases (e.g., CODIS).

18.2 MATERIAL COMPOSITIONS

In analogous fashion, a forensic scientist may thoroughly characterize a questioned material, perhaps by measuring trace-element concentrations or by determining the chemical speciation of the components that make up the sample. Such signature data comprise a material profile or "fingerprint" of the specimen, suitable for comparison against other samples to evaluate any possibilities of common origin. In essence, a material fingerprint is a distinctive profile or blueprint that elementally, and/or isotopically, and/or chemically embodies the composition of the specimen, and which generally conveys as much information as possible. These measurements are often quite valuable for forensic comparisons of different specimens. If two samples differ appreciably in a number of material comparison points, it is probable that they did not have a common origin. However, if they are very alike in composition, they may have had a common origin, but a potential accidental match must be assessed via an appropriate database of known materials measurements. The nuclear forensic analyst may well use any of the conventional assay techniques for an investigation, as well as comparisons of radioactive species and isotope-ratio values. The specific nature of questioned evidence will drive the choices for appropriate applied technologies.

As opposed to dermatoglyphic and DNA comparisons, a designated match in materials signatures must be tempered somewhat in its resulting interpretation and influence. In the first place, comparison databases for many questioned materials pertinent to nuclear forensic investigations, if they exist at all, are generally far inferior to those for human fingermarks and DNA. Further, a match of material profiles means only that no statistically significant discrepancies were measured between two specimens by whatever technical means were employed. Although the samples may have in fact been different, the conducted examinations merely failed to disclose the disparities.

Historically, in materials composition analyses, the final evaluation was subjective and based on the experience and judgment of the investigator. Typically, bar graphs of element concentrations, or scatter plots of isotope-ratio data, might be inspected side by side, and statements such as "the two samples look consistent to me" or "they appear too different" would be an end result. When the number of points of comparison is few, this range-overlap tactic could be a reasonable one, and for interpretation or courtroom presentation, the forensic result would usually be prefaced with "in my expert opinion." However, with modern analytic instrumentation such as ICP-MS or the particle microprobes, rigorously quantitative elemental and isotopic measurements of as many as 40–60 different species are routine. In such instances, an "eyeball" approach to meaningful match decisions is considerably more problematic.

However, for contemporary nuclear forensic analysis (NFA) methods that normally entail only qualitative measurements (e.g., GC–MS), visual inspection of larger data tables may still prove valuable. For example, Table 18.1 gives the results of signature species determined by SPME/GC–MS route analyses (see Chapter 16) for two questioned high-enriched uranium (HEU) samples. In this particular investigation, only toluene was measured in both specimens, and even with two dozen comparison points for consideration, a qualitative conclusion of dissimilar sample histories was straightforward and could be made subjectively with confidence. Moreover, as

TABLE 18.1
Organic Species in Questioned HEU Samples That Were Absent in Blank Analyses

t_R (min)	Tentative i.d. (Fit Quality)	Q1	Q2
1.51	Methyl propene (87)		X
1.83	Pentene (49)		X
2.31	Butanal (87)		X
3.81	Pentenal (43)		X
5.12	Tetramethyl tetrahydrofuran (53)		X
5.22	Dimethyl hexane (64)	X	
5.40	Pentanal (64)		X
5.41	Methyl heptene (76)	X	
5.57	Toluene (93)	X	X
5.93	Octane (70)		X
6.16	Hexenal (70)		X
7.03	Substituted cyclohexane		X
7.28	Methyl cyclopentyl ethanone (80)	X	
7.86	Dimethyl heptene (62)		X
8.00	Ethyl benzene/xylene (97)	X	
8.27	Heptanone (64)		X
8.41	Ethyl heptene (53)	X	
9.69	Camphene (96)		X
9.96	Benzaldehyde (97)		X
10.50	Methyl methylethyl cyclohexene (97)		X
10.68	Phenol (93)	X	
11.63	Methyl methylethyl benzene (97)	X	
14.11	Camphor (86)	X	

discussed later, elevating such data to semiquantitative value allows nonparametric statistical calculation of a Spearman coefficient to numerically assess correlation. Such computation is important when the degree of overlap between comparison species is not as clear-cut as depicted by Table 18.1.

18.3 CALCULATIONS

A modern tactic for analytic complexity is multivariate statistical analysis via one or more techniques in a diverse array of chemometric methods [5–7]. Thus, different investigators have variously applied principal component analysis (PCA), factor analysis, nearest-neighbor analysis, Hotelling's T^2 test, artificial neural networks, Hierarchical clustering, the Bayesian model, and other approaches to materials profiling for determinations of common provenance. These methods have been implemented in an assortment of subject areas, including glass [8,9], cartridge casings [10], minerals [11], heroin [12–14], environmental studies [15], amphetamines [16], and accelerants [17].

At Livermore, chemometrics have been effectively applied to measurements of comparative signature species for UOC samples (uranium ore concentrate or "yellowcake") [18]. Elemental, isotopic, and some speciation data are collected from both known and questioned UOC specimens, the former providing a large, vendor-linked training set in a relational database for geographic sourcing. Multidimensional data visualization and potential clustering are first constructed by dimension reduction with a PCA transformation. Attribution of samples of unknown or unconfirmed origin is then performed by autoscaling the data and narrowing possible production locations via an iterative, partial-least-squares discriminant analysis. The search engine for such database comparisons at LLNL is the Internet Discriminant Analysis Verification Engine (iDAVE) program.

18.3.1 Quantitative Data with Uncertainties

Nearly 50 years ago, Parker published two papers on this problem that were relatively lucid and statistically rigorous [19,20]. Although questioned-specimen comparisons to a large database are efficiently performed via chemometrics, Parker's systematics are valuable in NFA for comparing two specific samples for commonality through precise quantitative measurements of identical analytes. Parker's treatment also weights the multivariate input data by their empirical uncertainties. Such error propagation is a worthy tactic for modern forensic measurements, which often encompass major, minor, trace, and ultratrace species that can span widely varying relative-error values. The situation is often one in which material attributes (e.g., element concentrations) are numerically measured experimentally, with the resulting data subject to disparate relative standard deviations. These quantified variables must be completely independent of each other for correct treatment, however, and covariance information can be obtained by Pearson product-moment correlation statistics [9]. If two or more input values are found to be correlated, all but one are eliminated from the data set before the multivariate calculations are performed.

Parker's treatment divides the computation into two parts: the similarity problem and the identification problem. The former is the quantitation of the hypothesis that two specimens may have the same origin. The latter quantifies a probability that the overlap of the measured features does, in fact, imply a common origin, and it is dependent on the development of appropriate comparison databases. Two samples must be sufficiently similar before database considerations need be contemplated, however.

For direct comparison of two samples, the similarity problem considers whether both items might have a common origin. Let a questioned specimen (q) be characterized by n measurements, q_i, for $i = 1 \rightarrow n$ (i.e., a total of n attributes, i, have been measured for the specimen). Similarly, let a known comparison specimen (k) be characterized by k_i, $i = 1 \rightarrow n$. Let the difference between a given attribute measured in both q and k be denoted by d:

$$d_i = q_i - k_i$$

with error

$$\sigma_{di} = \left(\sigma_{qi}^2 + \sigma_{ki}^2\right)^{1/2}$$

For a common origin ($q = k$),

$$\sum_{i=1}^{n} d_i$$

would be Gaussian with mean = 0.

Parker defines a Discrepancy Index, C:

$$C \equiv \sum_{i=1}^{n} f_i^2 = \sum_{i=1}^{n} \left(d_i/\sigma_{d_i}\right)^2,$$

which is the sum of squares of n random normal deviates (and the square of the weighted, n-dimensional Euclidean distance). As all measured i are mutually independent, C is distributed as χ^2 with n degrees of freedom, small if $q \simeq k$, and large if $q \neq k$.

The variable C is then compared to a threshold value C_0, which is a function of the number of comparison attributes, where $C < C_0$ corresponds to a similarity match. The object is to select a value of C_0 with an associated small probability ε (denoted P in some texts), such that if q and k have the same origin, the probability that $C > C_0$ by chance is ε. That is, if q and k truly derive from a common source, the probability of obtaining a value C (or larger), that is greater than C_0, is less than ε. Therefore, $(1 - \varepsilon)$ is the probability that $C < C_0$ corresponds to a true similarity match. The critical values C_0 are given in statistical tables as the χ^2 distribution as functions of n and ε. A perfect similarity match would have probability $\varepsilon = 100\%$, and the most conservative similarity calculations derive from measurements having the highest precisions and accuracies.

However, $C < C_0$ is a necessary but insufficient condition for common origin. The identification problem must then be addressed, and comparison with a database is necessary. Parker thus defines a sensitivity criterion, Q, as an index that measures attribute efficiencies and a figure-of-merit ($\sigma_{di}\ g_i$), where a small value denotes good sensitivity for analyte or attribute i. Ensuing mathematics incorporates the Gaussian probability distribution and population statistics, a ($\sigma_{di}\ g_i$) product formulation, and the Γ function. Parker provides all individual data values for a detailed example problem in Tables 1 and 2 of his second paper (treating $n = 10$ trace-element concentrations from 30 individuals) [20].

An interesting, alternate approach to materials comparisons was suggested by Koons and Buscaglia [9]. In a study of 204 glass fragments via refractive index and the concentrations of 10 elements, the authors concluded that implementing a large number of highly discriminating measurements for trace-evidence characterization results in such low probability of two unrelated samples being indistinguishable that extensive comparison databases for exact probability calculations are unnecessary. Although this concept seemed intuitively reasonable, it was soon called into question by others [21].

18.3.2 Semiquantitative Data

Routine GC–MS analysis in NFA is qualitative in nature, and an analyst is generally satisfied with dependable identifications of detected species based on comparisons of their mass-spectral data to those in reliable libraries. Of course, quantitative analyses by GC–MS are also possible. They require a known analytic reference material and the measurement of GC retention time and (at least) three dominant MS ions for each chemical species of interest. Empirical calibration curves may then be determined by gravimetry, serial dilution, and the GC peak areas for each compound to be quantitated. However, as there may be hundreds of signature analytes in a given NFA GC–MS profile, quantitative calibrations for each species in these situations are generally prohibitive.

An established statistical method exists for the potential connection between two samples, one that requires only relative GC peak areas of semiquantitative value. These data are routinely measured in standard qualitative GC–MS runs. Under identical experimental conditions, rigorous statistical analysis of the similarity between two specimens may be performed by means of the nonparametric and distribution-independent Spearman correlation coefficient. The method is a semiquantitative comparison that examines the statistical dependence among variables with respect to the intensity and direction of their associations. Using the empirical analytes determined in the samples, and their relative peak areas, the Spearman ρ is calculated as

$$\rho = 1 - \left[6 \sum d_i^2 / n\left(n^2 - 1\right) \right],$$

where d is here the difference between the *ranks* of each measurement on the two samples (after converting the n raw data to rank scores). The ranked data are then compared to assess correlation between the variables. Similar (and related) to the Pearson coefficient, $\rho = +1$ indicates a perfect positive correlation, while $\rho = 0$ signifies no correlation ($\rho = -1$ would designate the identical chemical compounds with perfect reverse-order rankings). Reference [22] provides a good example of the application of the Spearman coefficient to relative GC–MS data.

REFERENCES

1. H.C. Lee and R.E. Gaensslen, *Advances in Fingerprint Technology*, 2nd ed., CRC Press, Boca Raton, FL, 2001.
2. A.J. Jeffreys, V. Wilson, S.L. Thein et al., Individual specific "fingerprints" of human DNA, *Nature*, 316, 76, 1985.
3. K. Inman and N. Rudin, *An Introduction to Forensic DNA Analysis*, CRC Press, Boca Raton, FL, 1997.
4. B.S. Weir, ed., *Human Identification: The Use of DNA Markers*, Kluwer Academic, Dordrecht, the Netherlands, 1995.
5. K. Esbensen, S. Schonkopf, and T. Midtgaard, *Multivariate Analysis in Practice*, Camo AS, Trondheim, Norway, 1994.
6. K.R. Beeve, R.J. Pell, and M.B. Seasholtz, *Chemometrics: A Practical Guide*, Wiley, New York, 1998.
7. D.L. Massart, B.G.M. Vandeginste, S.N. Deming et al., *Chemometrics: A Textbook*, Elsevier, New York, 1999.
8. J.M. Curran, C.M. Triggs, J.R. Almirall et al., The interpretation of elemental composition measurements from forensic glass evidence: I and II, *Sci. Justice*, 37(4), 241, 245, 1997.

9. R.D. Koons and J. Buscaglia, The forensic significance of glass composition and refractive index measurements, *J. Forensic Sci.*, 44(3), 496, 1999.
10. C.L. Heye and J.I. Thornton, Firearm cartridge case comparison by graphite furnace atomic absorption spectrochemical determination of Ni, Fe, and Pb, *Anal. Chim. Acta*, 288, 83, 1994.
11. R.J. Watling, Novel application of laser ablation inductively coupled plasma mass spectrometry in forensic science and forensic archaeology, *Spectroscopy*, 14(6), 16, 1999.
12. R. Myors, R.J. Wells, S.V. Skopec et al., Preliminary investigation of heroin fingerprinting using trace element concentrations, *Anal. Commun.*, 35, 403, 1998.
13. R.B. Myors, P.T. Crisp, S.V. Skopec et al., Investigation of heroin profiling using trace organic impurities, *Analyst*, 126, 679, 2001.
14. S. Klemenc, In common batch searching of illicit heroin samples—Evaluation of data by chemometrics methods, *Forensic Sci. Int.*, 115, 43, 2001.
15. G.W. Johnson and R. Ehrlich, State of the art report on multivariate chemometric methods in environmental forensics, *Environ. Forensics*, 3, 59, 2002.
16. M. Praisler, I. Dirinck, J. Van Bocxlaer et al., Pattern recognition techniques screening for drugs of abuse with gas chromatography-Fourier transform infrared spectroscopy, *Talanta*, 53, 177, 2000.
17. B. Tan, J.K. Hardy, and R.E. Snavely, Accelerant classification by gas chromatography/mass spectrometry and multivariate pattern recognition, *Anal. Chim. Acta*, 422, 37, 2000.
18. M. Robel, M.J. Kristo, and M.A. Heller, Nuclear forensic inferences using iterative multidimensional statistics, *LLNL-CONF-414001*, Livermore, CA, 2009.
19. J.B. Parker, A statistical treatment of identification problems, *J. Forensic Sci. Soc.*, 6(1), 33, 1966.
20. J.B. Parker, The mathematical evaluation of numerical evidence, *J. Forensic Sci. Soc.*, 7(3), 134, 1967.
21. J.M. Curran, J.S. Buckleton, and C.M. Triggs, Commentary on Koons, RD, Buscaglia J (Ref. [9]), *J. Forensic Sci.*, 44(6), 1324, 1999.
22. A.M. Curran, P.A. Prada, and K.G. Furton, The differentiation of the volatile organic signatures of individuals through SPME-GC/MS of characteristic human scent compounds, *J. Forensic Sci.*, 55(1), 50, 2010.

19 Source and Route Attribution

19.1 INTRODUCTION

Attribution is the end product of nuclear forensic analysis (NFA)—the integration of all relevant technical, analytical, and other (e.g., human intelligence) information about an incident into a consistent and insightful interpretation, leading to the identification and apprehension of the perpetrators [1–4]. Nuclear attribution integrates all relevant forms of evidence and information about a nuclear smuggling or illicit trafficking incident to produce a report that can be readily analyzed and interpreted and that will form the basis of a confident and meaningful response to the incident. Nuclear attribution uses inputs from many sources, including results from forensic sample analyses; an understanding of radiochemical, geochemical, and environmental signatures; knowledge of the methods of Special Nuclear Material (SNM) production and nuclear weapons development pathways; and information from intelligence sources and law enforcement agencies. The objective of attribution assessment is to satisfy policy makers' information needs, requirements, and questions within the framework of any given incident.

Nuclear attribution allows government agencies to discover and understand illicit trafficking in nuclear/radiological materials before perpetrators' plans progress to a point of actually developing, deploying, and exploding a nuclear weapon or radiologic device. An early clue would be the discovery of efforts to obtain nuclear or radiological material. It is important to pursue cases that at first glance do not appear to be especially serious, such as those involving, for example, small quantities of material or low-enriched uranium (LEU), because they may be linked to more serious threats that could emerge later. That is, such specimens could be nascent precursors. The early discovery of such efforts may represent the optimum opportunity, allowing sufficient time to stop nefarious plans before they progress to completion with potential attendant damage to life, property, and national security.

Attribution begins with NFA, the process by which intercepted illicit nuclear materials and any associated evidence (such as containers, documents, and packaging) are analyzed to provide clues to the origin and source of the interdicted material. The goal of NFA is to identify attribution indicators in an interdicted nuclear or radiologic sample, or in its surrounding environment. These indicators arise from known relationships between material characteristics and illicit activities. Thus, NFA entails more than merely characterizing the material and more than just a simple determination of its physical nature; successful NFA requires understanding the relationships between measured properties and the production and use-history of nuclear materials.

Attribution is best understood by separately considering two distinct but interrelated components: source and route evidence. Source attribution addresses questions related to the origin and original intended use of the interdicted radioactive material, and is based on analyses of the physical, isotopic, and chemical properties of the nuclear matter. Route attribution addresses questions related to the history of the material once it was diverted from the legitimate owners (e.g., how did it get from the point at which control was lost to the point of interdiction, and who were involved?) and relies on analyses of isotopic, chemical, and biological properties of the associated nonnuclear material. It is important to note that a number of first-order questions draw on both source and route attribution (e.g., identifying the point of loss of control of the nuclear material). Both route and source attribution are commonly referred to as "geolocation."

Analyses of the nuclear materials themselves are expected to provide information most directly related to the source of the evidence (i.e., production methods and processes), whereas analyses of the associated collateral materials are expected to be linked more closely to route considerations. The questions that may be asked by decision-making bodies can also be related to both source and route, as indicated in the following text.

19.1.1 Source Attribution Questions

1. Where did the materials originate?
2. What are the materials (SNM, weapons grade, etc.)?
3. Were the materials diverted from a legitimate pathway?
4. Where were the materials obtained?
5. Where was legitimate custody lost?
6. What is the potential for more illicit material from this source?

19.1.2 Route Attribution Questions

1. Are the characteristics of the route unique?
2. What was the route taken by the interdicted materials?
3. Was this an isolated event, or one in a series of shipments?
4. Who are the perpetrators?
5. What is the potential end-use (e.g., nation-state weapons program, subnational terrorist group, organized crime, etc.)?

Forensic analyses alone cannot answer all of these questions. Rather, providing the most complete answer requires that the results of forensic analyses be integrated with the knowledge of radiochemical and environmental signatures, a broad understanding of nuclear materials and weapons production, intelligence leads, and all other available information. This ensemble of information must be interpreted from several perspectives to obtain the most consistent and meaningful attribution assessment. Attribution estimation comprises the integration of all relevant forms of information (including nuclear forensic data) about a nuclear smuggling incident into a consistent and meaningful view, and this assessment forms the basis of a confident response to the incident. The goal of attribution assessment is to answer critical questions for policy makers within the timeframe of a given incident [1].

19.2 FORENSIC ANALYSIS OF INTERDICTED NUCLEAR MATERIALS

Two important keys to success in a forensic investigation are the ability to measure data both accurately and timely, and the ability to interpret those data in terms of a plausible scenario. The forensic process begins with detection of the incident and an on-site evaluation. The importance of on-site assessments, field analyses, and proper sample collection cannot be overstated. Given the suspected presence of radioactive materials, it is essential that safety and security be given top priority as well at the recovery site. Effective investigation in the presence of nuclear materials requires the capability to handle potentially large quantities of radioactivity while preserving trace quantities of nonnuclear materials of possibly high forensic value. Critical information linking the source and route of illicit materials to the responsible parties may be lost at this point if the collection opportunity is not exploited fully. Forensic analysis of evidence collected at a crime scene provides the only direct source of information to the attribution assessment. As with conventional forensic criminal investigations, the existence of a critical piece of evidence leading to source and route determination may not be initially apparent, but must be developed during the course of an investigation. Prescreening and general examination of materials collected from the scene of an illicit nuclear materials case are therefore critical steps for every forensic analysis. To obtain maximum information, experienced nuclear forensic investigators should be engaged to participate at the site evaluation whenever possible.

Once field samples are properly collected and packaged, they must be transported to a receival location at an appropriate facility capable of satisfying the requirements for both nuclear safety and preservation of forensic evidence. Prescreening by expert investigators will be used by the nuclear forensic team to determine the likely course of further analyses. Depending on the specific protocols deemed necessary, subsamples of the specimen and pertinent associated materials will be sent to the appropriate technical analysts. After analysis, the technical results should be made available to the entire nuclear forensic team for discussion and interpretation.

Evaluation of techniques chosen to conduct forensic analyses for the illicit trafficking of nuclear materials should be performed within the context of the questions posed by law enforcement agencies, intelligence communities, and decision-makers. In other words, the value of a particular technique should be related to its utility in answering the various attribution questions. The task of providing decision-makers with a timely and accurate assessment of an illicit nuclear material incident requires that the general approach be flexible. Using forensic protocols under the direction of an appropriate team of knowledgeable professionals provides the necessary flexibility. This team must be prepared to use whatever materials or information become available. Therefore, an incident-specific analytical procedure should be used, rather than a prescribed, comprehensive one. A general, prearranged approach would provide neither a cost-effective nor a timely response to the critical questions posed after a nuclear event.

To respond to the incident-specific nature of nuclear forensic investigations, most laboratories follow a tiered approach in which sample interrogation proceeds in stages, from routine nondestructive analyses to highly sophisticated, microanalytical characterizations that apply state-of-the-art instrumentation. The analytical protocol is structured to ensure that each successive level of study is based on previously acquired results, allowing the interrogation process to be halted as soon as sufficient data have been acquired to meet the requesting agency's requirements [3–6]. Following is a broad-brush outline of the general analytical approach used for a nuclear forensic investigation; a specific set of analyses is fine-tuned at later time, on a sample-by-sample basis, to address particular needs or concerns. Additional discussion relating to procedures that a response team and nuclear forensic laboratory should reasonably consider, developed through an international effort, is contained in the nuclear forensic model action plan developed by the Technical Working Group and the International Atomic Energy Agency (IAEA) [5].

19.3 LABORATORY CHARACTERIZATION OF NUCLEAR MATERIALS FOR SOURCE SIGNATURES

In a stepped approach, the first-stage characterization of any interdicted nuclear or radiologic material is initiated by entering all submitted specimens into a chain-of-custody tracking system and photodocumenting all salient features. The initial characterization involves nondestructive analyses to determine the bulk chemical composition and radiological nature of the sample. Typically, these analyses encompass x-ray fluorescence (XRF) and α-particle, γ-ray, neutron, and visible/near-infrared (Vis/NIR) spectroscopies. If the presence of informative organic substances is suspected, solid-phase microextraction sampling (SPME), followed by gas chromatography–mass spectrometry (GC–MS), will be applied. SPME protocols generate an efficient, safe, and convenient extraction of volatile and semivolatile analytes for direct injection and analysis by GC–MS. These techniques allow rapid screening for signature species of all flavors of weapons of mass destruction, as well as for extraordinary industrial compounds (see Section 16.5). At the conclusion of stage 1 (typically within 1–2 days following receipt of the sample), the analytical team should have a semiquantitative, bulk chemical composition, an estimate of the abundances of major radioactive species, a preliminary assay of any organic compounds, and a detailed photographic representation. Stage 1 provides essential groundwork for the more detailed analyses conducted in stages 2 and 3.

Stage 2 begins by dissecting the sample and examining its constituent parts with techniques of successively higher sensitivity, greater accuracy, and finer spatial resolution. An important first step employs optical microscopy to examine and characterize features down to roughly 5–10 μm in size (i.e., one-tenth the diameter of a human hair); the optical microscopes are equipped with digital cameras for photodocumentation. At this point, we also locate and remove adhering particles, fibers, hair, and other trace evidence, which may carry information useful for route attribution and geolocation. (The actual analyses of these materials are conducted in stage 3.)

A scanning electron microscope equipped with an energy-dispersive x-ray detector is used to continue physical characterization down to a spatial scale of a few tens of nm, revealing such characteristics as intrinsic grain size and morphology. The main focus of stage 2 is a complete, quantitative analysis of the abundance of prominent radionuclides, as well as major and minor elements. An electron microprobe is used to determine the elemental compositions and molecular stoichiometries of elements from B to U, at concentrations down to about 100 ppmw (100 μg/g), whereas a variety of MS techniques (including glow-discharge mass spectrometry; quadrupole and magnetic-sector inductively coupled plasma mass spectrometry (ICP-MS); and secondary-ion mass spectrometry (SIMS)) are applied to measure trace-element abundances. This range of instrumentation allows forensic examiners to use optimal techniques, depending on the specific nature of a sample, to cover the full suite of elements from H to Pu, with detection limits approaching 10 parts per quadrillion (10 fg/g).

The abundances of major actinide isotopes are determined by thermal ionization mass spectrometry (TIMS) or inductively coupled plasma mass spectrometry (MC-ICPMS). Stage 2 concentrates primarily on U and/or Pu isotopic compositions, depending on the material, to establish the basic characteristics of the specimen. Stage 2 also assesses several radionuclide parent–daughter pairs, such as ^{241}Pu–^{241}Am, ^{235}U–^{231}Pa, and ^{234}U–^{230}Th, to determine the age of the sample (the time elapsed since the sample was last chemically processed). With the completion of stage 2, the analytical team should have a complete physical description of the sample on spatial scales ranging from μm to cm, a representative assay of major and trace-element concentrations, U and Pu isotopic compositions, and a preliminary age determination.

Stage 3 entails analyses of oxygen, key low-abundance isotopes (including fission products), Sr, Pb, Nd, trace radionuclides (e.g., ^{3}H, ^{129}I, ^{226}Ra, ^{231}Pa), and signature organic species indigenous to the sample. This stage also involves chemical, isotopic, and trace-element characterization of any extraneous (environmental) material, such as particles, fibers, and hair, removed and isolated during stage 2. The measurement of trace radionuclides, including fission products and activation products, requires highly specialized radiochemical separation procedures to isolate purified aliquots of each species of interest. Radiochemical methods have been developed for the isotopes of the heavy elements Ra, Ac, Th, Pa, U, Np, Am, and Cm, as well as for the common activation and fission-product nuclides. Purified fractions are analyzed using α-particle and γ-ray spectroscopy and mass spectrometry (either TIMS or MC-ICPMS). During stage 3, the full suite of actinide radionuclides (Ra–Cm), plus several fission products (e.g., Zr, Cs, Sm), should be measured. As required, the capabilities of accelerator mass spectrometry (AMS) to measure ultratrace, low-specific-activity radionuclides, such as ^{10}Be, ^{14}C, ^{26}Al, ^{36}Cl, and ^{129}I, can also be employed [7]. GC–MS interrogates questioned specimens for organic compounds indicative of nuclear processing (e.g., TBP and organic radiolysis products) or other weaponization activities (e.g., high-explosives testing). Ultrahigh-performance liquid chromatography (uHPLC), coupled with electrospray-ionization tandem MS, can also be used to search for evidence of large biomolecules, including biotoxins.

Stage 3 also engages the measurement of stable isotope abundances of elements commonly associated with geochemical studies: C, N, O, Sr, Nd, and Pb. These isotopic systems have the potential to provide a wealth of information on source locations by capitalizing on data collected over ~50 years of geochemical investigations of rocks and minerals. Variations in stable-isotope abundances reflect worldwide systematic characteristics driven by age differences between ore bodies, as well as by isotope fractionation effects typical of transfer reactions between liquid and gaseous reservoirs (e.g., oceans and the atmosphere [8,9]). Data obtained in stage 3 also allow the application of several different age-dating schemes and the reconstruction of the neutron-exposure history. At the conclusion of stage 3, sufficient information should have been collected to perform a credible assessment of source attribution.

The forensic investigation of the HEU sample seized in Bulgaria, described in the case study in Chapter 20, provides an excellent example of the manner in which information collected on both nuclear and nonnuclear materials, using a variety of analytical techniques, is assembled to perform an attribution assessment. This investigation focused principally on source attribution, by both nuclear and conventional forensic analyses, insofar as information on route was available from other venues. A second illustrative example, describing how NFA was used to reconstruct the history of a Pu sample found in a US waste dump, may be found in [10].

19.4 LABORATORY CHARACTERIZATION OF NUCLEAR MATERIALS FOR ROUTE SIGNATURES

The characterization of any nonnuclear collateral material obtained in an interdiction is directed toward a fundamentally different goal than the analyses of nuclear material described earlier. Here, the focus is on route attribution, gathering clues to decipher the history of a sample from the time it passed from legitimate control until it was intercepted by a law-enforcement agency. Route attribution is based on the fact that the earth is heterogeneous with respect to the distribution of elements, isotopes, flora, and fauna, and that correlations between these signatures can be used to identify local geographic regions with high specificity. In its simplest form, route attribution relies on the identification of unique or unusual material or species that point unambiguously to a specific geographic location. For example, the rare mineral hibonite is mined almost exclusively on the island of Madagascar [11]. The observation of hibonite in dust or particulate debris in packaging associated with interdicted nuclear material would establish with a high degree of confidence that the sample had come from, or passed through, Madagascar. Similar specificity can be obtained in favorable cases through observations of pollen and microfossils [12].

Although potentially very powerful, this approach to route attribution is limited by the small number of unique, site-specific materials (minerals, pollens, microfossils). More commonly, route attribution uses analyses of major and trace elements, as well as stable and radiogenic isotopes, in soil particles, hair, fibers, vegetation, and pollen—basically anything that can be linked to specific geographic locations or manufacturing/production facilities. These analyses use many of the same analytical tools described for source attribution, but now applied to nonnuclear materials. Techniques such as the scanning electron microscope equipped with an energy-dispersive x-ray detector, SIMS, laser-ablation ICP-MS, and laser-ablation ion-trap MS are especially useful because of their ability to determine the composition of signature species in situ (e.g., on the exterior surfaces of shipping containers or wrapping paper). For more subtle and indirect investigations, techniques for the analyses of ultratrace chemical and radionuclear analytes in human and animal hair may be applied. Hair specimens can provide a faithful record of exposure to a number of weapons-of-mass-destruction signature species [13–15].

The success of geolocation and route attribution depends to a large degree on the ability to associate characteristic signatures of a sample, revealed through detailed analyses, with specific geographic locations. Merely collecting data is insufficient. Access to comprehensive relational databases containing information on a variety of environmental parameters (e.g., stable isotope abundances [H, C, N, O, S, Sr, and Pb], trace-element concentrations, pollen and microfossil habitats), as well as commercial and manufacturing signatures (paint compositions, common metal alloys, paper compositions, etc.), is required in order for geolocation and route attribution to succeed. The importance of choosing optimal analytical tools and identifying suitable database resources is discussed in the following paragraphs.

First, a set of forensic tools for route attribution is identified. This ensemble of techniques, which may be determined initially by a specific case, may then be viewed as a basic toolkit. As experience is gained, the toolbox may be subsequently expanded. A forensic tool is categorized in terms of a particular analytical measurement on a particular type of material, and data provided through application of the tool can be interpreted within the context of other information to help answer questions of route and source attribution [1].

The primary aspect of a forensic tool is the type of material to which the tool can be applied. Four broad categories of materials central to route attribution may be identified: biologic, geologic, industrial, and packaging. The biological group includes microscopic entities (pollen, spores), vegetation, and animal and human materials. The geological category includes aerosol particles and deposits, soil, and rock fragments. Industrial materials may be subdivided into fluids and particles, and they include all possible indicators of various industrial processes and materials. In this category are included such diverse constituents as lubricants, explosive residues, industrial plant effluents, detritus from commercial products, and so forth. Packaging materials include bulk items, exterior container and all nonnuclear materials contained within, and particles and fibers adhering to the packaging materials.

The second aspect of a forensic tool is the type of laboratory analytical measurement that would be performed on the interdicted materials. The primary categories of analyses are isotopes, major and trace-element compositions, organic and inorganic species, DNA and other biologics, and physical/structural characteristics. The degree to which these various categories are relevant will vary according to the specifics of a particular investigation. The category of physical/structural traits includes any characterization that would identify the nature of the materials collected. For example, a physical/structural characterization of microscopic materials would help identify the types of vegetation and animal detritus that are present. It also includes specific characteristics of the collected material that would constrain the origin; fiber analysis is a well-known criminalistics example. It is important to emphasize that the tools categorized here include the traditional forensic techniques that have been developed and used extensively by law-enforcement-focused forensic labs.

Using this scheme for categorization, a matrix of forensic tools has been generated. Even with these very broad categories, there are some 60 considered forensic tools, and if the specifics of the types of materials and analytic techniques are included, this number becomes much larger. Accordingly, some type of prioritization and discretion must be used to select the most useful tools. This topic is addressed in the next section.

19.5 PRIORITIZATION OF FORENSIC TOOLS FOR ROUTE ATTRIBUTION

There are relatively few case studies of intercepted illicit nuclear materials, and still fewer cases in which many of the possible route forensic tools have been applied. Accordingly, past experience provides little guidance regarding which route tools would be most useful. Of course, experience in forensic criminalistics would be extremely valuable for evaluating the nuclear forensic applications of traditional law-enforcement tools. However, the relative weighting assigned to these criminal forensic techniques, compared to more novel approaches (e.g., isotopic measurements on geological and industrial materials), is little more than educated guessing at this point. It is again important to emphasize that the value of these forensic tools depends on considerations about which scenarios are considered most likely and most important, and the implementation of any of the nuclear forensic tools will be highly dependent on policy makers' questions after a specific incident. Moreover, the ability to use a particular forensic tool in a specific case requires that the material actually be present and effectively collected.

The considered forensic tools are given in Table 19.1. Four categories of prioritization—essential, important, specialized, and not relevant—are shown. Essential forensic tools are those that could be implemented in numerous cases, would answer the largest number of relevant questions, and would provide timely answers. They would be expected to be applied during the first round of forensic analyses. Specialized tools provide specific information, but are unable to provide timely answers. Important tools fall between these two categories, and may be extremely valuable in answering questions for route attribution, but should be used only after the initial set of essential forensic analyses has been completed. Specialized tools are projected to be less useful across a broad spectrum of scenarios, but it should be noted that in some cases they may be of great importance (e.g., if highly unusual materials are present). Case histories in criminalistics provide many examples in which material initially considered incidental proved to be pivotal for an eventual successful conclusion to an investigation.

No single measurement is likely to provide geolocation information that is unique or even very specific. However, in many cases, the evidence provided by a suite of measurements can be interpreted to provide very strong constraints. One way of viewing this interpretational approach is to use each forensic geolocation measurement to successively narrow the possible areas of the world that are consistent with the measurement. The final solution would then be the area of overlap between all geolocation indicators. Even if each marker was individually consistent with a substantial portion of the earth, the intersection between just several of them may define a very small regional area.

TABLE 19.1
Route Forensic Toolbox

	Isotope Abundances	Major Element Composition	Trace Element Composition	Organic Constituents	DNA	Physical or Structural Characteristics
Biological						
Pollen, spores	I		S		S	E
Vegetation	S	S	S	S	S	E
Animal/human	I	S	S	S	E	E
Geological						
Aerosols	S	I	S	S		S
Soils	S	I	S	I		S
Rock fragments	I	S	S			E
Industrial						
Fluids	S	E	S	E		S
Particles	I	E	S	I		E
Bulk	S	E	S	I		E
Packaging materials						
Bulk	S	S	S	S		E
Particles/fibers	I	S	I	S		E

Note: E, essential; I, important; S, specialized; blank, not relevant.

19.6 ANALYTIC TECHNIQUES FOR NUCLEAR FORENSIC INTERROGATION

The instrumentation listed here represents a proposed set that could be used in nuclear forensic investigations, but we emphasize that the actual selection of measurements will depend on specific considerations within an individual case. Moreover, the actual materials present (biologic, geologic, industrial, and packaging) will determine which techniques should be used. The following list is a set of techniques that have actually been implemented for nuclear forensic applications. However, it does not distinguish utility as a function of material type, nor was any attempt made to prioritize the various analytical methods (e.g., to indicate which isotopic or chemical measurement might be most important). Also listed is a brief description of the type of materials best suited to each system. More extensive discussion of the most important instrumentation is presented elsewhere.

19.6.1 Isotopes

- Alpha spectrometry: actinides
- Beta spectrometry: fission and activation products
- Gamma spectrometry: actinides; fission and activation products
- SIMS: in situ microanalysis of all condensed samples and particles

- TIMS: high-precision analysis of metals and actinides; age-dating
- Multicollector plasma-source mass spectrometry (MC-ICPMS): high-precision analysis of metals and actinides; age dating
- Noble-gas mass spectrometry: low-level ^{3}H; headspace gas analysis
- Stable-isotope-ratio mass spectrometry (IRMS): H, C, N, O, and S
- AMS: cosmogenic radionuclides; trace U and Pu isotopes

19.6.2 Elemental Composition/Major and Trace Elements

- ICP-MS: Li through Pu; isotope dilution option
- Inductively coupled plasma, optical-emission spectroscopy (ICP-OES): generally less sensitive than ICP-MS
- XRF spectroscopy: Na through Pu; nondestructive
- Atomic absorption spectroscopy (AAS): Li through U
- Scanning electron microscopy (SEM) with x-ray analysis: sub-μm-scale imaging; element concentrations >0.05%
- Electron microprobe microanalysis: quantitative analysis at concentrations >0.005%; stoichiometry
- SIMS: in situ microanalysis; particle analysis; H through Pu; ng/g detection limits
- Particle-induced x-ray emission (PIXE): Ca through U, μg/g detection limits
- Chemical separation techniques: uHPLC, ion-exchange chromatography, ion chromatography (IC), thin-layer chromatography (TLC), capillary electrophoresis (CE)
- Electron spectroscopy for chemical analysis (ESCA): chemical speciation

19.6.3 Organic Species

- Gas chromatography–mass spectrometry (GC–MS): organic and organometallic compounds; 1–1000 amu
- Ion-trap mass spectrometry with MS-MS (ITMS); GC-ITMS with thermal-energy analyzer (nitrogen-specific analytes)
- uHPLC/electrospray-ionization/triple-quadrupole MS: analytes to 500,000 amu

19.6.4 DNA

- Polymerase chain reaction (PCR) amplification
- DNA sequencing
- Electrophoresis
- Immunoassay

19.6.5 Physical and Structural Characteristics

- Optical microscopy and spectroscopy
- SEM
- Transmission electron microscopy (TEM)

- X-ray diffraction (XRD)
- Neutron and x-ray radiography
- Metallurgical techniques
- Fourier-transform infrared (FTIR), Raman, nuclear magnetic resonance (NMR) spectroscopies

19.7 GEOLOCATION AND ROUTE ATTRIBUTION: REAL-WORLD EXAMPLES

19.7.1 Pb-Isotope Fingerprinting

For use as a geolocation indicator, the minute quantities of Pb present in a questioned sample are collected and analyzed by high-precision mass spectrometry (TIMS or MC-ICPMS). Contamination at some level of any material exposed to environmental windborne Pb is unavoidable, and essentially any material may consequently be sampled for its Pb isotopic composition. Forensic investigations have measured Pb within matrices such as newspaper, currency, shipping containers, ammunition, galvanized coatings from fences, and dust sequestered in clothing. The concentrations of Pb can be quite low (µg/g), and special clean-room procedures to isolate and purify Pb may be required. The Forensic Science Center at Lawrence Livermore National Laboratory, for example, developed a method for extracting Pb that had been trapped within paper when it was manufactured. Figure 19.1 compares the Pb-isotopic compositions of three paper samples investigated at LLNL with the isotopic composition of aerosols and gasoline. The paper specimens were packaged together with suspect nuclear material, and the question was posed whether any characteristics of the paper could be used to constrain the location where the paper was made. The distinct, anthropogenic Pb-isotope signatures of aerosols and gasoline from the United States, Cuba, Europe, and Southeast Asia reflect the variety of Pb ores used for different industrial activities and in different countries [16]. The Pb-isotopic composition of the questioned paper samples clearly excluded the possibility that the paper was produced in the United States or Western Europe, but was fully consistent with an origin in Southeast Asia. The location of the paper production site could be further refined by applying additional geolocation signatures, such as trace-element abundances or Sr- or O-isotopic compositions. Another use of Pb isotopes to identify the geographic origin of nuclear material, and the application of O isotopes to geolocation, are discussed in greater detail later.

19.7.2 O-Isotope Fingerprinting

The potential of O isotopes as a geolocation signature was first suggested in an investigation of uranium oxide ores and reactor fuel elements [17]. The O-isotopic composition of U ore reflects the local geological environment during ore formation, whereas O isotopes in uranium oxide fuel elements primarily carry information related to U refining and enrichment processes. Uranium oxide reactor fuel is produced from LEU (typically 2%–4% ^{235}U), which in turn is derived from U ore

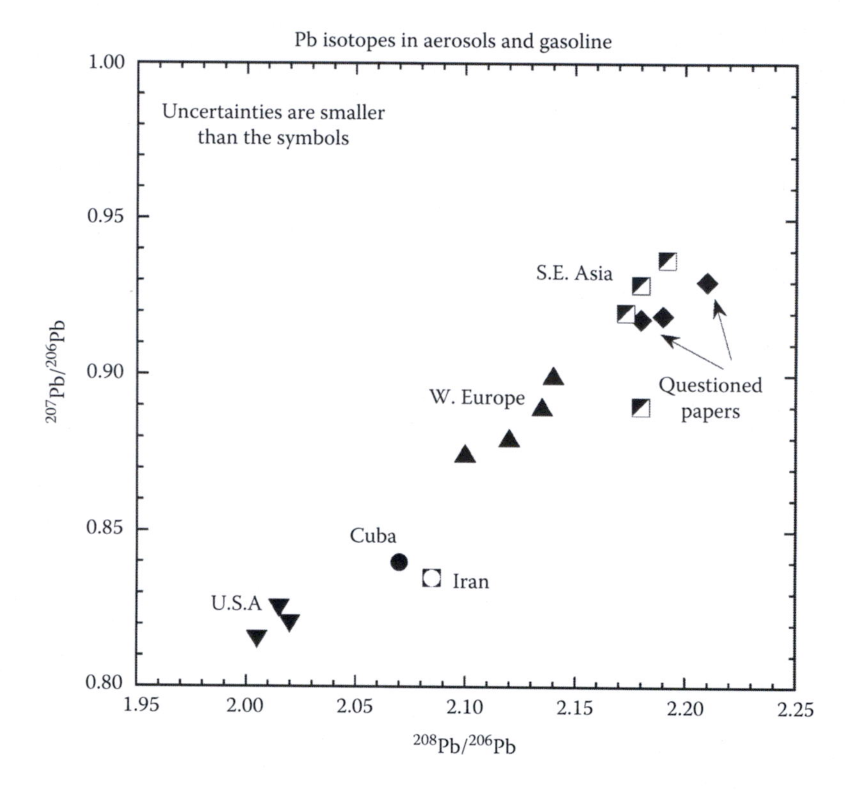

FIGURE 19.1 Comparison of Pb-isotope compositions ($^{207}Pb/^{206}Pb$ and $^{208}Pb/^{206}Pb$) for aerosol particles and gasoline from different geographic locations, and for questioned paper specimens collected together with nuclear material. On a global scale, industrial Pb signatures for aerosols and gasoline plot along linear trend-lines defined by Mississippi Valley Pb ore (very low $^{207}Pb/^{206}Pb$ and $^{208}Pb/^{206}Pb$ ratios), pre-Cambrian Pb ore (high $^{207}Pb/^{206}Pb$ and $^{208}Pb/^{206}Pb$ ratios), and Phanerozoic Pb ore (intermediate Pb isotope ratios). The Pb contained in the questioned paper samples had an isotopic composition very similar to aerosols in Southeast Asia and distinctly different from aerosols generated in the United States or Western Europe. On the basis of this comparison, the paper was assessed as most plausibly manufactured in Southeast Asia.

concentrate (UOC), commonly known as "yellowcake." Yellowcake is converted through UF_4 to UF_6 for U isotope enrichment, and then to UO_2 for fuel fabrication [18]. The conversion to UO_2 involves hydrolysis of the UF_6, commonly using water from sources close to the conversion facility. In this process, the UO_2 product assumes the O-isotope composition of the water. Thus, as U ore is converted to enriched UO_2 reactor fuel, the original O-isotopic composition of the ore is changed, step by step, until no memory of the starting composition remains. The $^{18}O/^{16}O$ ratio in the UO_2 fuel should reflect that of the water or acid used in the hydrolysis reaction.

Data collected by Pajo et al. [17] agree nicely with this simple model. Pajo studied five samples of U_3O_8 and six UO_2 fuel pellets, and found that the O-isotope

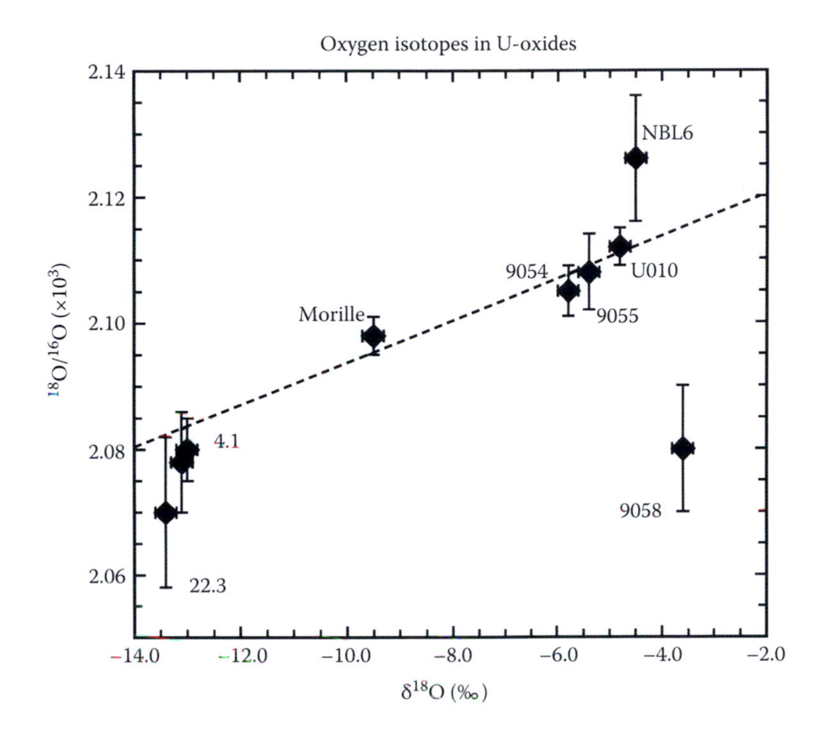

FIGURE 19.2 Comparison of O-isotope compositions ($^{18}O/^{16}O$) measured in uranium oxide samples with O-isotope compositions ($\delta^{18}O$, the ‰ deviation from mean ocean water) of meteoric water (e.g., rain water) characteristic of the geographic locations where the U oxides were obtained. (Based on data reported by Pajo, L. et al., *Fresenius J. Anal. Chem.*, 371, 348, 2001; Rozanski, K. et al., Climate change in continental isotopic records, Geophysical Monograph 78, American Geophysical Union, Washington, DC, 1993.) Samples NBL6 and U010 are from the United States, 9054 and 9055 are from Australia, 9058 is from Gabon, 22.3 is Russian, and 1 and 4.1 are from Kazakhstan. The correlation between $\delta^{18}O$ in meteoric water and $^{18}O/^{16}O$ in uranium oxide is very good for all samples except 9058 from Gabon. The correlation line is a weighted least-squares fit to the data and has an uncertainty in slope at the 95% confidence level of 15%. This correlation suggested that O-isotope compositions in uranium oxide may not be affected by secondary processing and may, in some cases, be a useful geolocation fingerprint in nuclear forensic investigations (see text).

ratios of the materials varied with their geographic origin and correlated with the respective O-isotope ratios of the local precipitation (meteoric water). Different samples from the same location had similar isotope compositions [17,19] (see Figure 19.2). Recently, Fayek [21] reviewed data on the composition of more than 250 samples of U ore and observed a significant geographic variation in O-isotope ratios. On the basis of these differences, Fayek suggested that O isotopes may provide a useful forensic signature, particularly when combined with other indicators such as lanthanide ratios.

A careful investigation by Plaue [22] confirmed the variation in $^{18}O/^{16}O$ observed by [17], but failed to find a correlation between O-isotope ratios in U_3O_8 and those of local meteoric waters. Oxygen-isotope ratios were measured for 15 samples of U_3O_8 from documented geographic locations. The samples were obtained from ore milling facilities, increasing the likelihood that they originated from nearby deposits and utilized local meteoric water for processing.

A resolution of the discrepancy between these studies is not immediately apparent. Limited analyses of oxygen self-diffusion rates have highlighted the sensitivity of diffusion rates to particle size and morphology, as well as to time at temperature. Diffusion processes during thermal oxidation of U oxides for production of U-oxide nuclear fuel may have constrained the extent of O-isotope exchange such that the $^{18}O/^{16}O$ values measured in real-world samples fail to reflect the conditions of origin or process. Oxygen-isotope ratios may yet provide a powerful new signatures technique for nuclear forensic science, but additional research into the subject is required.

19.7.3 Trace Elements and Other Isotopic Ratios

Uranium ore concentrate is an intermediate product of the front-end of the nuclear fuel cycle. According to ASTM standard C967-02a [23], UOC must contain at least 65 wt.% U, meaning that impurity elements often account for several tens of percent. The isotopic and chemical composition of yellowcake, particularly impurities, is highly variable and reflects the raw materials used for production, the milling processes, and the extent of refinement. The overall impurity content can range from tens of percent to as little as a few hundred ppm. For example, the abundances of Al, Zr, and the rare-earth elements (REEs; La–Lu) span a range of more than 100,000 in concentration and display distinctive patterns related to the origin and type of ore body, the milling process, and the details of chemical processing [24–26]. Stable-isotope ratios (e.g., C, N, O, Sr, and Pb) also show variations of several orders of magnitude between samples of UOC from different locations [22,27]. This large range of chemical and isotopic concentrations, coupled with their systematic relationship to origin and production, make UOC materials a model substance for illustrating the ability of NFA and signatures assessment to contribute to source attribution.

The chemical properties of the REEs are very similar (with the exceptions of Ce and Eu, and involving the +4 and +2 valence states, respectively); their relative abundances are typically unaffected by metamorphic processes. Consequently, the REEs frequently provide information on the source of U-bearing ores and ore bodies, in a manner analogous to their applications to substances as diverse as fossils and wines [28,29]. The systematic behavior of the REE is in accord with well-understood geochemical principles, such that REE abundance patterns may be used as predictive signatures should direct-comparison samples be unavailable. The predictive approach is more challenging than the frequently used comparative method, as it requires a firm theoretical foundation to interpret forensic signatures with confidence. Fortunately, the majority of the chemical and physical processes leading to the formation of U ore bodies and U-rich minerals are well understood.

As illustrative examples, consider UOC processed from unconformity-related U ores. Most samples display humped REE patterns (chondrite normalized) enriched in the central REEs (Sm–Dy), while UOC from quartz-pebble conglomerate deposits exhibits complementary patterns showing depletions in the middle REE [24,26]. UOCs derived from quartz-pebble conglomerates and deposits containing intrusive rocks have relatively flat REE patterns, with strong negative Eu anomalies, while yellowcake produced from phosphorite ores or shale deposits are characterized by enrichments in the heavy REE (HREE). This HREE enrichment is also found in seawater, controlled by pH and the abundance of colloids [26], as is the negative Ce anomaly found in the majority of these UOC samples.

The relationship between REE abundance patterns and the provenance of UOC samples is supported by a number of statistical comparisons [3,4,26]. They support the conclusion that REE abundances (as well as those of other trace elements) reflect the formation conditions of the parent U ore and are not significantly altered by milling and conversion processes.

Isotopic fractionation of the heavy elements is expected to be negligible during the production of UOC and suggests that isotope ratios may serve as a powerful tool for source attribution, providing isotope ratios characteristic of parental U-ore bodies can be identified and quantified. As the radiogenic lead isotopes (^{206}Pb, ^{207}Pb, and ^{208}Pb) are produced by the decays of parent U and Th, their relative abundances will depend upon the age and history of the ore deposit. One drawback of Pb isotope ratios as nuclear signature profiles is that radiogenic Pb in U ore is continually separated from U during chemical purification and is gradually diluted with common (nonradiogenic) Pb introduced as a contaminant. The Pb isotopic composition of a UOC sample may, therefore, differ significantly from that of the primary U ore, depending on the degree of processing. However, the Sr and Nd isotopic systems provide complementary geolocation tools that avoid much of the uncertainty contributed by sample contamination with common Pb.

As with Pb, variations in Sr and Nd isotope abundances correlate with the age of an ore body. The ^{87}Sr/^{86}Sr ratio reflects the ingrowth from the decay of ^{87}Rb (half-life = 48 Gy), while the ^{143}Nd/^{144}Nd ratio is produced by ingrowth from decay of ^{147}Sm (half-life = 106 Gy). The Rb–Sr system is easier to apply because of the higher concentrations of Rb and Sr in UOC, but it is more susceptible to modification due to metamorphism. Although the ^{87}Sr/^{86}Sr and ^{143}Nd/^{144}Nd ratios are widely exploited for geochronology, they return a much smaller range of values than the Pb isotopes [30].

The ability to measure multiple isotope systems markedly enhances the success and accuracy of source attribution. Not only do the different systems have different sensitivities to effects of metamorphism, but Sr and Nd appear much less sensitive to intra-mine variability and to contamination introduced during milling and processing. Figure 19.3 provides an example of a mutlielement isotope comparison plot, ^{206}Pb/^{204}Pb *vs* ^{87}Sr/^{86}Sr. Additional discriminatory power may be provided by multiple isotopic systems.

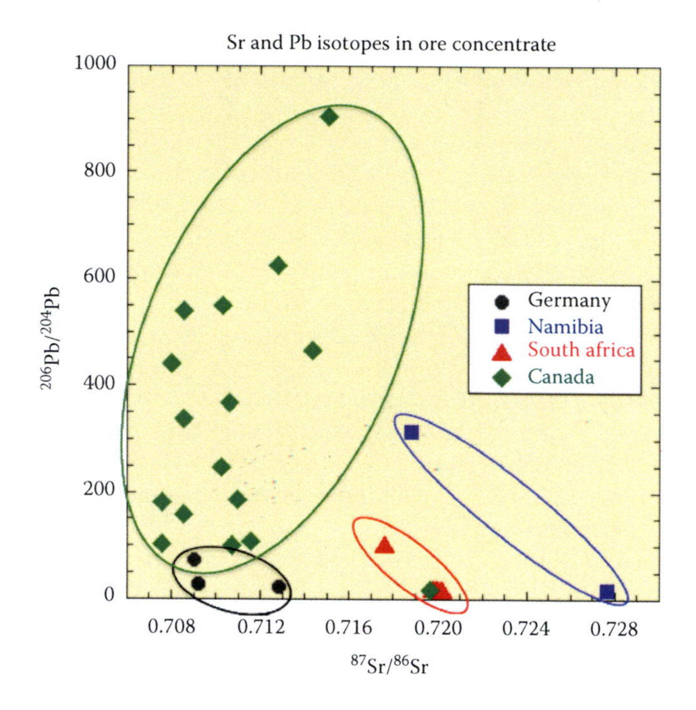

FIGURE 19.3 Diagram of $^{206}Pb/^{204}Pb$ ratios *vs* $^{87}Sr/^{86}Sr$ ratios for uranium ore concentrate samples (UOC; yellowcake) from four countries—Germany, Namibia, South Africa, and Canada. The distribution of the data into four distinct regions illustrates the utility of isotope-correlation diagrams for identifying source regions. (Uncertainties are smaller than the symbols.) However, note that some overlap does exist; for example, one UOC sample from Canada falls within the South African field. Data taken from the Uranium Sourcing database (see text).

19.8 REFERENCE DATA FOR ENHANCED INTERPRETATION: FORENSIC DATABASES

Driven by law enforcement desires to determine the origin of illicitly trafficked nuclear materials, as well as by concerns that a nonstate actor or terrorist organization could acquire sufficient material to assemble a nuclear explosive device, an international consensus to identify and characterize nuclear materials by NFA has recently emerged. By identifying material characteristics, process history, original intended use, production location, and route(s) that material traveled from loss of lawful control to point of interdiction, followed by collating this information into a nuclear forensic library (NFL), a powerful new resource for combating illicit nuclear trafficking is possible. Previously in this chapter, many of the techniques used to extract characteristic forensic signatures from questioned nuclear specimens for source and route attribution were discussed (see also the case studies in the final chapters of this book). Now we discuss how such signature profiles may be assembled and queried within developed forensic databases (i.e., NFLs) to improve the success of nuclear forensic investigations.

NFLs enhance the ability to determine whether nuclear material seized outside regulatory control originated from a particular state or from elsewhere. Much of the information that is, or should be, contained in NFLs already exists, having been collected at other times and/or for other purposes. An NFL collates this information within a single database to assist queries regarding a sample's origin, its history, intended use, and point at which legitimate control was lost. The creation and development of NFLs has recently been recognized as an important international issue by the IAEA, the Global Initiative to Combat Nuclear Terrorism, and the Nuclear Security Summits of 2010 and 2012 [31]. However, the development of NFLs has proven difficult, owing largely to issues about the types of information/data to be included, whether or not a sample archive is necessary, and what organization(s) will have access to, and control over, such libraries. The IAEA has provided guidance for the collection and organization of information related to nuclear material, as well as a context for comparing different types of materials, but has refrained from any direction regarding how to construct NFLs. Questions such as a best approach for developing a database and establishing a nuclear material archive, legal policies, and the utility and application of forensic databases are as yet unresolved.

One compilation that serves as a good example for the use of nuclear forensic databases for source attribution is the Uranium Sourcing Database, sponsored by the NNSA Office of Nuclear Controls. Motivated by the recovery of a large quantity of yellowcake in Libya in 2004, as well as a request to help verify Libyan declarations [32], the US Department of Energy initiated a program to develop improved methods for the determination of sources of UOC. Three DOE national laboratories (Livermore, Los Alamos, and Oak Ridge), collaborated to develop and implement a variety of analytical techniques to characterize the chemical, physical, and isotopic properties of UOC, then apply the data to populate the U-Sourcing Database. Today, this database contains information on >5000 samples originating from more than 30 countries. The information contained in this record may be used for source attribution through comparative analysis—methods generally described as *pattern classification*. The objective is to use the calibration data, consisting of the empirical characteristics of samples from known sources, to determine the most likely origin of an unknown sample via the measurement of variables that discriminate between sources. However, individual sources are typically represented by properties that span a range of values, especially for trace elements, and not by exact data. And more often than not, an array of values from different sources will overlap, creating a need for more information than is conveyed by any single measurement in order to perform an attribution assessment with high confidence.

Commonly used statistical approaches query databases in two dimensions (i.e., two variables) and look for clustering (e.g., principal components analysis). This method is not effective for UOCs, however, as different sources are too similar to distinguish using but two variables. Rather than down-selection, the DOE laboratories developed an alternative approach using latent variables, orthogonal combinations of measured variables, for multivariate regression [33,34]. The technique uses the information (i.e., variance) in UOC data and also has the benefit of signal averaging. This pattern classification approach, called the Discriminant Analysis Verification Engine (*DAVE*), utilizes the partial least squares discriminant analysis component of

PLS_Toolbox (Eigenvector Research, Inc. [35]). *DAVE* features an iterative process of elimination to predict the source of a sample based on information contained in the U-Sourcing Database [34]. Even should a unique source not be identified with high confidence, results from queries through *DAVE* may be used to corroborate other evidence, rule out potential sources, or inform follow-on investigations.

Information useful for nuclear forensic investigations is also contained in a database developed by the Nuclear Materials Information Program (NMIP). NMIP was established in 2006 under National and Homeland Security Presidential Directive, NSPD-48/HSPD-17 [36]. NMIP is an interagency effort managed by the US DOE, in close coordination with the Departments of State, Defense, Homeland Security, and Justice, the Nuclear Regulatory Commission, and the Director of National Intelligence. While many of the details about NMIP are classified, the goal is to create a database, consolidating information from multiple sources about worldwide nuclear material holdings and their security status, into an integrated and continuously updated information management system. NMIP is also developing a national registry for identifying and tracking nuclear material specimens throughout the United States to support US Government needs.

In 1998, the Institute for Transuranium Elements in Karlsruhe, Germany, and the All-Russia Research Institute for Inorganic Materials (VNIINM) in Moscow began the development of a database of civilian nuclear fuel materials [37,38]. The primary objective is to collect and manage data describing the characteristics of the civilian nuclear fuel cycle in Europe and the Commonwealth of Independent States, in order to support identification of the origin and the intended use of interdicted nuclear materials. This database contains distinguishing data for reactors, fresh fuel, and bulk nuclear materials. The general concept may be referred to as database-supported, case-specific analysis, where the database system is used for step-by-step narrowing of the range of potential sources by means of iterative exclusion. The relational database system structure was developed at ITU, with essential contributions from VNIINM experts, and this effort was recently complemented by an electronic literature archive on nonconventional nuclear fuels [39]. The database provides a logical path from measured characteristics of a nuclear material to targeted information about the origin and intended use of the material. Although originally intended for safeguards applications, the ITU-VNIINM work exhibits many of the features of databases designed for use in nuclear forensic investigations.

19.9 SOURCE + ROUTE ATTRIBUTION: TWO EXAMPLES

Independent consideration and comparisons of source and route forensic conclusions have been invaluable in real-world NFA for assessment and attribution of questioned SNM. Unfortunately, specific illustrations that can be described here are limited in number: first, by the relatively few genuine casework examples; second, by the fact that investigative route analyses are comparatively recent endeavors (20–25 years) within the current nuclear enterprise (75 years); and third, by existing classification restrictions on recent examinations. Nevertheless, two examples are accessible and are provided here, one real and one via an ITWG international exercise.

The interdicted Bulgarian HEU specimen is described in detail in Chapter 20. The source forensic result of an original 90% ^{235}U enrichment led to the limited conclusion of non-US material in the first edition of this book, with more specific identification of the likely source initially restricted information (a more focused conclusion attributed the probable origin of the HEU to the former Soviet Union [40,41]). The salient route analytic results included the $BaCrO_4$ loading of the wax, the fiber composition of the papers, and the Pb isotopic signature of the sample shielding, all consistent with an eastern European origin. For this investigation, the independent source and pathway aspects of the overall NFA were entirely complementary and served to strengthen the final assessment.

The second example was afforded by a round-robin exercise (RR#3) conducted by the nuclear ITWG in 2010 [42]. Formed in 1995–1996, the nuclear smuggling ITWG is a global association of interested parties with the noble vision of an idealistic, cooperative, nuclear-analytic world, working in close conjunction with the IAEA. Although ITWG members do not represent their respective countries, they promote multilateral international collaboration against nuclear proliferation and smuggling, and ITWG disseminates such NFA protocols as consensus guidelines, best practices, model action plans, and so forth. To date, however, shared nuclear databases and real casework investigations are not collective multilaboratory activities, but ITWG does sponsor analytic exercises with known nuclear materials for analyses by worldwide laboratories. Thus, in contrast to actual interdictions, correct answers are known a priori by a round-robin proctor. Moreover, the open-forum nature of these exercises precludes any effectual classification of methods and results by specific countries.

Livermore's activities in RR#3 marked our first measurements for both source and route signature species in an ITWG exercise. Two test samples of HEU metal (~92% ^{235}U) were processed by NFA. The nuclear isotopic and elemental analyses of the metal specimens conducted for source information resulted in a clear mismatch between the two samples, indicating that they were of different origins (e.g., a different production facility, different batches within a facility, etc.). Indeed, quantitative similarity comparisons (see Chapter 18) calculated from the weighted isotopic and elemental data measured at LLNL gave absolute 0% correlations between the two specimens.

Conversely, however, the route forensic examinations conducted by the LLNL FSC (Vis/NIR reflectance spectroscopy and SPME/GC–MS—see Chapter 16) concluded that those results were consistent with the two pieces having derived from a common source or initial processing (and/or an indistinguishable environmental history, such as handling, packaging, transport, storage, etc.).

Spectral comparisons of the two samples were performed in the visible and NIR wavelength regions. The results of visual and statistical assessment (discriminant analysis) of the data indicated that, from surficial spectral features, both samples were quite similar in nature and were consistent with having derived from a common source.

In addition, particularly compelling was the calculation of a Spearman rank coefficient (Chapter 18) $\rho = 0.88$ ($n = 28$) from the GC–MS data. It indicated a strong route correlation between the two metal specimens with respect to organic volatile/semivolatile species sorbed on their surfaces. Route conclusions, of course, are independent of any source trending.

Thus, the autonomous source and pathway forensic results for RR#3, rather than provide a consistent assessment, led to completely contradictory conclusions. And the correct answer was different batches of material out of the same facility. Thus, the ability to perform nonnuclear route analyses, although exceptional within nuclear-focused laboratories, allowed a final evaluation of all forensic aspects of this sample set that was completely consistent with the known history of the specimens. However, open knowledge of this result has been absent, as it was specifically omitted from the ITWG final report. The RR#3 summary document pointedly "excluded the consideration of traditional forensic evidence in order to focus efforts on advancing the relatively young science of nuclear forensics" (Lessons Learned, Technical Observation #11). Bizarrely, however, the same report expounds a recommendation to better engage law enforcement to "specifically target issues related to the integration of traditional and nuclear forensic evidence."

The ITWG has historically criticized the conduct of traditional forensic examinations in NFA as either overlooked or poorly executed. Yet, Livermore contributed productive general forensic measurements to NFA investigations for many years before any involvement, or even awareness of the problem, by conventional crime labs. We have actively engaged within mainstream criminalistics communities, as well as successfully performed the so-termed "law-enforcement" aspects of nuclear forensic investigations, for more than 20 years. For actual casework, of course, it is unreasonable to exclude any pertinent examination that could contribute to the overall assessment. Although curious that an expert group advancing forensic inquiry of any nature would disregard extant capabilities in the interest of endorsing some arbitrary level of functional parity, effective practitioners should recognize that any investigation embodying limited practices may not adequately reflect the full potential, scope, and evolution of pragmatic NFA.

REFERENCES

1. S. Niemeyer and I.D. Hutcheon, Geolocation and route attribution in illicit trafficking of nuclear materials, *Proceedings of the ESARDE 21st Annual Meeting Symposium on Safeguards and Nuclear Materials Management*, Seville, Spain, May 4–6, 1999, UCRL-JC-132398, 2000.
2. K. Mayer, M. Wallenius, and T. Fanghanel, Nuclear forensic science—From cradle to maturity, *J. Alloys Compd.*, 444–445, 50, 2007.
3. M.J. Kristo, Nuclear forensics, in *Handbook of Radioactivity Analysis*, 3rd ed., M.F. L'Annunziata, ed., Elsevier, Oxford, U.K., 2012, pp. 1281–1304.
4. I.D. Hutcheon, M.J. Kristo, and K.B. Knight, Nonproliferation nuclear forensics, in *Uranium: Cradle to Grave*, Mineralogical Society of Canada, Quebec City, Quebec, Canada, 2013, pp. 377–394.
5. International Atomic Energy Agency, *IAEA Nuclear Security Series #2, Nuclear Forensics Support Technical Guidance*, STI/PUB/1241, IAEA, Vienna, Austria, ISBN 92-0-100306-4, 2006.
6. International Atomic Energy Agency, Advances in destructive and nondestructive analysis for environmental monitoring and nuclear forensics, *Proceedings of the International Conference on Advances in Destructive and Nondestructive Analysis for Environmental Monitoring and Nuclear Forensics*, Karlsruhe, Germany, October 21–23, 2002, IAEA, Vienna, Austria, 2003.

7. M.A.C. Hotchkis, D.P. Child, and K. Wilchen, Applications of accelerator mass spectrometry in nuclear verification, *J. Nucl. Mater. Manag.*, XL, 60, 2012.
8. W.I. Finch, Uranium provinces of North America—Their definition, distribution, and models, U.S. Geological Survey Bulletin #2141, U.S. Geological Survey, Denver, CO, 1995.
9. R.F. Keeling, The atmospheric oxygen cycle: The oxygen isotopes of atmospheric CO_2 and O_2 and the O_2/N_2 ratio, *Rev. Geophys. Suppl.*, 1253, 1995.
10. J.M. Schwantes, M. Douglas, S.E. Bonde, J.D. Briggs, O.T. Farmer, L.R. Greenwood, E.A. Lepel, C.R. Orton, J.F. Wacker, and A.T. Luksic, Nuclear archeology in a bottle: Evidence of pre-Trinity U.S. weapons activities from a waste burial site, *Anal. Chem.*, 81, 1297, 2009.
11. P. Korbel and M. Novak, *The Complete Encyclopedia of Minerals*, Chartwell Books, New York, 2002.
12. C.E. Blome, P.M. Whalen, and K.M. Reed, Siliceous microfossils, Short Courses in Paleontology No. 8, Paleontological Society, Lancaster, PA, 1995.
13. I. Rodushkin and M.D. Axelsson, Application of double focusing sector field ICP-MS for multielemental characterization of human hair and nails. Part I. Analytical methodology, *Sci. Total Environ.*, 250, 83, 2000.
14. I. Rodushkin and M.D. Axelsson, Application of double focusing sector field ICP-MS for multielemental characterization of human hair and nails. Part III. Direct analysis by laser ablation, *Sci. Total Environ.*, 305, 23, 2003.
15. P.M. Grant, A.M. Volpe, and K.J. Moody, Forensic analysis of hair for inorganic and actinide signature species indicative of nuclear proliferation, *J. Intel. Comm. R&D*, Intelink, 10pp, 2009.
16. W.T. Sturges and L.A. Barrie, The use of stable lead 207/206 isotope ratios and elemental composition to discriminate the origins of lead in aerosols at a rural site in eastern Canada, *Atmos. Environ.*, 23, 1645, 1989.
17. L. Pajo, K. Mayer, and L. Koch, Investigation of the oxygen isotopic composition of oxidic uranium compounds as a new property in nuclear forensic science, *Fresenius J. Anal. Chem.*, 371, 348, 2001.
18. N.E. Lerch and R.E. Norman, Uranium oxide conversion, *Radiochim. Acta*, 36, 75, 1984.
19. C. Kendall and T.B. Coplen, Distribution of oxygen-18 and deuterium in river waters across the United States, *Hydrol. Process.*, 15, 1363, 2001.
20. K. Rozanski, L. Araguas-Araguas, and R. Gonfiantini, Climate change in continental isotopic records, Geophysical Monograph 78, American Geophysical Union, Washington, DC, 1993.
21. M. Fayek, J. Horita, and E. Ripley, The oxygen isotopic composition of uranium minerals: A review, *Ore Geol. Rev.*, 41, 1–21, 2011.
22. J. Plaue, Forensic signatures of chemical process history in uranium oxides, PhD thesis, Paper 1873, University of Nevada, Las Vegas, NV, 2013, http://digitalscholarship.unlv.edu/thesesdissertations/1873, accessed July 2, 2014.
23. ASTM International, Standard specification for uranium ore concentrate, ASTM Standard C967-13, ASTM International, West Conshohocken, PA, 2003, 19428, doi:10.1520/C0967-13, http://enterprise.astm.org/filtrexx40.cgi?+REDLINE_PAGES/C967.htm, accessed July 2, 2014.
24. E. Keegan, S. Richter, I. Kelly, H. Wong, P. Gadd, H. Kuehn, and A. Alonso-Munoz, The provenance of Australian uranium ore concentrates by elemental and isotopic analysis, *Appl. Geochem.*, 23, 765, 2008.
25. E. Keegan, M. Wallenius, K. Mayer, Z. Varga, and G. Rasmussen, Attribution of uranium ore concentrates using elemental and anionic data, *Appl. Geochem.*, 27, 1600, 2012.

26. Z. Varga, M. Wallenius, and K. Mayer, Origin assessment of uranium ore concentrates based on their rare-earth elemental impurity pattern, *Radiochim. Acta*, 98, 771, 2010.
27. Z. Varga, M. Wallenius, K. Mayer, E. Keegan, and S. Millet, Application of lead and strontium isotope ratio measurements for the origin assessment of uranium ore concentrates, *Anal. Chem.*, 81, 8327, 2009.
28. N. Jakubowski, R. Brandt, D. Stuewer, H.R. Eschnauer, and S. Görtges, Analysis of wines by ICP-MS: Is the pattern of the rare-earth elements a reliable fingerprint for the provenance? *Fresenius J. Anal. Chem.*, 364, 424, 1999.
29. D.E. Grandstaff and D.O. Terry, Rare-earth element composition of Paleogene vertebrate fossils from Toadstool Geologic Park, Nebraska, USA, *Appl. Geochem.*, 24, 733, 2009.
30. G. Faure, *Isotopes: Principles and Applications*, John Wiley & Sons, Hoboken, NJ, 2005.
31. IAEA, Board of Governors Nuclear Security Report, GOV/2013/36-GC(57)/16, 2013, www.iaea.org/About/Policy/GC/GC57/GC57Documents/English/gc57-16_en.pdf, accessed July 2, 2014.
32. IAEA, Implementation of the NPT safeguards agreement of the Socialist People's Libyan Arab Jamahiriya, Board of Governors Report, GOV/2004/59, 2004, http://www.iaea.org/Publications/Documents/Board/2004/gov2004-59.pdf, accessed July 2, 2014.
33. M. Robel and M.J. Kristo, Discrimination of source reactor type by multivariate statistical analysis of uranium and plutonium isotopic concentrations in unknown irradiated reactor fuel, *J. Environ. Radioact.*, 99, 1789, 2008.
34. M. Robel, M.J. Kristo, and M.A. Heller, Nuclear forensic inferences using iterative multidimensional statistics, *Institute of Nuclear Materials Management, 50th Annual Meeting*, Tucson, AZ, July 12–16, 2009.
35. B. Wise, N. Gallagher, R. Bro, J. Shaver, W. Windig, and R. Koch, PLS_Toolbox 4.0 for use with MATLAB™, Eigenvector Research, Inc., Wenatchee, WA, 2006.
36. J. Watts, J.D. Olivas, M.E. Cournoyer, and J.A. Judkins, Materials archive and materials data activities, Los Alamos Technical Report, LA-UR-12-26218, Los Alamos, NM, 2012, permalink.lanl.gov/object/tr?what=info:lanl-repo/lareport/LA-UR-12-26218.
37. A. Schubert, G. Janssen, L. Koch, P. Peerani, Yu.K. Bibilashvili, N.N. Chorokov, and Yu.N. Dolgov, A software package for nuclear analysis guidance by a relational database, *Proceedings of ANS International Conference on Physics of Nuclear Science and Technology*, New York, October 5–8, 1998.
38. Yu.N. Dolgov, Yu.K. Bibilashvili, N.N. Chorokov, A. Schubert, G. Janssen, K. Mayer, and L. Koch, Installation of a database for the identification of nuclear materials of unknown origin at VNIINM, Moscow, *Proceedings of ESARDE 21st Annual Meeting Symposium on Safeguards and Nuclear Materials Management*, Seville, Spain, May 4–6, 1999.
39. Y. Dolgov, Development of an electronic archive on nonconventional fuels as integral part of a nuclear forensic laboratory, *International Conference on Advances in Destructive and Nondestructive Analysis for Environmental Monitoring and Nuclear Forensics*, Karlsruhe, Germany, October 21–23, 2002.
40. I.D. Hutcheon, P.M. Grant, and K.J. Moody, Nuclear forensic materials and methods, in *Handbook of Nuclear Chemistry*, 2nd ed., Vol. 6, A. Vertes, S. Nagy, Z. Klencsar, R.G. Lovas, and F. Rosch, eds., Springer Science & Business Media, New York, 2011, Chap. 62, pp. 2837–2891.
41. B. Hart and P. Grant, Forensic analysis capabilities for investigating CBRN threats, IHS Jane's Defense, Risk, & Security Consulting, IHS Global Limited, Englewood, CO, September 2012, Issue 11, pp. 18–25.
42. M.J. Kristo and S.J. Tumey, The state of nuclear forensics, *Nucl. Instrum. Methods Phys. Res. B*, 294, 656, 2013.

20 Forensic Investigation of a Highly Enriched Uranium Sample Interdicted in Bulgaria

Just after midnight on May 29, 1999, alert border security guards stopped Urskan Hanifi, a Turkish national, at a border crossing in Ruse, Bulgaria, returning from a business trip to Turkey. Alerted by his suspicious behavior and lack of luggage, the guards searched Mr. Hanifi's car and discovered a bill of lading, written in Cyrillic, for a quantity of "99.99% uranium-235." They also found a 2.5-kg metallic Pb container (Figure 20.1) hidden within a portable air compressor in the trunk of the car. Inside the Pb shielding "pig" was a glass ampoule filled with several grams of fine black powder (Figure 20.2) that scientists from the Bulgarian National Academy of Science later determined to be highly enriched uranium (HEU). Mr. Hanifi told police that he had purchased the material in Moldova for sale to a client in Turkey. When the transaction fell through, Mr. Hanifi was forced to return to Moldova with the HEU, whereupon he was subsequently interdicted and arrested.

Approximately 9 months later, after learning that Bulgarian scientists had not yet carried out a thorough forensic investigation and had been unable to identify the source of the HEU, the US Department of State arranged to have the U, still contained in its unopened glass ampoule, transferred to the Lawrence Livermore National Laboratory, in Livermore, California. Over approximately the next 9 months, a team of forensic scientists from Livermore and three other Department of Energy laboratories performed an exhaustive study of the HEU and its associated packaging materials, revealing a wealth of information and providing important clues about the origin and original use of the HEU. This chapter relates the story of this investigation, which, at the time, was the most thorough and far-reaching analysis of illicit nuclear material ever conducted [1,2].

The sample received from the Bulgarian authorities included not only the HEU, but also the glass ampoule and the Pb shield, which was filled with a distinctive yellow substance used to cushion the ampoule (Figure 20.3). The first step in this and, in fact, any investigation of unknown nuclear material was to survey the contents with sensitive γ-ray spectrometers before opening the ampoule to ensure that the material contained no surprises in the form of undeclared radioactivities. Such nondestructive analysis (NDA) also provides an important first indication of the isotopic abundances of all actinides present, allowing forensic

FIGURE 20.1 Lead container found hidden within an air-compressor unit in the trunk of Mr. Hanifi's car. The container is ~6 cm in diameter. (Reproduced from Lawrence Livermore National Laboratory, Forensic analyses of a smuggled HEU sample interdicted in Bulgaria, UCRL-ID-143216, U.S. Department of Energy, 2001.)

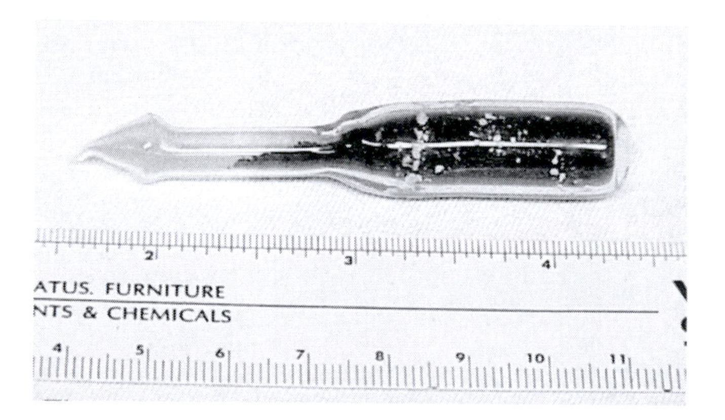

FIGURE 20.2 Glass ampoule containing the HEU powder. The ampoule was originally enclosed within the Pb container, and the HEU is visible as the dark powder in the ampoule. The ampoule is ~7 cm in length. (Reproduced from Lawrence Livermore National Laboratory, Forensic analyses of a smuggled HEU sample interdicted in Bulgaria, UCRL-ID-143216, U.S. Department of Energy, 2001.)

FIGURE 20.3 Interior of the Pb shield after the ampoule was removed. The yellow material, with a crudely fashioned, cylindrical hole used to hold the ampoule, is clearly visible. (Reproduced from Lawrence Livermore National Laboratory, Forensic analyses of a smuggled HEU sample interdicted in Bulgaria, UCRL-ID-143216, U.S. Department of Energy, 2001.)

radiochemists to lay the groundwork for follow-on work to completely characterize all radionuclides. The NDA of the Bulgarian sample confirmed its identification as HEU, containing no detectable Pu or other actinides, and also revealed the presence of three fission products, ^{134}Cs, ^{137}Cs, and ^{125}Sb. Results of the NDA are shown in Table 20.1.

TABLE 20.1
Preliminary Characterization of Highly Enriched Uranium by Nondestructive Analysis

Species[a]	Concentration (atom%)[b]
^{239}Pu	<0.002
^{238}U	14 ± 1
^{237}Np	<0.0002
^{236}U	13 ± 5
^{235}U	72 ± 4
^{234}U	1.3 ± 0.1
^{232}U	$(9 \pm 1) \times 10^{-7}$
^{231}Pa	$(0.12 \pm 0.03) \times 10^{-6}$

Source: Lawrence Livermore National Laboratory, Forensic analyses of a smuggled HEU sample interdicted in Bulgaria, UCRL-ID-143216, U.S. Department of Energy, 2001.

[a] Additionally, ^{134}Cs, ^{137}Cs, and ^{125}Sb were positively identified in the sample.

[b] Uncertainties are two standard deviations.

20.1 ANALYSES OF URANIUM OXIDE

Once the NDA results confirmed that the sample did not present a serious radiological hazard, work began in earnest. The Pb pig was opened and the ampoule removed; all of the steps in the disassembly procedure were extensively photodocumented, and individual evidentiary specimens were entered into forensic chain-of-custody. The ampoule was housed inside a cavity in the center of a block of yellow cushioning material that filled the interior of the pig. A piece of heavy paper surrounded the ampoule, completely separating it from the yellow substance. The ampoule was removed from the shielding container and examined under a stereomicroscope; the fine-grained nature of the HEU was confirmed. The ampoule was then opened and the powder removed, weighed, and divided into aliquots for additional distribution. As part of the comprehensive forensic study, we also prepared for examination aliquots of the Pb shielding container, the yellow material lining the interior of the pig, and two paper samples—an adhesive label applied to the pig exterior and the paper specimen surrounding the ampoule. Because nearly a year had elapsed since Mr. Hanifi was stopped at the border, the forensic team decided that full and complete characterization was more important than the speed of analysis in this instance and chose to apply a comprehensive suite of analytical tools to interrogate the questioned samples. The following techniques were applied: optical microscopy; scanning (SEM) and transmission electron microscopy (TEM), both with energy-dispersive x-ray analysis; x-ray and electron diffraction; radiochemistry followed by α- and γ-spectrometry and mass spectrometry (MS); optical emission spectrometry; ion-, gas-, and gel-permeation chromatography; gas chromatography–mass spectrometry (GC–MS); infrared spectrometry; x-ray photoelectron spectroscopy; x-ray fluorescence (XRF) spectroscopy; and metallurgical analysis. The following discussion focuses first on the uranium oxide sample itself, and then on the Pb pig and associated packaging materials.

Analyses by SEM, TEM, and x-ray diffraction revealed that the HEU was very fine-grained powder containing two major components: uranium oxide (U_3O_8) and a hydrated uranyl fluoride compound ($UO_2F_{2-x}(OH)_x \cdot 2H_2O$, with $0.3 < x < 1.2$). The uranium oxide was the dominant component, making up more than 90% of the sample. The oxide powder formed loosely compacted clumps ranging up to 100 μm in size (Figure 20.4a). These clumps were extremely friable and disaggregated into series of smaller clumps when touched. Individual particles were irregularly shaped (Figure 20.4b), and distinctive particle morphologies were absent at the resolution provided by the SEM images. The individual particles ranged in size from <0.1 μm to about 1 μm, with the smaller particles being much more abundant; the larger particles were perhaps aggregates themselves. Fragments of glass up to 10 μm in size were visible in several SEM images. These glass shards were most likely residue from the ampoule, introduced when the ampoule was cracked open.

TEM was performed on an aliquot of the uranium oxide sample to characterize the morphology and crystallography with a spatial resolution of ~2 Å, roughly 100 times better than that of the SEM. Small samples of the powder were ground with a clean mortar and pestle and dispersed onto carbon-coated Cu grids. A second grid was then placed on top to prevent grain dispersal during observations. The TEM images revealed two distinctive classes of particles: equant-to-slightly ovoid grains dominated the population, making up ~90% of the total, with rod-shaped and plate-shaped grains making up the remainder (Figure 20.5). A size-frequency analysis of

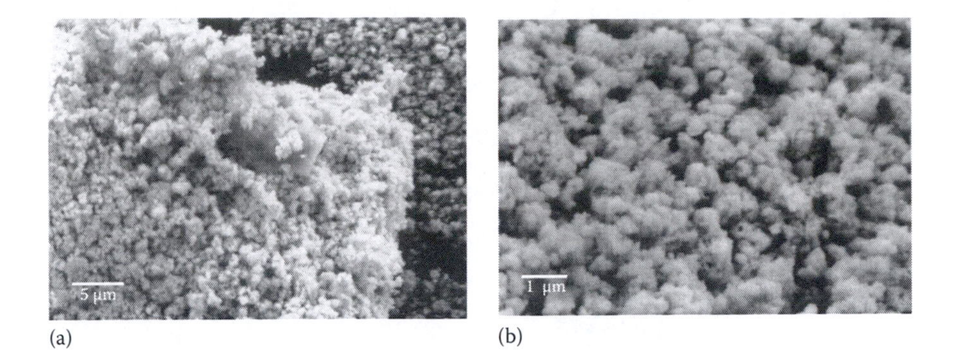

FIGURE 20.4 Two photomicrographs of the HEU powder taken with a scanning electron microscope (SEM). (a) The corner of a clump of HEU showing the very-fine-grained nature of the material. The smooth-surfaced, angular object in the upper right-hand corner of the clump is a shard of ampoule glass. The magnification is 1600×; the scale-bar represents 5 μm. (b) Close-up view of the HEU powder showing the irregular shape and absence of specific morphology. The magnification is 6500×; the scale-bar represents 1 μm. (Reproduced from Lawrence Livermore National Laboratory, Forensic analyses of a smuggled HEU sample interdicted in Bulgaria, UCRL-ID-143216, U.S. Department of Energy, 2001.)

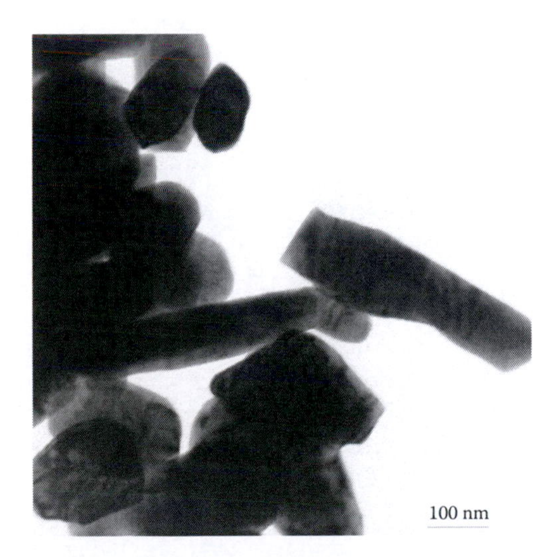

FIGURE 20.5 A photomicrograph of the HEU powder taken with a transmission electron microscope (TEM). The higher magnification of the TEM reveals the two distinctive shapes of grains found in the HEU: oval-to-equant grains making up ~90% of the sample and elongated rod- or plate-shaped grains. Both types of grains are predominantly U_3O_8. The magnification is 120,000×, and the scale-bar represents 100 nm. (Reproduced from Lawrence Livermore National Laboratory, Forensic analyses of a smuggled HEU sample interdicted in Bulgaria, UCRL-ID-143216, U.S. Department of Energy, 2001.)

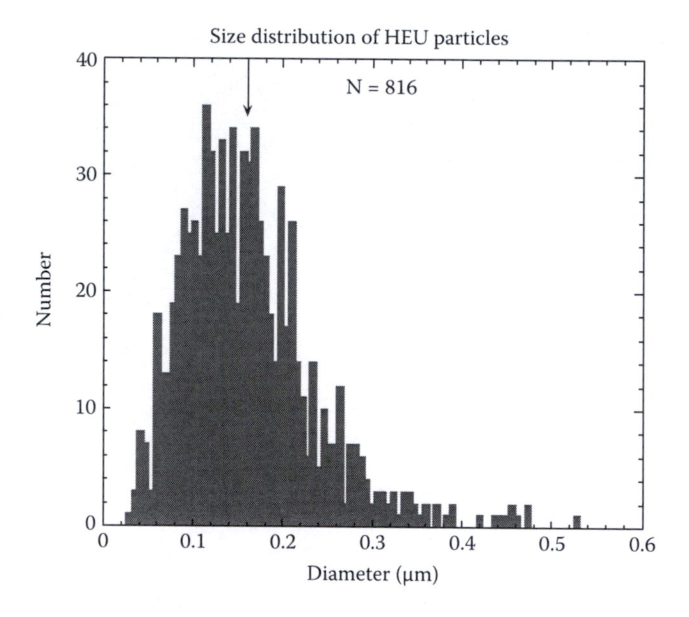

FIGURE 20.6 The grain size distribution of the HEU powder, determined from the TEM images. A total of 816 grains were measured. The arrow at the top of the figure indicates the mean size, ~160 nm. This very small average size suggests that the HEU was not subjected to conventional mechanical grinding and milling. (Reproduced from Lawrence Livermore National Laboratory, Forensic analyses of a smuggled HEU sample interdicted in Bulgaria, UCRL-ID-143216, U.S. Department of Energy, 2001.)

grains showed a wide distribution of sizes, spanning the interval from 30 to 550 nm, with a mean diameter of ~160 nm (Figure 20.6). The abundance of very small grains, with diameters <300 nm, provided an important clue to the manufacturing process used to make the HEU, as such small sizes cannot be generated in commercially available mechanical grinding and milling operations.

The major portion of the sample, ~2 g, was dissolved in nitric acid for radiochemical analyses. The concentrations of 35 individual radionuclides, spanning 15 orders of magnitude in chemical concentration, were determined by α- and γ-spectrometry following radiochemical separations to eliminate spectral overlaps. The major constituents were the six U isotopes—^{238}U, ^{236}U, ^{235}U, ^{234}U, ^{233}U, and ^{232}U—plus ^{230}Th (produced by the decay of ^{234}U). Also evident were ^{241}Am, five Pu isotopes (^{242}Pu, ^{241}Pu, ^{240}Pu, ^{239}Pu, and ^{238}Pu), ^{237}Np, and the residual fission products, ^{125}Sb, ^{134}Cs, and ^{137}Cs. The three fission-product species provide positive proof that the sample was derived from reactor-irradiated and reprocessed U. The empirical concentrations of the most abundant radionuclides are summarized in Table 20.2.

Following the methodology detailed earlier for age-dating radioactive materials, we applied nine radiochronometers, based on decays of U and Pu, to determine the age of the Bulgarian HEU. (To reiterate, age in this context refers to the time since the material was last chemically processed, separating daughter isotopes from their

TABLE 20.2
Concentrations of Radionuclides in the Bulgarian Highly Enriched Uranium

Nuclide	Concentration (wt.%)
^{242}Pu	4.6×10^{-9}
^{241}Pu	7.9×10^{-9}
^{240}Pu	2.6×10^{-8}
^{239}Pu	2.2×10^{-7}
^{238}Pu	1.1×10^{-8}
^{237}Np	6.6×10^{-7}
^{238}U	13.9
^{236}U	11.9
^{235}U	72.7
^{234}U	1.1
^{233}U	2.9×10^{-5}
^{232}U	1.1×10^{-6}
^{231}Pa	4.6×10^{-7}
^{230}Th	2.0×10^{-5}
^{229}Th	9.9×10^{-10}
^{228}Th	2.6×10^{-8}
^{227}Ac	3.3×10^{-11}
^{226}Ra	5.3×10^{-10}
^{137}Cs	4.6×10^{-10}
^{134}Cs	3.3×10^{-13}
^{125}Sb	5.3×10^{-11}

Source: Lawrence Livermore National Laboratory, Forensic analyses of a smuggled HEU sample interdicted in Bulgaria, UCRL-ID-143216, U.S. Department of Energy, 2001.

radioactive parents.) The mean age of the HEU was ~6.5 years, relative to the date the radiochemical separations were carried out at Livermore (April 17, 2000). This age indicated that final chemical processing of the HEU occurred on approximately October 30, 1993, with an uncertainty of <1 month. The fact that the nine radiochemical chronometers all provided ages in good agreement with each other indicated that the trace quantity of Pu contained in the sample was introduced during reprocessing and was not a contaminant added afterward. In principle, if the HEU were diverted from a facility subjected to the safeguards controls of the International Atomic Energy Agency, and if the records of fuel reprocessing were complete and available for inspection, the accuracy of this age determination alone could provide sufficient information to identify the facility.

The isotopic composition of U in the HEU was determined by three different MS techniques: secondary-ion MS (SIMS); thermal-ionization MS (TIMS); and multicollector inductively coupled plasma MS (MC-ICPMS). In addition, the isotopic composition of Pu was determined by TIMS. SIMS was used to provide a rapid (within 24 h of sample aliquoting), reasonably accurate, first analysis of the major U isotopes. TIMS and MC-ICPMS provided data on all six U isotopes, including the very low abundance isotopes ^{232}U and ^{233}U, with much higher accuracy than was available from SIMS or NDA, using purified aliquots of the pot solution prepared by radiochemical analysis. The analyses by all of the instrumental methods yielded a consistent U-isotopic composition for the HEU powder: $^{232}U = 1 \times 10^{-6}$, $^{233}U = 3 \times 10^{-5}$, $^{234}U = 1.2$, $^{235}U = 72.7$, $^{236}U = 12.1$, and $^{238}U = 14.0$ (all values in atom%). The uncertainties in these values are <0.5%, except for the very low-abundance isotope ^{232}U, with an error of ~10%. The trace amount of Pu contained in the HEU was enriched in ^{239}Pu: $^{238}Pu = 3.7$, $^{239}Pu = 82$, $^{240}Pu = 9.9$, $^{241}Pu = 2.8$, and $^{242}Pu = 1.7$ (all values in atom%). These data, together with those generated by the radiochemical analyses discussed earlier, are summarized in Table 20.2.

The U isotope abundances clearly identified the material as HEU containing ~73% ^{235}U; U of this composition is weapons usable. The empirical U isotope abundances designated that the HEU had an initial enrichment of ~90% and had been used as fuel in a nuclear reactor. The high concentration of ^{236}U indicated a prolonged irradiation history that burned about 50% of the usable ^{235}U initially present. The isotopic composition of the Pu was consistent with weapons-usable material, but

TABLE 20.3
Concentrations of Trace Elements in the Bulgarian Highly Enriched Uranium

Element	Concentration (ppm)[a]
B	2.8 ± 1.5
F	5.0 ± 0.4
Na	3.0 ± 1.5
Mg	28 ± 10
Al	8 ± 4
Si	21 ± 10
P	3.0 ± 1.5
S	130 ± 65
Cl	55 ± 26
K	20 ± 5
Cr	4.6 ± 0.5
Mn	3.1 ± 0.5
Fe	52 ± 15
Ni	3.4 ± 0.5
Cu	2.3 ± 1.1
Br	47 ± 6

Source: Lawrence Livermore National Laboratory, Forensic analyses of a smuggled HEU sample interdicted in Bulgaria, UCRL-ID-143216, U.S. Department of Energy, 2001.

[a] Uncertainties are one standard deviation.

the concentration (2 ng/g) is much too low for the Bulgarian HEU to be considered a significant source of weaponizable Pu.

The final step in the source characterization of the HEU sample was the complete determination of trace-element content. The concentrations of 72 elements, ranging from Li to Th, were measured using a variety of analytical techniques, including ICP-MS, spark-source MS, optical emission spectroscopy, and ion chromatography. The concentrations of individual elements varied widely, from <2 ppbw (1 ppbw = 1 ng/g) to ~200 ppmw. The total inventory of trace elements was fairly high, between 500 and 800 ppmw, with four elements (Cl, S, Fe, and Br) accounting for nearly 60% of the total inventory. The concentrations of the most abundant trace elements are given in Table 20.3. These trace-element concentrations were much higher than expected for laboratory-scale reprocessing and indicated that the material was an aliquot of a batch reprocessing operation. The pattern of relative elemental abundances, such as the enrichment of volatile, electronegative elements (S, Cl, and Br), is most readily interpreted as a signature of chemical reprocessing and can, in principle, be used to identify the specific chemical treatments used during reprocessing.

20.2 ANALYSES OF COLLATERAL (ROUTE) EVIDENCE

20.2.1 Lead Container

Once the characterization of the uranium oxide was underway, attention also focused on the nonnuclear components—the Pb shielding container (pig) with its distinctive yellow material, the glass ampoule, the label from the top of the Pb pig, and the paper inner-liner separating the ampoule from the yellow substance. These examinations used sophisticated conventional forensic techniques, some of which were similar to those employed by state and federal law enforcement agencies.

The Pb shield was examined via optical and scanning electron microscopy. The exterior of the container was weathered in a fashion consistent with a high-Pb alloy that had been exposed to elevated humidity for an extended period of time. Marks visible on the outer surface provided clear evidence of the use of a coarse hand file for shaping and smoothing. In contrast, marks indicative of the use of machine tools, such as an end mill or lathe, were absent, so that forming by swaging or extrusion was definitely excluded. The overall appearance of the container, and particularly a general lack of rectilinearity, indicated that the container was cast in a crude mold, probably made of sand, and shaped by hand. This supposition was confirmed by a careful metallurgical examination of the container.

A small fragment was cut from the container, mounted in epoxy and polished, and then etched with a solution of dilute acetic acid and hydrogen peroxide to reveal the microstructure. The structure, consisting of Pb dendrites surrounded by a two-phase eutectic region, was characteristic of common, cast-Pb metal parts. Examination by SEM equipped with energy-dispersive x-ray spectrometry (SEM/EDX) showed that approximately 5 wt.% Sb had been added to the initial Pb material to produce an alloy with greater malleability than simple cast Pb. The SEM/EDX study also revealed remnants of an aluminosilicate material, similar in composition to kyanite (Al_2SiO_5), trapped between folds of Pb. An aluminosilicate such as kyanite is commonly used for high-temperature insulation, and in this case appeared to have been used as a mold wash or liner, or as a release agent around a mold insert in the center of the casting. Kyanite is a naturally occurring mineral and is consistent with the type of material expected in a homemade device.

20.2.2 Yellow Wax

The yellow substance filling the interior of the Pb shield was analyzed by Fourier-transform infrared spectrometry (FTIR) to identify the molecular compounds present. Thin sections of several flakes of the material were optically scanned over areas ~5 mm × 5 mm in size to select target sites; background-corrected spectra were then obtained at those locations. On the basis of these FTIR spectra, the substance was identified as a paraffin wax (i.e., a mixture of solid hydrocarbons, derived from petroleum, with the general formula of C_nH_{2n+2}). Determining the specific brand of wax would require detailed peak-matching of the FTIR spectra to spectra from comparison samples of known manufacture and composition. Although this level of detail was beyond the scope of the investigation, a comparison to reference standards in the Sadtler database library indicated that the composition of the questioned yellow wax was inconsistent with many commercial waxes (e.g., chlorinated paraffin, carnauba, carbowax 300, shellac, or beeswax), but was strikingly similar to the paraffin-based wax, Parowax.

To identify the component imparting the distinctive yellow color to the wax, we attempted to extract a yellow dye with a variety of organic solvents. Although the wax itself was soluble, all of these attempts proved unsuccessful, and the residue remaining after dissolution retained its characteristic yellow color. The coloring agent was ultimately identified by extracting the paraffin away from the inorganic material via methylene chloride, and then using XRF to examine the remaining material. The XRF data identified the inorganic residue as barium chromate, $BaCrO_4$, a compound once generally used as yellow pigment in paints, glass, and ceramic overglazes; as an oxidizer in pyrotechnics; and as an oxidizer in heat powders and igniters. Barium chromate is rarely used in the United States or in most Western countries today because of environmental and health concerns about Cr as a carcinogen. However, widespread use of $BaCrO_4$ persists in Brazil, China, India, and many eastern European countries.

The next series of examinations were designed to characterize the wax by searching for any unusual trace compounds, either as wax by-products or as environmentally specific contaminants, and by determining the molecular weight of the wax. The yellow wax (as well as the HEU powder) was qualitatively analyzed for volatile and semivolatile organic compounds by GC–MS using headspace sampling by solid-phase microextraction (SPME). For each analysis, ~0.5 g of material was transferred into a 6-mL glass vial. The transfer was executed within an N_2-filled glovebag to prevent contamination by atmospheric gases. The vial was capped with a Teflon-barrier silicone septum and heated for about 1 h with a SPME fiber positioned in the vial headspace. This approach is advantageous for sampling trace species from solids because preconcentration is achieved in two steps: during equilibration (as analytes outgas from the solid into the headspace) and as the analyte is collected by the SPME fiber.

Table 20.4 summarizes the organic compounds measured in the wax. For material individualization purposes, several relatively unusual species (naphthalenic compounds)

TABLE 20.4
Significant Organic Species Identified in the Yellow Wax

Compound	CAS#	MW	Sig. Ion (m/z)
2-Methyl butane	78-78-4	72	43
Methylene chloride	75-09-2	84	49
Phenol	108-95-2	94	94
C_7H_{14}	NA	98	41
Heptane	142-82-5	100	43
Hexanal	66-25-1	100	44
Methyl isobutyl ketone	108-10-1	100	43
Benzyl alcohol	100-51-6	108	79
Trimethyl benzene	NA	120	105
C_9H_{20}	NA	128	43
Naphthalene	91-20-3	128	128
Decane	124-18-5	142	43
Methyl naphthalene	NA	142	142
Undecene	821-95-4	154	41
Dimethyl naphthalene	NA	156	156
Undecane	1120-21-4	156	43
1-Chloronaphthalene	25586-43-0	162	162
Dodecene	112-41-4	168	41
Dodecane	112-40-3	170	57
Trimethyl naphthalene	NA	170	155
2-Undecanone	112-12-9	170	43
Tridecene	2437-56-1	182	41
Tridecane	629-50-5	184	43
Tetradecene	1120-36-1	196	41
6,10-Dimethyl-2-undecanone	1604-34-8	198	58
Tetradecane	629-59-4	198	43
2-Tridecanone	593-08-8	198	58
Pentadecene	13360-61-7	210	57
Pentadecane	629-62-9	212	57
2-Tetradecanone	2345-27-9	212	58
Hexadecene	629-73-2	224	41
2-Pentadecanone	2345-28-0	226	58
Hexadecane	544-76-3	226	57
Heptadecene	2579-04-6	238	41
Heptadecane	629-78-7	240	57
Octadecene	112-88-9	252	56
Octadecane	593-45-3	254	57
Nonadecane	629-92-5	268	57
Eicosane	112-95-8	282	57
Heneicosane	629-94-7	296	57
Docosane	629-97-0	310	57
Tricosane	638-67-5	324	57

(Continued)

TABLE 20.4 (*Continued*)
Significant Organic Species Identified in the Yellow Wax

Compound	CAS#	MW	Sig. Ion (m/z)
Tetracosane	646-31-1	338	57
Pentacosane	629-99-2	352	57
Hexacosane	630-01-3	366	57
Heptacosane	593-49-7	380	57
Octacosane	630-02-4	394	57

Source: Lawrence Livermore National Laboratory, Forensic analyses of a smuggled HEU sample interdicted in Bulgaria, UCRL-ID-143216, U.S. Department of Energy, 2001.

Note: MW, molecular weight; Sig. Ion, significant ion of greatest response that distinguishes the chemical from interferences.

were identified. However, no volatile or semivolatile compounds were detected above background levels in the headspace gas collected from the HEU sample.

The melting point of the wax was determined with a Mel-Temp capillary melting-point instrument. Three replicate measurements indicated a melting point between 53°C and 55°C. The molecular weight was determined by gel-permeation chromatography. Two aliquots with different concentrations of isolated wax were prepared using tetrahydrofuran as solvent; each aliquot was analyzed three times. The six measurements yielded a weighted-average molecular weight of (393 ± 2.3) Da. These results were in reasonable agreement with a less accurate GC–MS datum, estimated by analysis of the total-ion chromatogram of the wax. The empirical molecular weight corresponds to a polymer with an average of 26 carbon atoms.

Only a few types of commercially available wax have melting points similar to that of the yellow wax (Table 20.5). Cresin and paraffin are petroleum-based waxes encompassing a broad range of products, whereas Japan wax is a vegetable wax composed of long-chain esters. The closest match is for paraffin waxes consisting

TABLE 20.5
Melting Points of Waxes Similar to the Questioned Yellow Wax

Wax	Melting-Point Range (°C)	Source
Cresin	53–85	Petroleum
Paraffins	44–74	Petroleum
127/129 AMP	51–54	Petroleum
USP petrolatum	53.3	Petroleum
Japan	46–52	Fruit of *Rhus succedanea* L., Anacardiaceae

Source: Lawrence Livermore National Laboratory, Forensic analyses of a smuggled HEU sample interdicted in Bulgaria, UCRL-ID-143216, U.S. Department of Energy, 2001.

principally of straight-chain, saturated hydrocarbons. The *n*-alkane content of the wax usually exceeds 75%, but can be as high as 100%. The molecular weights of these paraffin waxes range from ~280 to 560. A more specific identification of the yellow wax would require quantitative formulation analysis and detailed comparison to known wax standards, which were beyond the scope of this investigation.

20.2.3 Paper Liner and Label

The two paper specimens isolated from the Pb pig were characterized using forensic microscopy techniques to determine the composition of the wood fibers making up the paper. Fibers from the inner paper liner separating the ampoule from the paraffin wax consisted of 61% bleached softwood and 39% bleached hardwood. The softwood species were ~80% *Pinus sylvestris* (Scotch pine) and 20% of a *Picea* (spruce) variant; hardwood species were ~70% *Populus tremula* (European aspen) and 30% *Alnus glutinosa* (common alder). Fibers from the label removed from the cap of the shield consisted of 38% bleached softwood, 23% semi-bleached softwood, and 39% bleached hardwood. The softwood species were 80% *Pinus sylvestris* and 20% *Picea abies* (white spruce), and the semi-bleached softwood was 100% *Picea abies.* The hardwood species were ~70% *Populus tremula* and 30% *Alnus glutinosa* (white birch).

Both the softwood and the hardwood fibers in both paper samples were produced using the Kraft pulping process. Based on the use of the Kraft process, the paper was either fabricated in a large-scale, integrated manufacturing setting or in a smaller operation using purchased fiber. The semi-bleached fiber from the label argues for production in a large integrated facility, as there is generally no market demand for semi-bleached paper pulp. The fibers used in these papers are not found in North America or Scandinavia. The quality of the paper is that of commercial office paper, and assuming that the primary use was as packaging material, it is likely to have been obtained close to the point of manufacture, most likely within the greater European area.

20.2.4 Glass Ampoule

The quality of the seal and the absence of bubbles and macroscopic flaws indicated that the ampoule represents the work of a skilled glass blower. The form of the seal specified that the ampoule was sealed at atmospheric pressure and never evacuated.

A millimeter-size fragment of the ampoule was embedded in epoxy, polished, and analyzed via electron microprobe (using a 15-keV, 5-nA electron beam defocused to a diameter of 20 μm) to measure the chemical composition. The concentrations of C, Na, Mg, Al, Si, K, Ca, and Fe were determined by wavelength-dispersive x-ray spectrometry, whereas O was determined by stoichiometry. The fact that these elements summed to only ~84% indicated that the glass contained a light element for which we did not analyze as a major constituent (i.e., H, Li, Be, B, or N). A quick survey of common laboratory glasses indicated that the only reasonable choice was boron (B), and the boron oxide content was calculated by difference. The final composition of the ampoule glass was very similar to that of commercial, soft borosilicate glass. The major-element composition of this glass does not vary significantly between different sources, and additional data on trace-element abundances would be needed to identify possible glass manufacturers or suppliers.

20.3 ATTRIBUTION

The primary goal of an attribution analysis is to identify the original source of the material; its intended, or original, legitimate application; the point at which the material was diverted from lawful use; and, of course, the perpetrators. It should be understood that identifying the primary source (i.e., source attribution) does not necessarily indicate where the material passed from valid ownership, nor the identity of the individuals responsible for the theft or diversion. Source attribution is a necessary prerequisite, but does not lead directly to the perpetrators. Typically, an attribution analysis proceeds in stages, first focusing on unambiguous signatures (such as U isotopic composition), then proceeding to more subtle signatures (such as trace elements and physical properties), and finally consideration of collateral signatures found, for example, on packaging and containers.

The dominant signature of the Bulgarian material is the U-isotopic composition. The U isotope abundances, especially the unusually high concentration of ^{236}U, clearly indicate that the sample is HEU reactor fuel that was heavily irradiated and then reprocessed. The fuel was HEU with an initial ^{235}U isotope abundance of ~90%. Approximately 50% of the ^{235}U usable as reactor fuel was burned (consumed) during the irradiation, resulting in the measured U-isotopic abundances. The fact that the initial ^{235}U loading was ~90% immediately excludes fuel manufactured in the United States, as US HEU used for research and propulsion reactors has a ^{235}U content of 93%. Although a difference of 3% may appear small, it is much greater than the analytical uncertainty and provides compelling evidence of a non-US origin.

Determining the specific type of reactor in which the HEU was irradiated is a much more involved process, requiring intimate knowledge of reactor designs and operating conditions, as well as sophisticated computer modeling of fuel burn-up. This investigation focused on research and propulsion reactors and, although successfully excluding many common types of reactors, has yet to identify the specific reactor in question.

The accuracy to which the age was determined indicates that final chemical reprocessing of the irradiated fuel was completed on November 1, 1993, with an uncertainty of ~1 month. Careful examination of the information provided by the different isotopic parent–daughter pairs used for chronometry indicates that the PUREX process, which uses organic, phosphate-based reagents, was the primary reprocessing protocol. The PUREX process individually separates U, Np, and Pu from highly radioactive fission products contained in irradiated fuel, allowing the recycle of U and Pu as reactor fuel. As PUREX is popularly used in nuclear reprocessing facilities on a worldwide basis, identification of this process alone did not narrow the search for the source of the HEU.

Other physical characteristics of the fuel and the associated packaging also point to an origin outside of the United States. The extremely fine grain size of the powder is unlike that found in US facilities, where coarser sizes are produced in order to minimize the potential health hazards created by the respiration of fine, α-emitting dust. The size of the Bulgarian HEU grains is more characteristic of material prepared for specialized use, for example, in powder metallurgy at a fuel-fabrication facility. In terms of grain size and overall consistency, the Bulgarian sample has the characteristics of feedstock used for the production of fuel pellets or blending with other batches of uranium oxide at U fuel-conversion facilities. Multiple samples, approximately the same size as this specimen, are commonly aliquoted from process batches of uranium oxide product for laboratory analysis and archival purposes. The Bulgarian HEU appears to have the characteristics of such material.

The ampoule strongly resembles glass containers used at nuclear fuel reprocessing centers to preserve subsamples of production runs as archival material. Dozens of these ampoules, of uncertain accountability, have been reported at some facilities. The barium chromate, used to give the wax its distinctive yellow color, has been banned in the United States as a carcinogen, but it is still in widespread use in Brazil, China, India, and many European states. In similar vein, the composition of the wood fibers clearly reflect paper products derived from mixtures of hardwood and softwood trees not found in the United States, but which are common in eastern Europe. Finally, the isotopic signature of the Pb used in the shielding container is inconsistent with Pb mined in the United States, but is compatible with Pb ores from Asia and Europe. These results from the nonnuclear materials consistently reinforce the conclusion that the HEU sample is not of US origin and was most plausibly produced somewhere in Europe.

Attribution of the Bulgarian HEU thus remains somewhat incomplete. Despite the comprehensive forensic investigation and wealth of data, neither the original source of the HEU, nor the point at which legitimate control was lost, has yet been unambiguously identified. The possibility that the HEU represents material produced in the United States has been eliminated, and efforts to further refine the attribution analysis are continuing. This lack of definitive answers underscores the complexity of nuclear forensic investigations and highlights the need to develop wide-ranging databases linking measurable properties of nuclear materials to specific production facilities (see Chapter 19 for additional discussion of this point).

Even after the laboratory analyses had been completed for some time, the saga of the Bulgarian HEU continued. The smuggler, Mr. Hanifi, was tried and convicted of trafficking in controlled nuclear materials. At his sentencing, the judge attempted to determine the commercial value of the HEU sample. When experts from the Bulgarian Institute for Nuclear Research and Nuclear Energy testified that a uranium oxide sample weighing 5–10 g would normally sell for <$10,000 (far below the value of illicit drugs in the many cases presided over by this judge), the judge sentenced Mr. Hanifi to time already served, fined him a few thousand dollars, and released him. After his release, Mr. Hanifi was thought to have died in Moldova several months later, with his official cause of death remaining unresolved. However, recently uncovered information indicated that this was a case of mistaken identity.

Although the Bulgarian seizure was the first instance of a comprehensive nuclear forensic investigation of smuggled HEU by the United States, it has not been an isolated incident. Since the interdiction in Ruse in 1999, at least five additional seizures of weapons-grade or weapons-usable HEU have been reported [3–9]. Some reports suggested that organized crime may be seeking an entrée into the illicit trafficking of nuclear materials, and, in toto, these reports indicate that the black market in nuclear materials may be far more extensive than previously recognized [6,7,10]. Illicit trafficking in nuclear materials remains an important area of concern for the IAEA,

Europol, and national law enforcement agencies [11]. It has also gained increased attention in the context of the 2010 and 2012 Nuclear Security Summits [12,13].

One incident, in particular, had striking similarities to the Bulgarian seizure. In July 2001, the French national security service obtained information that an individual had a large quantity of HEU for sale, possibly as much as 30 kg. A successful police operation led to the arrest of two men and the recovery of a small sample of HEU [7]. The seized materials had a striking similarity to materials interdicted in Bulgaria and described in this chapter. The HEU was a black powder, sealed inside a borosilicate glass ampoule, and confined within a cylindrical lead container lined with a yellow wax. Analyses of this HEU showed a U isotope composition indistinguishable from that of the Bulgarian HEU for all measured isotopes. In particular, the presence of trace amounts of ^{232}U, ^{233}U, ^{239}Pu, and ^{137}Cs indicated that the material was irradiated in a reactor and then reprocessed, just as was the Bulgarian HEU. Baude [7] concluded that the HEU samples interdicted in Bulgaria and Paris very likely had the same origin, and efforts to evaluate potential linkage between these and other interdicted materials remain active in several countries.

REFERENCES

1. Lawrence Livermore National Laboratory, Forensic analyses of a smuggled HEU sample interdicted in Bulgaria, UCRL-ID-143216, U.S. Department of Energy, Livermore, CA, 2001.
2. G. Vogel, Crime and (puny) punishment, *Science*, 298, 952, 2002.
3. Police nab three suspects in France's first U-235 smuggling, *Nucleonics Week*, 42(30), 13, July 26, 2001.
4. T. Berthemet and M. Bietry, Five grams HEU seized in Paris, FBIS document EUP20010723000399, *Le Figaro*, Paris, France, July 23, 2001.
5. M. Bronner, 100 grams (and counting …), notes from the nuclear underworld, Project on managing the atom, Harvard University, Cambridge, MA, 2008.
6. E.K. Sokova and W.C. Potter, The 2003 and 2006 enriched uranium seizures in Georgia: Some answers and possible lessons, *Illicit Nuclear Trafficking: Collective Experience and the Way Forward*, IAEA Publication 1316, Vienna, Austria, 2008, pp. 405–424.
7. S. Baude, HEU seized in July 2001 in Paris: Analytical investigations performed on the material, *Illicit Nuclear Trafficking: Collective Experience and the Way Forward*, IAEA Publication 1316, Vienna, Austria, 2008, pp. 397–399.
8. M.J. Kristo, Nuclear forensics, in *Handbook of Radioactivity Analysis*, 3rd ed., M.F. L'Annunziata, ed., Elsevier, Oxford, U.K., 2012, pp. 1281–1304.
9. I.D. Hutcheon, M.J. Kristo, and K.B. Knight, Nonproliferation nuclear forensics, in *Uranium: Cradle to Grave*, P.C. Burns and G.E. Sigmon, eds., Vol. 43, Mineralogical Society of Canada, Winnipeg, Manitoba, Canada, 2013, p. 377.
10. W.H. Webster and the CSIS Task Force, The nuclear black market: global organized crime project, CSIS Panel Report, Center for Strategic and International Studies, Washington, DC, 1996, p. 47.
11. International Atomic Energy Agency, Illicit nuclear trafficking: Collective experience and the way forward, *IAEA Proceedings of the International Conference*, Edinburgh, U.K., International Atomic Energy Agency, Vienna, Austria, 2008.
12. Communiqué, Washington Nuclear Security Summit, 2010, http://www.whitehouse.gov/the-press-office/communiqu-washington-nuclear-security-summit, accessed July 2, 2014.
13. Communiqué, Seoul Nuclear Security Summit, 2012, http://www.cfr.org/proliferation/seoul-communiqu-2012-nuclear-security-summit/p27735, accessed July 2, 2014.

21 Counterforensic Investigation of U.S. Enrichment Plants

21.1 BACKGROUND

There was a possibility that the US gaseous diffusion facilities at Portsmouth, Ohio, and Paducah, Kentucky, would be subject to international inspections for safeguards and accountability purposes. The question arose of whether an inspector with a covert agenda could surreptitiously learn classified information from the analysis of environmental materials found at these sites. To explore that possibility, the Livermore Forensic Science Center (FSC) was contacted by a program sponsor to undertake "counterforensic" sampling to evaluate the potential magnitude of the risk to classified technology.

Unclassified information that was available before the sampling trip was the fact that the barrier consisted of tubes of porous, sintered Ni powder with a specified maximum length and pore diameter. It is designed to function in a dry fluoride-rich environment at a temperature above that of the triple point of UF_6. As a ferromagnetic metal, Ni can be separated from diverse mixtures of collected materials in any given dry sample through application of a permanent magnet, and this property helped guide the creation of a sampling plan.

21.2 SAMPLING

Three forensic analysts traveled to the Portsmouth plant and collected a total of 18 samples at that site in July 1994. The same sampling team then continued to Paducah and obtained 21 samples at that facility later in the month. A wide spectrum of sampling tools was available to the collection team; however, the environment at the plants precluded particle collection, and brute-force tools such as hacksaws and hatchets were found to be unnecessary. Although some swipe specimens were taken, the most useful sampling devices were spatulas and horseshoe magnets. The latter were covered with plastic wrap during the collection of magnetic debris, which was then deposited into a zip-locked plastic bag by physically separating the magnet from the plastic wrap.

Collected samples were placed in either high-purity, polyethylene liquid scintillation vials or in zip-locked polyethylene bags. A second zip-locked bag was employed for secondary sample containment, and double-bagged samples were placed in a plastic box. The package was then sealed with vinyl tape. Tamper-indicating seals were employed at appropriate places on the exterior of all of the packaged samples. The collections from each plant were returned to the Livermore FSC by classified mail. On receipt, the samples were photomicrographed, screened for radioactivity, and examined for Ni content by both magnetic screening and x-ray fluorescence (XRF).

Figures 21.1 through 21.3 are photographs of material in three of the samples before their radiochemical analysis. Figure 21.1 shows a sample containing fairly representative magnetic debris from the Paducah plant, mostly Fe, but containing other transition metals and organic matter (sample 94-7-45). Figure 21.2 shows a sample of mixed magnetic materials, including Ni shards, collected at the Portsmouth plant (sample 94-7-17). Figure 21.3 shows a sample containing almost pure Ni shards, which are visibly porous, collected with a spatula from an out-of-the-way location at the Portsmouth plant (sample 94-7-14). Samples such as those shown in Figure 21.3 were studied extensively and subjected to metallurgical analysis.

Five samples were subjected to a full radiochemical analysis. The samples were selected on the basis of their radioactivity content, as determined with a β–γ survey instrument; only one of them (that associated with Figure 21.3) was composed of prime candidate barrier material. In addition, a blank sample was constructed from the employed analytic reagents, and it was subjected to the same radiochemical procedures as were the inspection samples. The samples taken for radiochemical analyses are described in Table 21.1.

Sample 94-7-45

FIGURE 21.1 Photomicrograph of materials from sample 94-7-45, collected at Paducah.

FIGURE 21.2 Photomicrograph of materials from sample 94-7-17, collected at Portsmouth.

FIGURE 21.3 Photomicrograph of materials from sample 94-7-14, collected at Portsmouth; consists largely of barrier material.

TABLE 21.1
Samples Selected for Radiochemical Analyses from the Portsmouth and Paducah Site Inspections

Sample I.D.	Weight (g)	Description
94-7-14	7.0	Ni flakes collected with a spatula into a polyvial; from an elevated catwalk, Portsmouth (Figure 21.3).
94-7-17	1.4	Magnetic flakes, both rust and Ni, collected from the top of an I-beam about 3 m from the floor, Portsmouth (Figure 21.2).
94-7-19	3.3	Magnetic grit, mostly Fe, collected from a groove in the floor around railroad tracks, Portsmouth.
94-7-38	4.9	Loose debris, perhaps some Ni flakes, collected with a spatula from a bracket 5 m from the floor, Paducah.
94-7-45	20.5	Magnetic grit, with most of the Ni removed for metallurgical analysis, collected along a gap in construction materials, about 2.5 m from the floor, Paducah (Figure 21.1).
Blank	4.7	Radiochemical blank, constructed from Fe_2O_3, Ni_2O_3, and CaO.

21.3 RADIOCHEMISTRY

The creation of stable, initial pot solutions was a formidable problem. The dissolution of samples containing Ni, Fe, dirt, paint flakes, plastic shards, oil, broken glass, hair, and (in one case) a dead insect was accomplished as follows: the samples were poured into Erlenmeyer reaction flasks, and the original sample containers were sequentially washed with water and with 8 M HNO_3 to ensure quantitative transfer. Samples were heated gently until reaction was complete. A few drops of 9 M HCl were added to each reaction flask, and the samples were gently evaporated to moist residues. Several times over the next day, 9 M HCl was added to the samples, which were then evaporated to moist residues. To each sample, 4 M HCl was added, and the supernatant solutions were filtered through Whatman #42 paper into fresh Erlenmeyer flasks, which were to serve as primary containers for the analytical solutions. Several times over the next day, 8 M HNO_3 was added to the residues in the reaction flasks, which were then evaporated to moist deposits. Next, 2 mL of 7.5 M $HClO_4$ and several milliliters of 8 M HNO_3 were added to each flask, and the resultant mixtures were evaporated until the evolution of copious perchloric fumes was observed. After that, 4 M HCl was added to the samples, and the supernatant solutions were filtered into the Erlenmeyer primary containers.

Several milliliters of 8 M HNO_3 and 1 mL of concentrated HF were added to the residues in the reaction flasks, which were then evaporated to moist deposits. Three milliliters of 7.5 M $HClO_4$ were then added to each reaction flask, and the samples were evaporated through perchloric fumes to dryness. The reaction flasks were then heated vigorously with a Bunsen burner to volatilize SiF_4. Several milliliters of 4 M HCl were added to each flask, and the resultant solutions were filtered into the Erlenmeyer pots. The filter papers were washed with 4 M HCl and were removed from the funnels and placed into the appropriate reaction flasks. Several milliliters of 8 M HNO_3 and 2 mL of 7.5 M $HClO_4$ were added to each flask, and the samples were heated until the filter papers were destroyed with the evolution of copious perchloric fumes. The contents of the flasks were quantitatively transferred to centrifuge cones with 4 M HCl, solutions were centrifuged, and supernates were carefully decanted into primary containers.

Four of the six samples were completely dissolved, but several milligrams of intractable white powder remained of samples 94-7-38 and 94-7-45 (the two Paducah specimens). XRF showed that this material was predominantly Ti, with a trace of Zr, and was therefore probably residue from paint chips. It was decided to treat these residues separately from bulk samples, under the assumption that only a small fraction of the total radionuclide content was entrained in the precipitates. The residues were transferred to Ni crucibles with a minimum quantity of dilute HCl and were evaporated to dryness. Several grams of NaOH and 1 g of Na_2O_2 were added to the crucibles, which were heated to melt the mixtures. Twenty milliliters of water were added to the molten samples; after cooling, volumes were doubled with 9 M HCl, and supernatant liquids were transferred to fresh Erlenmeyer flasks. Several milliliters of 4 M HCl were added to residues in crucibles; after heating for several minutes, supernates were decanted into flasks containing the previous solutions. Residues were heated to dryness, 0.5 g of $(NH_4)_2SO_4$ and 4 mL of concentrated H_2SO_4 were added to each crucible, and the mixtures were heated for several hours. Resulting solutions were diluted with 4 M HCl and transferred quantitatively to flasks containing previous supernates. Only a trivial amount of residue was left in the crucibles. Ti solutions were analyzed with an abbreviated version of the analytical procedure given later and were found to contain <1% of the Th, U, Np, and Pu measured in the main samples. Analytical results given in Tables 21.5 and 21.6 for samples 94-7-38 and 94-7-45 have been corrected for these losses.

The final analytical working solutions are described in Table 21.2. Aliquots of 60 μL of each solution were dried on 1-in.-diameter Whatman #541 filter papers and subjected to XRF analysis. The results for major constituents, scaled to the total solution volume, are also given in Table 21.2. In many cases, the sum of the measured masses of these constituents is significantly lower than that of the original sample; the reason is that XRF is insensitive to low-*Z* materials such as C, Al, and Si, particularly in the presence of significant amounts of Cl from the solvent. In the case of the blank, however, a direct comparison of measured XRF values with the known materials input to the dissolution procedure gave good agreement (1940 mg Fe measured by XRF versus 2050 mg added, and 1240 mg Ni measured versus 1330 mg added). The explanation for the minor quantities of Cu and Zn measured in the blank is their presence as contaminants in reagent compounds used for solution synthesis.

Aliquots of each pot solution were taken for different aspects of the radiochemical analyses. The 60-μL aliquots for XRF analysis were indicated earlier, and other aliquots are listed in Table 21.3. Gamma-spectrometry aliquots were transferred to plastic counting vials and sealed in Saran wrap. All other samples were delivered into 40-mL centrifuge cones.

Unspiked samples were not traced with radionuclides. "Spike mix S" solutions were traced with standard aliquots containing about 175 dpm ^{236}Pu, 30 dpm ^{232}U (with ^{228}Th and daughters in near equilibrium), and 440 dpm ^{243}Am (with ^{239}Np in equilibrium). "Spike mix N" solutions were traced with 820 dpm ^{237}Np (with ^{233}Pa

TABLE 21.2
Description of Dissolved Samples and Results of an X-Ray Fluorescence Analysis

Sample I.D.	Total Volume (mL)	Color	X-Ray Fluorescence Composition
94-7-14	70	Green	6550 mg Ni, 6 mg Cu
94-7-17	75	Yellow	725 mg Fe, 225 mg Ni, 4 mg Cu, 3 mg Zn, 6 mg U
94-7-19	100	Orange-tan	2220 mg Fe, 145 mg Ni, 68 mg Cu, 8 mg Zn, 4 mg Pb, 14 mg U
94-7-38	150	Orange	1500 mg Fe, 108 mg Ni, 13 mg Cu, 75 mg Zn, 16 mg Pb, 32 mg U
94-7-45	215	Dark brown	8600 mg Fe, 1680 mg Ni, 204 mg Cu, 337 mg Zn, 9 mg Pb, 183 mg U
Blank	75	Olive	1940 mg Fe, 1240 mg Ni, 7 mg Cu, 1 mg Zn

TABLE 21.3
Aliquot Information for Radiochemical Analysis

Sample I.D.	γ-Spec (mL)	Unspiked (U) (mL)	Spiked (S) (mL)	Spiked (N) (mL)
94-7-14	25	15	10	10
94-7-17	25	20	10	10
94-7-19	25	20	20	20
94-7-38	25	25	25	25
94-7-45	25	30	25	25
Blank	25	20	10	10

in equilibrium). The comparison of measurements of both spiked and unspiked samples results in chemical yield determinations and quantitative, absolute data for radionuclide atoms/mL of pot solution.

The γ-spec aliquots were counted with large-volume Ge detectors for emitted photons having energies between 50 and 2000 keV. Each data acquisition period was between 3 and 6 days in length. Using detector calibration parameters, γ-ray peaks were identified with proper energies and were corrected for detector efficiency to yield resultant photon intensities. The only radionuclides detected in the samples with this technique were ^{235}U, ^{238}U, and ^{237}Np. The numbers of atoms thus determined in each γ-spec sample are listed in Table 21.4.

It was somewhat surprising that the ^{237}Np contamination level was so high, particularly in the Paducah samples, where the radioactivity resulting from decaying ^{237}Np was comparable to that from U isotopes. From the radionuclide assay of the blank specimen, no significant cross-contamination among the samples was evident during dissolution. The U measurements allow the calculation of enrichment values, given in the last column of Table 21.4. All of the Portsmouth samples were somewhat higher in ^{235}U than is natural uranium (0.71%), whereas both the Paducah samples were depleted in ^{235}U.

TABLE 21.4
Results of γ-Spec Analysis of Radiochemical Pot Aliquots

Sample I.D.	U-238 Atoms	U-235 Atoms	Np-237 Atoms	U-235 Enrich.
94-7-14	$(9.05 \pm 2.00) \times 10^{17}$	$(1.75 \pm 0.07) \times 10^{16}$	$(1.29 \pm 0.27) \times 10^{13}$	$(1.90 \pm 0.43)\%$
94-7-17	$(5.39 \pm 0.79) \times 10^{18}$	$(5.35 \pm 0.13) \times 10^{16}$	$(3.07 \pm 0.48) \times 10^{13}$	$(0.98 \pm 0.15)\%$
94-7-19	$(1.10 \pm 0.14) \times 10^{19}$	$(3.67 \pm 0.06) \times 10^{17}$	$\leq 9 \times 10^{12}$	$(3.23 \pm 0.41)\%$
94-7-38	$(1.81 \pm 0.23) \times 10^{19}$	$(5.50 \pm 0.18) \times 10^{16}$	$(1.31 \pm 0.02) \times 10^{15}$	$(0.30 \pm 0.04)\%$
94-7-45	$(6.92 \pm 0.80) \times 10^{19}$	$(2.46 \pm 0.05) \times 10^{17}$	$(2.99 \pm 0.04) \times 10^{15}$	$(0.35 \pm 0.04)\%$
Blank	$\leq 4 \times 10^{16}$	$\leq 5 \times 10^{14}$	$\leq 9 \times 10^{12}$	—

Gamma-spec results are presented here separately, instead of being combined with the more precise results of the radioanalytical chemistry procedure described later, for two reasons: The results were available before chemical processing was begun and were the first clear indication that procedures developed for the measurement of tracer-level actinide nuclides would need modification to accommodate the multi-milligram quantities of U present in analytical samples, and the data presented in Table 21.4 also give an indication of the limitations of the precision of radionuclide determinations via nondestructive γ-ray spectrometry.

The analytical samples described in Table 21.3 were evaporated to dryness. Each sample was then processed by the methods outlined in Chapter 10. After two oxidation/reduction cycles, the samples were dissolved in 6 M HCl, and Fe was extracted with hexone. The aqueous fractions were evaporated to dryness and redissolved in 1 M HCl. Fe^{3+}, La^{3+}, and Al^{3+} carriers were added (~1 mg each), the solutions were thoroughly mixed, and the volumes were doubled by the addition of concentrated NH_4OH, after which the samples were digested in a warm water bath for several hours. After centrifugation, the supernatant liquids were decanted to waste, the precipitates were contacted with water, and 1 mL of concentrated NH_4OH was added to each. After digesting for an hour, the samples were centrifuged, and the supernatant liquids were decanted to waste. Hydroxide precipitates were dissolved in a minimum volume of 9 M HCl, solution volumes were doubled by the addition of concentrated HCl, two drops of 8 M HNO_3 were added to each solution, and the solutions were loaded onto the first anion-exchange column described in Chapter 10.

According to the prescription given there, fractions containing (Ra, Ac, Th, Am, Cm), Pu, Pa, Np, and U were eluted sequentially from the column. Pu, Pa, Np, and U fractions were further purified with anion-exchange steps similar to those used with the original column to generate the fractions. The (Ra, Ac, Th, Am, Cm) fractions were evaporated to dryness and dissolved in 8 M HNO_3, to which a drop of 18% NH_4NO_2 had been added; after warming briefly, these solutions were passed through anion-exchange columns that adsorbed Th from the mixture. Th was eluted from the columns with 9 M HCl and was further purified via anion exchange. The residual (Ra, Ac, Am, Cm) fractions were separated with DOWEX-50X4 columns, from which (Am, Cm), Ra, and Ac fractions were eluted with increasing molarities of HCl, as described in Chapter 10. Finally, the (Am, Cm) fractions, which also contained La carrier, were dissolved in saturated HCl and purified with AG MP-50 columns, again following the Chapter 10 prescription.

Radium fractions from samples 94-7-38 and 94-7-45 contained ^{226}Ra, as determined by α spectrometry (see following). Unfortunately, the chemical yield as determined from short-lived ^{224}Ra in the S samples was not valid; early precipitation steps that may have had poor efficiencies for Ra resulted in intermediate samples containing high yields of the ^{228}Th precursor nuclide, which would have caused ^{224}Ra to grow back into the samples. For the two samples with indications of Ra (again, the Paducah samples), special analytical aliquots were constructed from 15 ml of working solution and three times the standard S-Mix aliquot of ^{236}Pu and daughters. After a single oxidation/reduction cycle and several hexone extractions as described earlier, the samples were evaporated to dryness, dissolved in 9 M HCl, and loaded onto anion-exchange columns subsequently washed with 9 M HCl. Load and wash solutions were collected as Ra fractions, evaporated to dryness several times with the addition of HNO_3, and then dissolved in 8 M HNO_3. These solutions were loaded onto anion-exchange columns, which were then washed with more 8 M HNO_3 (this is the Th–Ra separation time). The load and wash solutions were collected as the Ra fractions and evaporated to dryness several times with the addition of 6 M HCl. The Ra fractions were then purified with the same cation-exchange procedure outlined earlier for the (Ra, Ac, Am, Cm) fractions.

Final Ra, Ac, Th, Pa, Np, and (Am, Cm) samples were dissolved in minimum volumes of 6 M HCl and transferred to W filaments, where they were re-evaporated. Counting samples were prepared by volatilizing the samples onto 1-in.-diameter Pt disks through the rapid heating of the filaments. Uranium fractions were subdivided into several samples of different strengths, some of which were volatilized onto Pt, whereas others were subjected to isotopic analysis by ICP-MS. Pu fractions from the S and N samples were volatilized onto Pt, as were all of the unspiked fractions except those from samples 94-7-38 and 94-7-45, which were analyzed for isotopic composition by thermal ionization mass spectrometry.

Neptunium and Pa samples were mounted in standard-geometry γ-spec holders and were counted close to the faces of large-volume Ge photon detectors; γ-ray intensities were used to measure the activities of the ^{239}Np and ^{233}Pa chemical-yield indicators. Gross α-decay rates of these samples were measured with calibrated 2π gas-proportional counters to complete the link to chemical yields. All Pt-backed samples were pulse-height counted for α-particle emission with surface-barrier or Frisch-grid gas-filled detectors.

In total, more than 130 samples were generated and counted in the course of these radiochemical investigations.

21.4 RESULTS

Results of the analyses of the six working solutions, scaled to the masses of the original samples, are given in Table 21.5. These data were derived from a combination of α-particle, γ-ray, and MS analyses. Data are given in units of atoms, except for ^{239}Pu and ^{240}Pu in those samples where Pu fractions were not subjected to MS analysis. Because ^{239}Pu and ^{240}Pu are not resolvable by α spectrometry (due to their

TABLE 21.5
Results of the Analysis of the Heavy-Element Contents of the Analytical Samples

Nuclide	94-7-14	94-7-17	94-7-19
^{241}Am	$(5.22 \pm 1.43) \times 10^{9}$	$(1.60 \pm 0.20) \times 10^{9}$	$(9.14 \pm 3.28) \times 10^{8}$
^{242}Pu			
^{241}Pu			
^{240}Pu			
^{239}Pu	$\sim 9 \times 10^{10}$	$\sim 2 \times 10^{11}$	$\sim 1 \times 10^{11}$
^{238}Pu	$(4.43 \pm 0.47) \times 10^{7}$	$(1.57 \pm 0.21) \times 10^{7}$	$(4.58 \pm 0.77) \times 10^{7}$
^{237}Np	$(2.96 \pm 0.29) \times 10^{13}$	$(8.46 \pm 0.50) \times 10^{13}$	$(3.57 \pm 1.93) \times 10^{13}$
^{238}U	$(3.23 \pm 0.13) \times 10^{18}$	$(1.63 \pm 0.05) \times 10^{19}$	$(4.23 \pm 0.07) \times 10^{19}$
^{236}U	$(2.24 \pm 0.14) \times 10^{14}$	$(9.54 \pm 0.67) \times 10^{14}$	$(1.75 \pm 0.08) \times 10^{15}$
^{235}U	$(5.61 \pm 0.23) \times 10^{16}$	$(1.80 \pm 0.06) \times 10^{17}$	$(1.66 \pm 0.03) \times 10^{18}$
^{234}U	$(3.36 \pm 0.14) \times 10^{14}$	$(1.03 \pm 0.03) \times 10^{15}$	$(1.71 \pm 0.03) \times 10^{16}$
^{233}U	$\leq 1.3 \times 10^{11}$	$\leq 1.5 \times 10^{11}$	$\leq 2.3 \times 10^{12}$
^{232}U	$\leq 1.5 \times 10^{7}$	$(2.07 \pm 1.79) \times 10^{7}$	$(5.86 \pm 0.40) \times 10^{8}$
^{231}Pa	$\leq 6.3 \times 10^{10}$	$\leq 1.3 \times 10^{11}$	$(5.27 \pm 0.31) \times 10^{11}$
^{232}Th	$(3.59 \pm 2.17) \times 10^{15}$	$(3.77 \pm 2.09) \times 10^{15}$	$(2.08 \pm 1.29) \times 10^{15}$
^{230}Th	$(1.54 \pm 0.09) \times 10^{12}$	$(7.14 \pm 0.38) \times 10^{12}$	$(5.33 \pm 0.28) \times 10^{12}$
^{229}Th	$\leq 9.0 \times 10^{8}$	$\leq 1.4 \times 10^{9}$	$\leq 1.0 \times 10^{9}$
^{227}Ac	$(4.05 \pm 2.08) \times 10^{7}$	$\leq 2.5 \times 10^{7}$	$\leq 2.9 \times 10^{7}$
^{226}Ra			
Activity ^{239}Pu + ^{240}Pu	(7.10 ± 0.20)	(13.65 ± 0.30)	(8.40 ± 0.26)

Nuclide	94-7-38	94-7-45	Blank
^{241}Am	$(3.89 \pm 0.06) \times 10^{10}$	$(1.23 \pm 0.02) \times 10^{11}$	$\leq 7.4 \times 10^{8}$
^{242}Pu		$(1.06 \pm 0.11) \times 10^{10}$	
^{241}Pu		$(3.79 \pm 0.15) \times 10^{10}$	
^{240}Pu	$(5.59 \pm 0.19) \times 10^{11}$	$(1.81 \pm 0.02) \times 10^{12}$	
^{239}Pu	$(8.12 \pm 0.10) \times 10^{12}$	$(2.79 \pm 0.03) \times 10^{13}$	$\leq 6.0 \times 10^{9}$
^{238}Pu	$(9.70 \pm 0.28) \times 10^{8}$	$(2.36 \pm 0.06) \times 10^{9}$	
^{237}Np	$(7.81 \pm 0.08) \times 10^{15}$	$(2.56 \pm 0.02) \times 10^{16}$	$\leq 6.0 \times 10^{11}$
^{238}U	$(9.70 \pm 0.33) \times 10^{19}$	$(5.25 \pm 0.15) \times 10^{20}$	$\leq 8.0 \times 10^{15}$
^{236}U	$(3.85 \pm 0.17) \times 10^{15}$	$(2.08 \pm 0.09) \times 10^{16}$	
^{235}U	$(3.22 \pm 0.11) \times 10^{17}$	$(2.03 \pm 0.06) \times 10^{18}$	
^{234}U	$(1.90 \pm 0.07) \times 10^{15}$	$(1.24 \pm 0.04) \times 10^{16}$	$\leq 4.4 \times 10^{10}$
^{233}U	$\leq 1.5 \times 10^{10}$	$(1.77 \pm 0.73) \times 10^{11}$	
^{232}U	$\leq 9.6 \times 10^{7}$	$\leq 2.5 \times 10^{8}$	
^{231}Pa	$(2.82 \pm 0.26) \times 10^{11}$	$(8.53 \pm 0.58) \times 10^{11}$	$\leq 5.0 \times 10^{10}$
^{232}Th	$(1.49 \pm 0.13) \times 10^{17}$	$(3.28 \pm 0.26) \times 10^{17}$	$\leq 3.0 \times 10^{15}$
^{230}Th	$(1.29 \pm 0.07) \times 10^{14}$	$(5.15 \pm 0.27) \times 10^{14}$	$\leq 3.0 \times 10^{10}$
^{229}Th	$\leq 1.3 \times 10^{9}$	$(1.16 \pm 0.44) \times 10^{10}$	
^{227}Ac	$\leq 3.5 \times 10^{7}$	$(1.24 \pm 0.46) \times 10^{8}$	$\leq 4.0 \times 10^{7}$
^{226}Ra	$(1.26 \pm 0.15) \times 10^{10}$	$(3.23 \pm 1.13) \times 10^{10}$	
Activity ^{239}Pu + ^{240}Pu	(556.3 ± 5.5)	(1886 ± 16)	≤ 0.30

Note: Data reported are atoms in each sample, decay corrected to day 260, 1994. Also given are the disintegrations per minute of the sum of ^{239}Pu and ^{240}Pu α decays in each sample.

similar α-decay energies), the sum of their activities (in disintegrations/minute) is given in the table. The assessments of ^{239}Pu content are derived from $^{240}Pu/^{239}Pu$ mass ratios estimated from $^{238}Pu/(^{239}Pu + ^{240}Pu)$ activity ratios (see below). Values listed for ^{233}U are calculated from the ^{234}U content and the relative amounts of their decay daughters (^{229}Th and ^{230}Th, respectively); as will be seen later, this tactic may yield an ambiguous calculation.

The distribution of U isotopes in each sample is given in Table 21.6 in the form of atom ratios to ^{238}U content. These are substantially the same data given in Table 21.5, but do not include propagated errors on chemical yield.

Observables in the Pu fractions are given in Table 21.7. The $^{238}Pu/(^{239}Pu + ^{240}Pu)$ activity ratios were obtained from α-particle spectrometry, whereas the atom ratios are the results from thermal ionization mass spectrometry. The data for the higher-mass Pu isotopes in sample 94-7-38 were compromised by a contamination in the instrument; data beyond the $^{240}Pu/^{239}Pu$ ratio for that sample should be treated with suspicion.

TABLE 21.6
U Isotopic Ratios of the Site-Inspection Samples (Atom Ratios)

Ratio	94-7-14	94-7-17	94-7-19
$^{232}U/^{238}U$	$\leq 5 \times 10^{-12}$	$(1.3 \pm 1.1) \times 10^{-12}$	$(1.38 \pm 0.10) \times 10^{-11}$
$^{233}U/^{238}U$	$\leq 4 \times 10^{-8}$	$\leq 9 \times 10^{-9}$	$\leq 6 \times 10^{-8}$
$^{234}U/^{238}U$	$(1.040 \pm 0.011) \times 10^{-4}$	$(6.30 \pm 0.06) \times 10^{-5}$	$(4.034 \pm 0.020) \times 10^{-4}$
$^{235}U/^{238}U$	$(1.738 \pm 0.013) \times 10^{-2}$	$(1.108 \pm 0.009) \times 10^{-2}$	$(3.927 \pm 0.016) \times 10^{-2}$
$^{236}U/^{238}U$	$(6.92 \pm 0.29) \times 10^{-5}$	$(5.86 \pm 0.37) \times 10^{-5}$	$(4.14 \pm 0.04) \times 10^{-5}$
Ratio	**94-7-38**	**94-7-45**	**Natural**
$^{232}U/^{238}U$	$\leq 1 \times 10^{-12}$	$\leq 5 \times 10^{-13}$	0
$^{233}U/^{238}U$	$\leq 2 \times 10^{-10}$	$(3.4 \pm 1.4) \times 10^{-10}$	0
$^{234}U/^{238}U$	$(1.960 \pm 0.013) \times 10^{-5}$	$(2.358 \pm 0.017) \times 10^{-5}$	5.47×10^{-5}
$^{235}U/^{238}U$	$(3.320 \pm 0.021) \times 10^{-3}$	$(3.873 \pm 0.027) \times 10^{-3}$	7.25×10^{-3}
$^{236}U/^{238}U$	$(3.97 \pm 0.12) \times 10^{-5}$	$(3.96 \pm 0.15) \times 10^{-5}$	0

Note: Natural U values are included for comparison.

TABLE 21.7
Observables in Site-Inspection Samples, Pu Fractions

	Activity	Atom Ratios		
Sample	$^{238}Pu/(^{239}Pu + ^{240}Pu)$	$^{240}Pu/^{239}Pu$	$^{241}Pu/^{239}Pu$	$^{242}Pu/^{239}Pu$
94-7-14	0.0936 ± 0.0089			
94-7-17	0.0173 ± 0.0023			
94-7-19	0.0816 ± 0.0131			
94-7-38	0.0262 ± 0.0007	0.069 ± 0.002		
94-7-45	0.0188 ± 0.0004	0.0648 ± 0.0001	0.00136 ± 0.00005	0.00038 ± 0.00004

Results for the radioanalytical blank were consistent with negligible radionuclide content, giving no evidence for significant cross-contamination among the samples during dissolution and subsequent analytical procedures. The data in the last column of Table 21.5 also give a sense of the sensitivity of the technique that would be of potential use for planning future site inspections.

21.5 INTERPRETATION

The basic assumption underlying all data interpretation that follows is that, unless otherwise indicated, the radionuclide content of a given sample is characteristic of a specific location in the enrichment cascade and that there was a single release time associated with that sample. Contributions from multiple releases of material at different cascade process times would compromise many of the following conclusions. Information about operations at the Paducah and Portsmouth plants is available in the unclassified literature [1,2], and these resources were the starting point of the ensuing forensic analysis. The main function at both plants is the enrichment of U by means of gaseous diffusion of warm UF_6; however, other enrichment techniques had also been explored (e.g., gas centrifugation) and might have contributed to the isotopic content of specific samples, but not significantly to the general level of contamination at these sites. Operations at both plants started in the mid-1950s and continued past the time that the samples were collected (1994). During much of this time, the Paducah plant produced U of low ^{235}U enrichment (1%–5%), some of which was used as feedstock at the Portsmouth plant. Portsmouth made several different products at ^{235}U enrichments of 97% or less.

This last statement is confirmed by the data in Table 21.6. It is much more likely that feed or tails materials would be represented in the site-collection samples than would be product materials, because of both their relative masses and their relative financial values. Paducah samples (94-7-38 and 94-7-45) were distinctly depleted in ^{235}U relative to natural U ($^{235}U/^{238}U_{nat}$ = 0.00725), but those from Portsmouth were all distinctly enriched. The fact that each of the five samples represents different ^{235}U enrichments is confirmation that these are factories that are actively engaged in changing the isotopic composition of U, and not just using standard materials modified at other facilities.

Most of the diffusion barrier in both plants was replaced between 1975 and 1983 in a Cascade Improvement Program (CIP). Because of the enormous volume of contaminated material handled during the CIP, it seems likely that most of the site contamination, particularly in sheltered or out-of-the-way places, dates from that time.

An examination of the data in Tables 21.5 and 21.6 shows that there is a significant quantity of ^{236}U present in the samples. The mass-236 isotope is not a component in natural U, arising only through the interaction of neutrons with ^{235}U. This measurement proves that at least some of the cascade feed material had been irradiated in a reactor and that the Pu and Np in the samples probably came into the plants as contamination in the U, rather than as unrelated entities. Furthermore, the mass-240 content of the Pu in the Paducah samples (Table 21.7) is not what one would expect of a material that had been used in high-burnup power applications, but is characteristic of the enrichments found in standard weapons-grade materials. Therefore, even though we did not find any direct evidence of HEU, sample analyses clearly indicated that the plants have some tie to the weapons complex.

The ORIGEN2 code [3,4] was used to understand the isotopic compositions of the site-inspection samples. ORIGEN2 calculates the nuclide content of nuclear fuels at specified power levels and burnup intervals for several types of reactor. The calculated correlations between the mass ratio $^{240}Pu/^{239}Pu$ and the activity ratio $^{238}Pu/(^{239}Pu + ^{240}Pu)$ for three standard reactor types and an external target (blanket) were given earlier. Reactors that are unlikely to have contributed to weapons production in the 1960s (e.g., mixed-fuel or liquid-metal cooled) were not considered. A reactor power level of 37.5 MW_t/t was assumed for calculations. The measured data from the Paducah samples fall between the curves given for CANDU and boiling water reactors (BWR). ^{238}Pu is a fairly short-lived nuclide ($t_{1/2}$ = 88 years); because the samples were likely at least 15 years old at the time of the analysis (midpoint of the CIP), allowance for radioactive decay occurring since the material was discharged from a reactor would lead to a corrected value for $^{238}Pu/(^{239}Pu + ^{240}Pu)$ that is significantly closer to the BWR curve. Moreover, the mass diffusion principle works on Pu isotopes in PuF_6 as readily as it does on U isotopes in UF_6. The Paducah samples consisted of plant tails materials, depleted in low-mass isotopes; consequently, a correction for the action of the diffusion process would also lead to Pu data closer to the BWR curve (see Figure 21.4).

Feed material to the Paducah plant was probably not significantly enriched (or depleted) in ^{235}U. If the Pu component of the plant feed had been produced in a reactor with enriched fuel, a much higher $^{240}Pu/^{239}Pu$ value would have resulted from an irradiation that would have burned the ^{235}U content to near-natural levels. This implies that the production of weapons-grade Pu in the United States took place in

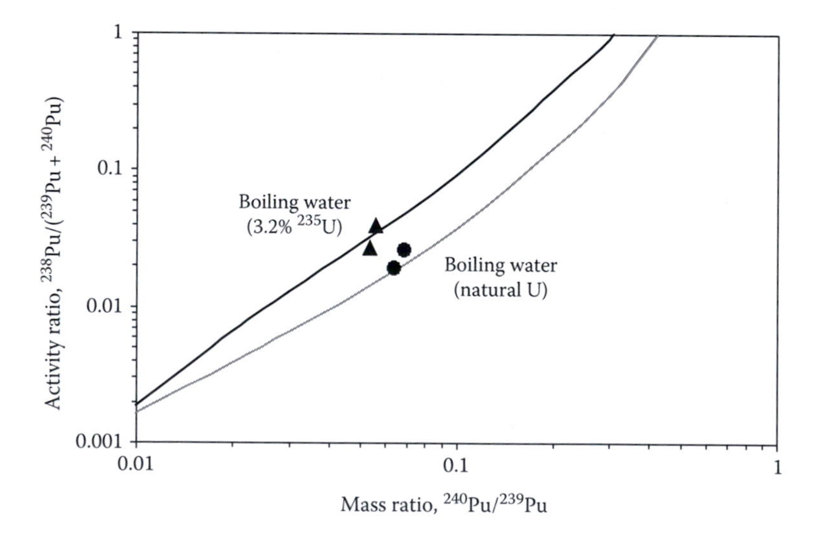

FIGURE 21.4 Results of ORIGEN2 calculations for boiling water reactors and the effect of variable fuel enrichment on Pu observables. Circles are derived from the data in Table 21.7 for samples 94-7-38 and 94-7-45. After correction for the effects of 20 years of radioactive decay (arbitrary) and mass fractionation in the cascade, adjusted points are represented by triangles.

reactors using U fuel of near-natural composition. However, the observables in the Paducah Pu samples do not match the ORIGEN2 calculations for either of the natural-U-fueled standard reactor types (CANDU and blanket), particularly after making the aforementioned corrections. It would be naïve to conclude that Pu production occurred in BWR or pressurized water reactors (PWRs) with lower fuel enrichment; however, an ORIGEN2 calculation for a hypothetical (unrealistic) natural-fuel BWR resulted in the curve shown in Figure 21.4, compared with the same calculation depicted for a 3.2% enrichment ^{235}U BWR in Figure 18 of Chapter 3. Some of the difference between calculations results from the fact that much of the production of ^{238}Pu is from the capture of a neutron by ^{237}Np. At medium-to-long irradiation times, ^{237}Np arises largely from sequential capture of two neutrons by ^{235}U, followed by β decay of 6.7-day ^{237}U. Enrichment in the ^{235}U content of the starting material thereby results in an increase in the production of ^{238}Pu.

Also shown in Figure 21.4 are measured Paducah Pu observables (given as circular data points). The result of correcting the Pu data for a 20-year decay period, and an enrichment (or depletion) of 33% of that defined by the ^{235}U content of the U component of the related samples, is denoted by triangular points (as the mass difference among the Pu isotopes is only ±1 unit, although it is –3 units in the U). Although there seems to be agreement between the original data and the hypothetical "natural-U BWR" curve, modified Pu isotopics very closely match those expected from an enriched-fuel BWR after realistic corrections are made for decay and the effects of mass separation.

There is thus an apparent conflict, in that a natural-U-fueled reactor is required by the comparison of the analytic data with the literature, but the only reactor types in the ORIGEN2 data library that fit the corrected measurements require enriched fuel. However, another path to ^{238}Pu is through the interaction of high-energy neutrons with ^{239}Pu, via the ^{239}Pu(n,2n) reaction. The production of the Pu isotopic mix in the Paducah samples from natural U requires a harder neutron spectrum than that generated by reactors in the calculations, implying a less-efficient (higher-Z) moderator than either light or heavy water. This nuance means that US weapons-grade Pu production took place in graphite-moderated reactors. Unfortunately, the ORIGEN2 libraries available to us did not contain a graphite-moderated, production-reactor dataset. Nevertheless, as most of the U and Pu isotopes are produced through neutron-capture reactions that are relatively spectrum insensitive, it seems likely that Pu and U production in the hypothetical "natural-U BWR" would closely approximate that resulting in the cascade feed material.

The isotopic composition of Pu as a function of burnup in a "natural-U BWR," which we assume is similar to that arising in a graphite-moderated reactor, is given in Figure 21.5, along with the production of ^{239}Pu relative to the unburned ^{238}U remaining in the fuel. The value of ^{239}Pu used in the calculation of these ratios is the sum of the number of atoms of ^{239}U, ^{239}Np, and ^{239}Pu given in the code output at the appropriate irradiation length. Again, a power level of 37.5 MW_t/t is assumed in the calculation. Isotopic ratios obtained in sample 94-7-45 (Table 21.7) are plotted as circular points at an irradiation time of 33 days (1250 MW_t-day/t), which is defined by the ^{240}Pu/^{239}Pu value. The value of ^{242}Pu/^{239}Pu for the sample falls very close to the appropriate curve. ^{241}Pu is a short-lived nuclide ($t_{1/2}$ = 14.4 years), and a decay correction of 41 years from the reference time is required to adjust the observed data to the 241/239 curve. This decay time (i.e., an average time of reactor discharge in 1953) is consistent with the time since the last chemical separation, calculated from the ^{241}Am content of the sample (see following).

The correction of the Pu isotope ratios for the action of the cascade, assuming a linear enrichment with a mass number as estimated from the $^{235}U/^{238}U$ value for sample 94-7-45 (Table 21.6), results in the data plotted as triangular points at an irradiation time of 28.5 days (1070 MW_t-day/t) in Figure 21.5. The fit of the corrected value of $^{242}Pu/^{239}Pu$ to the calculated curve becomes better. The decay correction required to make the adjusted value of $^{241}Pu/^{239}Pu$ fit the calculated curve increases slightly to 42 years. The assumption of a linear relationship between enrichment and isotope mass is probably reasonable for small enrichments (or depletions), such as those resulting in sample 94-7-45; for larger enrichments, it is no longer appropriate (see following).

The comparison of the Pu data from the Paducah sample with the curves given in Figure 21.5 indicates that the accuracy of the calculation for the production of Pu isotopes in a precascade reactor irradiation is good. Therefore, it can be assumed that the change of U isotope ratios from the same ORIGEN2 calculation are given in Figure 21.6, along with the production of ^{237}Np (including 6.7-day ^{237}U) relative to residual ^{238}U.

From Figure 21.5, the discharged reactor fuel that contributed to the Paducah feedstock could be expected to have a $^{239}Pu/^{238}U$ atom ratio of about 10^{-3}. From Table 21.5,

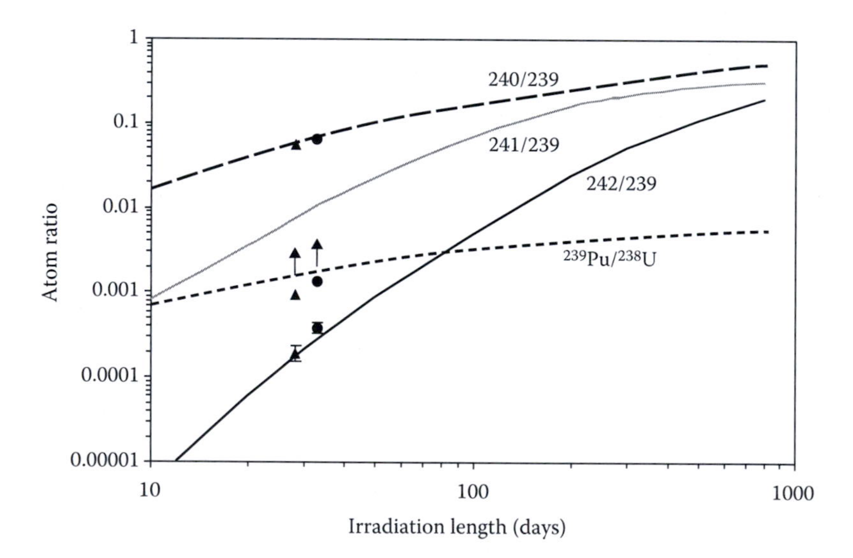

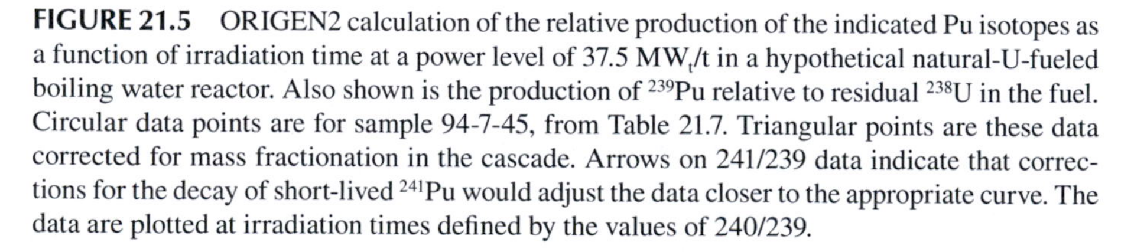
FIGURE 21.5 ORIGEN2 calculation of the relative production of the indicated Pu isotopes as a function of irradiation time at a power level of 37.5 MW_t/t in a hypothetical natural-U-fueled boiling water reactor. Also shown is the production of ^{239}Pu relative to residual ^{238}U in the fuel. Circular data points are for sample 94-7-45, from Table 21.7. Triangular points are these data corrected for mass fractionation in the cascade. Arrows on 241/239 data indicate that corrections for the decay of short-lived ^{241}Pu would adjust the data closer to the appropriate curve. The data are plotted at irradiation times defined by the values of 240/239.

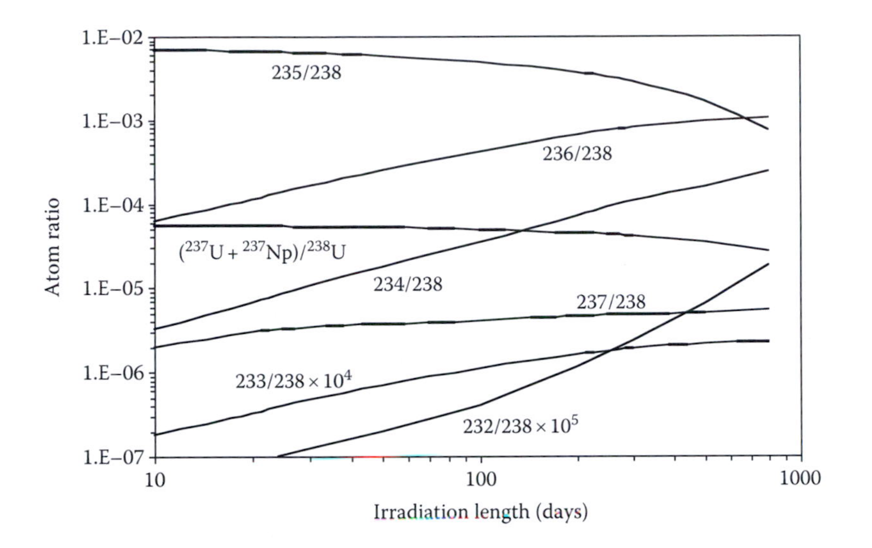

FIGURE 21.6 ORIGEN2 calculation of the relative production of the indicated U isotopes (and ^{237}Np) as a function of irradiation time at a power level of 37.5 MW_t/t in a hypothetical natural-uranium-fueled boiling water reactor. Computation is analogous to Figure 21.5.

the observed ^{239}Pu/^{238}U is about 10^{-7} in the Paducah samples and about 10^{-8} in Portsmouth samples. If the assumption is made that the source of the Portsmouth Pu is the Paducah product and that the mass fractionation of ^{239}Pu from ^{238}U is negligible, then loss of Pu in the Paducah cascade from chemical fractionation (or in the transportation containers) is on the order of 90%. This is not surprising, as volatile PuF_6 is much more chemically reactive than is UF_6 and is reduced to PuF_4 by residual air (both N_2 and O_2) and by small quantities of UF_4 contaminating the UF_6. If the further assumption is made that the irradiated U made up a large fraction of the Paducah feedstock, then the observed ^{239}Pu/^{238}U values are consistent with 10^{-5} to 10^{-4} of the weapons-grade Pu product being lost to the U fraction in the precascade fuel reprocessing.

From Figure 21.6, the ^{237}Np/^{238}U value in the discharged reactor fuel was about 10^{-5}. From Table 21.5, the observed ^{237}Np/^{238}U is on the order of 5×10^{-5} in the Paducah samples and on the order of 5×10^{-6} in the Portsmouth samples. If the same assumptions apply to ^{237}Np as they apply to ^{239}Pu, then loss of Np in the Paducah cascade is also on the order of 90%; volatile NpF_6 is intermediate in chemical reactivity between UF_6 and PuF_6, and a significant cascade loss would be expected. The fuel reprocessing was notably unsuccessful in removing ^{237}Np from the Paducah feedstock, leaving essentially all Np with the U. This last observation reveals a detail about the postprocessing conversion of the product to volatile hexafluorides. Although UF_6 can be produced in high yield by a variety of strong fluorinating agents (e.g., BrF_3, ClF_3, SF_4, etc.) reacting with several different U compounds at temperatures near 220°C, the formation of NpF_6 in high yields requires the reaction of F_2 with lower Np fluorides at higher temperatures. This means that the UF_6 production probably went through a UF_4 intermediate in reaction with elemental fluorine.

Although the relative concentrations of Np and Pu do not provide a conclusive signature, the reprocessing of spent fuel by means of solvent extraction with tributyl phosphate (TBP) (the PUREX process) provides an efficient recovery of Pu and a good decontamination from fission products, but leaves Np with the U. Residual TBP was identified in organic-rich samples collected at the Paducah site. It would have been interesting to look for long-lived ^{129}I or ^{99}Tc in the site-inspection samples, both of which have fluorides that are volatile under the same conditions as is UF_6 and whose contamination levels might provide further clues to the reprocessing technique.

The ^{237}Np and residual Pu comprised a very small mass fraction, and only approximately 1% each of the total α radioactivity, of the Paducah feed material. However, the ^{237}Np content approximately doubled the γ-ray dose from the material, requiring increased shielding, and the presence of Np and Pu in the final product would require additional chemical cleanup for most nonreactor applications. Rather than permanently contaminating the cascade and fouling the plant product, it would have been preferable to clean the material thoroughly before producing the UF_6 feed. There is a possible explanation for use of this less-than-ideal strategy: Once a UF_6-diffusion plant is in operation, it cannot be stopped and restarted without enormous effort and expense, as well as a large lead time. Feed of UF_6 must be continuous to match withdrawals. In the absence of feedstock, the cascade must be operated in a steady-state mode, with no product withdrawn, but with all of the expense of operation. It seems likely that the use of contaminated U in the cascade in the early 1950s (determined from ^{241}Pu decay) was a symptom of a chronic U shortage, where stockpiles were insufficient to support independent programs for the production of weapons-grade Pu and HEU, and the output required of the enrichment cascades by the growing nuclear arsenal [5] did not permit sufficient time for a lengthy decontamination of the feedstock.

The Pu activity ratios given in Table 21.7 can be corrected for enrichment in the cascade by assuming, once again, that the action of the cascade on the Pu isotopes is similar in magnitude to the enrichment of U isotopes and that the effect is roughly linear with mass. After each activity ratio was corrected for an arbitrary decay period of 20 years, the 3.2% enrichment ^{235}U BWR curve in Figure 21.2 was used to estimate the corresponding $^{240}Pu/^{239}Pu$ mass ratio. The ^{238}Pu and ^{240}Pu activities were corrected for the effect of the cascade on their enrichment (defined by ±33% of the ^{235}U enrichment of the sample), the activity ratio was recalculated, and Figure 21.2 was again used to estimate $^{240}Pu/^{239}Pu$. The process was iterated until the value of $^{240}Pu/^{239}Pu$ converged.

Results of the calculation for each sample are given in Table 21.8. The atom ratio, $^{240}Pu/^{239}Pu$, in the feedstock can be used with Figure 21.5 to determine the length of the reactor irradiation resulting in the Pu isotopic mix, in units appropriate for use with Figure 21.6. Both $^{240}Pu/^{239}Pu$ and $^{241}Pu/^{239}Pu$ (from Figure 21.5), corrected for the effect of the cascade on the samples, are given in Table 21.8. However, the ^{241}Pu data contain no correction for decay since the time of reactor discharge. Considering the approximations inherent in computing the data in Table 21.8, resultant values for reactor irradiation length are remarkably consistent, yielding an average exposure of 1100 ± 300 MW_t-day/t for the five samples.

TABLE 21.8
Pu Isotope Production and Reactor Irradiation Length (Assuming 37.5 MW_t/t Power Level)

	$^{238}Pu/(^{239}Pu + ^{240}Pu)$	Atoms $^{240}Pu/^{239}Pu$		Atoms $^{241}Pu/^{239}Pu$	
Sample	Activity, Corrected	Feed	Cascade	Cascade Action	Irradiation Length (d)
94-7-14	0.069	0.083	0.057	0.0083	43
94-7-17	0.017	0.036	0.031	0.0024	20
94-7-19	0.035	0.056	0.023	0.0018	28
94-7-38	0.039	0.060	0.073	0.0140	32
94-7-45	0.027	0.048	0.057	0.0081	25

Note: The $(^{240}Pu/^{239}Pu)_{feed}$ data are taken from the 3.2% enriched ^{235}U BWR curve in Figure 21.4; the $^{238}Pu/(^{239}Pu + ^{240}Pu)$ and $(^{240}Pu/^{239}Pu)_{cascade}$ data are corrected for mass fractionation in the cascade and a 20-year decay period; the $^{241}Pu/^{239}Pu$ data are taken from Figure 21.5, and are corrected for the action of the cascade, but not for decay since reactor discharge.

Irradiation lengths given in Table 21.8 can be used with Figure 21.6 to estimate the starting composition of the reactor-irradiated fraction of the cascade feed that resulted in analytical samples. It should be noted that at these low burnups, atom ratios of $^{234}U/^{238}U$ and $^{235}U/^{238}U$ have not changed much from their values in natural U. The apparent enrichment of each U isotope relative to the cascade feedstock can be calculated from the data in Table 21.6. Enrichment = $(^{x}U/^{238}U)_{observed}/(^{x}U/^{238}U)_{Figure\ 21.6}$, where x is the atomic mass of the isotope of interest. The enrichments of ^{234}U, ^{235}U, and ^{238}U, relative to $^{238}U_{feed}$ for each of the analytical samples, are plotted in Figure 21.7 (lines connecting the points are drawn freehand). It is observed that the enrichment as a function of mass number is not even approximately linear for the Portsmouth samples, casting considerable doubt on calculations resulting in their entries in Table 21.8. For the Paducah samples (used in most of the prior analyses), the linear-enrichment assumption is reasonable.

For each of the five samples, the calculated enrichment of ^{236}U is considerably less than that expected from the freehand curves of Figure 21.7. For samples 94-7-17, 94-7-38, and 94-7-45, the ^{236}U enrichment is approximately 30% of that expected from the curves. For sample 94-7-14, the ^{236}U enrichment is 10% of that expected, and for sample 94-7-19, it is only 4% of that expected. If it is assumed that the ORIGEN2 calculation of ^{236}U production is accurate, then about 30% of the Paducah feedstock was irradiated U and 70% was unirradiated, natural U. Sample 94-7-17 from Portsmouth exhibits the same isotopic fingerprint, thereby connecting it to the Paducah cascade. The other two Portsmouth samples originated from starting materials with a significantly smaller dilution of reactor fuel in natural U. This implies that Portsmouth received its feed from more than one source and that the cascade was subdivided into different sections that received variable proportions of the different materials. An alternative explanation would be that the fraction of reactor-irradiated materials in the feedstock changed over the years and that samples 94-7-14 and 94-7-19 represent materials from a different

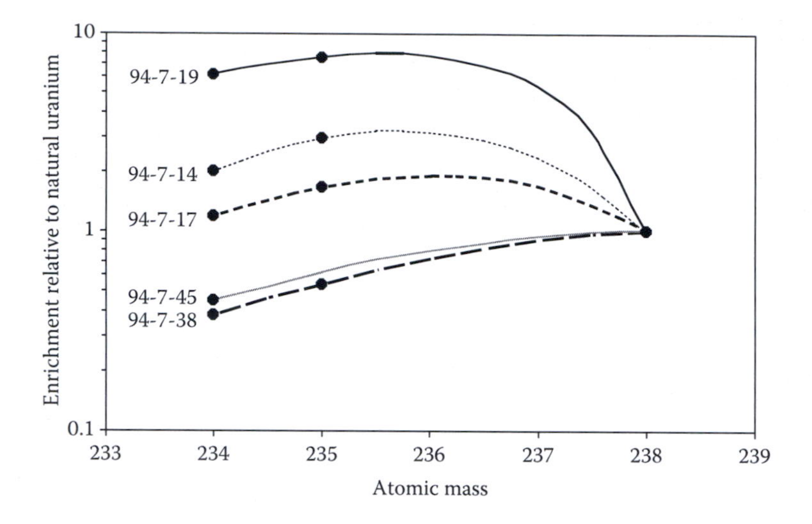

FIGURE 21.7 Enrichment of U isotopes relative to cascade feedstock, defined as $(^{x}U/^{238}U)_{obs}/(^{x}U/^{238}U)_{feed}$, for the five site-inspection samples. Curves are drawn freehand.

operational era at the plant than do the others. The ^{235}U enrichments of 94-7-14 and 94-7-19 are higher than in the other Portsmouth sample. Given that feed or tails materials are more likely to be released into the environment than would be the product, these two samples may be representative of tails from a separate cascade used for the production of HEU.

Samples can be age-dated using pairs of radionuclides that are linked by radioactive decay. It must be assumed that there existed a time in the past when the sample contained only parent activities, meaning that a chemical separation was performed to remove daughters. A gaseous-diffusion enrichment plant would appear to be an ideal facility for age determinations, as a likely mechanism for radioactive release into the environment would be the escape of volatile UF_6 (and NpF_6 and PuF_6) into the air. Such a process would leave refractory Pa, Th, Ac, Ra, and Am daughters and granddaughters behind.

Table 21.5 gives concentrations for several pairs of nuclides that can be used to calculate specimen ages, which are given in Table 21.9. Ages thus calculated from the measurements of relative quantities of volatile and refractory nuclides result in values of the age that are inconsistent and absurdly old. This implies that a chemical separation has indeed occurred, but that the refractory daughter nuclides were dispersed after most of the volatile parent activities were removed. For example, consider the case of a large quantity of UF_6 stored in a container. There is constant production of Th and Pa fluoride species through radioactive decay of the U isotopes. These daughter compounds are nonvolatile and tend to deposit on the walls of the container. If the UF_6 is pumped away (leaving a small residual), a sample of the mixture of nuclides on the walls of the container would appear too old. It would normally be assumed that the daughters arose from the long-term decay of a small mass of the parent, rather than from the short-term decay of a large (missing) mass of the parent. Another way that this mixture could be produced is through the enrichment

TABLE 21.9
Pseudo-Ages of the Site-Inspection Samples (Years)

	Sample				
Chronometer	**94-7-14**	**94-7-17**	**94-7-19**	**94-7-38**	**94-7-45**
$^{230}Th/^{234}U$	1640 ± 160	2490 ± 210	110 ± 8	28,600 ± 3,100	16,200 ± 1,500
$^{226}Ra/^{234}U$				754 ± 62	457 ± 93
$^{231}Pa/^{235}U$	≤1200	≤770	320 ± 25	900 ± 110	430 ± 40
$^{227}Ac/^{235}U$	1200 ± 600	≤250	≤53	≤200	130 ± 40
$^{241}Am/^{241}Pu$[a]					30.5 ± 0.9

Note: The data are calculated from the indicated nuclide pairs, assuming that at the time defining the age there was a complete absence of the radiogenic daughter. No corrections have been made for the radionuclide content of soil.

[a] Calculated from Table 21.5 data, no corrections.

process itself. The decay of U occurs throughout the enrichment cascade, and ThF_4 and PaF_5 would tend to accumulate in the diffusion barrier.

If the assumption is made that the ^{230}Th activity in each sample was produced from a steady-state ingrowth of 10 years (the proper order of magnitude), then 3.5×10^4 atoms of ^{234}U were required to account for each atom of ^{230}Th observed. If, at the end of the accumulation period, all volatile UF_6 was removed from the sample, the data in Table 21.5 can be used to estimate the loss in the process caused by chemical reactions producing nonvolatile uranium compounds (e.g., UO_2F_2). In Paducah samples, about 0.05% of the U remained with its Th daughter. In Portsmouth samples 94-7-14 and 94-7-17, 0.5% of the U remained, and sample 94-7-19 retained 10% of its original U content. It seems unlikely that the diffusion barrier would clog to this extent before it was replaced or that this magnitude of accumulation on the walls of a container would not be recovered. Sample 94-7-19 is clearly more volatile in nature than are the other four samples, and it may have had contributions through both a mechanical dispersal of daughters and a volatile gas release (see later).

The ages determined from the relative quantities of nonvolatile daughter and granddaughter activities (^{230}Th and ^{226}Ra, or ^{231}Pa and ^{227}Ac) are lower than expected, possibly as the result of the presence of a background caused by the incorporation of dirt within samples. The activity ratio of daughter/granddaughter is about 1.0 in soil, but is much higher for samples in which the daughter was recently produced. The ^{232}Th content of each sample provides a means of estimating the soil contribution to its radionuclide content. The earth's crust contains an average of 12 ppm ^{232}Th. If soil contributions to analytical samples contain Th in the same proportion, data in Table 21.5 can be used to calculate the soil components given in Table 21.10. (Also given are the original sample weights from Table 21.1.) Soil contributed only a small fraction to the masses of the Portsmouth samples, all of which were collected magnetically. Paducah sample 94-7-38 was collected with a spatula, and only about one-third of its mass could be accounted for by XRF analysis (Table 21.2). It was therefore comprised principally of low-*Z* dirt. Paducah sample 94-7-45 was part of

TABLE 21.10
Soil Contributions to the Radionuclide Contents of the Site-Inspection Samples, Calculated from ^{232}Th Contents Given in Table 21.5

	94-7-14	94-7-17	94-7-19	94-7-38	94-7-45
Sample weight (g)	7.0	1.4	3.3	4.9	20.5
Soil content (g)	0.12	0.12	0.07	4.8	10.5
Soil U atoms	9.1×10^{14}	9.1×10^{14}	5.1×10^{14}	3.6×10^{16}	8.0×10^{16}
^{231}Pa soil fraction	≥0.5%	≥0.2%	0.03%	4.3%	3.2%
^{230}Th soil fraction	1.0%	0.2%	0.2%	0.5%	0.3%
^{227}Ac soil fraction			≥0.4%	≥23%	15%
^{226}Ra soil fraction				100%	85%

Note: Uranium soil content is assumed to be 25% of the Th content.

a larger sample from which Ni flakes were segregated during prescreening. Even though it was collected magnetically, it is reasonable to believe that about one-half of the residue was dirt entrained during collection.

The ^{238}U content of soil is generally approximately 25% that of ^{232}Th. On that basis, an estimate of the soil contribution to the U in each sample is given in Table 21.10. Comparing these values to the data in Table 21.5, it was concluded that soil did not significantly perturb either the U content or the relative U isotopics of any of the samples. If it is assumed that the radiogenic daughters of the soil nuclides are in equilibrium with their parents, then corrections to data in Table 21.5 for ^{231}Pa, ^{230}Th, ^{227}Ac, and ^{226}Ra can be calculated. These corrections are also provided in Table 21.10. Any corrections to the ^{230}Th and ^{231}Pa data are small and would not be significant given the precision of the measurements. The soil contribution to the ^{226}Ra content of the Paducah samples is quite high, however—on the order of 100%.

If it is valid to assume that the mix of ^{231}Pa and ^{230}Th in each sample arose from the decay of U of the same composition as that in the samples, then the same (incorrect) ages should be calculated from both the ^{230}Th/^{234}U and ^{231}Pa/^{235}U nuclide pairs. This is clearly not the case. With the exception of sample 94-7-19, the ^{231}Pa ages are significantly lower than the ^{230}Th ages, indicating that ^{231}Pa was removed from the samples after release. Another way to look at this is as follows: The atom-ratio ^{235}U/^{234}U, in the material whose decay resulted in the nonsoil radionuclides in samples, can be calculated from the U half-lives and the ^{231}Pa/^{230}Th atom ratio. These data are given in Table 21.11, along with the direct measurement of ^{235}U/^{234}U derived from Table 21.6. Again, with the exception of sample 94-7-19, there is less ^{231}Pa in samples than there should be. One might consider a possible explanation to be that the U composition of the samples is unrelated to the daughter content. However, this could not explain either of the Paducah samples, for which the derived values of ^{235}U/^{234}U are an order of magnitude lower than would be produced by gaseous diffusion of the known starting materials.

An explanation for the loss of ^{231}Pa lies in its aqueous chemistry. PaF_5 is soluble in water (in fact, it is deliquescent), whereas AcF_3 and ThF_4 are not. The 94-7-14 and 94-7-17 samples from Portsmouth may have been moist, resulting in partial leaching

TABLE 21.11
Evidence of Chemical Activity in the Site-Inspection Samples after Release, from Chronometric Nuclides

	94-7-14	94-7-17	94-7-19	94-7-38	94-7-45
Direct	167 ± 2	176 ± 2	97 ± 1	169 ± 2	164 ± 2
Daughter	≤125	≤56	290 ± 20	6.3 ± 0.7	4.8 ± 0.4

Notes: Calculated atom ratio $^{235}U/^{234}U$ from Table 21.6 ("direct") and from the relative amounts of ^{231}Pa and ^{230}Th given in Table 21.5 ("daughter"). Natural uranium $^{235}U/^{234}U$ = 132.

of ^{231}Pa. The ^{231}Pa content of the Paducah samples is so severely reduced that it seems likely that water was used in the process that resulted in the dispersal of the material. Both samples were collected well above floor level, indicating the possible use of pressurized water. Because uranyl fluoride (UO_2F_2, the product of the reaction of UF_6 with moist air) is also water soluble, it is likely that the process U loss calculated earlier for the Paducah samples is actually an order of magnitude too low and that Portsmouth and Paducah both suffered from process losses on the order of 1%.

The mix of activities in sample 94-7-19 is difficult to explain. The relative proportions of ^{231}Pa and ^{230}Th in the material are inconsistent with U isotopics, even with a correction for different solubilities. Much of the radionuclide content of the sample may have been leached from other materials by water, which would explain the elevated levels of U and ^{231}Pa relative to other activities.

The only age in Table 21.9 that is consistent with samples being generated during the CIP is the $^{241}Am/^{241}Pu$ value given for sample 94-7-45. If the ^{241}Am component of the sample was the result of its long-term accumulation in barrier material while (partially) volatile PuF_6 flowed by, the computed age would be significantly higher than the actual age. In fact, it is impossible to calculate an age for probable-barrier samples 94-7-14 and 94-7-17 from the calculated ^{241}Pu data in Table 21.8 because there is more measured ^{241}Am in samples than can be attributed to full decay. However, Table 21.2 indicates that potential barrier material (Ni) made up less than 10% of sample 94-7-45. It seems likely that the radionuclides in this specimen were introduced with debris components that were not barrier related. These might be fragments from the walls of a container used for storage of tails material, on which a large fraction of the highly reactive PuF_6 had reduced to a nonvolatile compound. This would result in an age value that was only slightly too old. The midpoint in plant operations between startup (~1955) and the CIP (~1979) is 1967, or an age of 27 years relative to the analysis time.

21.6 SUMMARY

The radionuclide contents of samples taken during the site inspections of the Paducah and Portsmouth enrichment plants revealed not only information about operations at the plants, but also about the US weapons complex as a whole.

Feedstock to the enrichment plants contained a significant quantity of U that had been irradiated in a natural-U-fueled, graphite-moderated reactor for the purpose of producing weapons-grade Pu. The recovery of Pu from this fuel was likely accomplished with the PUREX process, leaving about $10^{-3}\%$ of the Pu and virtually all of the by-product ^{237}Np with the U. The use of this incompletely decontaminated material in the cascade feed was probably required by a U shortage occurring sometime after the plants started operations. The UF_6 cascade feedstock was produced through the action of free fluorine on UF_4, rather than with other fluorinating agents.

The same isotopic signatures exist in the Paducah and Portsmouth feedstocks. However, feed of a different signature was also used in Portsmouth, implying the presence of more than one enrichment cascade. This separate cascade had tails with a moderate enrichment in ^{235}U, as would be expected from a process for the production of HEU (weapons grade). Cascade losses of U by both Portsmouth and Paducah were on the order of 1%; about 90% of the Pu and Np introduced with the Paducah feedstock were lost in the cascade.

The site contamination at both plants took place by dispersion of nonvolatile radioactive compounds, rather than through leaks of volatile UF_6 into the gas phase. In the Portsmouth facility, this took place mechanically, whereas in the Paducah facility high-pressure water may have been used. It proved difficult to age-date any of the operations at the plant. However, the decay of ^{241}Pu relative to other Pu isotopes, and ingrowth of ^{241}Am, indicated enrichment operations in Paducah occurring during the 1950s and 1960s.

REFERENCES

1. J.R. Merriman and M. Benedict, eds., *Recent Developments in Uranium Enrichment*, AIChE Symposium Series 221, New York, 1982.
2. S. Villani, ed., *Uranium Enrichment*, Springer, Berlin, Germany, 1979.
3. A.G. Croff, Origen2: A versatile computer code for calculating the nuclide composition and characteristics of nuclear materials, *Nucl. Technol.*, 62, 335, 1983.
4. A.G. Croff, *A User's Manual for the ORIGEN2 Computer Code*, ORNL/TM-7175, RSICC Code Collection CCC 371, Oak Ridge National Laboratory, Oak Ridge, TN, 1980.
5. T.B. Cochran, W.M. Arkin, R.S. Norris, and M.M. Hoenig, *Nuclear Weapons Databook, Vol. II, U.S. Nuclear Warhead Production*, Ballinger Publishing, Cambridge, MA, 1987, Chapter 3.

22 Nuclear Smuggling Hoax
D-38 Counterweight

22.1 BACKGROUND AND ANALYSES

A dense, 8.96-kg metal specimen, dark gray in color, was involved in a sale of illicit nuclear materials in Hong Kong in 1988. It had been offered for sale as nuclear weapons material by a Southeast Asian military official. The part was subsequently "rediscovered" in a US consulate nearly 10 years later, and forensic characterization was requested.

The piece was indeed radioactive, and it had a complex shape that indicated it was intended to be bolted to another part (a photograph may be found in Ref. [1] and on the back cover of this book). Arc-shaped wear patterns centered on large, circular depressions on opposite sides of the part indicated that the specimen was possibly a component of an assembly that functioned as a type of hinge, perhaps in a shipping container.

Nondestructive γ analysis with an HPGe spectrometer system revealed that the main radioactive component of the specimen was uranium that was considerably depleted in ^{235}U. The surface α-emission rate measured with a portable survey instrument gave a maximum of approximately 1000 counts/min, which was 10 times lower than that expected from the surface of bare U metal. Bulk analysis of the sample resulted in a density of (17 ± 0.3) g/cm^3, somewhat lower than the density of pure U, and the surface of the questioned part was mottled and stained in a way that indicated it had not been stored in a protected environment. However, pure U metal would have oxidized to an extent that the surface would no longer have appeared metallic.

The reduced density of the part indicated that voids could be present or that it was composed of two or more inhomogeneous phases, but a 9-MeV x-ray radiograph of the piece with an electron linac revealed no apparent voids or discontinuities in the interior. Qualitative x-ray fluorescence spectrometry (XRF) with an ^{241}Am excitation source was complicated by the presence of photons from the specimen itself, as well as from the external source. Nevertheless, the method determined that the part was coated with Ni and that some of the internal U was exposed. However, although the Ni cladding had pulled away from the underlying U metal in spots, particle-microprobe microscopy indicated that it was still well attached in general. The nondestructive analysis γ-results revealed that the Ni coating had only a negligible effect on the self-absorption of low-energy photons, so it had to be <1 mm thick.

The sample was leached with ultrapure 0.4 M HNO_3 for 5 min to remove milligram quantities of material of the outermost layer. One corner of the specimen (~3 cm^2 of surface area) was then immersed in ultrapure 8 M HNO_3 for 5 min to dissolve some of the Ni cladding (but not penetrate to the underlying U). These two HNO_3 solutions were then subjected to forensic radiochemistry. Because of the pyrophoricity of U metal, the bulk specimen was placed in an Ar-filled glove bag, and a 3-g subsample was removed with a hacksaw. The cut surface of this analytical sample was polished and examined by electron and ion microprobes. It was subsequently dissolved in ultrapure 8 M HNO_3, to which a few drops of 4 M HCl and one drop of 6 M HF had been added, for isotopic and elemental assays.

22.2 RESULTS AND DISCUSSION

The results of these analyses showed that the material was predominantly depleted U (D-38) that contained approximately 0.3 wt.% ^{235}U, but was actually an alloy of 90% U with 10% Mo. The Ni surface layer had been electroplated and ranged in thickness between 85 and 150 μm, with about 90 μm being a reasonable average. It was >99% pure and <0.2 wt.% in both Cr and Fe. A crenulated outer margin of the U implied that the piece had been cast in shape and had not been machined before being plated with Ni. The ICP-MS analyses also detected a small quantity of Cd, but only in the surface-leach solution. As Cd is often used to coat moving parts due to its lubricating properties, this result reinforced the concept of a hinge-like apparatus.

Radiochemical analyses of the various analytic solutions identified, or set limits for, a number of long-lived actinide, fission-product, and activation-product species. However, no evidence for any use involving other radionuclides was found. This D-38 was not the modern standard, as levels of minor isotopes were too high, and the presence of ^{236}U precluded mixing with natural U. Radiochronometry measured the date of its last chemical purification as 1961 (±3 years).

With the aforementioned information obtained from the forensic analyses, conventional detective work was undertaken to formulate any valid conclusions for attribution purposes. One such finding was the absence of any evidence for an association of either fabrication or application by the nuclear industry. Telephone inquiries revealed that the part had been made by the National Lead Company of Albany, New York, which was now out of business, and that the historical records had been transferred to Nuclear Metals, Inc. The questioned D-38 specimen was most likely a piece of an aircraft counterweight assembly, probably military in origin.

In the nuclear smuggling world, this specimen was one of the earliest contraband items in what ultimately became known as the "Southeast Asian Uranium" scam (see Section 1.7). This hoax is a pervasive swindle, first reported in 1991 and particularly prevalent in Thailand, Vietnam, and Cambodia. Transactions are in specimens of irregularly shaped metal objects, alleged to be ^{235}U, with sales costs on the order of $10,000 (US) per piece. The smuggled U is also used for barter as substitute currency in drug trafficking operations.

REFERENCE

1. P.M. Grant, K.J. Moody, I.D. Hutcheon et al., Nuclear forensics in law enforcement applications, *J. Radioanal. Nucl. Chem.*, 235(1–2), 129, 1998.

23 Nuclear Smuggling Hoax
Sc Metal

23.1 BACKGROUND AND ANALYSES

In the mid-1990s, a confidential informant in Moldova provided law enforcement with a small sample of gray metal in a 35-mm film canister. It was said to be representative of a much larger quantity of material that was available for sale as a replacement for highly enriched uranium in nuclear weapons. The law enforcement agency requested comprehensive nuclear forensic analysis of the specimen. This investigation was strictly for intelligence information, requiring neither sample documentation nor adherence to federal rules of evidence for criminal prosecution (chain-of-custody protocols were followed nonetheless upon sample arrival at the laboratory). However, it was necessary to return at least one-half of the specimen to the informant within a few weeks, and the analyses were constrained to being as nondestructive as possible.

The submitted sample weighed <0.4 g and measured <1 cm in any direction. Multidisciplinary forensic analyses were conducted on the entire piece (nondestructive analysis only) and on milligram-quantity subsamples of the material. Nuclear counting for α- and β-particle surface emission, and HPGe spectrometry for γ-ray activity, were both negative. Qualitative XRF analysis indicated that the specimen consisted primarily of elemental Sc.

At the 2σ confidence level, diagnostic radiochemistry measured a solitary nuclear signature, ^{230}Th, at a trace level. The half-life of this nuclide is too short for its existence in nature from primordial nucleosynthesis, but it is found as a daughter product in the decay chain of naturally occurring U. The presence of ^{230}Th thus indicated that the sample was linked to uranium at some point in its history. It was quite possible that it originated in a U mine and was refined and isolated as an ancillary product by chemical processing.

The metallurgy of the Sc specimen indicated that it was derived from a casting pour. Electron-probe analyses showed differences in the concentrations of six elements at surface and interior locations and gave detailed depth profiles of Sc, Fe, and Cr within the sample. Quantitative measurements of 55 minor and trace elements were performed by SIMS and ICP-MS. The major impurities in the material, at several tenths of a weight %, were P, C, Cr, and Fe. This relatively large concentration of P similarly implicated U mining or processing in the history of the sample, as phosphoric acid refining methods have been used for the industrial recovery of U from ores. In addition, P-based reagents (e.g., TBP and di-2-ethylhexyl orthophosphoric acid [HDEHP]) are often the basis of U solvent-extraction procedures.

23.2 RESULTS AND DISCUSSION

The questioned specimen was thus determined to be relatively clean Sc metal that had been purified from U-rich materials. It was likely that the origin of the sample was a U mine and that it had been later refined by chemical methods. However, no fissionable species or any other weaponization signatures were detected. Although the chemical properties of Sc are analogous to those of actinides and lanthanides, its nuclear attributes are not. The alleged use of the sample as substitute bomb material was clearly a scam.

As is often the case with real-world unknowns, however, some puzzling characteristics of the specimen were also measured. The distribution of lanthanide concentrations was very strange, with the levels of Tb, Dy, Ho, and Er elevated by factors of 50–500 over normal terrestrial values. It was speculated that perhaps practical-grade erbia (Er_2O_3) had been used as a crucible mold wash before casting the Sc metal, but this conjecture was very tentative. In addition, organic analyses of surficial contamination on the Sc metal revealed a number of long, even- and odd-chain fatty acids. Even-chain acids were typical of human handling and were therefore expected. However, odd-chain acids were characteristic of synthetic processes and were unusual in that they are rarely found in questioned forensic specimens. Although some of the detected organic species were anthropogenic, others were artificial products of unidentified origin.

Detailed forensic protocols, specific data, and more extensive discussion can be found in the primary literature reference for this investigation [1].

REFERENCE

1. P.M. Grant, K.J. Moody, I.D. Hutcheon et al., Forensic analyses of suspect illicit nuclear material, *J. Forensic Sci.*, 43(3), 680, 1998.

24 Fatal "Cold Fusion" Explosion

24.1 BACKGROUND AND ANALYSES

On January 2, 1992, an explosion in an electrochemistry laboratory at SRI International in Menlo Park, California, killed scientist Andrew Riley, injured three others, and caused extensive lab damage. Riley and others at SRI were conducting experiments in a relatively new, but widely controversial, field of study termed "cold fusion." Beginning with a report by Pons and Fleischmann in 1989 [1], investigators around the world reported the production of small quantities of excess heat in electrolysis experiments with heavy water (D_2O). Some also measured the emission of conventional nuclear fusion signature species (such as neutrons, ^3H, ^4He, and prompt γ rays) accompanying the surplus heat, while others could not reproduce the phenomenon at all.

The Lawrence Livermore National Laboratory, through its Forensic Science Center, conducted forensic examinations of explosion debris collected by the local fire department and relinquished to California's Division of Occupational Safety and Health (Cal-OSHA). This investigation was likely the first comprehensive nuclear forensic endeavor, as developed in this book, of an unclassified nature, and it was designed to elucidate the cause or causes of the explosion. It was notable for its multidisciplinary character. The applied analytic methodologies encompassed the technical realms of nuclear science (γ-ray spectrometry and liquid scintillation counting), physical science (optical microscopy, scanning electron microscopy, and x-ray radiography), inorganic and organic chemistry (inductively coupled plasma mass spectrometry, gas chromatography–mass spectrometry, and ion chromatography), isotopic analysis (ion microprobe), materials science (x-ray fluorescence spectroscopy and inductively coupled plasma mass spectrometry), and engineering (metallurgy and weld inspection).

At the time of the incident, the tragedy of Dr. Riley's death was diluted somewhat by the intriguing speculation that finally, after several years of marginal experimentation on an international scale, the SRI experiment had somehow managed to find the Holy Grail and unleash the energy promise of "cold fusion," albeit in an uncontrolled excursion. The first examinations conducted at Livermore were nondestructive γ-ray interrogations of various specimens of metal debris to search for evidence of neutron activation products. Amid the heated disagreements over "cold fusion" during the early 1990s, neutron production as a fusion by-product managed to achieve some degree of credibility among the scientific skeptics, and approximately 50 laboratories worldwide had reported the simultaneous detection of neutron emission in their electrochemistry experiments. Indeed, an initial account of the fatal explosion reported that SRI had subsequently buried other "cold fusion" cells to safeguard against exposure to radioactivity induced by analogous experimentation [2].

24.2 RESULTS AND DISCUSSION

Had the SRI explosion also produced neutron emission, detectable radioactivation species could have been produced in the 1.8-kg stainless steel cell or the 0.5 kg of brass heat-exchanger fins that surrounded the explosion in close proximity. This debris was counted for several days with state-of-the-art Ge(Li) and HPGe γ-spectrometer systems, in ultralow background facilities, for approximately a dozen radioactive nuclides that possessed adequate γ intensities and sufficient half-lives to have survived from the blast to the count times. None were detected, however, and only natural background activities ($\gamma^{\pm}$, ^{40}K, U, and Th daughter products) were measured in the spectra. Thus, no evidence of signature species indicative of neutrons or any other orthodox nuclear events was found in the metal debris from the subject explosion.

However, the various nonnuclear analyses in the investigation provided a wealth of collateral information. Materials analyses and metallurgy indicated the type of stainless steel used in the electrolytic cell, the quality of the end-cap weld, and an absence of any corrosion or hydrogen embrittlement. Nondestructive radiography at 200-keV and 4-MeV energies revealed no apparent obstructions in the gas-phase pathways of the experiment. Analyses of recovered particles by ion and electron microprobes assayed explosion detritus to be an alumina matrix with approximately 10-μg Pt inclusions—likely remnants of the recombination catalyst in the electrochemical system. They also measured LiOH and LiOD with enriched D/H ratios. Inorganic analyses determined the presence of Li, Pd, Pt, and other analytes in the apparatus but did not detect nitrate, nitrite, or any other low-explosive signatures. The principal significance of these data was that, had the circumstances of the explosion been unknown a priori, the discovery of Pt, Pd, Li, and enriched D/H in the debris would have led to an identification of the device as a "cold fusion" experiment with a high degree of probability.

The most salient finding of this investigation resulted from a most unlikely protocol: The electrolysis cell was sampled for any organic compounds remaining on the interior wall after the blast. Because the various attempts at "cold fusion" implemented environments that were wholly inorganic, this tactic was questioned by some, but was performed nevertheless for forensic completeness. By means of modern gas chromatography–mass spectrometric analyses, nanogram detection limits were placed on diverse nefarious species, such as conventional high explosives, propellants, accelerants, oxidizers, and other exceptional industrial chemicals. However, an organic oil, likely residual lubricant from the extensive prior fabrication of the cell in a machine shop, was measured at levels estimated to be a few hundred micrograms and, among all of the submitted evidence, was detected only in the cell interior.

The importance of this finding derives from the material mechanisms at work during the "cold fusion" process. D_2O is electrolyzed to form gaseous $D_2\uparrow$ and $O_2\uparrow$ in the cell interior. As the D_2 absorbs into a Pd cathode to produce the anomalous heating effect (by whatever mechanisms), the cell headspace becomes increasingly enriched in pressurized O_2. It is well known that hydrocarbon oils in contact with high-pressure O_2 have been sources of significant energetic incidents over many years. For this reason, apparatus used with overpressures of O_2 (such as gas lines in hospitals) must be thoroughly cleaned with efficient solvents to remove all organic residues to avoid explosions and catastrophic pressures from the heats of combustion of the contaminants.

Residual oil present in the interior of the SRI cell when pressurized O_2 was generated by electrolysis could have resulted in an exothermic, explosive reaction. Various phthalates and silicones were also detected in the SRI cell by gas chromatography–mass spectrometry (at relatively lower levels), and the consumption of diverse organic residues as combustion sources appeared quite credible in the investigation. As a consequence, it was concluded that oxidation of the oil could have reasonably contributed to the overall energy release, as well as perhaps the initiation, of the SRI explosion. The total quantity of oil inside the cell at the time of the accident could not be projected, but it would be logical to infer it to be greater than the hundreds of micrograms remaining after the strong explosion. Cal-OSHA subsequently imposed effective cleaning procedures on SRI, as if for oxygen service in a health-care facility, for all internal cell surfaces and components of future "cold fusion" experiments. The requirement was part of an OSHA Special Order, issued on June 29, 1992, before the resumption of laboratory work was permitted.

24.3 COMMENTARY

Photographs and the forensic details of the nuclear [3] and conventional [4] aspects of this inquiry were published in refereed journals. They should have logically been incorporated into one paper, but one referee for the *Journal of Forensic Sciences* did not agree with reporting the nuclear forensic data. As a consequence, the aspect of the investigation that initially held the highest interest for a technical audience was excluded from publication in a premier forensic journal. However, nuclear forensic analysis was an unusual field of examination at that time, and with today's potential IND and RDD terrorist threats, such prejudice has not persisted.

An interesting sociologic and scientific postscript to this investigation has not been previously described. Cal-OSHA was mandated by law to issue any special order justified by an event within a period of 6 months after the incident. Negotiations between the legal staffs of SRI and OSHA delayed the onset of analyses at Livermore by 3 months and precluded assays of any potential short-lived neutron activation products. (Such radionuclides would have been the most sensitive indicators of a neutron pulse.) Nonetheless, Livermore completed the forensic work in a timely fashion, and the Forensic Science Center submitted a confidential, 96-page report to Cal-OSHA detailing the laboratory results and conclusions several weeks before the 6-month deadline. OSHA shared the report with SRI.

The Forensic Science Center director then received a telephone call from an SRI scientist troubled with the content of the report. He opined that the organic findings and inferences were surely wrong for the subject explosion and could not have been the sole source of the incident (a possibility never contended by Livermore). Further, the work was poorly done, was incorrect, and, most importantly, could have a serious, negative impact on the entire enterprise of "cold fusion" if made public. The director was urged to send a "clarification" letter to Cal-OSHA to further expand on and explain any implications of the organic oil for this incident.

Very soon after that, and before the OSHA Special Order, SRI convened a press conference and released an explanation for the explosion that derived from joint studies between themselves and their "cold fusion" sponsor, the Electric Power Research Institute (EPRI). SRI and EPRI had established several intramural subcommittees to investigate various facets of the experiment and the accident. An SRI Scientific Investigative Committee presented the resultant reports to an Independent Review Committee (also created by SRI and EPRI), composed of three extramural members. The independent committee completely agreed with all SRI findings and conclusions and did not contribute any original insight to the analysis. When, sometime later, SRI was asked to comment on the Livermore result of a possible contribution by organic oil, they dismissed the possibility entirely ("SRI spokeswoman Heather Page scoffed at the theory. She said SRI scientist Michael McKubre—a participant in the experiment—calls it 'completely erroneous.'" [5]). It appears that an evaluation of SRI's theory of the explosion, within the context of the actual forensic measurements performed by Livermore, was never made. The following is an attempt at such assessment.

The first steps in the SRI analysis were computations and theoretical modeling of the experimental configuration. Subsequent comparison with the explosion-induced deformation of the stainless-steel electrolysis cell implicated a blast in the vapor head space, as opposed to within the condensed heavy-water phase. Measurements of the dimensions of the distended vessel allowed a calculation of the internal detonation pressure induced by the explosion. The result was 300 atm. SRI then assumed that the sole source of the explosion was the $D_2 + O_2$ recombination reaction and, from knowledge of its chemistry and kinetics, calculated that the cell overpressure just prior to ignition was 30 atm (with a total energy inventory of 30–40 kJ).

However, a leak repair performed on the vessel the day before the explosion indicated that, to all appearances, the cell was venting freely to the ambient environment. With the experimental system parameters in effect from then until the time of the blast, it was not feasible to build up an internal pressure of 30 atm from a starting pressure of 1 atm. As a consequence, SRI speculated that the repair on January 1, 1992, did not completely vent the apparatus and that perhaps the Teflon insert in the vessel end-cap occluded the exit hole of the gas tube, thereby effectively sealing the cell interior.

Then, the 5 wt.% Pt-Al_2O_3 recombiner catalyst was to have failed catastrophically for the pressure to build to 30 atm (now from an arbitrary initial pressure). The review committee further speculated that the catalyst was poisoned, perhaps by frothing LiOD, D_2O condensation, or Pt sintering. The initiation of a 30 atm D_2 + O_2 runaway reaction was induced by removal of the cell from the calorimeter. This maneuver allegedly activated friable, fragmenting catalyst, or perhaps exposed a dry Pd cathode by tipping the vessel, to cause the ensuing deflagration. The latter postulate was considered more likely in the opinion of the review committee.

There were a number of difficulties with this theory when considered in conjunction with the results of the forensic measurements performed at Livermore. The conjecture that the Teflon insert of the cell end-cap somehow plugged the gas-tube exit at the top of the apparatus was unsupported by direct observation. Radiography revealed an unrestricted air gap between the insert and the metal cap. It also showed that the Teflon was skewed within the vessel cap; for an effective seal of the gas tube, however, it should have been in even contact with the metal cap. If the insert had sealed the gas tube from at least 1/1/92 to 1/2/92, it must have been dislodged by the pressure release or by postdetonation interactions. As for cell collisions potentially dislodging a tight Teflon plug, SRI's own kinetic-energy analyses indicated that most of the energy of the rocketing vessel was absorbed by nonrigid matter (Dr. Riley's head). Further, the cell vessel cap was reasonably undamaged by the incident and underwent disassembly at Livermore with minimal effort. The disassembly procedure gave no evidence of Teflon extruded into the gas tube or of any depressions or deformations of the Teflon piece at the entrance to the gas tube.

Presumption that the recombiner catalyst must have failed was also dubious because the catalyst should have become even more effective at increased gas pressure. According to an SRI personal communication, only two of the approximately two dozen catalyst balls in the electrochemical system were necessary to achieve complete $D_2 + O_2$ recombination and to maintain cell equilibrium. After the explosion, about half of the catalyst spheres were found reasonably intact. Moreover, one recovered ball (SRI debris specimen 36) was assessed to be undamaged by the detonation, to be uncontaminated by Li or anything else, and to actually exhibit more catalytic activity than an average, unused sphere.

The SRI explosion theory therefore required complete catalyst deactivation and a preexistence of pressure at the repair of the cell leak conducted at 845 operating hours (e.g., by [unapparent] blockage of the gas tube by the Teflon insert of the cap). However, these hypotheses were inconsistent with SRI's own documented considerations of approximately 10^5 operating hours of catalyst experience, the observed gas venting rate before the explosion, the measured cell voltage (and implied negligible pressure rise from 843 to 865 h), the recombination kinetics, mass balance, and the cell excess power readings from 844 to 865 h.

However, the Livermore-postulated mechanism of an O_2 reaction with lubricating oil as a possible factor was arguably more reasonable. Oil residues in the cell interior were definitely established by forensic analysis, and this oil provided an additional exothermic energy source within the electrolytic system. Moreover, the abundant Pt metal used in the experiment was also an excellent catalyst for carbon oxidation, and imposed conditions of complete catalyst failure and 30 atm of $D_2 + O_2$ overpressure would be unnecessary. The SRI scenario might therefore have realized greater credibility had the possibility of the oil contamination been admitted within their overall conjectures.

The potential contribution of any organic oil to the energy output of the explosion can be appreciated from the following considerations: Oils employed as common lubricants correspond to a general range of C_{20}–C_{38} branched hydrocarbons and have densities of about 0.8 g/mL. Empirical thermochemical measurements of, for example, $C_{24}H_x$ indicate that the enthalpy of combustion of 1 mg (~1.2 μL) of such materials is $\Delta H \sim -40$ J. Therefore, as little as 0.1 mL of lubricating oil in the vapor-phase reaction mixture of the SRI cell would have yielded about 3–4 kJ of energy if completely combusted. This value is 10% of the total energy generated in the accident, as calculated by SRI. One-tenth of a milliliter of oil is not a large volume. Although it represents a greater mass than the hundreds of micrograms estimated from the forensic analyses, it should be recognized that the latter was but a minimal value that reflected the residue remaining on the cell interior following a blast of considerable force.

REFERENCES

1. M. Fleischmann, S. Pons, and M. Hawkins, Electrochemically induced nuclear fusion of deuterium, *J. Electroanal. Chem.*, 261, 301, 1989.
2. C. Anderson, Cold fusion explosion kills one, *Nature*, 355, 102, 1992.
3. P.M. Grant, R.E. Whipple, F. Bazan et al., Search for evidence of nuclear involvement in the fatal explosion of a "cold fusion" experiment, *J. Radioanal. Nucl. Chem.*, 193, 165, 1995.
4. P.M. Grant, R.E. Whipple, and B.D. Andresen, Comprehensive forensic analyses of debris from the fatal explosion of a "cold fusion" electrochemical cell, *J. Forensic Sci.*, 40, 18, 1995.
5. K. Davidson, *San Francisco Examiner*, March 13,1994.

25 Questioned Sample from the U.S. Drug Enforcement Agency

25.1 BACKGROUND AND NONDESTRUCTIVE ANALYSIS

The Livermore Forensic Science Center was contacted by agents of the Drug Enforcement Agency (DEA), who had obtained a sample of radioactive material in the raid of a clandestine drug laboratory. They requested a nuclear forensic analysis of the confiscated sample, principally to complete their investigation and dispose of the radioactive material. We received the sample in September 1996, and the detailed results of the analysis are given in Ref. [1].

The sample consisted of a brown-tinted glass bottle, about one-fourth full of a white crystalline solid. The lid was threaded black plastic with a paper liner. The white solid was somewhat self-adhesive, which is consistent with what would be expected of a simple, hydrated salt, and the specimen had the acrid/musty smell of a nitrate. Interrogation with a β–γ survey instrument showed that the sample was significantly radioactive; placing an α survey instrument over the open mouth of the bottle resulted in the detection of α radioactivity.

The sample was placed whole in a clean plastic bag and was counted for γ-rays with a Ge semiconductor detector. The detected γ-rays were consistent with those expected to emit from a ^{232}Th sample with daughter activities in near equilibrium. Because of the poorly defined counting geometry for the initial assay, it was impossible to quantify any disequilibria among the daughter nuclides. There was no evidence for other radionuclides, including those from the naturally occurring U decay series. Using a spatula, enough of the sample was transferred to a standard γ-spec vial to cover the bottom of the container, and a polyethylene disk was placed over the material to keep it uniformly distributed. The second Ge count confirmed the earlier semiquantitative assessment that the radioactive constituents of the material consisted of those present in an aged ^{232}Th sample. On the basis of this information, and considering the description of thorium nitrate in the CRC Handbook [2], the initial working hypothesis was that the crystalline solid was aged, purified $Th(NO_3)_4 \cdot 4H_2O$.

Before proceeding with analytical work, an attempt was made to understand what a drug laboratory would be doing with thorium nitrate. The best conjecture was that it was used in an old German procedure [3] for the synthesis of P2P (1-phenyl-2-propanone, *aka* methylbenzylketone or phenylacetone) from acetic acid and phenylacetic acid (see Figure 25.1). P2P is a methamphetamine precursor, and the thorium oxide required by the procedure can be obtained by roasting thorium nitrate in a laboratory furnace.

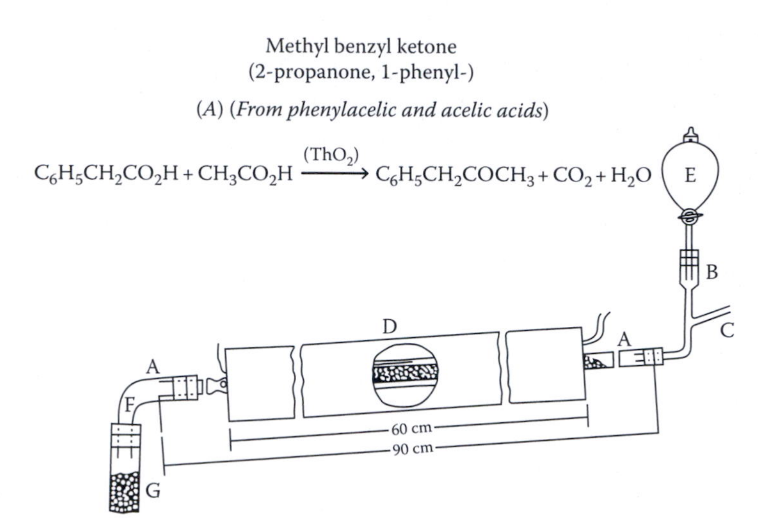

FIGURE 25.1 Synthesis of P2P from phenylacetic and acetic acids, with ThO_2 as catalyst. (From Blatt, A.H., ed., *Organic Syntheses: Collective Volume*, Vol. 2, Wiley & Sons, New York, 1943. With permission.)

Thorium and its compounds can be very difficult to dissolve, although in this case the solid was readily soluble in both water and dilute nitric acid. Nevertheless, a procedure was employed that was developed to dissolve any sample that is principally thorium or one of its salts. Thorium metal, thorium nitrate, thorium oxide (thoria), and thorium oxalate can all be dissolved with the following procedure.

25.2 RADIOCHEMISTRY AND RESULTS

As much as 1 g of sample is weighed into a 50 mL round-bottomed flask (the reaction vessel), which is placed in a heating mantle. Ten milliliters of water is added, and the sample is slowly evaporated to dryness under a gentle air stream. Care must be taken that no insoluble material is lost to the air, particularly in the case of fine-powder oxides or oxalates. Ten milliliters of 8 M HNO_3 are added, and the sample is gently evaporated to dryness under the air stream; this step is then repeated with more vigorous heating. The dry sample is slurried with 10 mL of 2 M HNO_3, warmed, and allowed to settle. The supernatant liquid is transferred to a 40-mL centrifuge cone, centrifuged, and decanted into a 50-mL volumetric flask, henceforth referred to as the analytical "pot."

Any precipitate remaining in the centrifuge cone is transferred back to the reaction vessel with 10 mL of concentrated HNO_3. A water-cooled condenser is joined to the reaction vessel, and the HNO_3 solution is refluxed for several hours (additional concentrated HNO_3 can be added as needed to keep the reaction vessel from going dry). Although the reaction vessel is still hot, a single drop of concentrated HF is added, and refluxing is continued. (Dissolution of Th metal or thoria is catalyzed by the presence of a small amount of fluoride ion [or fluosilicic acid]; however, the reaction proceeds at a negligible rate unless maintained at a temperature >110°C.)

The reflux is continued for 1 h beyond the point when no visible deposit remains in the reaction vessel. If loss of HF through evaporation is suspected, a second drop of HF can be added, but the presence of too much fluoride can cause complications in the subsequent chemical processing.

The condenser is removed from the reaction vessel, and the sample is evaporated to dryness. Ten milliliters of 2 M HNO_3 are added, the mixture is warmed and allowed to settle, and the liquid phase is transferred to the centrifuge cone used earlier. The sample is centrifuged, and the supernatant liquid is decanted to the pot. The centrifuge cone is washed with 5 mL of concentrated HNO_3, which is transferred back to the reaction vessel. Two milliliters of 7.5 M $HClO_4$ are added to the reaction vessel, which is evaporated past the point at which copious perchloric fumes are detected and to the point at which a moist deposit is obtained. This deposit is dissolved in 10 mL of 2 M HNO_3, and the resulting solution is warmed and transferred to the centrifuge cone. After centrifuging, the supernatant liquid is decanted to the pot.

At this time, there should be no visible deposit in either the reaction vessel or in the centrifuge cone. A radiation survey of both empty glass items with a β–γ survey instrument should also read only background. If not, the refluxing step should be repeated with an appropriate reduction in the volumes of reagents. Otherwise, wash the reaction vessel and centrifuge cone with small aliquots of 2 M HNO_3 and add them to the pot, with mixing and dilution to the 50-mL line. The pot solution, which is a quantitative dissolution of the original solid specimen, is stable indefinitely; we have seen no evidence of precipitation or turbidity after time periods as long as 6 months.

Accurate measurement of the concentrations of the heavy elements is accomplished through radiochemical separations—modifications of those presented earlier in this book. For each specimen, three samples are processed: a 10-mL aliquot of pot solution with nothing added (the unspiked sample), a 10-mL aliquot of pot solution to which a standard radionuclide tracer has been added (the spiked sample), and an aliquot of the tracer solution diluted to 10 mL with 2 M HNO_3 (the blank). The suggested radionuclide tracer aliquot contains about 10 dpm of ^{232}U, in which the daughter activities are in radioactive equilibrium, and ^{236}Pu. In the case of the DEA sample, a 10-mL aliquot was pipetted into a standard γ-spec vial for γ-ray spectrometry, a small aliquot was stippled onto a Pt disk (which was then heated in a Bunsen-burner flame) for gross-α decay-rate quantitation, and a few μL of solution were assayed by plasma-source mass spectrometry (ICP-MS) for trace-element analysis.

To ensure that the tracer activities exchange with the radionuclides in the aliquots, the samples are subjected to a single oxidation/reduction procedure with HI and HNO_3, as outlined elsewhere in the book. Fortunately, most of the radiogenic species produced by the decay of Th form single valence states in aqueous solution (e.g., Th(IV), Ac(III), and Ra(II)); the redox procedure is principally for the chemistry of potential contaminant species, such as U or Pu. Each valence-adjusted sample is then dissolved in 3 mL of 9 M HCl. One drop of 8 M HNO_3 is added to each of the samples, which are warmed before being loaded onto 12-mm diameter × 10-cm columns of AG1-X8 anion-exchange resin. The column eluents from the load and subsequent 9 M HCl wash solutions contain Th, Ac, and Ra, still in the same relative concentrations as in the pot solution, and are collected in 40-mL centrifuge cones. Uranium and Pu are then stripped from the columns with 0.5 M HCl. All fractions collected from these columns are evaporated to dryness under an air stream. The dry U, Pu samples are processed according to the procedures given earlier and result in fractions for both α- and mass-spectrometry analyses.

The dry Th, Ac, Ra samples are dissolved in a minimum volume of 8 M HNO_3 and reevaporated. The samples are then dissolved in 3 mL each of 8 M HNO_3, and the resulting solutions are loaded onto 12-mm diameter × 10-cm anion-exchange columns (as described earlier). The column eluents from the load and 8 M HNO_3 wash solutions are collected in 40-mL centrifuge cones; the time at which the elution is complete should be noted as the separation time of Ac and Ra (in the eluent) from Th (still on the column). Thorium is then eluted from the columns into fresh 40-mL cones with 9 M HCl. All fractions collected from these columns are evaporated to dryness as before. From this point on, the purification of the Ra/Ac fractions should take priority over that of the other fractions, as the chemical-yield indicators are ^{224}Ra, with a half-life of 3.66 days, and ^{228}Ac, with a half-life of 6.13 h (in equilibrium with 5.75-year ^{228}Ra in this fraction).

The dry Ra, Ac fractions should be examined. If a large quantity of white or off-white salt (residual thorium nitrate) is present, the previous paragraph should be repeated. If not, Ra and Ac are separated from each another using the cation-exchange procedures outlined in Chapter 10. Final Ra and Ac fractions are volatilized onto stainless-steel counting plates, as is a small aliquot of the chemically purified Th fraction. Samples are counted for γ-rays and α particles, the intensities of which are converted into atom concentrations. These concentrations must be corrected for decay taking place between the counting time and the separation time of the relevant fraction; they must then be further corrected for decay occurring between the separation time and the reference time for the sample.

For the sample obtained from the DEA, 5.439 g of the specimen was weighed into a 50-mL round-bottomed flask, and the protocol outlined earlier was followed. Virtually this entire specimen dissolved on the initial addition of water, but we followed the full procedure nevertheless. This solubility was another indication that the sample was thorium nitrate, as most Th salts have limited solubility in water unless accompanied by a vigorous reaction [2]. The resultant pot solution (quantitatively diluted to 50 mL) was aliquoted and analyzed as described earlier, though scaled to the larger sample size. The counting data were processed to give the radionuclide concentrations listed in Table 25.1, decay-corrected to noon on September 24, 1996.

In addition to the radiochemical measurements specified previously, 200 μL of the pot solution were diluted to 10 mL with water and subjected to ICP-MS analysis. The results, scaled by volume to the original pot solution, are given in Table 25.2.

25.3 DISCUSSION

From the weight of the dissolved sample and the assay given in Table 25.1, the original solid contained 0.443 ± 0.005 g of Th per gram of specimen. This result is supported by careful analysis of the NDA γ spectrum obtained from the solid sample. The ICP-MS measurement of the Th concentration (Table 25.2) was 0.47 g/g

TABLE 25.1
Results of the Analysis of a Sample of Thorium Nitrate Obtained from the DEA

Nuclide	Atoms/mL of Pot Solution	Nuclide	Atoms/mL of Pot Solution
$^{239,240}Pu$	$\leq 7.3 \times 10^7$	^{230}Th	$(8.71 \pm 0.15) \times 10^{13}$
^{238}Pu	$\leq 1.4 \times 10^5$	^{229}Th	$\leq 8.9 \times 10^8$
^{238}U	$\leq 1.7 \times 10^{14}$	^{228}Th	$(1.428 \pm 0.006) \times 10^{10}$
^{235}U	$\leq 1.6 \times 10^{13}$	^{227}Ac	$\leq 2.6 \times 10^6$
^{234}U	$\leq 2.8 \times 10^{10}$	^{228}Ra	$(4.43 \pm 0.06) \times 10^{10}$
^{233}U	$\leq 3.6 \times 10^{10}$	^{226}Ra	$(5.80 \pm 0.08) \times 10^{10}$
^{232}Th	$(1.251 \pm 0.014) \times 10^{20}$		

Note: Reference time is noon on September 24, 1996.

TABLE 25.2
Concentrations of Inorganic Constituents in the DEA Specimen

Element	ppm	Element	ppm	Element	Ppm
Li	n.d.	Rb	n.d.	Sm	n.d.
Be	n.d.	Sr	n.d.	Eu	n.d.
B	14	Y	n.d.	Dy	n.d.
Mg	n.d.	Zr	n.d.	Er	n.d.
Sc	n.d.	Nb	n.d.	Yb	n.d.
V	9	Mo	n.d.	Lu	n.d.
Cr	28	Cd	n.d.	Hf	n.d.
Mn	n.d.	Sb	n.d.	Ta	n.d.
Fe	46	Te	n.d.	W	n.d.
Co	n.d.	Cs	n.d.	Pb	n.d.
Ni	n.d.	Ba	360	Bi	n.d.
Cu	5	La	n.d.	Th	469,000
Zn	46	Ce	n.d.	U	n.d.
As	n.d.	Nd	n.d.		

Note: n.d. = "not detected" and is typically <5 ppm for these data.

of sample, with an uncertainty of about 10%; it is in reasonable agreement with the value obtained via radiation counting. From stoichiometry, 0.420 g of Th is expected per gram of thorium nitrate tetrahydrate. As mentioned earlier, the specimen had a distinct nitric acid odor, and the higher-than-theoretical Th concentration of the salt may be attributed to denitration or dehydration with age.

In performing the measurements and subsequent analyses that generated the data of Table 25.1, we found that, peculiar to samples consisting of mostly Th, most of our conclusions could be drawn from the α pulse-height count of the unspiked

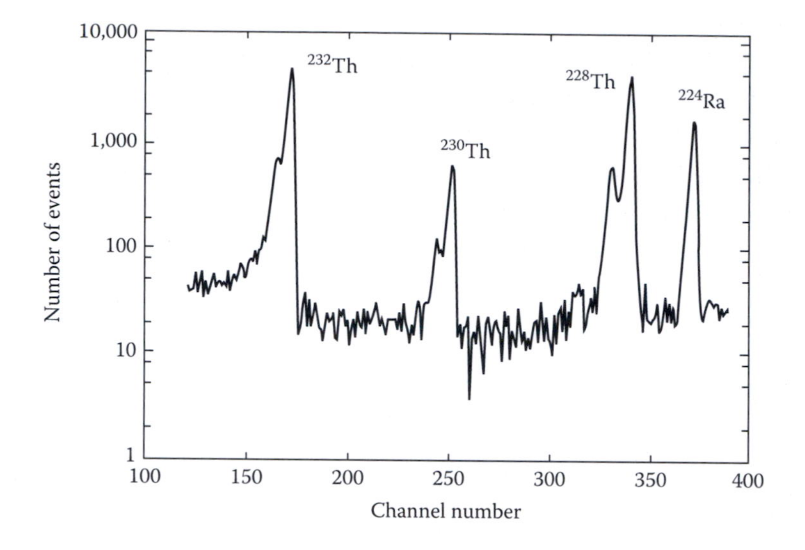

FIGURE 25.2 Alpha pulse-height spectrum from a volatilized aliquot of the purified Th fraction. Note the disequilibrium between ^{228}Th and its ^{224}Ra daughter, caused by partial ingrowth of ^{224}Ra following the separation. The volatilization process also contributes to the fractionation of Ra from Th.

Th sample, obtained with a Si surface-barrier detector. The spectrum obtained for the DEA sample is shown in Figure 25.2. Most of the nuclear forensic information determined from this sample arose from an interpretation of the data presented in this figure.

In Figure 25.2, the peak resulting from decay of 3.66-day ^{224}Ra derives from its ingrowth from ^{228}Th decay between the time the Th fraction was purified and the time the spectrum was acquired. The decay daughters of ^{224}Ra (55-s ^{220}Rn, 145-ms ^{216}Po, etc.) were also observed at higher energies, but are not shown in the figure. At lower energies, only the isotopes of Th are observed. The presence of 1.9-year ^{228}Th is the result of its production in the decay chain of the 4n-nuclide, ^{232}Th (see Figure 6.5). The daughter of ^{232}Th decay is 5.75-year ^{228}Ra, whose decay product (discounting a short-lived intermediate) is ^{228}Th. In an ore body, all of these species are in radioactive equilibrium with one another, and the decay rate of ^{228}Th is equal to that of ^{232}Th. When the sample is purified and Ra removed, the equilibrium is disrupted and the ^{228}Th begins to decay, relative to ^{232}Th, with the ^{228}Th half-life. However, simultaneously with the decay of ^{228}Th, the ^{232}Th is producing fresh ^{228}Ra, which in turn produces ^{228}Th. The ingrowth of ^{228}Ra in a purified sample of ^{232}Th is shown in Figure 25.3. Eventually, ^{228}Th builds back into the sample, and after a few decades, equilibrium is reestablished. In Figure 25.4, the calculated ratio of the activities of ^{228}Th and ^{232}Th, as a function of time after purification, is shown by the solid line. While Figure 25.3 describes a standard single-valued ingrowth curve, Figure 25.4 depicts a two-valued function (i.e., a single value of ^{228}Th/^{232}Th defines two ages). To decide which age is correct for a given sample requires collateral information from Table 25.1 and the use of Figure 25.3.

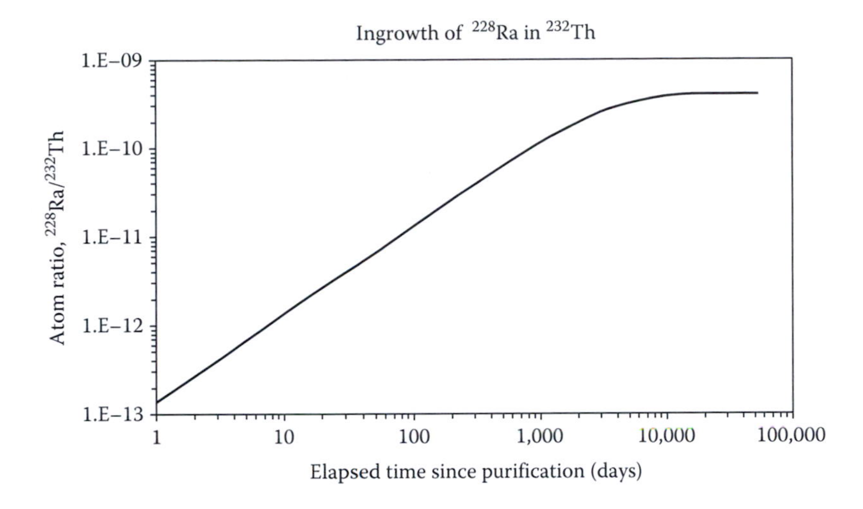

FIGURE 25.3 The ingrowth of ^{228}Ra in a purified ^{232}Th sample, expressed as a ratio of atoms.

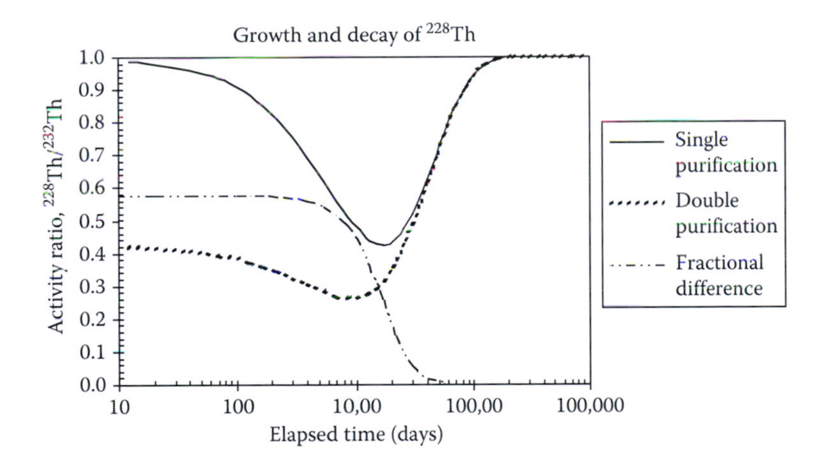

FIGURE 25.4 Ratio of activities of ^{228}Th and ^{232}Th in a sample of Th from which Ra was removed, as a function of time after separation. For the single-purification curve (solid line), it is assumed that the radionuclides were in equilibrium in the preseparation sample and that the separation took very little time relative to the ^{228}Th 1.9-year half-life. For the double-purification curve (dashed line), it is assumed that 5 years elapsed between the first separation and the second separation, which reset the clock. The fractional difference curve (dash-dot line) is described in the text.

We made the assumption that the time involved in recovering the Th from the ore, and its subsequent purification, was short compared to the ^{228}Th half-life. However, if one allows that this was not the case, and that a second purification from Ra was performed about 5 years after the first, the use of Figure 25.3 is still completely valid. The derived age would then be that associated with the second (final) separation, assuming that all Ra was removed from the Th sample. Following the first separation, ^{228}Th will decay relative to ^{232}Th according to the solid line in Figure 25.4. After approximately 5 years, it reaches 42% of its equilibrium concentration. This is the minimum value attained from single-pass chemistry; afterward, the ^{228}Th concentration begins to return to equilibrium. If ^{228}Ra is again removed from the sample at this point, ^{228}Th will continue to decay relative to ^{232}Th (following the dashed curve in Figure 25.4), relative to the new separation time. Clearly, at short ingrowth times, the age extracted from the activity ratio $^{228}Th/^{232}Th$ is very sensitive to how many separations were performed and the time interval between them. The ratio can even be unphysical (<0.42) if the analyst assumes that only a single purification step was performed.

The dot-dash curve shown in Figure 25.4 is the fractional difference between the single- and double-purification curves:

$$\text{Fractional difference} =$$

$$\frac{\left[\text{single pass}\left({}^{228}\text{Th}/{}^{232}\text{Th}\right) - \text{double pass}\left({}^{228}\text{Th}/{}^{232}\text{Th}\right)\right]}{\left[\text{single pass}\left({}^{228}\text{Th}/{}^{232}\text{Th}\right)\right]}.$$

As seen in the figure, the fractional difference between the two purification schemes becomes less important at long times. If the analyst assumes a single-purification procedure when, in fact, the double-purification procedure was followed, the extracted age from $^{228}Th/^{232}Th$ will be approximately correct if the age of the sample is sufficiently large. At 7 years following the second separation, the sample would appear to be 30% younger than it is; at 10 years, the sample would appear to be 3% younger than its actual age; and at 15 years, the inaccuracy in the determined age becomes less than 0.1 year (which is probably the limit of precision with which the activity ratio can be determined). If a Th sample has been purified several times, Figure 25.3 can always be used to extract the time since last purification, but Figure 25.4 can be used only for samples that are more than 12 years old.

In the DEA sample, the activity ratio of $^{228}Th/^{232}Th$ was (0.8415 ± 0.0059); the uncertainty is smaller than that expected from Table 25.1 because the error associated with establishing the absolute concentration does not propagate with the ratio. Using Figure 25.4, the activity ratio defines a specimen age of either (0.49 ± 0.02) year or (18.55 ± 0.30) years. From Figure 25.3, the $^{228}Ra/^{232}Th$ atom ratio gives an age of (16.8 ± 0.8) years, which is certainly more consistent with the longer ingrowth time. There is a third chronometric pair in the Table 25.1 data: the $^{226}Ra/^{230}Th$ value defines an age of 74 years. Although the $^{228}Th/^{232}Th$ and $^{228}Ra/^{232}Th$ ages are in rough agreement at the 2σ level, the $^{226}Ra/^{230}Th$ age is clearly inconsistent. An explanation is the incomplete separation of Ra from Th. If all of the long-lived ^{226}Ra was not removed at separation time, at a later time, there would be more ^{226}Ra present than could have grown in from ^{230}Th. This situation would result in an apparent age that is too old. The inference of incomplete separation from Ra is supported by the observation of a chemical homolog, Ba, in the ICP-MS analysis of the sample (Table 25.2). Purification from Ra is usually accomplished by adding a Ba carrier and precipitating the sulfate: $BaSO_4$ is a very effective coprecipitant of radium (see following). To rationalize the difference between the 18-year real age and 74-year apparent age, it must be assumed that the yield of Ra in the final specimen was about 2.4% that of Th, starting from the ore.

Of course, the same 2.4% incomplete separation factor also applies to ^{228}Ra relative to the other members of its decay chain. Fortunately, the 5.75-year half-life of ^{228}Ra is much smaller than that of 1600-year ^{226}Ra, so its equilibrium concentration in the ore is much lower relative to the parent Th isotopes and is much less than the age of the sample, so its effect is minimized through decay. Correcting for residual ^{228}Ra results in lowering the ^{228}Th/^{232}Th age to (18.40 ± 0.30) years and the ^{228}Ra/^{232}Th age to (16.6 ± 0.8) years.

The corrected ^{228}Th/^{232}Th and ^{228}Ra/^{232}Th ages should be the same value. The reason that they are significantly different results from the analytical methods used for the determination of the nuclide concentrations. The value of ^{228}Th/^{232}Th comes from a single measurement of an α spectrum, such as that shown in Figure 25.2. Errors in chemical yield and spectroscopic parameters do not affect the ratio. However, the determination of ^{228}Ra is via both γ spectroscopy of the ^{228}Ac daughter and observation of ^{228}Th ingrowth in the α spectrum of the separated Ra fraction. Correction for chemical yield of Ra relative to the pot solution must also be made to place ^{228}Ra and ^{232}Th on the same basis. Small systematic errors, in the spectroscopic quantities, the chemical yield, and the counting efficiencies, can exist and become particularly important as equilibrium is approached and all activity ratios converge to 1.00.

Long-lived (75,400-year) ^{230}Th is part of the $4n + 2$ decay chain and arises from the decay of ^{238}U and daughters. Because ^{230}Th is so long-lived, its concentration relative to 1.4×10^{10}-year ^{232}Th has not significantly changed as a result of decay processes following its separation from the parent ore. If it is assumed that the parent material was of natural origin, in which all radioactive species were in secular equilibrium with either ^{232}Th or ^{238}U before purification, the value of ^{230}Th/^{232}Th obtained from the α spectrum translates directly into the atom ratio, ^{238}U/^{232}Th = 0.0414 ± 0.0008 in the parent ore. This value is likely characteristic of a particular ore body. It is consistent with monazite (the principal ore from which Th is recovered), with nominal ^{238}U/^{232}Th = 0.03–0.05 [4,5]. However, attribution of the questioned specimen to a specific ore body would only be possible through comparison to an appropriate database.

Use of Th in nuclear reactors is becoming more widespread. India is actively working to establish the ^{232}Th/^{233}U fuel cycle [6], and reactors at Elk River, Minnesota, and Indian Point, New York, have used fuel in which Th was the fertile material [7]. The forensic assessment of a sample containing Th that was recovered from material used in nuclear-fuel applications is nontrivial. Neutron irradiation of Th modifies the isotopic composition of the sample and can lead to incorrect conclusions from the analytical data. Fortunately, previous neutronic use of Th in an analytical specimen can be established from the ^{229}Th content of the material. In a reactor, the ^{232}Th comprising the major component of the Th is irradiated with neutrons. During the course of this irradiation, some fraction of the ^{232}Th atoms capture neutrons, becoming 22-min ^{233}Th, which decays to 27-day ^{233}Pa, which in turn decays to 1.59×10^5-year ^{233}U, the desired product from nuclear fuel reprocessing [8]. The irradiated fuel is set aside for about 1 year, both to allow efficient ^{233}Pa decay and to let fission products that unavoidably arise in the neutron irradiation decay to acceptable levels. During this year, some of the ^{233}U product decays to 7900-year ^{229}Th, which follows the other Th isotopes through reprocessing and purification chemistries. The ^{229}Th isotope does not occur in nature in measurable amounts and is an unambiguous signature of neutron irradiation in a reactor. A reprocessed sample will have an atom ratio of $^{229}Th/^{232}Th$ on the order of 10^{-8} to 10^{-7}. By means of the equilibrium concentrations of ^{229}Th progeny, it is relatively simple to measure this ratio to as low as 10^{-11}, assessing the possibility of any significant prior use in a reactor.

The atom ratio of $^{229}Th/^{232}Th$ in the drug agency sample was $\leq 7.1 \times 10^{-12}$. This is several orders of magnitude lower than expected in Th samples recovered from spent fuel. The Pu isotopes were also absent. Because Pu has a 4+ oxidation state with chemistry similar to Th, residual Pu contamination in Th might be expected in reprocessed material, particularly from the earlier phases of establishing a $Th/^{233}U$ fuel cycle when the heavier U isotopes are added to the fuel. Similarly, $^{233}U/^{232}Th$ was $\leq 3 \times 10^{-10}$. Thus, the invalidation of data interpretation by modification of Th isotopics by neutron irradiation was not a factor in this investigation.

There are two principal industrial processes for extracting Th from phosphate ores (e.g., monazite) [5]. One relies on leaching the ore with sulfuric acid at elevated temperatures, followed by dilution with water to precipitate thorium pyrophosphate. This intermediate product is then dissolved with more H_2SO_4, from which Th is reprecipitated by addition of oxalic acid. The second method relies on leaching the ore with sodium hydroxide, followed by dilution with water, and reconcentration of the leachate through evaporation to a thorium hydroxide precipitate. The $Th(OH)_4$ is dissolved in HNO_3 and reprecipitated by adjusting the solution to near-neutrality via additional NaOH. Both processes use a final solvent extraction with tributyl phosphate (TBP) to effect a final purification of Th.

In the first process, a quantity of Ba carrier is added to the initial leach liquor to coprecipitate Ra; if this step is not quantitative, however, some fraction of the Ra (and Ba) would accompany Th through the oxalate precipitation. The second process is just as effective in removing Ra from Th as it is in removing Ac and the lanthanides, and no carrier addition is necessary. Similarly, the TBP step should be very effective for removing both Ra and Ac from Th. In addition to the excess Ra, significant Ba was also measured in the questioned DEA specimen (Table 25.2). If the overall Ba chemical yield was similar to that of Ra throughout the purification chemistry, about 30 mg of Ba carrier were used for each gram of Th in the process solution, and this value is a reasonable quantity for an industrial-scale process. The experimental data, therefore, indicate that Th was recovered from the ore through the acid-leach process, rather than by alkaline leach, and that a procedure other than TBP extraction was used for final purification of the product.

25.4 SUMMARY

In conclusion, we received an interdicted sample of a Th salt from the DEA and analyzed it according to established methodology. The specimen was derived from ore in which the U/Th ratio was 0.0414, consistent with the nominal concentrations of the elements in the mineral monazite. The Th was likely recovered from this ore through an acid-leach process, as residual Ra and added Ba carrier remained in the sample. No significant fraction of the material comprising the final specimen was ever incorporated in reactor fuel. The last chemical purification of the sample material occurred (18.4 ± 0.3) years before the analysis time, or May 1, 1978, ±110 days.

REFERENCES

1. K.J. Moody and P.M. Grant, Nuclear forensic analysis of thorium, *J. Radioanal. Nucl. Chem.*, 241, 157, 1999.
2. D.R. Lide, ed., *Handbook of Chemistry and Physics*, 74th ed., CRC Press, Boca Raton, FL, 1993.
3. A.H. Blatt, ed., *Organic Syntheses: Collective Volume*, Vol. 2, Wiley, New York, 1943, p. 389.
4. K.H. Wedepohl, ed., *Handbook of Geochemistry*, Vols. II/5, Springer, Berlin, Germany, 1978.
5. H.A. Wilhelm, ed., *The Metal Thorium*, American Society for Metals, Cleveland, OH, 1956, Chap. 1.
6. A. Palamalai, S.V. Mohan, M. Sampath, R. Srinivasan, P. Govindan, A. Chinnusamy, V.R. Raman, and G.R. Balasubramanian, Final purification of uranium-233 oxide product from reprocessing treatment of irradiated thorium rods, *J. Radioanal. Nucl. Chem.*, 177, 291, 1994.
7. J. Belle and R.M. Berman, eds., *Thorium Dioxide; Properties and Nuclear Applications*, DOE/NE-0060, U.S. Department of Energy, Washington, DC, 1984.
8. B.I. Spinrad, J.C. Carter, and C. Eggler, Reactivity changes and reactivity lifetimes of fixed-fuel elements in thermal reactors, *Proceedings of the International Conference on Peaceful Uses of Atomic Energy*, Vol. 5, United Nations, New York, 1956.

26 Radioactive Pillow Shipment

26.1 BACKGROUND AND ANALYSES

In 2006, a consignment of pillows from a foreign manufacturer triggered a radiation alarm at a US port, estimated at the facility to contain on the order of 10s–100s of μCi of penetrating γ radiation. Field x-rays of the isolated container verified that no apparently hazardous components were present. Scoping NDA field measurements of the questioned cardboard box were performed around its surface, with the following results:

- β–γ survey meter: >60k cpm at contact
- α survey meter: negative
- Presumptive ELITE kit for HE: negative

The questioned cardboard container was then opened in a hood in a radiochemistry lab at the Livermore Forensic Science Center. Detailed radiation sensing was performed with a β–γ survey meter and pancake probe of ~5-cm diameter active area. The wide probe tip and relatively slow response of the meter made localization of the radioactivity somewhat inefficient because of the low spatial resolution, and the isolation procedure consequently became iterative in nature. However, it was quickly determined that the entire radioactivity was on or in the cardboard box, with no contamination present on the pillows. Further, the radioactivity was localized on a small portion of the container and was not distributed throughout the cardboard.

The cardboard consisted of five separate layers glued together: three flat sheets with corrugation furrows between them. After rough localization of the activity with the β–γ probe, areas of the box were excised, and each layer of cardboard was separated with scalpel and forceps. Ultimately, the activity was isolated to a roughly trapezoidal section of the container. Subdivision of the active specimen was terminated after it had been segregated to an area of approximately 2.0 × 2.0 × 2.4 × 2.7 cm. This size allowed its placement with fixed geometry within a standard γ-spec counting vial.

The isolated substrate was then microscopically inspected for any unusual characteristics. While several interesting particles and features were observed, it could not be determined whether any were intimately associated with the radioactivity. An attempt at overnight autoradiography on a photographic film was attempted, but the activity was too small to induce exposure.

Qualitative and quantitative analyses of the sample were performed on a low-background, high-efficiency HPGe spectrometer system for all γ emissions between ~30 and 2000 keV. Two spectra were acquired and analyzed with the GAMANAL software program: a short, 10-min count for the identification of the principal radioactivity as ^{137}Cs and a longer, 24-h count for minor impurity activities.

26.2 RESULTS AND DISCUSSION

The radioactivity was determined to be on, or within, the interior flat layer of the cardboard container, with the hot spot near the top of a side wall, not on the opening lid. At a zero time of November 28, 2006, GAMANAL unambiguously identified two isotopes of Cs in or on the cardboard:

^{137}Cs 8.2 μCi ± 0.7% (1σ) $t_{1/2}$ = 30.17 year

^{134}Cs 0.11 nCi ± 14% $t_{1/2}$ = 2.065 year

The detection of ^{134}Cs was unanticipated, but its presence allowed considerations of potential chronometry for the isotope production. Parent–progeny relationships in U and Pu isotopic systems are routinely assessed to determine the time elapsed since the last chemical processing. However, the following application to Cs nuclides was novel to this particular investigation.

The empirical counting data also allowed an inference of the limit for another Cs isotope. From decay schemes and detector efficiencies, the detection limit for ^{136}Cs would have been approximately equivalent to that of ^{134}Cs. We assuredly could have detected half as much ^{134}Cs as we actually did, thereby making the presence of 13.16-day ^{136}Cs < 120 dpm. Thus, the $^{134}Cs/^{137}Cs$ atom ratio was 9.0×10^{-7} ± 14%, while the $^{136}Cs/^{134}Cs$ ratio was <0.009. No other radionuclides were observed, an indication that chemical processing of the isotope-production host material was effective and complete.

The consideration of various possible nuclide production mechanisms was important to this development. As ^{137}Cs is four neutrons heavier than stable ^{133}Cs, a simple accelerator irradiation of a Cs target was unfeasible. Similarly, while high-energy spallation production of radiocesium (*Cs) from a heavier element was possible, the relatively large quantity of ^{137}Cs (4.2×10^{14} atoms, or ~0.1 μg) would have required a long irradiation at high current, a much more expensive proposition than other alternatives. We therefore concluded that it was unlikely that a particle accelerator was involved in the production of this *Cs. Consequently, the material was generated by nuclear fission.

The fission production routes considered were the following:

Thermal neutrons + ^{235}U, ^{239}Pu, ^{233}U
Fission-spectrum neutrons + ^{235}U, ^{238}U, ^{239}Pu, ^{233}U, ^{232}Th
14-MeV neutrons + ^{235}U, ^{238}U, ^{239}Pu, ^{233}U
Spontaneous fission of ^{252}Cf

The cumulative yields of the products of all of these fission systems can be found in the comprehensive compilation by England and Rider. Using those data to calculate $^{134}Cs/^{137}Cs$ atom ratios, and assuming negligible differential decay of the *Cs nuclides, only two systems were close to the measurements of the present sample: thermal neutrons + ^{235}U and fission neutrons + ^{235}U. The *Cs ratios from all other fission systems were disparate by more than a factor of 8, and all but one (fission-spectrum neutrons + ^{238}U) were dissimilar by >24×.

Further, the *Cs isotopic mixture of the specimen was unlikely to have arisen from nuclear testing. The measured ratio was consistent with bomb debris no more than 19–20 years old for simple devices. However, more sophisticated devices would be compatible only with younger debris and would be inconsistent with the Chinese test program.

The questioned *Cs mixture must therefore have arisen from a reactor application, either fuel reprocessing or isotope production from a fissile target. The presence of shielded ^{134}Cs in the mixture indicated that the *Cs could not have been recovered from gas-phase precursors during the reactor operation. If the mixture had been derived from fuel reprocessing for Pu, the calculation of $^{134}Cs/^{137}Cs$ would have been determined by the initial $^{235}U/^{238}U$ ratio, the duration of the irradiation, and the quantity of ingrown ^{239}Pu. Even in fresh debris (no decay after discharge), a mixture of fission products from ^{235}U and ^{238}U alone would not have produced the measured cesium ratio; this situation would have been made even worse had a significant cooling period been required before processing. Appreciable admixture of ingrown ^{239}Pu fissions would have been required, a scenario incompatible with the production of weapons-grade plutonium, which involves low grams/ton of the material. Had the radionuclides in the questioned sample arose from fuel reprocessing, it must have been from higher burn-up civilian fuel.

Of the two production modes, long irradiation and reprocessing of U in the civil fuel cycle is a nonlinear problem that could be addressed through ORIGIN calculations and more complete reactor analysis. However, as this investigation was not a particularly urgent national problem, the sponsor did not authorize such efforts here. (Actually, we could not strictly eliminate the Th fuel cycle or MOX fuel from the *Cs ratio alone, but both were assessed to be lower probability production reactors.)

Limited-activity *Cs production in a reactor thermal port is a much more tractable chronometry problem. HEU would be the preferred target, since minimizing the quantity of ^{238}U minimizes the production of dangerous α emitters, and the target could be reconstituted and reused for additional activations. Further, it is unlikely that saturation-factor considerations would be important for this production method, as the irradiation time would almost certainly be shorter than 2 years. From the England and Rider tables, if ^{235}U is irradiated in an isotope-production reactor with near-thermal neutrons, two chronometers apply via the empirical measurements of the present specimen. Relative to November 28, 2006,

$$^{134}\text{Cs}/^{137}\text{Cs} \quad \text{age} = (357 \pm 50) \text{ days}$$

$$^{136}\text{Cs}/^{134}\text{Cs} \quad \text{age} > 215 \text{ days}$$

Reprocessing an irradiated fuel element for ^{137}Cs would be realistic only if the commercial production of MCi of activity were desired; for example, for sterilization irradiators. In the absence of that scenario, it would be more economical to discard the mixed fission products and spent reprocessing chemicals as waste and produce the *Cs material from a smaller reactor target.

There exists yet another radionuclide in the sample that could potentially provide additional information if assayed. Although ^{135}Cs should be present at approximately the same level as ^{137}Cs, its very long half-life precludes a radiometric measurement. However, it may be that $^{135}Cs/^{137}Cs$ could be determined by mass spectrometry.

In summary, several types of isotope production scenarios could be excluded based on the quantity of material in the sample (e.g., spontaneous fission), and others could be eliminated based on the $^{134/137}Cs$ isotope ratio (e.g., spallation and nuclear testing). Of the remaining scenarios, the most likely (in our opinion) was the irradiation of highly enriched uranium with low-energy neutrons, with the subsequent chemical recovery of radiocesium. If this were true, then the target was irradiated during the first week of December 2005 (with a 50-day error), and the *Cs was relatively recent material. If untrue, however, reprocessed civilian fuel would have been the next likely production mode, but then all bets were off.

27 Afghanistan Scam Specimens

27.1 BACKGROUND AND QUESTIONED SPECIMENS

The trafficking of nuclear hoax items was introduced in Section 1.7. Of the five persistent regional scams, recovered Afghan objects have been resurgent of late, due in part to the presence of allied forces in Afghanistan for Operation Enduring Freedom. Since the mid-1990s, nuclear smuggling incidents in both Afghanistan and Pakistan have displayed a regular theme that has been labeled the "Afghanistan uranium scam."

This scam variation attempts to sell materials, advertised as highly enriched uranium (HEU), that are claimed to have been abandoned by Russian military forces upon their withdrawal from Afghanistan in 1989. The alleged weapons materials in these sales have since diversified to also include Pu. Discriminating traits of the contraband objects include their external markings and the conical or cylindrical shapes of the containers, with Cyrillic characters sometimes comingled with English letters. Perhaps based on some original Russian artifact, large numbers of forged variants of these items were subsequently produced for nuclear smuggling scam activities.

Historically, none of the interrogated items were found (solely by NDA γ-spec or field radiation detection instrumentation) to contain any radioactive substance, let alone weapons-grade U or Pu. However, with the current increased frequency of their discovery, more thorough forensic investigation was considered prudent.

Examples of two such specimens are shown in Figure 27.1. The "billet" was a nonright circular cylinder, painted brown and weighing 3.98 kg. External labeling on the side wall included the notations "U-235 150 Г" (the Cyrillic "ghe" presumably specifying grams), "Pu-239 50 Г," "A○12 KT," and "1988 CCCP." The "hourglass" weighed 4.16 kg and was painted olive green, with three stars on its upper surface sloppily painted bright red. It had the same ^{235}U and ^{239}Pu designations as the billet, but "CCCP RX" rather than a date.

27.2 INITIAL NDA

Nondestructive analyses for safety information are always the first protocols for questioned specimens of this nature. Both samples assayed nominally nonradioactive via survey meters, but long counts by HPGe γ-spectrometry identified three naturally occurring radionuclides at activities above background: ^{40}K, ^{226}Ra, and ^{232}Th. However, ^{226}Ra was anomalous, and it presented the first mystery of this investigation

FIGURE 27.1 Examples of Afghan scam objects. The cylindrical-shaped "billet" is on the left and the conical-shaped "hourglass" on the right.

because its activity depended on which end of an object was closest to the detector. Indeed, the end-to-end Ra inhomogeneity of the billet was greater than a factor of 10.

Simple water-displacement experiments approximated the densities of the two items. Both objects had the heft and feel of Pb, and XRF analyses at their surfaces indicated such. Assay of the hourglass estimated $\rho \sim 10$ g/cm^3, sufficiently close to Pb ($\rho = 11.3$) to pose little concern. However, the billet $\rho \sim 3.8$ g/cm^3 implied that the metal container was either a metal other than Pb (with XRF merely reflecting the composition of the surface paint) or that the object contained significant voids. Radiography and computed tomography (CT) of the specimens were the logical next steps.

Both x-ray and neutron irradiations were conducted. At LLNL, an electron linac irradiated a W target to produce bremsstrahlung and generate high-energy (9-MeV) photons. Neutron transmission was measured at a thermal beam port of the UC Davis 2-MW TRIGA reactor.

Mild neutron activation of the devices induced by the reactor irradiations produced additional forensic information. Gamma-spec analyses of the objects after neutron exposure identified the following elements in their compositions: Mn, Fe, Co, Zn, Sr, Ag, Sb, Ba, and Hg. The Sr and Ba, in conjunction with ^{226}Ra, encompassed the suite of heavy alkaline-earth elements, and the significance of this result will become apparent later.

A neutron radiograph of the hourglass is shown in Figure 27.2. The spine of the object consisted of a large, inverted hex-head bolt and an externally threaded pipe. The screw end of the bolt had been forcibly inserted into the interior of the pipe, and the void space in the pipe was clearly visible. Some lower-Z debris was also present at the bottom, in proximity to the bolt head, but the remainder of the specimen appeared to be solid Pb molded around the makeshift spine. The hourglass was assessed to be considerably less interesting than the billet,

FIGURE 27.2 Digital neutron radiograph of questioned hourglass specimen.

FIGURE 27.3 Afghan scam billets: the primary questioned specimen of this investigation.

however, and no further description of the former is given here. The remainder of this investigation was devoted to the billet (Figure 27.3), which was completely sealed by virtue of a crudely welded base.

Digital radiographs of the billet structure (Figure 27.4) indicated a hollow container with primary Pb walls that were lined with a thinner metal insert. A number of

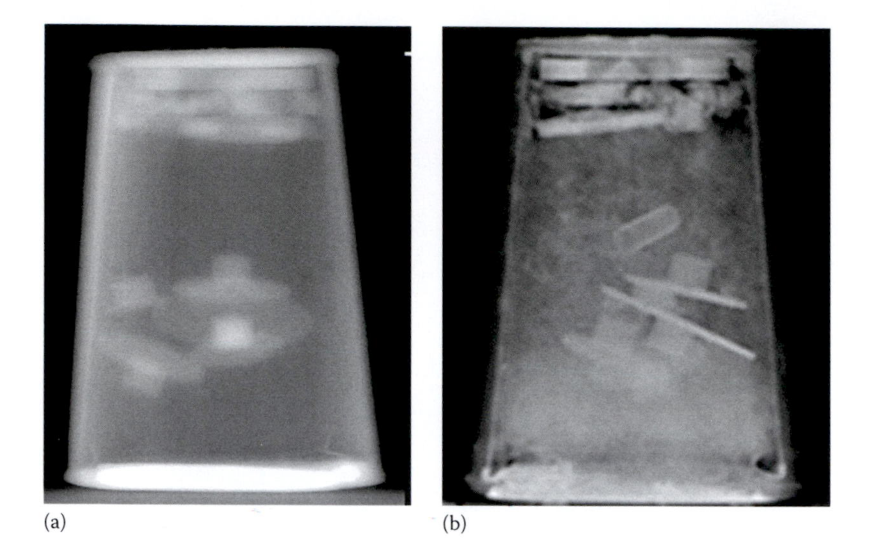

FIGURE 27.4 (a) Digital x-ray and (b) neutron radiographs of Afghan scam billet. A number of disk-like objects were visible near the center of the billet, while fragments and an intact metal washer were positioned near the top. The neutron measurements provided better spatial resolution, depicted the packing material as a hazy presence, and gave better indication of a thin metal liner inside the Pb walls.

strange, disk-like objects were supported in the center of the device, and the neutron radiography revealed a low-Z object immediately above them that had not been visible by x-ray. Axial CT imagery showed that the metal items at the very top of the interior were an intact circular washer (uppermost), along with fragments of broken washers directly beneath it. An interior, lower-Z packing and support material was clearly evident in the neutron images. The Pb sidewall of the billet was cast unevenly, with the thickness varying from ~0.5 to 0.8 cm near the base. All radiography and CT data were obtained through these Pb walls.

27.3 FORENSIC ANALYSES AFTER BILLET B&E

Breaking and entering the billet was effected by sawing through the sidewall just above the welded base plate. Generous utilization of interior packaging support material by the maker was immediately evident, as was the severed edge of the metal inner liner of the wall (see Figure 27.5).

The packing material was extremely inhomogeneous and contained hairs, fibers, assorted particles, and diverse debris. An initial prescreen by FTIR-ATR identified a number of interesting compounds through comparisons to known specimens in established IR spectral libraries (our adopted figure of merit for reliable identifications was >75% match similarity). These analytes included asbestos, cocaine, wool flour, and tetramine, the last being a generalized chemical name for compounds containing four amine groups. For example, the "tren" tetramine variant is exploited in coordination chemistry as a chelating ligand

FIGURE 27.5 Billet opened at base revealing packing material and thinner metal liner along inner sidewall.

for complexation with transition metals, while the hydrochloride salt of trien (*aka* TETA, triethylenetetramine) is used in medical therapy to remove Cu from patients with Wilson's disease. Not all tetramines are benevolent or benign, however, as TETS is a disulfonated analog that has been used as a rodenticide but is now banned in most countries.

A larger sampling of the packing material was interrogated later and more thoroughly in this investigation by the same technique. The additional chemical species identified by these subsequent analyses are summarized in Table 27.1.

The cylindrical, open-ended, lower-Z item visible in the neutron radiographs, just above the circular disks suspended within the interior packing material, was a white, nondescript plastic cap with no markings. It likely originated from a writing pen or color marker, but the value of any subsequent forensic analyses was not apparent, and no further examinations of it were conducted.

The disks were metallic and magnetic, and consisted of five randomly oriented, well-worn objects of similar design but varying dimensions. XRF analyses assayed their elemental compositions to be approximately 90% Fe, Zn, Si, and Al. They were identified as different brands of driver components that are used in audio speakers, one manufacturer of which was Energy EAS (Figure 27.6).

The intact metal washer and broken pieces at the top of the billet (Figure 27.7) were very strongly magnetic and composed of ferrite (a ceramic material containing Fe_2O_3). The most abundant elements in the metal were Fe (~65%), Ba (~17%), Sr (~1%), and Si (~1%). Permanent magnets are made of hard ferrites, the most common of which is $MFe_{12}O_{19}$ ($\equiv MO \cdot 6Fe_2O_3$), where M = Sr, Ba, or Co. The ferrite in the billet was thus predominantly $BaFe_{12}O_{19}$, which provided a plausible explanation for the ^{226}Ra counting data. Exceptional chemical separation would be required to completely isolate Ba from other alkaline earths in any industrial process, a tactic unnecessary for this particular application. Consequently,

TABLE 27.1
Additional Chemical Compounds Sequestered in the Billet Packing Material and Identified through FTIR-ATR

Tentative ID	Match (%)	Use/Function
Cellulose	89	Wood pulp; paper
Cellulose-N,N-diethyl-aminoethyl ether	90	Cellulose-based adsorbent polymer
Dextran	85	Bacterial product; dental plaque; medicinal uses
Hydrocodone bitartrate hemipentahydrate	77	Opioid
Elixophyllin-SR	85	Respiratory drug (~caffeine)
Methyclothiazide	83	Pharmaceutical; diuretic
Fucose	86	Sugar in cosmetics, pharmaceuticals, dietary supplements
Troykyd Defoamer 262	87	Defoaming agent
Bentone 27	85	Organoclay in cosmetics and toiletries
Goulac	81	Leather filler
Conoco 100	83	Defoamer
Polygalacturonic acid (K salt)	89	Plant matter
Polypropylene yarn	87	Plastic
Veegum T	87	Mg-Al-Si clay
Polyester	86	Plastic; fibers
Van-amid 340	87	Polymer
Polypropylene	83	Plastic
Poly(ethylene terephthalate)	82	Plasticizer; synthetic fibers
Others	—	—

small impurities of Sr and Ra in $BaFe_{12}O_{19}$ are not surprising, and the ^{226}Ra assay anomaly resulted from the asymmetric loading of ferrite at the top of the billet.

Because of the concerted effort to place magnetic objects within the interior, magnetometry was also conducted on the item. For context, background measurements of the magnetic field at the surface of the earth give a range of values over approximately ±0.5 gauss. At the top, external surface of the billet, center-to-edge magnetometer measurements varied from 80 to −10 G. The interior (without the speaker drivers, but with the ferrite) registered ~−700 G, while an individual ferrite fragment was ~400–900 G at contact.

Other material analyses of the structure of the billet included the composition of the stiff inner liner of the Pb wall. It was a 65% Zn alloy containing Si (12%), Al (6%), Fe (3%), and Mg (3%) as the major constituents. In addition, an attempt was made at geolocation analyses of the paint and exterior wall of the billet by means of Pb-isotope measurements. Six subsamples of the surface paints and metal were analyzed for 204/206/207/208 isotopic distributions (see Section 19.7.1), and most of the samples matched best to database entries from Kazakhstan. However, this estimation is not considered reliable at the present time and would not contribute to any overall assessment by this investigation.

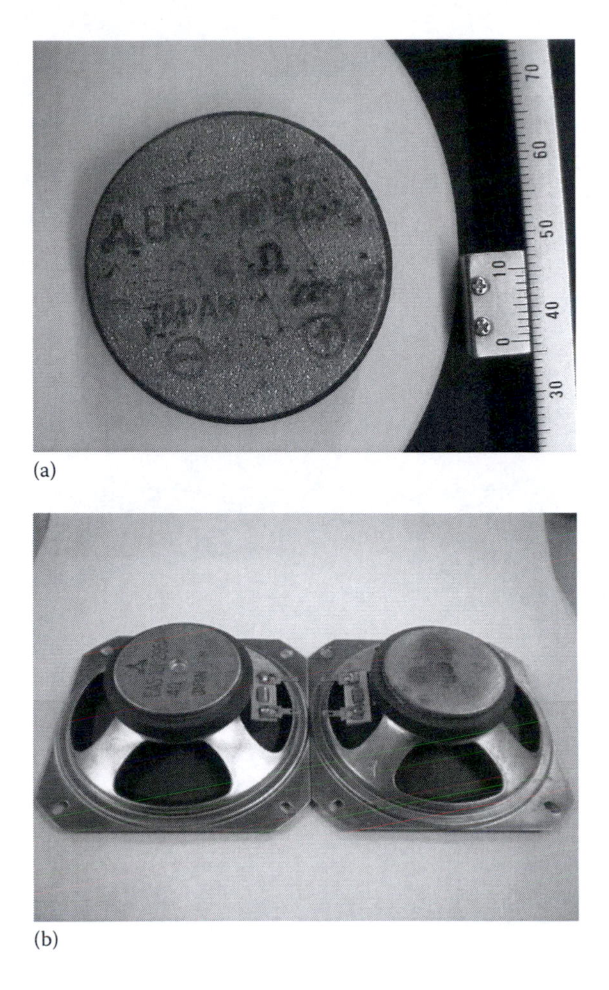

(a)

(b)

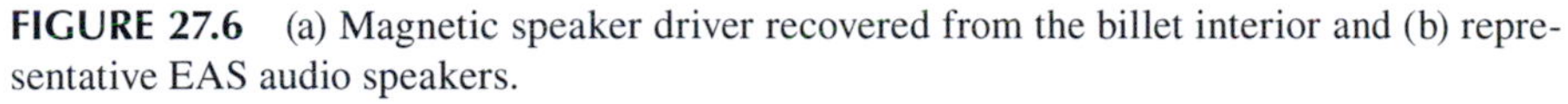

FIGURE 27.6 (a) Magnetic speaker driver recovered from the billet interior and (b) representative EAS audio speakers.

The older entries in this specific database were likely biased due to incorrect fractionation corrections, and a large effort beyond the practical scope (and budget) of this study would be required for a dependable result.

27.4 VERY UNUSUAL INCORPORATED OBJECTS

A number of strange and unexpected artifacts were also recovered from within the packing material of the billet (Figure 27.8). The two most interesting (and forensically challenging for a nuclear forensic investigation) were a plain white tablet having no identifying descriptors on its surface and a seeming mineral grain, rocky in appearance.

GC–MS analyses (Section 16.5) of the tablet identified sulfamethoxazole and trimethoprim as the principle components, with the relative intensity ratio of their peaks measured at ~5:1. Aniline and sulfanilamide were also detected at lesser concentrations.

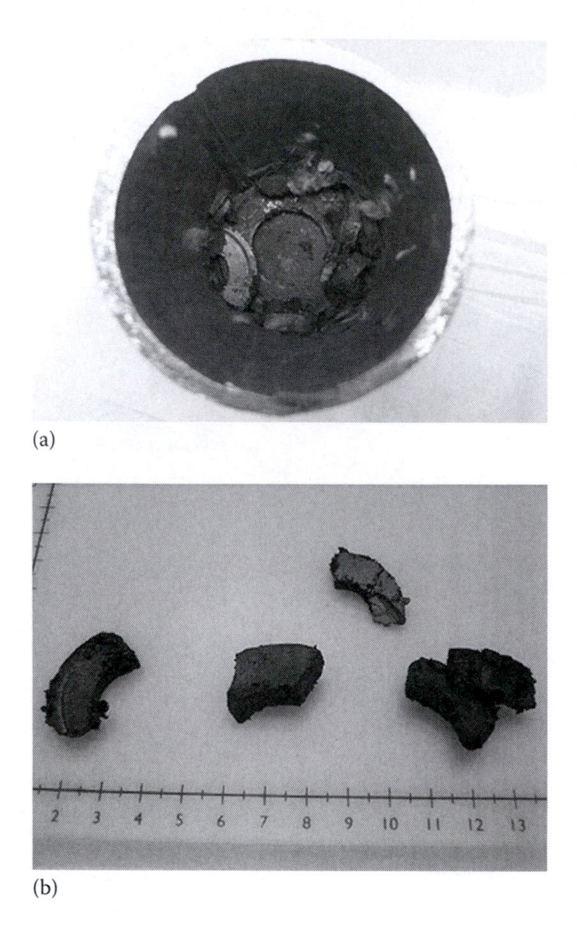

FIGURE 27.7 (a) View from the base of the opened billet toward the top. The intact washer (uppermost) and strongly adherent broken ferrite pieces are clearly visible. (b) Ferrite fragments removed from the billet.

From these measurements, the pill was identified as a Bactrim tablet, whose active ingredients are 400 mg of sulfamethoxazole and 80 mg of trimethoprim. Inactive ingredients in the formulation include docusate sodium, sodium benzoate, glycolate, magnesium stearate, and starch. Bactrim is a prescription drug used therapeutically to treat urinary tract infections, bronchitis, shigellosis, pneumonia, traveler's diarrhea, and other bacterial infections. The minor compounds measured in the GC–MS data were consistent with by-products or degradation of the sulfamethoxazole.

The apparent mineral grain (Figure 27.9) was interrogated in detail by SEM and XRD for potential geolocation information. It was very finely grained, with clay-like morphology, and was only weakly crystalline. XRD identified quartz, feldspar, and minor Fe_2O_3 in the material, but the conclusion from geochemical analyses was that it was composed predominantly of "organic" material (i.e., 95% consisted of elements with $Z < 11$ (Na)), with embedded quartz and feldspar grains.

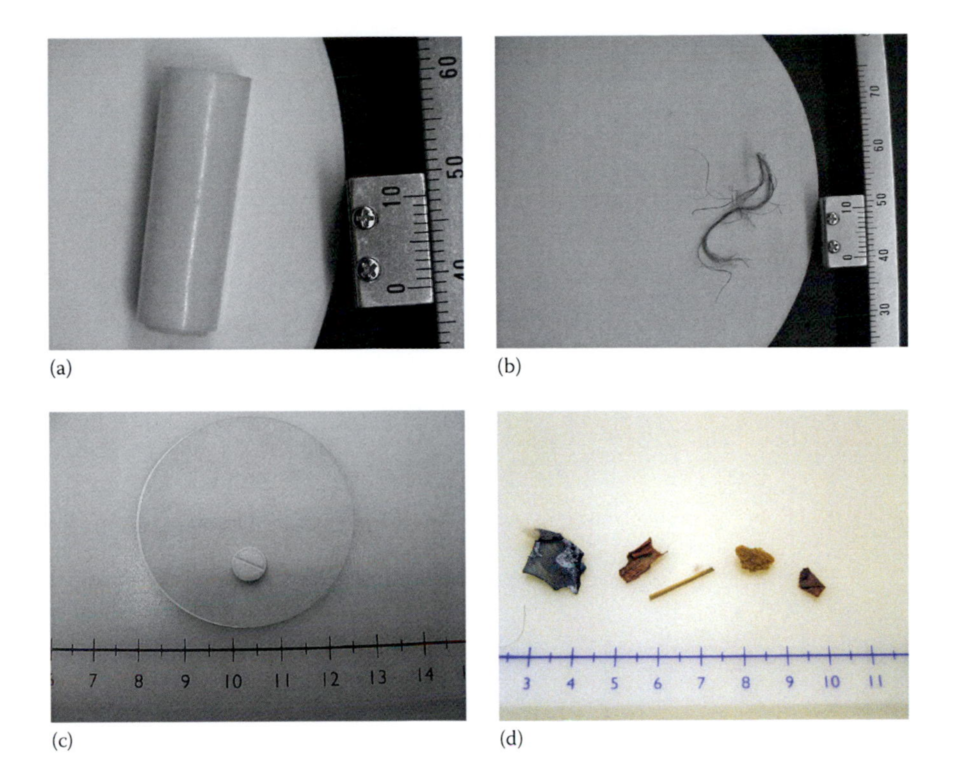

FIGURE 27.8 Diverse specimens recovered from the interior packing material of the billet. (a) The plastic cap, (b) a tuft of brown hair, (c) an unidentified white tablet, (d) mélange of ferrite fragment, pieces of wood, and an apparent mineral grain (#4 of 5 from left).

Organic analyses were therefore conducted. Strangely, the grain was ostensibly insoluble in both CH_2Cl_2 and BSTFA derivatizing agent, as well as in various aqueous media. It did burn briefly in a Bunsen-burner flame and decomposed to a black, ash-like color, but the pyrolysis was incomplete. Residue on the tip of a spatula reflected a visible, enduring framework of the lattice of the original material, an observation unique within our long experience of flame tests for suspected organic materials.

Subsequent analysis by ATR-FTIR, in conjunction with the Sadtler polymer database, identified the grain (with 85% match) as having a ceramic component. Further analyses and the Sigma-Aldrich library identified the principal composition of the grain as Lipase AYS Amano, with a 95% match probability. Amano Lipase is a hydrolase enzyme used in synthesis and diagnostic procedures (CAS # 9001-62-1). It catalyzes the hydrolytic cleavage of proteins, nucleic acids, starch, fats, and other species. For example, hydrolysis of oils and fats (i.e., triglycerides) gives free fatty acids and glycerol, a valuable commercial product. In addition, applications of the lipase within biocatalysis can produce enantioselective syntheses.

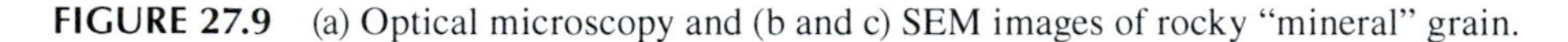

FIGURE 27.9 (a) Optical microscopy and (b and c) SEM images of rocky "mineral" grain.

Amano Lipase has no human use, but it was commercially available, immobilized on ceramic or diatomite substrates, from Sigma-Aldrich. It has since been discontinued.

27.5 DISCUSSION AND ASSESSMENT

The Afghan billet explored in this investigation was clearly a scam in the nuclear smuggling world. Despite external markings to the contrary, there were no SNM or other extraordinary radioactive substances associated with the object. However, interesting and unexpected items that were most unusual for a nuclear forensic inquiry were discovered within the interior.

The best rationalization for the various magnetic items in the object to date is entirely speculative and was proposed by the LLNL Nuclear Assessment group. The reasoning was that, during the 1980s and 1990s, legitimate radiation-detection equipment was not as readily available in black-market circles as it perhaps is today. To convince a potential buyer that something emitting no radiation was actually radioactive, a seller might hold the object close to a television or microwave oven and induce adverse effects. The claim that any disturbance of the appliance due to the magnetic field was indicative of the presence of ionizing radiation might then be plausible in certain scenarios.

We surmise that the inhomogeneous packing material may have been bulk stuff lying about an ad-hoc assembly facility, ready for insertion into a cast billet whenever a need arose. Consequently, it could have integrated a diverse assortment of environmental artifacts that were subsequently incorporated within the finished object. The prevalence of miscellaneous biochemical and pharmaceutical analytes was interesting from an attribution viewpoint. They could be indicative of a (perhaps legitimate) function of the facility used to fabricate the scams, albeit now a 20-year-old clue of dubious value.

This object was certainly one of the more interesting questioned specimens investigated by nuclear forensic activities at Livermore. There were no dangerous materials in this particular item, but clearly the potential for mischief exists within the hollow device designs. Future collectors would be advised to exercise more initial prudence than has perhaps been considered in the past.

ACKNOWLEDGMENTS

Such multidisciplinary technical endeavors as required in this case have been indicative of LLNL forensic investigations for more than 20 years, but the Afghan billet was more wide ranging than many and surely the most extensive within the nuclear forensic efforts. The Livermore scientists who produced the results were Armando Alcaraz, Bill Brown, Gary Eppich, Pat Grant, Ian Hutcheon, Ron Kane, Kim Knight, Harry Martz, Ken Moody, Sarah Roberts, Paul Spackman, Audrey Williams, and Ross Williams.

Finally, to end this book, special acknowledgement and gratitude go to the LLNL Nuclear Assessment Program (NAP; now renamed Nuclear Assessment Operations) for critical support during the initial development of the modern nuclear forensic enterprise at Livermore. Although the discipline was underway before its involvement, NAP was instrumental in providing real-world casework for ground-truth assessments during the formative years of NFA, as well as being the initiator of this latest investigation of the Afghan billet. The analyses of just some of the NAP-facilitated questioned specimens at LLNL, beginning in the mid-1990s, may be found in Chapters 20, 22, 23, 26, and 27 of this work.

Accordingly, special thanks to Fred Jessen, Shawn Cantlin, Rob Allen, Mike Frank, Kim Budil, Bill Buckley, Yvonne Winger, Alex Loshak, and Lori Grant.

Index

D

G

H

N

O

P

Q

R

S

T

U

V

W

X